NUREG-1910
Supplement 1

Environmental Impact Statement for the Moore Ranch ISR Project in Campbell County, Wyoming

Supplement to the Generic Environmental Impact Statement for *In-Situ* Leach Uranium Milling Facilities

Final Report

Manuscript Completed: August 2010
Date Published: August 2010

Prepared by:

U.S. Nuclear Regulatory Commission
Office of Federal and State Materials and
Environmental Management Programs

ABSTRACT

The U.S Nuclear Regulatory Commission (NRC) issues licenses for the possession and use of source material provided that proposed facilities meet NRC regulatory requirements and would be operated in a manner that is protective of public health and safety and the environment. Under the NRC environmental protection regulations in the *Code of Federal Regulations* (CFR), Title 10, Part 51, which implement the National Environmental Policy Act (NEPA) of 1969, issuance of a license to possess and use source material for uranium milling requires an environmental impact statement (EIS) or a supplement to an EIS.

In May 2009, NRC issued NUREG–1910, "*Generic Environmental Impact Statement for In-Situ Leach Uranium Milling Facilities*" (the GEIS). In the GEIS, NRC assessed the potential environmental impacts from the construction, operation, aquifer restoration, and decommissioning of an *in-situ* leach uranium recovery facility [also known as an *in-situ* recovery (ISR) facility] located in four specified geographic regions of the western United States. As part of that assessment, NRC determined which potential impacts would be essentially the same for all ISR facilities and which would result in varying levels of impacts for different facilities, thus requiring further site-specific information to determine potential impacts. The GEIS provides a starting point for the NRC NEPA analyses for site-specific license applications for new ISR facilities, as well as for applications to amend or renew existing ISR licenses.

By letter dated October 2, 2007, Energy Metals Corporation submitted a license application to the NRC for a new source material license for the Moore Ranch Project, to be located in Campbell County, Wyoming, which is in the Wyoming East Uranium Milling Region identified in the GEIS. In October 2009, the name of Energy Metals Corporation U.S. was changed to Uranium One Americas, Inc. (Uranium One). The NRC staff prepared this SEIS to evaluate the potential environmental impacts from Uranium One's proposal to construct, operate, conduct aquifer restoration, and decommission an ISR facility at the proposed Moore Ranch Project. This SEIS describes the environment potentially affected by the proposed site activities, presents the potential environmental impacts resulting from reasonable alternatives to the proposed action, and describes Uranium One's environmental monitoring program and proposed mitigation measures. In conducting its analysis in this SEIS, the NRC staff evaluated site-specific data and information to determine whether the applicant-proposed activities and site characteristics were consistent with those evaluated in the GEIS. The NRC staff then determined relevant sections, findings, and conclusions in the GEIS that could be incorporated by reference in this SEIS, and areas that required additional analysis. Based on its environmental review, the NRC staff recommends that, unless safety issues mandate otherwise, that the source material license be issued as requested.

Paperwork Reduction Act Statement

This NUREG contains and references information collection requirements that are subject to the Paperwork Reduction Act of 1995 (44 U.S.C. 3501 et seq.). These information collections were approved by the Office of Management and Budget, approval numbers 3150-0014, 3150-0020, 3150-0021, and 3150-0008.

Public Protection Notification

The NRC may not conduct or sponsor, and a person is not required to respond to, a request for information or an information collection requirement unless the requesting document displays a currently valid OMB control number.

CONTENTS

Section	Page
ABSTRACT	iii
FIGURES	xiv
TABLES	xv
EXECUTIVE SUMMARY	xvii
ABBREVIATIONS/ACRONYMS	xxxi
SI (MODERN METRIC) CONVERSION FACTORS	xxxv

1 INTRODUCTION 1-1
   1.1 Background 1-1
   1.2 Proposed Action 1-1
   1.3 Purpose of and Need for the Proposed Action 1-1
   1.4 Scope of the Supplemental Environmental Analysis 1-3
      1.4.1 Relationship to the GEIS 1-3
      1.4.2 Public Participation Activities 1-4
      1.4.3 Issues Studied in Detail 1-5
      1.4.4 Issues Outside the Scope of the SEIS 1-6
      1.4.5 Related NEPA Reviews and Other Related Documents 1-6
   1.5 Applicable Regulatory Requirements 1-7
   1.6 Licensing and Permitting 1-7
      1.6.1 NRC Licensing Process 1-8
      1.6.2 Status of Permitting with Other Federal, Tribal, and State Agencies 1-8
   1.7 Consultations 1-10
      1.7.1 Endangered Species Act of 1973 Consultation 1-10
      1.7.2 National Historic Preservation Act of 1966 Consultation 1-10
      1.7.3 Coordination with Other Federal, Tribal, State, and Local Agencies 1-11
        1.7.3.1 Coordination with the Bureau of Land Management 1-11
        1.7.3.2 Coordination with the Bureau of Indian Affairs 1-11
        1.7.3.3 Interactions with Tribal Governments 1-12
        1.7.3.4 Coordination with the Wyoming Department of Environmental Quality 1-12
        1.7.3.5 Coordination with the Wyoming Game and Fish Department ... 1-13
        1.7.3.6 Coordination with the Wyoming State Engineer's Office 1-13
        1.7.3.7 Coordination with the Wyoming Governor's Planning Office 1-13
        1.7.3.8 Coordination with the Wyoming Community Development Authority 1-14
        1.7.3.9 Coordination with Localities 1-14
   1.8 Structure of the SEIS 1-14
   1.9 References 1-14

2 IN-SITU URANIUM RECOVERY AND ALTERNATIVES 2-1
   2.1 Alternatives Considered for Detailed Analysis 2-1
      2.1.1 The Proposed Action (Alternative 1) 2-1
        2.1.1.1 Disposal Via Class I Injection Well 2-1
          2.1.1.1.1 Site Description 2-1

CONTENTS (CONTINUED)

Section		Page
2.1.1.1.2	Construction Activities	2-2
2.1.1.1.2.1	Central Plant	2-4
2.1.1.1.2.2	Access Roads	2-7
2.1.1.1.2.3	Wellfields	2-7
2.1.1.1.2.3.1	Injection and Production Wells	2-8
2.1.1.1.2.3.2	Monitoring Wells	2-9
2.1.1.1.2.3.3	Well Construction and Testing	2-9
2.1.1.1.2.3.4	Pipelines	2-10
2.1.1.1.2.4	Other Structures and Systems	2-10
2.1.1.1.2.5	Schedule	2-10
2.1.1.1.3	Operation Activities	2-10
2.1.1.1.3.1	Uranium Mobilization	2-11
2.1.1.1.3.2	Lixiviant Chemistry	2-11
2.1.1.1.3.3	Lixiviant Injection and Production	2-11
2.1.1.1.3.4	Excursion Monitoring	2-13
2.1.1.1.3.5	Uranium Processing	2-13
2.1.1.1.3.5.1	Ion Exchange	2-14
2.1.1.1.3.5.2	Elution	2-14
2.1.1.1.3.5.3	Precipitation, Drying, and Packaging	2-14
2.1.1.1.3.6	Management of Production Bleed and Other Liquid Effluents	2-15
2.1.1.1.3.7	Schedule	2-15
2.1.1.1.4	Aquifer Restoration Activities	2-15
2.1.1.1.4.1	Groundwater Restoration Methods	2-16
2.1.1.1.4.1.1	Groundwater Transfer	2-17
2.1.1.1.4.1.2	Groundwater Treatment	2-18
2.1.1.1.4.2	Schedule	2-19
2.1.1.1.5	Decontamination, Decommissioning, and Reclamation Activities	2-19
2.1.1.1.5.1	Wellfield Decommissioning	2-19
2.1.1.1.5.2	Topsoil Handling and Replacement	2-20
2.1.1.1.5.3	Final Contouring and Revegetation	2-20
2.1.1.1.5.4	Procedures for Removing and Disposing of Structures and Equipment	2-21
2.1.1.1.5.4.1	Preliminary Radiological Surveys and Contamination Control	2-21
2.1.1.1.5.4.2	Removal of Process Buildings and Equipment	2-21
2.1.1.1.5.4.3	Building Materials, Equipment, and Piping Released for Unrestricted Use	2-21
2.1.1.1.5.5	Schedule	2-22
2.1.1.1.6	Effluents and Waste Management	2-22
2.1.1.1.6.1	Gaseous and Airborne Particulate Emissions	2-22
2.1.1.1.6.2	Liquid Effluents	2-24
2.1.1.1.6.3	Solid Wastes	2-26
2.1.1.1.7	Transportation	2-27
2.1.1.1.8	Financial Surety	2-28
2.1.1.2	Alternative Wastewater Disposal Options	2-28
2.1.1.2.1	Evaporation Ponds	2-32

CONTENTS (CONTINUED)

Section				Page
		2.1.1.2.2	Land Application	2-33
		2.1.1.2.3	Surface Water Discharge	2-34
		2.1.1.2.4	Class V Injection Well	2-35
		2.1.2	No-Action (Alternative 2)	2-35
	2.2	Alternatives Eliminated from Detailed Analysis		2-35
		2.2.1	Conventional Mining and Milling	2-36
		2.2.2	Conventional Mining and Heap Leaching	2-37
		2.2.3	Alternate Site Location	2-38
		2.2.4	Alternate Lixiviants	2-38
		2.2.5	Alternate Wastewater Treatment Methods	2-38
		2.2.6	Comparison of the Predicted Environmental Impacts	2-39
	2.3	Final Recommendation		2-44
	2.4	References		2-44
3	AFFECTED ENVIRONMENT			3-1
	3.1	Introduction		3-1
	3.2	Land Use		3-1
		3.2.1.1	Existing Uses	3-4
		3.2.1.2	Rangeland and Pastureland	3-4
		3.2.1.3	Minerals and Energy	3-4
	3.3	Transportation		3-4
	3.4	Geology and Soils		3-6
		3.4.1	Geology	3-8
		3.4.2	Soils	3-13
	3.5	Water Resources		3-14
		3.5.1	Surface Waters and Wetlands	3-14
		3.5.1.1	Drainage Basins	3-14
		3.5.1.2	Surface Water Features	3-14
		3.5.1.2.1	Intermittent Streams	3-17
		3.5.1.2.2	Ponds	3-17
		3.5.1.3	Surface Water Flow	3-18
		3.5.1.4	Surface Water Quality	3-18
		3.5.1.5	Wetlands	3-19
		3.5.2	Groundwater	3-24
		3.5.2.1	Regional Groundwater Resources	3-24
		3.5.2.2	Local Groundwater Resources	3-24
		3.5.2.3	Uranium-Bearing Aquifers	3-25
		3.5.2.3.1	Hydrogeologic Characteristics	3-25
		3.5.2.3.2	Level of Confinement	3-25
		3.5.2.3.3	Groundwater Quality	3-26
		3.5.2.3.4	Current Groundwater Uses	3-28
		3.5.2.4	Surrounding Aquifers	3-29
	3.6	Ecology		3-29
		3.6.1	Terrestrial Ecology	3-30
		3.6.1.1	Vegetation	3-30
		3.6.1.2	Wildlife	3-32
		3.6.1.2.1	Big Game	3-32

CONTENTS (CONTINUED)

Section			Page
	3.6.1.2.2	Avian Species	3-33
	3.6.1.2.3	Other Mammals, Reptiles, and Amphibians	3-34
	3.6.2	Aquatic Ecology	3-34
	3.6.3	Protected Species	3-35
3.7	Meteorology, Climatology, and Air Quality		3-40
	3.7.1	Meteorology and Climatology	3-40
	3.7.1.1	Temperature	3-41
	3.7.1.2	Wind	3-41
	3.7.1.3	Precipitation	3-42
	3.7.1.4	Evaporation	3-42
	3.7.1.5	Climate Change and Greenhouse Gases	3-42
	3.7.2	Air Quality	3-45
3.8	Noise		3-46
3.9	Historical and Cultural Resources		3-48
	3.9.1	Cultural History	3-49
	3.9.1.1	Prehistoric Era	3-49
	3.9.1.2	Protohistoric/Historic Era	3-50
	3.9.2	Historic and Cultural Resources Identified and Places of Cultural Significance	3-50
	3.9.2.1	Previous Cultural Resources Investigations	3-50
	3.9.2.1.1	Archaeology–Identification and Evaluation	3-50
	3.9.2.1.2	Ethnology–Identification and Evaluation	3-52
	3.9.3	Historic Properties Listed in the National Register of Historic Places	3-52
	3.9.4	Tribal Consultation and Places of Cultural Significance	3-53
3.10	Visual and Scenic Resources		3-53
3.11	Socioeconomics		3-54
	3.11.1	Demographics	3-55
	3.11.2	Income	3-57
	3.11.3	Housing	3-57
	3.11.4	Employment Structure	3-57
	3.11.5	Local Finance	3-57
	3.11.6	Education	3-58
	3.11.7	Health and Social Services	3-58
3.12	Public and Occupational Health and Safety		3-58
	3.12.1	Background Radiological Conditions	3-59
	3.12.2	Public Health and Safety	3-62
	3.12.3	Occupational Health and Safety	3-62
3.13	Waste Management		3-63
	3.13.1	Liquid Waste Disposal	3-63
	3.13.2	Solid Waste Disposal	3-63
3.14	References		3-64
4	ENVIRONMENTAL IMPACTS AND MITIGATIVE ACTIONS		4-1
4.1	Introduction		4-1
4.2	Land Use Impacts		4-1

CONTENTS (CONTINUED)

Section			Page
	4.2.1	Proposed Action (Alternative 1)	4-2
	4.2.1.1	Construction Impacts	4-2
	4.2.1.2	Operation Impacts	4-3
	4.2.1.3	Aquifer Restoration Impacts	4-3
	4.2.1.4	Decommissioning Impacts	4-4
	4.2.2	No-Action (Alternative 2)	4-4
4.3	Transportation Impacts		4-5
	4.3.1	Proposed Action (Alternative 1)	4-5
	4.3.1.1	Construction Impacts	4-5
	4.3.1.2	Operation Impacts	4-7
	4.3.1.3	Aquifer Restoration Impacts	4-10
	4.3.1.4	Decommissioning Impacts	4-11
	4.3.2	No-Action (Alternative 2)	4-12
4.4	Geology and Soils Impacts		4-12
	4.4.1	Proposed Action (Alternative 1)	4-12
	4.4.1.1	Construction Impacts	4-12
	4.4.1.2	Operation Impacts	4-14
	4.4.1.3	Aquifer Restoration Impacts	4-15
	4.4.1.4	Decommissioning Impacts	4-16
	4.4.2	No-Action (Alternative 2)	4-17
4.5	Water Resources Impacts		4-17
	4.5.1	Surface Water and Wetlands Impacts	4-17
	4.5.1.1	Proposed Action (Alternative 1)	4-17
	4.5.1.1.1	Construction Impacts	4-18
	4.5.1.1.2	Operation Impacts	4-19
	4.5.1.1.3	Aquifer Restoration Impacts	4-20
	4.5.1.1.4	Decommissioning Impacts	4-21
	4.5.1.2	No-Action (Alternative 2)	4-21
	4.5.2	Groundwater Impacts	4-22
	4.5.2.1	Proposed Action (Alternative 1)	4-22
	4.5.2.1.1	Construction Impacts	4-22
	4.5.2.1.2	Operations Impacts	4-23
	4.5.2.1.2.1	Operations Impacts to Shallow (Near-Surface) Aquifers	4-24
	4.5.2.1.2.2	Operations Impacts to Production and Surrounding Aquifers	4-25
	4.5.2.1.2.3	Operations Impacts to Deep Aquifers Below the Production Aquifers	4-31
	4.5.2.1.3	Aquifer Restoration Impacts	4-32
	4.5.2.1.4	Decommissioning Impacts	4-33
	4.5.2.2	No-Action (Alternative 2)	4-34
4.6	Ecological Resources Impacts		4-34
	4.6.1	Proposed Action (Alternative 1)	4-34
	4.6.1.1	Construction Impacts	4-35
	4.6.1.1.1	Impacts to Vegetation	4-35
	4.6.1.1.2	Impacts to Wildlife	4-36
	4.6.1.1.2.1	Impacts to Big Game	4-37

CONTENTS (CONTINUED)

Section			Page
	4.6.1.1.2.2	Impacts to Other Mammals	4-38
	4.6.1.1.2.3	Impacts to Avian Species	4-38
	4.6.1.1.2.4	Impacts to Reptiles and Amphibians	4-39
	4.6.1.1.3	Impacts to Aquatic Resources	4-39
	4.6.1.1.4	Impacts to Threatened and Endangered Species	4-39
	4.6.1.1.4.1	Impacts to Species of Concern	4-40
	4.6.1.2	Operations Impacts	4-40
	4.6.1.2.1	Impacts to Vegetation	4-41
	4.6.1.2.2	Impacts to Wildlife	4-41
	4.6.1.2.2.1	Impacts to Big Game	4-41
	4.6.1.2.2.2	Impacts to Other Mammals	4-41
	4.6.1.2.2.3	Impacts to Avian Species	4-42
	4.6.1.2.2.4	Impacts to Reptiles and Amphibians	4-42
	4.6.1.2.3	Impacts to Aquatic Resources	4-42
	4.6.1.2.4	Impacts to Threatened and Endangered Species	4-42
	4.6.1.2.4.1	Impacts to Species of Concern	4-42
	4.6.1.3	Aquifer Restoration Impacts	4-43
	4.6.1.4	Decommissioning Impacts	4-43
	4.6.2	No-Action (Alternative 2)	4-44
4.7	Air Quality Impacts		4-44
	4.7.1	Proposed Action (Alternative 1)	4-46
	4.7.1.1	Construction Impacts	4-46
	4.7.1.2	Operation Impacts	4-48
	4.7.1.3	Aquifer Restoration Impacts	4-48
	4.7.1.4	Decommissioning Impacts	4-49
	4.7.2	No-Action (Alternative 2)	4-49
4.8	Noise Impacts		4-50
	4.8.1	Proposed Action (Alternative 1)	4-50
	4.8.1.1	Construction Impacts	4-50
	4.8.1.2	Operation Impacts	4-51
	4.8.1.3	Aquifer Restoration Impacts	4-51
	4.8.1.4	Decommissioning Impacts	4-52
	4.8.2	No-Action (Alternative 2)	4-52
4.9	Impacts to Historical and Cultural Resources		4-53
	4.9.1	Proposed Action (Alternative 1)	4-53
	4.9.1.1	Construction Impacts	4-53
	4.9.1.2	Operation Impacts	4-54
	4.9.1.3	Aquifer Restoration Impacts	4-55
	4.9.1.4	Decommissioning Impacts	4-55
	4.9.2	No-Action (Alternative 2)	4-55
4.10	Visual and Scenic Resources Impacts		4-56
	4.10.1	Proposed Action (Alternative 1)	4-56
	4.10.1.1	Construction Impacts	4-56
	4.10.1.2	Operation Impacts	4-57
	4.10.1.3	Aquifer Restoration Impacts	4-58
	4.10.1.4	Decommissioning Impacts	4-58
	4.10.2	No-Action (Alternative 2)	4-59

CONTENTS (CONTINUED)

Section			Page
4.11	Socioeconomic Impacts		4-59
	4.11.1	Proposed Action (Alternative 1)	4-60
	4.11.1.1	Construction Impacts	4-60
	4.11.1.1.1	Demographics	4-60
	4.11.1.1.2	Income	4-60
	4.11.1.1.3	Housing	4-60
	4.11.1.1.4	Employment Structure	4-61
	4.11.1.1.5	Local Finance	4-61
	4.11.1.1.6	Education	4-61
	4.11.1.1.7	Health and Social Services	4-61
	4.11.1.2	Operation Impacts	4-61
	4.11.1.2.1	Demographics	4-62
	4.11.1.2.2	Income	4-62
	4.11.1.2.3	Housing	4-62
	4.11.1.2.4	Employment Structure	4-62
	4.11.1.2.5	Local Finance	4-62
	4.11.1.2.6	Education	4-63
	4.11.1.2.7	Health and Social Services	4-63
	4.11.1.3	Aquifer Restoration Impacts	4-63
	4.11.1.4	Decommissioning Impacts	4-64
	4.11.2	No-Action (Alternative 2)	4-64
4.12	Environmental Justice Impacts		4-64
	4.12.1	Methodology	4-65
	4.12.2	Proposed Action (Alternative 1)	4-67
	4.12.3	No-Action (Alternative 2)	4-68
4.13	Public and Occupational Health and Safety Impacts		4-68
	4.13.1	Proposed Action (Alternative 1)	4-68
	4.13.1.1	Construction Impacts	4-68
	4.13.1.2	Operation Impacts	4-69
	4.13.1.2.1	Radiological Impacts to Public and Occupational Health and Safety From Normal Operations	4-69
	4.13.1.2.2	Radiological Impacts to Public and Occupational Health and Safety from Accidents	4-72
	4.13.1.2.3	Nonradiological Impacts to Public and Occupational Health and Safety From Normal Operations	4-74
	4.13.1.2.4	Nonradiological Impacts to Public and Occupational Health and Safety From Accidents	4-76
	4.13.1.3	Aquifer Restoration Impacts	4-77
	4.13.1.4	Decommissioning Impacts	4-77
	4.13.2	No-Action (Alternative 2)	4-78
	4.14	Waste Management Impacts	4-78
	4.14.1	Proposed Action (Alternative 1)	4-79
	4.14.1.1	Disposal Via Class I Injection Well	4-79
	4.14.1.1.1	Construction Impacts	4-79
	4.14.1.1.2	Operation Impacts	4-79
	4.14.1.1.3	Aquifer Restoration Impacts	4-81
	4.14.1.1.4	Decommissioning Impacts	4-81

CONTENTS (CONTINUED)

Section				Page
		4.14.1.2	Alternative Wastewater Disposal Options	4-83
		4.14.1.2.1	Evaporation Ponds	4-83
		4.14.1.2.2	Land Application	4-84
		4.14.1.2.3	Surface Water Discharge	4-86
		4.14.1.2.4	Class V Injection Well	4-88
		4.14.2	No-Action (Alternative 2)	4-89
	4.15	REFERENCES		4-89
5	CUMULATIVE IMPACTS			5-1
	5.1	Introduction		5-1
		5.1.1	Other Past, Present, and Reasonably Foreseeable Future Actions	5-1
		5.1.1.1	Uranium Recovery Sites	5-2
		5.1.1.2	Coal Mining	5-5
		5.1.1.3	Oil and Gas Production	5-5
		5.1.1.4	Coal Bed Methane Development Projects	5-5
		5.1.1.5	Other Mining	5-7
		5.1.1.6	EISs as Indicators of Past, Present, and Reasonably Foreseeable Actions	5-7
		5.1.2	Methodology	5-7
	5.2	Land Use		5-11
	5.3	Transportation		5-15
	5.4	Geology and Soils		5-16
	5.5	Water Resources		5-17
		5.5.1	Surface Waters and Wetlands	5-17
		5.5.2	Groundwater	5-20
	5.6	Ecological Resources		5-22
		5.6.1	Terrestrial Ecology	5-22
		5.6.2	Aquatic Resources	5-24
		5.6.3	Threatened and Endangered Species	5-24
	5.7	Air Quality		5-24
		5.7.1	Global Climate Change and Greenhouse Gas Emissions	5-27
		5.7.1.1	Greenhouse Gas (GHG) Emissions in the Region	5-28
		5.7.1.2	GHG Emissions from the Proposed Moore Ranch Project	5-29
		5.7.1.3	Moore Ranch ISR Facility GHG Emissions Impact	5-30
		5.7.1.4	Effect of Climate Change on the Moore Ranch ISR Facility	5-30
		5.7.1.5	GHG Mitigation Measures	5-31
		5.7.1.6	Other Mining Activities in the Powder River Basin	5-31
	5.8	Noise		5-33
	5.9	Historical and Cultural Resources		5-34
	5.10	Visual and Scenic Resources		5-34
	5.11	Socioeconomics		5-35
	5.12	Environmental Justice		5-38
	5.13	Public and Occupational Health and Safety		5-39
	5.14	Waste Management		5-40

CONTENTS (CONTINUED)

Section			Page
	5.15	References	5-42
6	ENVIRONMENTAL MEASUREMENTS AND MONITORING PROGRAMS		6-1
	6.1	Introduction	6-1
	6.2	Radiological Monitoring	6-1
		6.2.1 Airborne Radiation Monitoring	6-2
		6.2.2 Soils and Sediment Monitoring	6-3
		6.2.3 Vegetation, Food, and Fish Monitoring	6-3
		6.2.4 Surface Water Monitoring	6-3
		6.2.5 Groundwater Monitoring	6-3
		6.3 Physiochemical Monitoring	6-3
		6.3.1 Wellfield Groundwater Monitoring	6-4
		6.3.1.1 Preoperational Groundwater Sampling	6-4
		6.3.1.2 Groundwater Quality Monitoring	6-5
		6.3.2 Wellfield and Pipeline Flow and Pressure Monitoring	6-7
		6.3.3 Surface Water Monitoring	6-7
		6.3.4 Meteorological Monitoring	6-7
	6.4	Ecological Monitoring	6-8
		6.4.1 Vegetation Monitoring	6-8
		6.4.2 Wildlife Monitoring	6-8
	6.5	References	6-8
7	COST-BENEFIT ANALYSIS		7-1
	7.1	No-Action Alternative	7-1
	7.2	Benefits from Proposed Action in Campbell County	7-1
		7.2.1 Benefits From Potential Production	7-2
		7.2.2 Costs to the Local Communities	7-3
	7.3	Evaluation Findings of the Proposed Moore Ranch Project	7-4
	7.4	References	7-4
8	SUMMARY OF ENVIRONMENTAL CONSEQUENCES		8-1
	8.1	Proposed Action (Alternative 1)	8-2
	8.2	No-Action (Alternative 2)	8-2
	8.3	Reference	8-3
9	LIST OF PREPARERS		9-1
	9.1	U.S. Nuclear Regulatory Commission Contributors	9-1
	9.2	Center for Nuclear Waste Regulatory Analyses (CNWRA®) Contractor Contributors	9-2
	9.3	CNWRA Consultants and Subcontractors	9-4
10	DISTRIBUTION LIST		10-1
	10.1	Federal Agency Officials	10-1
	10.2	Tribal Government Officials	10-2
	10.3	State Agency Officials	10-3
	10.4	Local Agency Officials	10-3
	10.5	Other Organizations and Individuals	10-4

CONTENTS (CONTINUED)

Section	Page
APPENDIX A CONSULTATION CORRESPONDENCE	A-1
APPENDIX B PUBLIC COMMENTS ON THE DRAFT ENVIRONMENTAL IMPACT STATEMENT FOR THE MOORE RANCH IN-SITU RECOVERY PROJECT IN CAMPBELL COUNTY, WYOMING AND U.S. NUCLEAR REGULATORY COMMISSION RESPONSES	B-1
APPENDIX C ALTERNATE CONCENTRATION LIMITS	C-1
APPENDIX D NONROAD COMBUSTION ENGINE EMISSIONS ESTIMATES	D-1

FIGURES

Figure		Page
1-1	Location of the Proposed Moore Ranch Project	1-2
2-1	Site Layout at the Proposed Moore Ranch Project	2-3
2-2	Wellfield Patterns and Infrastructure for the Proposed Moore Ranch Project	2-5
2-3	Layout of the Central Plant	2-6
2-4	Water Balance for the Proposed Moore Ranch Project	2-12
3-1	Surface Ownership at the Proposed Moore Ranch Project	3-2
3-2	Mineral Rights Ownership at the Proposed Moore Ranch Site	3-3
3-3	Land Use in the Vicinity of the Proposed Moore Ranch Project	3-5
3-4	Generalized Stratigraphy Sequence Showing the 70 Sand Ore Production in the Wasatch Formation	3-9
3-5	Isopach Map Showing the Thickness of the 70 Sand	3-10
3-6	West-East Stratigraphic Cross Section of Wellfield 1	3-11
3-7	West-East Stratigraphic Cross Section of Wellfield 2	3-12
3-8	Drainage Basin and Sub-Watersheds at the Proposed Moore Ranch Project	3-15
3-9	CBM Production Near the Proposed Moore Ranch Project	3-16
3-10	Wetlands Located on the Proposed Moore Ranch Project	3-23
3-11	Vegetation Communities on the Proposed Moore Ranch Project	3-31
3-12	Seasonal Wind Roses for the Antelope Coal Company	3-44
4-1	Difference in Size and Type of Drawdown in an Unconfined Aquifer and Confined Aquifer from an Extraction Well Operating at a Same Rate	4-26
5-1	Conventional and ISR Uranium Facilities, BLM Pasture Allotments and Croplands Near the Proposed Moore Ranch ISR Facility	5-12
5-2	Oilfields, Coalfields, CBM Project Areas, Uranium Occurrences, and Uranium Facilities near the Proposed Moore Ranch ISR Facility	5-14
5-3	Energy Developments Within 50-Mile Radius of Proposed Moore Ranch ISR Project	5-18

FIGURES (CONTINUED)

Figure		Page
5-4	Wyoming, Campbell Country, Johnson Country, and Natrona County Population, 1970–2020	5-36
5-5	Projected Campbell County Population and Employment to 2020	5-37
6-1	Proposed Moore Ranch Project Operational Environmental Monitoring Locations	6-2

TABLES

Table		Page
1-1	ISL GEIS Range of Expected Impacts in the Wyoming East Uranium Milling Region	1-4
1-2	Environmental Approvals for the Moore Ranch Uranium Project	1-9
2-1	Soil Cleanup Criteria and Goals	2-20
2-2	Summary of Anticipated Liquid 11e.(2) Water Quality Parameters	2-25
2-3	Comparison of Different Liquid Wastewater Disposal Options	2-30
2-4	Comparison of Predicted Environmental Impacts	2-40
3-1	Annual Average Daily Traffic Counts	3-7
3-2	Surface Water Quality at the Proposed Moore Ranch Project	3-20
3-3	Average Preoperational Baseline Groundwater Quality for Site Aquifers	3-26
3-4	Areal Distribution of Vegetation Communities within the Proposed License Area	3-32
3-5	Migratory Bird Species of Management Concern Observed on the Proposed Moore Ranch Project	3-37
3-6	Federally- and State-Listed Species	3-38
3-7	Climate Data for Midwest, Wyoming, Climate Station	3-43
3-8	Existing Conditions—2007 Ambient Air Quality Monitoring Data	3-46
3-9	Noise Abatement Criteria: One Hour, A-Weighted Sound Levels in Decibels (dBA)	3-47
3-10	Newly Identified and Relocated Archaeological Sites by Location and Site Characteristics	3-51
3-11	Population and Percent Growth in Campbell County, Wyoming, From 1980 to 2050	3-55
3-12	Demographic Profile of the Population in Campbell County in 2000	3-56
3-13	Demographic Profile of the Population in Campbell County, 2006–2008 Three-Year Estimate	3-56
3-14	Income Information for the Region of Influence 2006–2008 American Community Survey 3-Year Estimates	3-57
3-15	Housing in Campbell County, Wyoming	3-58
4-1	Estimated Annual Average Daily Traffic on State Highway 387 for the Construction Phase of the Proposed Moore Ranch Project	4-6
4-2	Estimated Annual Average Daily Traffic on State Highway 387 for the Operations Phase of the Proposed Moore Ranch Project	4-8

TABLES (CONTINUED)

Table		Page
4-3	Estimated Annual Average Daily Traffic on State Highway 387 for the Aquifer Restoration Phase of the Proposed Moore Ranch Project	4-10
4-4	Estimated Annual Average Daily Traffic on State Highway 387 for the Decommissioning Phase of the Proposed Moore Ranch Project	4-11
4-5	Percent Living in Poverty and Percent Minority in 2000	4-67
4-6	Estimated Radon-222 Releases (Ci yr-1)* from the Proposed Moore Ranch	4-71
4-7	Generic Accident Dose Analysis for ISR Operations	4-72
5-1	Uranium Mining and Milling Projects—Distance and Direction From the Proposed Moore Ranch Project	5-3
5-2	Coal Mining and Milling Projects—Distance and Direction From the Proposed Moore Ranch Project	5-6
5-3	Cumulative Impacts on Environmental Resources	5-9
5-4	Comparison of Carbon Dioxide Emissions by Source	5-28
5-5	Wyoming Historical and Reference Case GHG Emissions	5-28
5-6	Annual Greenhouse Gas Emissions at the West Antelope II Mine	5-32
5-7	Estimated Annual Equivalent CO_2 Emissions at the Black Thunder, Jacobs Ranch, and North Antelope Rochelle Mines	5-33
7-1	Towns Near to the Proposed Moore Ranch Project	7-3
7-2	Summary of Costs and Benefits of the Proposed Moore Ranch Project	7-4
8-1	Summary of Environmental Consequences	8-4

EXECUTIVE SUMMARY

BACKGROUND

By letter dated October 2, 2007, Energy Metals Corporation (EMC), a wholly-owned subsidiary of Uranium One Americas, submitted an application to the U.S. Nuclear Regulatory Commission (NRC) for a new source material license for the Moore Ranch Uranium Project (Moore Ranch Project), located in Campbell County, Wyoming. As of August 10, 2007, Energy Metals Corporation was acquired by Uranium One, Inc., therefore, the applicant for the proposed Moore Ranch Facility is Uranium One Americas, Inc. (Uranium One). Uranium One is proposing to recover uranium using the *in-situ* leach (ISL) recovery process [also known as the *in-situ* recovery (ISR)]. The proposed Moore Ranch Project includes a central processing plant, two wellfields, two to four deep disposal wells for liquid effluents, and the attendant infrastructure (e.g., pipelines).

The Atomic Energy Act of 1954, as amended by the Uranium Mill Tailings Radiation Control Act of 1978, authorizes the NRC to issue licenses for the possession and use of source material and byproduct material. The NRC must license facilities, including ISR operations, in accordance with NRC regulatory requirements to protect public health and safety from radiological hazards. Under the NRC environmental protection regulations in the *Code of Federal Regulations*, Title 10, Part 51 (10 CFR Part 51), that implement the National Environmental Policy Act of 1969 (NEPA), preparation of an environmental impact statement (EIS) or supplement to an EIS is required to issue a license to possess and use source material for uranium milling [see 10 CFR 51.20(b)(8)].

In May 2009, the NRC staff issued NUREG–1910, "*Generic Environmental Impact Statement for In-Situ Leach Uranium Milling Facilities*" (herein after referred to as the GEIS). In the GEIS, NRC assessed the potential environmental impacts from the construction, operation, aquifer restoration, and decommissioning of an ISR facility located in four specified geographic regions of the western United States. The proposed Moore Ranch Project is located within the Wyoming East Uranium Milling Region identified in the GEIS. The GEIS provides a starting point for the NRC NEPA analysis for site-specific license applications for new ISR facilities, as well as for applications to amend or renew existing ISR licenses. This supplemental environmental impact statement (SEIS) incorporates by reference from the GEIS and uses information from the applicant's license application and other independent sources to fulfill the requirements in 10 CFR 51.20(b)(8).

This SEIS includes the NRC staff analysis that considers and weighs the environmental effects of the proposed action, the environmental impacts of alternatives to the proposed action, and mitigation measures to either reduce or avoid adverse effects. It also includes the NRC staff's recommendation regarding the proposed action.

PURPOSE AND NEED OF THE PROPOSED ACTION

NRC regulates uranium milling, including the ISR process, under 10 CFR Part 40, "Domestic Licensing of Source Material." Uranium One is seeking an NRC source material license to authorize commercial-scale ISR uranium recovery at the Moore Ranch site. The purpose and need for the proposed federal action is to either grant or deny the Uranium One license application to use ISR technology to recover uranium and produce yellowcake at the proposed Moore Ranch Project. Yellowcake is the uranium oxide product of the ISR milling process that is used to produce fuel for commercially-operated nuclear power reactors. Based on the application, the NRC's federal action is the decision whether to issue the license to Uranium One.

This definition of purpose and need reflects the Commission's recognition that, unless there are findings in the safety review required by the Atomic Energy Act or findings in the NEPA environmental analysis that would lead the NRC to reject a license application, the NRC has no role in a company's business decision to submit a license application to operate an ISR facility at a particular location.

THE PROJECT AREA

The Moore Ranch Project is located in southwest Campbell County, in south-central Wyoming, about halfway between the Towns of Wright, located 40 km [25 mi] to the northeast, and Midwest-Edgerton, located 39 km [24 mi] to the southwest, with populations of 1,604 and 439 people. The City of Gillette is located approximately 80 km [50 mi] to the northeast; the City of Casper is located approximately 85 km [53 mi] to the southwest. Planned facilities associated with the proposed project include a central plant with processing capabilities; two wellfields with injection, production, and monitor wells, header houses, and pipeline to connect the wellfields with the central plant; and an access road network.

The proposed license area consists of approximately 2,879 ha [7,110 ac] and is remotely located on private land with about 14 percent of the surface rights being administered by the State of Wyoming. The U.S. Department of the Interior, U.S. Bureau of Land Management does not administer surface rights within the proposed Moore Ranch license area.

IN-SITU RECOVERY PROCESS

During the ISR process, an oxidant-charged solution, called a lixiviant, is injected into the production zone aquifer (uranium ore body) through injection wells. Typically, a lixiviant uses native groundwater (from the production zone aquifer), carbon dioxide, and sodium carbonate/bicarbonate, with an oxygen or hydrogen peroxide oxidant. As the lixiviant circulates through the production zone, it oxidizes and dissolves the mineralized uranium, which is present in a reduced chemical state. The resulting uranium-rich solution is drawn to recovery wells by pumping, and then transferred to a processing facility via a network of pipes, which may be buried just below the ground surface. At the processing facility, the uranium is removed from the solution. The resulting barren solution is then recharged with the oxidant and reinjected to recover more uranium from the wellfield.

During production, the uranium recovery solution continually moves through the aquifer from outlying injection wells to internal recovery wells. These wells can be arranged in a variety of geometric patterns depending on the ore body configuration, aquifer permeability, and operator preference. Wellfields are often designed in a five-spot or seven-spot pattern, with each recovery (i.e., production) well being located inside a ring of injection wells. In the case of the Moore Ranch Project, the applicant has proposed a design based on a five-spot pattern. Monitoring wells completed in the production zone aquifer surround the wellfield pattern area; monitoring wells are also completed in the overlying and underlying aquifers. These monitoring wells are screened in the appropriate stratigraphic horizons to detect lixiviant in case it migrates out of the production zone. The uranium that is recovered from the solution would be processed, dried into yellowcake and packaged into NRC- and U.S. Department of Transportation (USDOT)-approved 205-L [55-gal] steel drums, and trucked offsite to a licensed conversion facility.

ALTERNATIVES

The NRC environmental review regulations in 10 CFR Part 51 require NRC to consider reasonable alternatives, including the No-Action alternative, before acting on a proposal. The NRC staff considered a range of alternatives that could fulfill the underlying purpose and need for the proposed action. Based on this screening analysis, a set of reasonable alternatives was developed, and the impacts of the proposed action were compared with the impacts that would result if a given alternative were implemented. This SEIS evaluates the potential environmental and public health and safety impacts of the proposed action, the No-Action alternative, and considers alternative wastewater disposal options. Under the No-Action alternative, Uranium One would not be issued a license to construct and operate an ISR facility at the proposed site. Alternatives considered but eliminated from detailed analysis include conventional mining and milling, conventional mining and heap leach processing, alternate site location, alternate lixiviants, and alternate wastewater treatment methods.

SUMMARY OF THE ENVIRONMENTAL IMPACTS

This SEIS includes the NRC staff analysis that considers and weighs the environmental impacts resulting from the proposed construction, operation, aquifer restoration, and decommissioning of an ISR facility at the proposed Moore Ranch Project site and the No-Action alternative. This SEIS also provides mitigation measures to either reduce or avoid potential adverse impacts from the proposed action. This SEIS uses the assessments and conclusions reached in the GEIS, combined with site-specific information to assess and categorize impacts.

As discussed in the GEIS and consistent with NUREG–1748 (NRC, 2003), the significance of potential environmental impacts is categorized as follows:

SMALL: The environmental effects are not detectable or are so minor that they will neither destabilize nor noticeably alter any important attribute of the resource.

MODERATE: The environmental effects are sufficient to alter noticeably but not destabilize important attributes of the resource.

LARGE: The environmental effects are clearly noticeable and are sufficient to destabilize important attributes of the resource.

Chapter 4 provides the NRC evaluation of the potential environmental impacts of the construction, operation, aquifer restoration, and decommissioning of the proposed Moore Ranch Project. A list of the significance level of impacts by phase of the ISR facility lifecycle is provided next, followed by a brief summary of impacts by environmental resource area and ISR facility lifecycle phase.

Impacts by Resource Area and ISR Facility Phase

Land Use

Construction: Impacts would be SMALL. Approximately 60 ha [150 ac] of the 2,879 ha [7,110 ac] or 2 percent of the proposed license area would be disturbed by the construction phase of the proposed Moore Ranch Project. Topsoil would be stripped and stockpiled to build the central plant, develop two wellfields and the attendant infrastructure, and to construct

Executive Summary

access roads. Livestock grazing and natural resources extraction (e.g., coal-bed methane development) would be excluded from the fenced areas surrounding the wellfields and the central plant during the life of the project.

Operation: Impacts would be SMALL. Livestock grazing and natural resources extraction would continue to be limited from the wellfields and the central plant during the ISR lifecycle, limiting access to approximately 2 percent of the proposed license area for the life (approximately 12 years) of the project. As all facilities would be constructed, the direct impacts to land from earthmoving activities would be less than that during the construction phase.

Aquifer Restoration: Impacts would be SMALL. The impact to land use would either be similar to, or less than that described for the operations phase. Access to wellfields would continue to be restricted from other uses as described for the operations phase.

Decommissioning: Impacts would be SMALL. The impact on land use from decontaminating and decommissioning the proposed Moore Ranch Project would be similar to that experienced during the construction phase. Decommissioning the buildings, wellfields, access roads, and removing potentially contaminated soil would result in a temporary increase in land-disturbing activities. Upon completion of the plugging and abandonment of wells, soil would be reseeded and reclaimed in areas where it had been removed. At the completion of decommissioning activities, because the reclaimed land would be released for other uses and no longer restricted, the land use impacts for disturbed areas would be MODERATE until vegetation in seeded areas becomes established. Once vegetation is established in reclaimed areas, the land would be returned to a condition that would support a variety of land uses; therefore, the impact would be SMALL.

Transportation

Construction: Impacts would be SMALL. Truck traffic during construction activities would result in about a 9 percent increase in local traffic. Localized fugitive dust emissions, noise from traffic, and incidental wildlife or livestock kills could potentially occur but would be limited by the short access road distance.

Operation: Impacts would be SMALL. Transportation impacts would be less than those during the construction phase because fewer trucks would be on the road. The probability of an accident was determined to be low. To minimize the risk of an accident, materials would be packed and shipped in accordance with NRC and the U.S. Department of Transportation regulations, and the applicant would develop emergency response plans.

Aquifer Restoration: Impacts would be SMALL. Transportation impacts would be less than those during the construction phase and comparable to that during the operations phase. The need to transport hazardous materials and uranium-loaded resins would decrease as aquifer restoration proceeded; therefore, the potential for accidents resulting in spills or leaks would also decrease.

Decommissioning: Impacts would be SMALL. Transportation impacts would be less than those during the construction and operations phases. Transport of hazardous materials would cease during decommissioning, and access roads would either be reclaimed or left in place for future use.

Geology and Soils

Construction: Impacts would be SMALL. Within the 61 ha [150 ac] that would be directly affected by ISR activities, most earthmoving activities would be limited to the construction of the central plant facilities and access roads, drilling of wells and installation of well header houses, and excavation of trenches to lay and bury pipelines. Topsoils removed during these activities would be saved and reused later to restore disturbed areas. Implementation of best management practices, the short duration of the construction phase, and mitigative measures such as reestablishing temporary native vegetation as soon as possible after implementation, would further minimize the potential impact on soils.

Operation: Impacts would be SMALL. The removal of uranium mineral coatings on sediment grains in the target sandstones during ISR operations would result in a small permanent change to the mineralogical composition of the uranium-producing formations. However, the rock matrix of sediment grains would continue to form the support structure for the rock layers, and no significant matrix compression or ground subsidence would be expected to occur. The potential for spills during transfer of uranium-bearing lixiviant to and from the central plant would be further mitigated by implementing onsite standard operating procedures and complying with NRC and WDEQ requirements for spill response and reporting of surface releases.

Aquifer Restoration: Impacts would be SMALL. During aquifer restoration, the process of groundwater transfer would not remove rock matrix or structure. The formation pressure would be decreased during restoration to ensure that the direction of groundwater flow was into wellfields to reduce the potential for lateral migration of constituents; however, the change in pressure would not result in collapse of overlying rock strata into the extraction zone of the aquifer. The potential for and response to spills would be comparable to that described for the operations phase.

Decommissioning: Impacts would be SMALL. Disruption or displacement of soils would occur during dismantling of the facilities and reclamation of the land; however, the disturbed lands would be restored to their preextraction land use. Topsoil would be reclaimed and regraded to the original topography.

Surface Waters and Wetlands

Construction: Impacts would be SMALL. The occurrence of surface water at the Moore Ranch Project is limited, and surface water flow in channels is intermittent. Although the proposed construction activities such as laying pipeline and drilling wells could generate surface water runoff, implementation of best management practices and mitigative measures such as working in channels when they were dry would further minimize potential impacts. Well construction would avoid channels whenever possible. Temporary disturbances to the soil from traffic during construction could result in surface water runoff and sediment transport during periods of surface flow. Wetland areas would be avoided.

Operation: Impacts would be SMALL. The applicant would construct the central plant and support facilities landward of intermittent channels and above peak flood elevations. Furthermore, the central plant and chemical and fuel storage tanks would also have secondary containment. Spills and leaks could potentially impact surface waters, but the implementation of best management practices would minimize the potential impact. A storm water management plan would detain or treat runoff. Routine maintenance of Wellfield 2 would require vehicular crossing of an intermittent channel; however, the applicant would implement sedimentation and erosion control measures to further minimize surface water runoff from such temporary disturbances.

Aquifer Restoration: Impacts would be SMALL. There would be no impact to surface water or groundwater during aquifer restoration because waste water generated during this ISR phase would be disposed of via deep well injection. Automated sensors would monitor the injection pressure of the deep disposal wells to detect potential leaks or pipeline/well ruptures that could result in a discharge.

Decommissioning: Impacts would be SMALL. The impact from decommissioning would be similar to that during the construction phase. Land recontouring would restore areas to their preexisting condition to minimize the long-term impact to intermittent streams that were traversed during well maintenance.

Groundwater

Construction: Impacts would be SMALL. The primary impact to groundwater during the construction phase of the proposed Moore Ranch Project would be from the consumptive use of groundwater, injection of drilling fluids and muds during well installation, and from surface spills that could potentially migrate to groundwater. Groundwater for consumptive use would be from an aquifer located deeper than the proposed uranium recovery zone and the volume would be small and temporary relative to the water supply in the affected aquifer. The introduction of drilling or production fluid into a wellbore could impact the groundwater quality in the aquifers that would be penetrated; however, the use of drilling muds designed to seal the wellbore to set the casing would mitigate this impact. The applicant would use best management practices during facility construction and wellfield installation, including the implementation of a spill prevention and cleanup program, to prevent soil contamination that would require an immediate cleanup response to prevent soil contamination or infiltration to groundwater.

Operation: Impacts on water levels in local wells and on groundwater quality would be SMALL. The operations phase of the proposed Moore Ranch Project could impact shallow (near-surface) aquifers, the aquifer containing the ore body and surrounding aquifers, and deep aquifers below the ore production zone that are used for the disposal of liquid effluent. One domestic well, two stock wells, and one miscellaneous water well drilled by EMC occur within the proposed license area. The stock wells are not permitted through the State Engineers Office. No irrigation groundwater wells occur within the proposed license area. If the domestic well is completed in the exempted portion of the ore-bearing aquifer, the well cannot be used as a source of drinking water, in compliance with 40 CFR 146. If industrial or livestock wells are completed in the exempted portion of the aquifer, these wells are required to be properly plugged and abandoned to both avoid potential negative impacts on targeted bleed rates during ISR operations and to minimize potential adverse impacts on the environment. Domestic wells outside the proposed license area would be protected through the excursion monitoring and remediation requirements in 40 CFR 144.54 and 40 CFR 144.55 during ISR operation. Therefore, the potential impacts to the surficial aquifer would be SMALL. The potential impact to groundwater supplies in the ore production zone and surrounding aquifers is related to the consumptive use of groundwater and groundwater quality. Groundwater modeling, performed by the applicant and verified by the NRC staff, of the ore production zone predicted that the potential drawdown in private wells located within 3.2-km [2-mi] radius surrounding the proposed license area would experience a nominal drawdown in their private wells, but well yield would not be impacted. ISR operations would degrade groundwater quality in the ore production zone. However, the establishment of an inward hydraulic gradient, as well as the applicant-installed groundwater monitoring network by the applicant to detect potential vertical and horizontal excursions, would limit the potential for undetected groundwater excursions that could degrade groundwater quality. Because the ore production zone is overlain and underlain by impermeable shale layers, this further ensures hydraulic isolation of the ore production zone to minimize potential groundwater contamination above and below the production zone.

Liquid effluent generated from operation of the proposed Moore Ranch Project would be disposed of via deep well injection into Class I disposal wells permitted by the Wyoming Department of Environmental Quality (WDEQ). The groundwater in the formations being considered for deep well disposal must not be a potential underground source of drinking water and must comply with the WDEQ Water Quality and Regulations for Underground Management of Hazardous or Toxic Waste (Chapter 8, Section 6).

Aquifer Restoration: Impacts would be SMALL. Groundwater restoration activities occur when a wellfield is no longer used to produce uranium. During aquifer restoration, there are potential impacts to groundwater quality and water levels. Groundwater modeling performed by the applicant and verified by NRC estimated groundwater drawdown in the ore production zone in Wellfields 1 and 2. The results of this modeling showed that the potential drawdown at the boundary of the proposed license area could range from 1 to 9 ft and from 1 to 6 ft at Wellfields 1 and 2, from aquifer restoration activities. A drawdown contour of 1 ft extended from 1 to 4 mi from the proposed license area to the north, northwest, west and southwest. Only one private well was identified that could potentially be affected by the drawdown; however, the predicted drawdown would have a negligible impact on well yield.

Because the ore production zone coalesces with an underlying aquifer in a portion of Wellfield 2, the applicant performed groundwater modeling to estimate the impact of aquifer restoration activities on water levels in the underlying aquifer and to surrounding users. The results of this analysis showed that private wells within a 3.2-km [2-mi] radius surrounding the proposed license area would experience a nominal 0.012 to 0.4-m [0.04- to 1.2]-ft drawdown, which would not be expected to impact well yields. Therefore, the impact on groundwater levels would be SMALL. Furthermore, the hydrologic test package for Wellfield 2 will be reviewed and approved by the NRC because of the more complex geology at that location.

Impacts on groundwater quality would be SMALL. Postrestoration groundwater quality would be protective of the public and the environment. The goal of aquifer restoration is to restore groundwater quality in the ore production zone to preextraction baseline conditions. If the aquifer cannot be restored to baseline conditions, then NRC requires that either the production zone be returned to maximum contaminant levels in Table 5C of 10 CFR Part 40, Appendix A, or to NRC-approved alternate concentration limits.

Decommissioning: Impacts would be SMALL. The potential impact to groundwater quality during decommissioning and reclamation would be comparable to that described previously for the construction phase of the proposed Moore Ranch Project.

Ecological Resources

Construction: Impacts would be SMALL. The impact would be further reduced by implementing the mitigative measures discussed in Section 4.6 of this SEIS. An estimated 61 ha [150 ac] of land would be disturbed during construction activities to build the central plant, develop wellfields, and lay pipeline, which would result in some habitat loss or alteration, displacement of wildlife, and injury or mortality from encounters with vehicles or heavy equipment, although wildlife species would generally be expected to disperse from the area when construction activities begin. The applicant could mitigate these impacts by observing Wyoming Fish and Game Department (WFGD) guidelines regarding noise, vehicular traffic, and human proximity during the construction phase. No threatened or endangered species are known to occur within the proposed license area.

Operation: Impacts would be SMALL. The impact could be further reduced by implementing the mitigative measures discussed in Section 4.6 of this SEIS. Impacts would be similar to but less than those experienced during the construction phase because fewer earthmoving activities would occur. The applicant would reseed disturbed areas with WDEQ-approved seed mixtures.

Aquifer Restoration: Impacts would be SMALL. Impacts would be similar to those experienced during the operations phase with no major differences in type or degree of impact.

Decommissioning: Impacts would be SMALL. Temporary disturbances to land and soils could displace vegetation and wildlife species that recolonized the proposed license area after the construction phase. Revegetation and recontouring would restore habitat previously altered during construction and operations.

Meteorology, Climatology, and Air Quality

Construction: Impacts would be SMALL. Combustion engine exhausts from non-road mobile diesel equipment used during construction would generate air emissions. The magnitude of these emissions would be well below Clean Air Act thresholds for major stationary sources of air pollution. Considered along with meteorological conditions that are often favorable for dispersion, the emissions would be unlikely to change the present status of attainment with the National Ambient Air Quality Standards (NAAQS) nor impact the air quality of the nearest Class I Prevention of Significant Deterioration area. Fugitive dust emissions from road travel would be mitigated by the applicant's plans for wetting and stabilizing unpaved roads.

Operation: Impacts would be SMALL. Impacts would be similar to, but less than those experienced during construction. Operating ISR facilities are not major point source emitters of regulated nonradiological pollutants and emissions would be well below Clean Air Act thresholds for major sources of air pollution and therefore would be unlikely to change the present status of attainment with the National Ambient Air Quality Standards (NAAQS).

Aquifer Restoration: Impacts would be SMALL. Impacts would be similar in type and degree as those experienced during the operations phase. Less vehicular traffic would be required during the aquifer restoration phase than during operations because there would be fewer yellowcake shipments than during operations. The use of existing infrastructure and reduced traffic volume would reduce fugitive dust and road vehicle exhaust emissions. Fugitive dust emissions from road travel would be further mitigated by the applicant plans for wetting and stabilizing unpaved roads.

Decommissioning: Impacts would be SMALL. Impacts to air quality would be similar to that experienced during construction since the same type of activities would occur (e.g., earthmoving activities that generate fugitive dust and combustion engine emissions). The emissions would decrease as decommissioning progressed.

Noise

Construction: Impacts would be SMALL. Increased traffic and the use of drill rigs, heavy trucks, bulldozers, and other heavy equipment to construct and operate the wellfields, drill wells, construct access roads, and build the central plant would generate noise audible above the undisturbed background levels. The sound from construction activities would return to preexisting conditions at a distance of approximately 300 m [1,000 ft]. Therefore, there would be no audible noise at the location of the nearest resident approximately 4.5 km [2.8 mi] east of the center of the proposed Moore Ranch licensed area.

Operation: Impacts would be SMALL. Traffic would be the primary noise-generating activity that could be heard offsite. The central plant and other processing activities would generate indoor noise audible to workers. The nearest resident would not notice a change in noise at their location approximately 4.5 km [2.8 mi] east of the center of the proposed project.

Aquifer Restoration: Impacts would be SMALL. Noise impacts would be similar to, or less than, those experienced during the operations phase. Pumps and other wellfield equipment contained in buildings would reduce the potential sound impact to an offsite individual. Because the location of the nearest resident is approximately 4.5 km [2.8 mi] east of the center of the proposed project, there would be no change in background noise.

Decommissioning: Impacts would be SMALL. Noise impacts would either be similar to, or less than, those experienced during the construction phase. Noise during this phase would be temporary, and when decommissioning and reclamation activities were complete, the noise level would return to baseline. At the location of the nearest resident approximately 4.5 km [2.8 mi] east of the center of the proposed project, there would be no change in background noise.

Historical and Cultural Resources

Construction: Impacts would be SMALL. No National Register of Historic Places (NRHP) eligible properties would be affected by construction, and any potentially eligible archaeological sites would be avoided. Should historical or cultural resources be encountered during construction, work would stop and appropriate federal and state officials would be notified.

Operation: Impacts would be SMALL. No properties recommended eligible for listing to the NRHP would be affected by facility operations. None of the potentially eligible sites are located in areas affected by operations. As noted above, should historical or cultural resources be encountered during the plant operations, work would stop, and appropriate federal and state officials would be notified per the license condition.

Aquifer Restoration: Impacts would be SMALL. Impacts to historical and cultural resources during the aquifer restoration phase would be similar to operations. Should historical or cultural resources be encountered during aquifer restoration, work would stop and the appropriate federal and state officials would be notified.

Decommissioning: Impacts would be SMALL. No properties recommended eligible for listing to the NRHP would be affected during the decommissioning phase. Should historical or cultural resources be encountered during decommissioning, work would stop and appropriate federal and state officials would be notified.

Executive Summary

Visual and Scenic Resources

Construction: Impacts would be SMALL. The existing land use surrounding the proposed Moore Ranch Project has pipelines, wellfields for CBM production, and utility lines that disturb the landscape; implementing the proposed action would not change the character of the landscape. Temporary and short-term visual impacts during the construction period in each wellfield would result from header house construction, well drilling, and construction of access roads and electrical distribution lines.

Operation: Impacts would be SMALL. The wellfields would operate for approximately 3.25 years, and they would be similar in visual impact to the CBM installations that occur in the area. The central plant would remain operational for approximately 12 years. The proposed operations are consistent with the BLM visual classification for this area.

Aquifer Restoration: Impacts would be SMALL. The visual impact would be the same as described for the operations phase. No modifications to either scenery or topography would occur during aquifer restoration, which is estimated to last from 3.5 to 5 years. There would also be less vehicular traffic during aquifer restoration, creating less of a visual impact.

Decommissioning: Impacts would be SMALL. Temporary impacts to the visual landscape would be comparable to those during the construction phase. Reclamation would return the visual landscape to baseline contours and would reduce the visual impact by removing buildings and the associated infrastructure.

Socioeconomics

Construction: Overall impacts would be SMALL. Most of the construction work force is expected to be found locally, so the short-term increased demand for housing would be a SMALL impact. Housing construction workers in nearby towns would have no demographic impacts. Workers would be paid the regional rates typical of the area; therefore income impacts would be SMALL. Local workers and contractors would be employed whenever possible, which would have a SMALL impact on employment rates. The local economy could experience a SMALL beneficial impact from the purchasing of local goods and services and an increase in sales and income tax revenues, however an increased short-term demand for social services would have a SMALL impact on these resources.

Operation: Impacts could range from SMALL to MODERATE. The in-migration of workers and their families to nearby towns would have a SMALL impact on demographics. Workers would be paid similar rates to the average income in Wyoming; therefore income impacts would be SMALL. Housing demand would increase in local areas with low vacancy rates. The impact on housing could range from SMALL to MODERATE because of limited housing in the immediate area around the proposed ISR facility. Operation of the ISR facility at Moore Ranch would create new jobs, but because of the small size of the workforce, impacts on employment would be SMALL. The local economy would experience a SMALL beneficial impact from the purchasing of local goods and services and an increase in sales and income tax revenues. An increased demand for schools would have a SMALL impact on education because the current school system is not at full capacity. Increased demand for education and social services would have a SMALL impact.

Aquifer Restoration: Impacts would be SMALL. Impacts would be similar to, but less than, those during the operations. Fewer workers would be required, thus reducing the potential pressure on housing, education, and health and social services.

Decommissioning: Impacts would be SMALL. Impacts would be similar to those during the construction phase. By this stage of the project, local governments would have adapted to the changes brought on by the project years earlier, and thus, housing, education, and health and social services demand would be more likely to be met.

Environmental Justice

All Phases: No minority or low-income block groups were identified in the vicinity of the proposed Moore Ranch Project. Therefore, there would be no disproportionately high and adverse impacts to minority and low-income populations from the construction, operation, and decommissioning of the proposed ISR facility at Moore Ranch.

Public and Occupational Health and Safety

Construction: Impacts would be SMALL. Construction activities, including the use of construction equipment and vehicles, could disturb the topsoil and create fugitive dust emissions. Radiological environmental monitoring data indicate that radioactivity levels in the topsoil at the proposed Moore Ranch Project are at background levels. Therefore, the inhalation of these concentrations of residual radioactivity would pose a radiological dose comparable to that from natural background exposure.

Operation: The radiological impacts from normal operations would be SMALL. Public and occupational exposure rates at ISR facilities during normal operations have historically been well below regulatory limits. The remote location of the proposed Moore Ranch Project, in addition to the proposed technology coupled with the procedures to be implemented by the applicant, indicate that public and occupational health impacts from the operation of the facility would be consistent with historic observations. The radiological impacts from accidents would be SMALL for workers if the applicant's radiation safety and incident response procedures in its NRC-approved radiation protection plan are followed, and SMALL for the public due to the facility's remote location. The nonradiological public and occupational health impacts from normal operations and accidents, due primarily to risk of chemical exposure, would be SMALL if handling and storage procedures were followed.

Aquifer Restoration: Impacts would be SMALL. Impacts would be similar to, but less than, those during the operations phase. The reduction or elimination of some operational activities further limits the relative magnitude of potential worker and public health and safety hazards.

Decommissioning: Impacts would be SMALL. Impacts would be similar to, but less than, those experienced during construction. Soil and facility structures are decontaminated and lands are restored to preoperational conditions.

Waste Management

Construction: Impacts would be SMALL. Small-scale and incremental wellfield development would generate low volumes of construction waste consisting primarily of building materials, piping, and other solid wastes. No byproduct material would be generated during construction. Nonhazardous solid waste would be disposed of at a nearby municipal solid waste landfill.

Hazardous construction wastes, such as organic solvents, paints, used oil and paint thinners would be disposed of in accordance with the requirements in the Resource Conservation and Recovery Act (RCRA). The nearby landfill and associated construction and demolition pit are not at capacity and would be able to continue receiving municipal solid waste and construction and demolition waste.

Operation: Impacts would be SMALL. Liquid waste, including process bleed, restoration water, resin-transfer wash, filter washing, brine, and plant washdown, would be disposed of according to applicable NRC, federal, and state permits. Applicable permit requirements would mitigate potential adverse impacts from liquid waste management. From two to four Class 1 deep disposal wells permitted by the WDEQ and reviewed by the NRC would be drilled onsite for disposal of liquid effluent. Solids classified as Atomic Energy Act section 11e.(2) byproduct material (herein called "byproduct material") would be sent to a licensed disposal facility. Some contaminated materials would be decontaminated and disposed of in accordance with applicable NRC regulations.

Aquifer Restoration: Impacts would be SMALL. Waste decontamination and disposal procedures would be the same as those during the operation phase, resulting in similar impacts. Wastewater generated may increase but would be offset by the reduction in production capacity from the removal of wellfields.

Decommissioning: Impacts would be SMALL. All process or potentially contaminated equipment and materials including tanks, filters, pumps, and piping would be inventoried and designated for removal to a new location for future use; removed to another licensed facility; disposal as byproduct material at a licensed facility; or decontaminated to meet unrestricted release criteria. The process building would be decontaminated, dismantled, and released for use at another location. Safe handling, storage, and disposal of decommissioning wastes would be addressed in a decommissioning plan, which would be approved by the NRC prior to initiating decommissioning activities. A preoperational agreement with a licensed disposal facility to accept radioactive wastes would ensure that sufficient disposal capacity would be available for byproduct material generated by decommissioning activities.

Cumulative Impacts

The cumulative impact on the environment that results from the incremental impact of the proposed licensing action when added to other past, present, and reasonably foreseeable future actions was also considered, regardless of what agency (Federal or non-Federal) or person undertakes such other actions. The NRC staff determined that the SMALL to MODERATE impacts from the proposed Moore Ranch Project are not expected to contribute perceptible increases to the MODERATE cumulative impacts, due primarily to; concurrent CBM activities at the proposed Moore Ranch Project in conjunction with other oil and gas exploration and mining activities occurring throughout the Powder River Basin.

Summary of the Costs and Benefits of the Proposed Action

The implementation of the proposed action would generate primarily regional and local costs and benefits. The regional benefits of building the proposed project would be increased employment, economic activity, and tax revenues in the region around the proposed site. Costs associated with the proposed Moore Ranch ISR Project are, for the most part, limited to the immediate area surrounding the site.

Comparison of Alternatives

Under the No-Action alternative, NRC would not issue a license to Uranium One to construct, operate, restore the aquifer, or decommission the proposed Moore Ranch Project. The land would be available for other uses. There would be no incremental increase in traffic on local roads attributable to the proposed action. No land disturbance from earthmoving activities that could disrupt vegetation or current grazing patterns would occur nor would increased surface water runoff result from such activities. CBM operations in the area would continue, resulting in surface water discharges to the local intermittent drainages. There would neither be an impact on groundwater quality nor to the water levels in surrounding private wells from operating the ISR facility. There would be no noise-generating activity nor increased fugitive dust or exhaust emissions from either earthmoving activities or increased commuter traffic to the site. If the No-Action alternative were implemented, no new jobs would be created from the proposed action nor would additional tax revenue accrue to the local economy. There would be no affect on housing availability, the education system, or public services. There would be no disproportionate high and adverse impacts to minority or low-income populations under either alternative, nor would there be any generation of byproduct material requiring disposition.

Final Recommendation

After weighing the impacts of the proposed action and comparing the alternatives, the NRC staff, in accordance with 10 CFR 51.91(d), sets forth its NEPA recommendation regarding the proposed action. Unless safety issues mandate otherwise, the NRC staff's recommendation to the Commission related to the environmental aspects of the proposed action is that the source material license be issued as requested. This recommendation is based upon (1) the license application, including the environmental report submitted by Uranium One and the applicant supplemental letters and responses to the staff requests for additional information; (2) consultation with federal, state, tribal, and local agencies; (3) the NRC staff's independent review; (4) the NRC staff's consideration of comments received on the draft SEISs; and (5) the assessments summarized in this SEIS.

ABBREVIATIONS/ACRONYMS

AADT	annual average daily traffic count
ADAMS	Agency Wide Documents Access and Management System
ACL	alternate concentration limit
AEA	Atomic Energy Act
ALARA	as low as reasonably achievable
AMSL	above mean sea level
APE	area of potential effect
APLIC	Avian Power Line Interaction Committee
AQD	Air Quality Division
ARPA	Archaeological Resources Protection Act of 1979
bgs	below ground surface
BIA	Bureau of Indian Affairs
BLM	U.S. Bureau of Land Management
BMP	best management practice
B.P.	before present
CAA	Clean Air Act
CBM	coal bed methane
CCESC	Campbell County Educational Services Center
CCS	Center for Climate Strategies
CDNR	Colorado Department of Natural Resources
CEQ	Council on Environmental Quality
CERCLA	Comprehensive Environmental Response, Compensation, and Liability Act
CESQG	Conditionally Exempt Small Quantity Generator
CFR	Code of Federal Regulations
CO	carbon monoxide
CWA	Clean Water Act
dBA	decibels
DOC	U.S. Department of Commerce
DOE	U.S. Department of Energy
EMC	Energy Metals Corporation
EA	Environmental Assessment
EIS	Environmental Impact Statement
ENSR	ENSR Corporation
E.O.	Executive Order
EPA	U.S. Environmental Protection Agency
ER	Environmental Report
ERP	emergency response plan
ESA	Endangered Species Act of 1973
FCR	fire-cracked rock
FHWA	Federal Highway Administration
FONSI	finding of no significant impact
FR	Federal Register

Abbreviations/Acronyms

FSME	Office of Federal and State Materials and Environmental Management Programs
FWS	U.S. Fish and Wildlife Service
GCRP	U.S. Global Change Research Program
GHG	greenhouse gas
GEIS	Generic Environmental Impact Statement
gpm	gallons per minute
HDPE	high-density polyethylene
HKM	HKM Engineering, Inc.
I	Interstate
ISL	*in-situ* leach
ISR	*in-situ* recovery
IX	ion exchange
JCSD	Johnson County School District
LQD	Land Quality Division
Lpm	liters per minute
MBHFI	Migratory Birds of High Federal Interest
MCL	maximum contaminant level
MIT	mechanical integrity test
MOA	Memorandum of Agreement
MOU	Memorandum of Understanding
MSDS	material safety data sheets
NAAQS	National Ambient Air Quality Standards
NCDC	National Climatic Data Center
NCRP	National Council for Radiation Protection
NCTHPO	Northern Cheyenne Tribal Historic Preservation Office
NEPA	National Environmental Policy Act
NHPA	National Historic Preservation Act of 1966, as amended
NMSS	Nuclear Materials Safety and Safeguards
NOAA	National Oceanographic and Atmospheric Association
NOI	Notice of Intent
NPDES	National Pollutant Discharge Elimination System
NRC	U.S. Nuclear Regulatory Commission
NRCS	Natural Resource Conservation Service
NRHP	National Register of Historic Places
NWI	National Wetlands Inventory
OSHA	Occupational Safety and Health Administration
PA	Programmatic Agreement
PDR	Public Document Room
PM	particulate matter
PRI	Power Resources Inc.
PRRCT	Powder River Regional Coal Team

PSD	Prevention of Significant Deterioration
psig	pounds per square inch gauge
PSM	Process Safety Management
PVC	plastic polyvinyl chloride
RAI	Request for Additional Information
RCRA	Resource Conservation and Recovery Act
RFFA	reasonably feasible future action
RO	reverse osmosis
ROD	Record of Decision
ROI	region of influence
RQ	Reportable Quantity
RTV	Restoration Target Value
SDWA	Safe Drinking Water Act
SEIS	Supplemental Environmental Impact Statement
SER	Safety Evaluation Report
SHPO	State Historic Preservation Office
SR	State Highway
TCP	traditional cultural property
TEDE	total effective dose equivalent
TDS	total dissolved solids
THPO	Tribal Historic Preservation Office
TNM	Traffic Noise Model Version 2.5
TPQ	Threshold Planning Quantity
TQ	Threshold Quantity
TR	Technical Report
TSCA	Toxic Substances Control Act
TSS	total suspended solids
UCL	upper control limits
UIC	underground injection control
UMTRCA	Uranium Mill Tailings Radiation Control Act
U.S.	United States (or) United States Highway
USACE	U.S. Army Corps of Engineers
USDA	U.S. Department of Agriculture
USDOT	U.S. Department of Transportation
USFS	U.S. Forest Service
USC	United States Code
USCB	U.S. Census Bureau
USDW	Underground Source of Drinking Water
USGS	U.S. Geological Survey
VRM	Visual Resource Management
WBC	Wyoming Business Council
WDE	Wyoming Department of Education
WDEQ	Wyoming Department of Environmental Quality
WDOE	Wyoming Department of Employment, Research, and Planning
WDOR	Wyoming Department of Revenue

Abbreviations/Acronyms

WGFD	Wyoming Game and Fish Department
WLS	Western Land Services
WNDD	Wyoming Natural Diversity Database
WQD	Water Quality Division
W.S.	Wyoming Statute
WSEO	Wyoming State Engineer's Office
WYDOT	Wyoming Department of Transportation
WYNDD	Wyoming Natural Diversity Database
WYPDES	Wyoming Pollutant Discharge Elimination System

SI* (MODERN METRIC) CONVERSION FACTORS

\multicolumn{5}{c}{Approximate Conversions From SI Units}				
Symbol	When You Know	Multiply By	To Find	Symbol
\multicolumn{5}{c}{Length}				
cm	centimeters	0.39	inches	In
m	meters	3.28	feet	ft
m	meters	1.09	yards	yd
km	kilometers	0.621	miles	mi
\multicolumn{5}{c}{Area}				
mm^2	square millimeters	0.0016	square inches	in^2
m^2	square meters	10.764	square feet	ft^2
m^2	square meters	1.195	square yards	yd^2
Ha	hectares	2.47	acres	ac
km^2	square kilometers	0.386	square miles	mi^2
\multicolumn{5}{c}{Volume}				
mL	milliliters	0.034	fluid ounces	fl oz
L	liters	0.264	gallons	gal
m^3	cubic meters	35.314	cubic feet	ft^3
m^3	cubic meters	1.307	cubic yards	yd^3
m^3	cubic meters	0.0008107	acre-feet	acre-feet
Mass				
G	grams	0.035	ounces	oz
Kg	kilograms	2.202	pounds	lb
Mg (or "t")	megagrams (or "metric ton")	1.103	short tons (2,000 lb)	T
\multicolumn{5}{c}{Temperature (Exact Degrees)}				
°C	Celsius	1.8C + 35	Fahrenheit	°F

*SI is the symbol for the International System of Units. Appropriate rounding should be performed to comply with Section 4 of ASTM E380 (ASTM International. "Standard for Metric Practice Guide." West Conshohocken, Pennsylvania: ASTM International. (Revised 2003).

1 INTRODUCTION

1.1 Background

The U.S. Nuclear Regulatory Commission (NRC) prepared this Supplemental Environmental Impact Statement (SEIS) in response to an application submitted by Energy Metals Corporation (EMC) on October 2, 2007, to develop and operate the Moore Ranch Uranium Project (referred to herein as the proposed Moore Ranch Project), located in Campbell County, Wyoming, by *in-situ* recovery (ISR) methods (EMC, 2007) (Figure 1-1). EMC is a wholly owned subsidiary of Uranium One, Inc. as of August 10, 2007, Energy Metals Corporation was acquired by Uranium One, Inc, therefore, the applicant for the proposed Moore Ranch Facility is Uranium One Americas, Inc. (Uranium One). This site-specific SEIS supplements the *Generic Environmental Impact Statement for In-Situ Leach Uranium Milling Facilities* (referred to herein as the GEIS) in accordance with the process described in Section 1.8 of the GEIS (NRC, 2009a) and as detailed in Section 1.4.1 of this chapter. The NRC's Office of Federal and State Materials and Environmental Management Programs prepared this SEIS as required by Title 10, "Energy," of the *U.S. Code of Federal Regulations* (10 CFR) Part 51. These regulations implement the requirements of the *National Environmental Policy Act of 1969 (NEPA)*, as amended (Public Law 91-190) which requires the Federal Government to assess the potential environmental impacts of major federal actions that may significantly affect the human environment.

The GEIS used the terms "in-situ leach (ISL) process" and "11e.(2) byproduct material" to describe this uranium milling technology and the wastestream generated by this process. The SEIS replaces the term "in-situ leach (ISL)" with "in-situ uranium recovery (ISR)" to be consistent with the terminology used by the mining industry. The SEIS also uses the term "byproduct material" instead of "11e.(2) byproduct material" to describe the wastestream generated by this milling process to be consistent with the definition in 10 CFR 40.4.

1.2 Proposed Action

On October 2, 2007, EMC initiated the proposed federal action by submitting an application for an NRC source material license to construct and operate an ISR facility at the proposed Moore Ranch Project, and to conduct the consequent aquifer restoration and site decommissioning and reclamation activities. Based on the application, the NRC's federal action is the decision whether to either grant or deny the license to Uranium One. Uranium One's proposal is discussed in detail in Section 2.1.1 of the SEIS.

1.3 Purpose of and Need for the Proposed Action

NRC regulates uranium milling, including the ISR process, under 10 CFR Part 40, "Domestic Licensing of Source Material." Uranium One is seeking an NRC source material license to authorize commercial-scale ISR uranium recovery at the proposed Moore Ranch Project. The purpose and need for the proposed Federal action is to either grant or deny Uranium One's license application to use ISR technology to recover uranium and produce yellowcake at the proposed Moore Ranch Project. Yellowcake is the uranium oxide product of the ISR milling process that is used to produce fuel for commercially-operated nuclear power reactors. This definition of purpose and need reflects the Commission's recognition that, unless there are either findings in the safety review required by the Atomic Energy Act or findings in the NEPA

Introduction

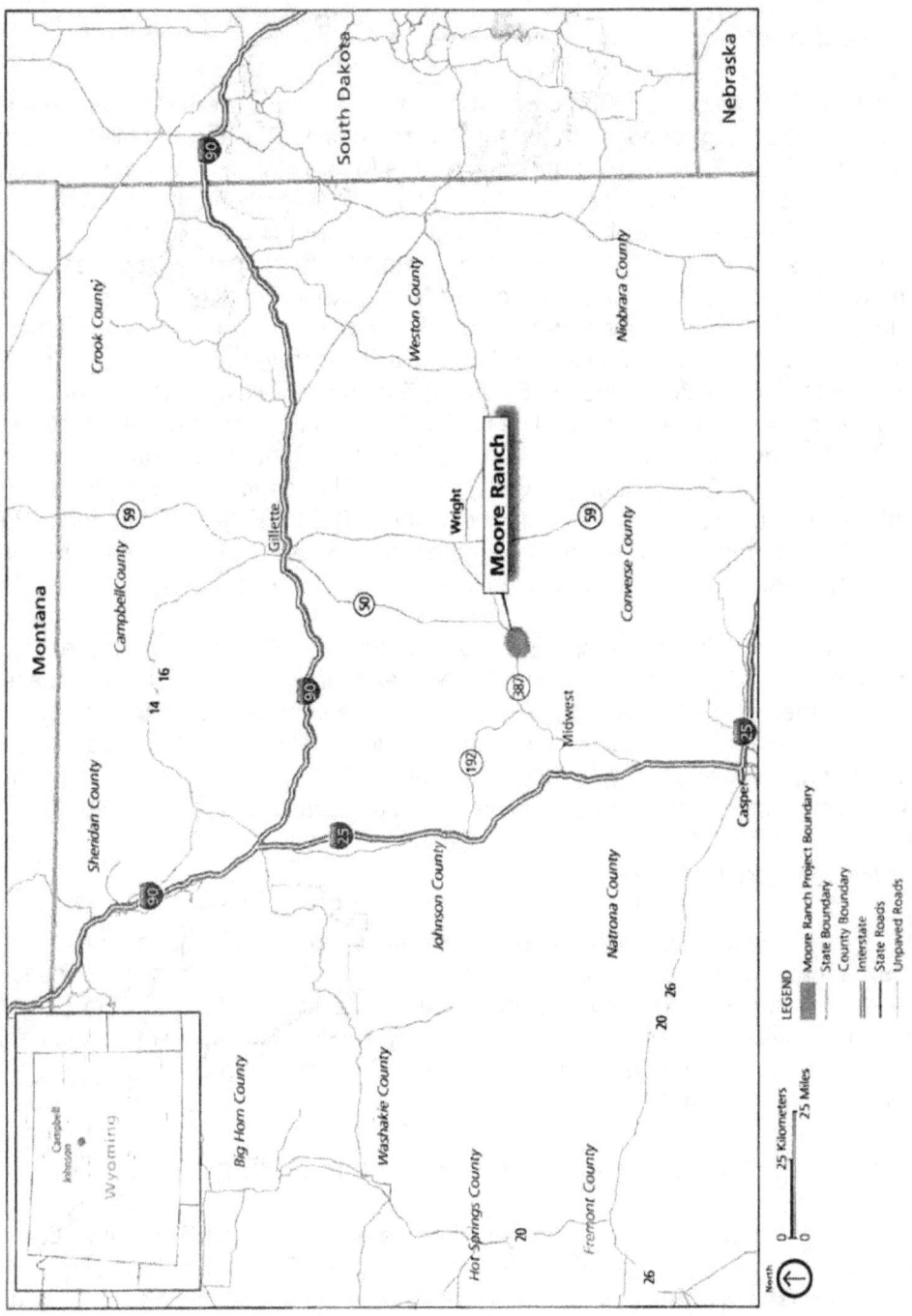

Figure 1-1. Location of the Proposed Moore Ranch Project

environmental analysis that would lead the NRC to reject a license application, the NRC has no role in a company's business decision to submit a license application to operate an ISR facility at a particular location.

1.4 Scope of the Supplemental Environmental Analysis

The NRC prepared this SEIS to analyze the potential environmental impacts (i.e., direct, indirect, and cumulative impacts) of the proposed action and of reasonable alternatives to the proposed action. The scope of this SEIS considers both radiological and nonradiological (including chemical) impacts associated with the proposed action and its alternatives. This SEIS also considers unavoidable adverse environmental impacts, the relationship between short-term uses of the environment and long-term productivity, and irreversible and irretrievable commitments of resources.

1.4.1 Relationship to the GEIS

As discussed previously, this SEIS supplements the GEIS, published as a final report in May 2009 (NRC, 2009a). The GEIS assessed the potential environmental impacts associated with the construction, operation, aquifer restoration, and decommissioning of an ISR facility located in four specific geographic regions of the western United States. The proposed Moore Ranch Project is located in the Wyoming East Uranium Milling Region. Table 1-1 summarizes the expected environmental impacts by resource area in the Wyoming East Uranium Milling Region based on the GEIS analyses.

The NRC staff considers the scope of the GEIS to be sufficient for the purposes of defining the scope of this SEIS. NRC accepted public comments on the scope of the GEIS from July 24 to November 30, 2007, and held three public scoping meetings, one of which was in the State of Wyoming. Additionally, NRC held eight public meetings to receive comments on the draft GEIS, published in July 2008. Three of these meetings were held in the State of Wyoming. Comments on the draft GEIS were accepted between July 28 and November 8, 2008. Comments received during scoping and on the draft GEIS are available through NRC's Agencywide Documents Access and Management System (ADAMS) database on the NRC's website (http://www.nrc.gov/reading-rm/adams.html). Transcripts of the scoping meeting and draft GEIS comment meetings in Wyoming are available at http://www.nrc.gov/materials/uranium-recovery/geis/pub-involve-process.html. A scoping summary report is provided as Appendix A to the GEIS (NRC, 2009a).

This SEIS was prepared to fulfill the requirement at 10 CFR 51.20(b)(8) to prepare either an EIS or supplement to an EIS for the issuance of a source material license for an ISR uranium recovery facility (NRC, 2009a). The GEIS provides a starting point for NRC's NEPA analyses for site-specific license applications for new ISR facilities, as well as for applications to amend or renew existing ISR licenses. As discussed in the GEIS, the GEIS provides criteria for each environmental resource area to help determine the significance level for potential impacts (e.g., SMALL, MODERATE, or LARGE). The NRC staff applied these criteria to the site-specific conditions at the proposed Moore Ranch Project.

This SEIS tiers and incorporates by reference from the GEIS relevant information, findings and conclusions concerning potential environmental impacts. The extent to which NRC incorporates GEIS impact conclusions depends on the consistency between Uranium One's proposed facility and activities and conditions at the proposed Moore Ranch Project and the reference facility

Introduction

Table 1-1. ISL GEIS Range of Expected Impacts in the Wyoming East Uranium Milling Region				
Resource Area	Construction	Operation	Aquifer Restoration	Decommissioning
Land Use	S to L	S	S	S to M
Transportation	S to M	S to M	S to M	S
Geology and soils	S	S	S	S
Surface Water	S	S to M	S to M	S to M
Groundwater	S	S to L	S to M	S
Terrestrial Ecology	S to M	S	S	S
Aquatic Ecology	S	S	S	S
Threatened and Endangered Species	S to L	S	S	S
Air Quality	S	S	S	S
Noise	S to M	S to M	S to M	S to M
Historical and Cultural Resources	S to L	S	S	S
Visual and Scenic Resources	S	S	S	S
Socioeconomics	S to M	S to M	S	S to M
Public Health and Safety	S	S to M	S	S
Waste Management	S	S	S	S
S: SMALL impact M: MODERATE impact L: LARGE impact Source: NRC, 2009a				

description and activities and information or conclusions in the GEIS. NRC's determinations regarding potential environmental impacts and the extent to which GEIS impact conclusions were incorporated by reference are discussed in Chapter 4 of this SEIS. Section 1.8.3 of the GEIS describes in detail the relationship between the GEIS and the conduct of site-specific reviews as documented in this SEIS (NRC, 2009a).

1.4.2 Public Participation Activities

As part of the preparation of this SEIS, NRC staff met with federal, state, and local agencies and authorities during the course of an expanded visit to the proposed Moore Ranch Project and site vicinity in January 2009. The purpose of these meetings was to gather additional site-specific information to assist in the NRC staff's environmental review and to aid the staff in its determination of the consistency between site and local information and similar information in the GEIS. As part of this effort to gather additional site-specific information, the NRC staff also contacted potentially interested Native American tribes and local authorities, entities, and public interest groups in person and via email and telephone.

NRC published a Notice of Opportunity for Hearing in the *Federal Register* on January 25, 2008 (see 73 FR 4642) related to the Moore Ranch license application (NRC, 2008a). No hearing requests were received. NRC also published a Notice of Intent to prepare this SEIS on August 21, 2009 (see 74 FR 42332) (NRC, 2009b).

1.4.3 Issues Studied in Detail

To meet its NEPA obligations related to its review of the Moore Ranch license application, the NRC staff has conducted an independent, comprehensive evaluation of the potential environmental impacts from construction, operation, aquifer restoration, and decommissioning of an ISR facility at the proposed Moore Ranch Project. As discussed in Section 1.8.3 of the GEIS, the GEIS (1) evaluated the types of environmental impacts that may occur from ISL uranium milling facilities, (2) identified and assessed impacts that are expected to be generic (the same or similar) at all ISR facilities (or those with specified facility or site characteristics), and (3) identified the scope of environmental impacts that needed to be addressed in site-specific environmental reviews. Therefore, although all of the environmental resource areas identified in the GEIS will be addressed in site-specific reviews, certain resource areas would require a more detailed analysis, because the GEIS analysis concluded there could be a range in potential impacts (e.g., SMALL to MODERATE, SMALL to LARGE) depending upon site-specific conditions (see Table 1-1). Based on the results of the GEIS analyses, this SEIS provides a more detailed analysis of the following resource areas:

- Land Use
- Transportation
- Geology and Soils
- Surface Water
- Groundwater
- Terrestrial Ecology
- Threatened and Endangered Species
- Meteorology, Climatology, and Air Quality
- Noise
- Historical and Cultural Resources
- Visual and Scenic Resources
- Socioeconomics
- Public Health and Safety
- Waste Management

Furthermore, certain site-specific analyses not conducted in the GEIS (e.g., assessment of cumulative impacts, analysis of environmental justice) were considered in this SEIS.

Additionally, the NRC also includes a brief discussion about the effect of global climate change on the potential impacts from the proposed action based on a 10 year licensing period and the effect of the facility on global climate change.

Introduction

1.4.4 Issues Outside the Scope of the SEIS

Some issues and concerns raised during the scoping process on the GEIS (NRC, 2009a; Appendix A) were determined to be outside the scope of the GEIS. These issues and concerns, (e.g., general support or opposition for uranium milling, potential impacts associated with conventional uranium milling, comments regarding the alternative sources of uranium feed material, comments regarding energy sources, requests for compensation for past mining

impacts, and comments regarding the credibility of NRC) were also determined to be outside the scope of this SEIS.

1.4.5 Related NEPA Reviews and Other Related Documents

The following NEPA and other related documents were reviewed as part of the development of this SEIS to obtain information relevant to the issues raised:

NUREG–1910, Generic Environmental Impact Statement for *In-Situ* Leach Uranium Milling Facilities, Final Report (NRC, 2009a). As discussed previously, the GEIS was prepared to assess the potential environmental impacts from the construction, operation, aquifer restoration, and decommissioning of an ISR facility located in four different geographic regions of the western United States, including the Wyoming East Uranium Milling Region where the Moore Ranch Project is located. The environmental analysis in this SEIS both tiers from and incorporates by reference the GEIS.

NUREG–0706, Final Generic Environmental Impact Statement on Uranium Milling (NRC, 1980). This EIS provided a detailed evaluation of the impacts and effects of anticipated conventional milling operations in the U.S. through the year 2000 including analysis of tailings disposal programs. The environmental impacts of underground mining and conventional milling would be more severe than using ISR technology. As discussed in Section 2.2.1 of the final SEIS, conventional mining and milling were considered, but eliminated from detailed analysis at the proposed Moore Ranch Project.

NUREG–1508, Final Environmental Impact Statement to Construct and Operate the Crownpoint Uranium Solution Mining Project, Crownpoint, New Mexico (NRC, 1997). This EIS evaluated the use of ISR technology at the Church Rock and Crownpoint sites at Crownpoint, New Mexico. Alternative uranium mining methods were not evaluated because the proposed sites were too deep to be extracted economically and underground mining would have more significant environmental impacts than ISR recovery.

Final Environmental Impact Statement for the West Antelope II Coal Lease Application WYW163340 (BLM, 2008). The U.S. Bureau of Land Management (BLM) prepared this EIS for the West Antelope Coal Lease Application WYW163340, located approximately 24.6 km [15 mi] east of the Moore Ranch Project. The document evaluates the environmental impacts of leasing and mining coal on approximately 1,664 ha [4,109 ac].

NRC's Safety Evaluation Report. The NRC staff is preparing a safety evaluation report (SER) for the Moore Ranch license application. In the SER, the NRC staff evaluates whether the licensee's proposed action can be accomplished in accordance with the applicable provisions of 10 CFR Part 20 and 10 CFR Part 40, Appendix A. The SER evaluates the licensee's proposed facility design, operational procedures, and radiation protection program to ensure that the applicable requirements in 10 CFR Part 20 and 10 CFR Part 40 would be met by the applicant.

The SER also provides the staff's analysis of the initial estimate from the applicant of the funding needed to complete site decommissioning and reclamation.

Final Environmental Impact Statement and Proposed Amendment of the Powder River Basin Oil and Gas Project, WY–070–02–065 (BLM, 2003). The BLM prepared this EIS to evaluate the potential effect from drilling, completing, operating and reclaiming 39,400 new natural gas wells and their associated infrastructure (roads, pipelines for gathering gas and produced water, electrical utilities, and compressors). This EIS evaluated potential effects to subwatersheds within the proposed Moore Ranch Project as well as to the aquifer that would be mined as part of the proposed action being considered in this SEIS.

NUREG–0889, Draft Environmental Statement related to the operation of Sand Rock Mill Project, Docket No. 40-8743, Conoco, Inc. (NRC, 1982). NRC evaluated the potential environmental impact from issuing a license to Conoco, Inc. to construct and operate a uranium mill, associated with an open-pit mine in the same geographic area now being considered for *in-situ* recovery of uranium. This environmental statement evaluated alternatives for tailings management including the use of evaporation ponds, which were considered but eliminated from detailed analysis in this SEIS.

NRC's Environmental Review for the Nichols Ranch ISR Project. The NRC is reviewing a license application from Uranerz Energy Corporation for an ISR project located on about 1,365 ha [3,371 ac] about 32 km [20 mi] northwest of the proposed Moore Ranch Project.

NRC's Environmental Review for the Irigaray and Christensen Ranch ISR Project License Renewal. The NRC is reviewing a license application from COGEMA Mining, Inc. for the renewal of Source Material License SUA–1341, which is located in Campbell and Johnson Counties about 30 km [19 mi] north-northwest of the proposed Moore Ranch Project. The Irigaray project was commercially licensed for ISR operations in 1978. The license was amended in 1987 to include the Christensen Ranch satellite facility. In June 2000, production ended and the site has been undergoing wellfield restoration and site decommissioning.

1.5 Applicable Regulatory Requirements

NEPA establishes national environmental policy and goals to protect, maintain, and enhance the environment. NEPA provides a process for implementing these specific goals for those Federal agencies responsible for an action. This SEIS was prepared in accordance with NEPA requirements and NRC's implementing regulations in 10 CFR Part 51, and consistent with other regulations that were in effect at the time of writing. Appendix B of the GEIS summarizes other Federal statutes and implementing regulations and Executive Orders that are potentially applicable to environmental reviews for the construction, operation, aquifer restoration, and decommissioning of an ISR facility.

Sections 1.6.3.1 and 1.7.5.1 of the GEIS provide a summary of the State of Wyoming's statutory authority pursuant to the ISR process, relevant state agencies that are involved in the permitting of an ISR facility, and the range of state permits that would be required (NRC, 2009a).

1.6 Licensing and Permitting

NRC has statutory authority through the *Atomic Energy Act* as amended by the *Uranium Mill Tailings Radiation Control Act* (UMTRCA) to regulate uranium ISR facilities. In addition to obtaining an NRC license, uranium ISR facilities must also obtain the necessary permits from

Introduction

the appropriate federal, state, local and tribal governmental agencies. The NRC licensing process for ISR facilities was described in Section 1.7.1 of the GEIS. Sections 1.7.2 through 1.7.5 of the GEIS describe the role of the other Federal, tribal, and state agencies in the ISR permitting process.

This section of the SEIS summarizes the status of the NRC licensing process at the proposed Moore Ranch Project site and the status of Uranium One's permitting with respect to other applicable Federal, tribal, and state requirements.

1.6.1 NRC Licensing Process

By letter dated October 2, 2007, Uranium One submitted a final license application to NRC for the Moore Ranch Project (EMC, 2007). As discussed in Section 1.7.1 of the GEIS, NRC initially conducts an acceptance review of a license application to determine whether the application is complete enough to support a detailed technical review. The NRC staff accepted the Moore Ranch license application for detailed technical review by letter dated December 20, 2007 (NRC, 2007).

The NRC's detailed technical review of the Moore Ranch license application is comprised of both a safety review and an environmental review. These two reviews are conducted in parallel (see Figure 1.7-1 of the GEIS). The focus of the safety review is to assess compliance with the applicable regulatory requirements in 10 CFR Part 20, 10 CFR Part 40, and Part 40, Appendix A. The environmental review is conducted in accordance with the regulations in 10 CFR Part 51.

The NRC hearing process (10 CFR Part 2) applies to proposed licensing actions and offers stakeholders a separate opportunity to raise concerns associated with the proposed action. NRC published a Notice of Opportunity for Hearing in the *Federal Register* on January 25, 2008 (see 73 FR 4642) related to the Moore Ranch license application (NRC, 2008a). No request for a hearing was received on the Moore Ranch license application.

1.6.2 Status of Permitting with Other Federal, Tribal, and State Agencies

In addition to obtaining a source material license from NRC prior to conducting ISR operations at the Moore Ranch Project, Uranium One is also required to obtain necessary permits and approvals from other federal, tribal, and state agencies. These permits and approvals would address issues such as (1) the underground injection of solutions and wastewater associated with the ISR process; (2) the exemption of all or a portion of the extraction zone aquifer from regulation under the Safe Drinking Water Act; and (3) the discharge of stormwater during construction and operation of the ISR facility.

Table 1-2 provides the permit status for Uranium One's proposed Moore Ranch Project.

Table 1-2. Environmental Approvals for the Moore Ranch Uranium Project		
Issuing Agency	Description	Status
Wyoming Department of Environmental Quality (WDEQ)	Underground Injection Control Class III Permit (WDEQ Title 35-11)	Class III UIC Permit application under review; anticipated approval by WDEQ in June 2010
	Aquifer Exemption (WDEQ Title 35-11)	Aquifer exemption application under preparation; anticipated WDEQ review complete in June 2010
	Underground Injection Control Class I (WDEQ Title 35-11)	Class I UIC Permit application under review; anticipated approval by WDEQ in second quarter 2010
	Industrial Stormwater NPDES Permit (WDEQ Title 35-11)	An Industrial Stormwater NPDES will be required for the Central Plant area. Expected submittal in second quarter 2011
	Construction Stormwater NPDES Permit (WDEQ Title 35-11)	Construction Stormwater NPDES authorizations are applied for and issued annually under a general permit based on projected construction activities. The Notice of Intent will be filed at least 30 days before construction activities begin in accordance with WDEQ regulations.
	Mineral Exploration Permit (WDEQ Title 35-11)	Mineral Exploration Permit 342DN Approved: August 22, 2006
	Underground Injection Control Class V (WDEQ Title 35-11)	The Class V UIC permit will be applied for following installation of an approved site septic system during facility construction
	Air Quality Permit	Air Quality Permit Application (AP-10490) submitted March 2010, under technical review
U.S. Nuclear Regulatory Commission	Source Materials License (10 CFR Part 40)	Application under review
U.S. Environmental Protection Agency (EPA)	Aquifer Exemption (40 CFR 144, 146)	Aquifer exemption application will be forwarded to EPA following WDEQ action
U.S. Army Corps of Engineers	Nationwide Permit (NP) #12 Authorization	Activities authorized on May 10, 2010 per NP #12 as defined in Part II of the *Federal Register* (Vol. 72 No. 47, March 12, 2007).
Source: Winter, 2010		

Introduction

1.7 Consultations

As a Federal agency, the NRC is required to comply with consultation requirements in Section 7 of the *Endangered Species Act of 1973* (ESA), as amended, and Section 106 of the *National Historic Preservation Act of 1966* (NHPA), as amended. The GEIS took a programmatic look at the environmental impacts of ISR uranium recovery within four distinct geographic regions and acknowledged that each site-specific review would include its own consultation process with relevant agencies. Section 7 and Section 106 consultation conducted for the proposed Moore Ranch Project is summarized in Sections 1.7.1 and 1.7.2 below. Copies of the correspondence for this consultation are provided in Appendix A of this SEIS. Section 1.7.3 discusses NRC coordination with other federal, state, and local agencies that was conducted during the development of the SEIS.

1.7.1 Endangered Species Act of 1973 Consultation

The ESA was enacted to prevent the further decline of endangered and threatened species and to restore those species and their critical habitats. Section 7 of the Act requires consultation with the U.S. Fish and Wildlife Service (FWS) to ensure that actions they authorize, permit or otherwise carry out will not jeopardize the continued existence of any listed species or adversely modify designated critical habitats.

By letter dated April 9, 2008, NRC staff initiated consultation with the FWS, requesting information on endangered or threatened species or critical habitat on the Moore Ranch Project area (NRC, 2008b). NRC received a response from the Ecological Services Wyoming Field Office of the FWS, dated May 7, 2008, that (1) provided a list of the threatened or endangered species that may occur in the project area; (2) discussed obligations to protect migratory birds; (3) noted the negative impacts that can result from the land application of ISR wastewater; and (4) recommended avoidance of wetland and riparian areas and protection of sensitive species, such as the mountain plover and sage grouse (FWS, 2008).

NRC staff also met with the FWS Buffalo office on January 14, 2009 to discuss site-specific issues. The main concern expressed by the Buffalo office was potential impacts to sage grouse and typical mitigation measures were discussed (NRC, 2009c).

1.7.2 National Historic Preservation Act of 1966 Consultation

Section 106 of the NHPA requires that federal agencies take into account the effects of their undertakings on historic properties and allow the Wyoming State Historic Preservation Office (SHPO) to comment on such undertakings.

NRC initiated consultation with the Wyoming SHPO via a letter dated April 9, 2008 (NRC, 2008c), requesting information from the SHPO to facilitate the identification of historic and cultural resources that could be affected by the proposed project. NRC staff also met with members of the SHPO's office on January 12, 2009 (NRC, 2009c), to discuss site-specific issues, including Wyoming SHPO's review process, cumulative impacts to historic sites, and best management practices.

By letter dated October 22, 2009, the NRC solicited comment from the Wyoming SHPO regarding Uranium One's cultural resources survey reports and related documentation (NRC, 2009d). In a letter dated November 3, 2009, the Wyoming SHPO concurred that the sites located within the proposed Moore Ranch project area are ineligible for listing on the

National Register of Historic Places. In addition, the letter states: "We recommend the U.S. Nuclear Regulatory Commission allow the project to proceed in accordance with state and federal laws subject to the following stipulation: If any cultural materials are discovered during construction, work in the area shall halt immediately, the federal agency and SHPO staff be contacted, and the materials be evaluated by an archaeologist or historian meeting the Secretary of the Interior's Professional Qualification Standards (48 FR 22716, September 1983). Additionally, if any future disturbance is planned at the locations of sites 48CA964, 48CA6694 or 48CA6696 that evaluative testing be completed and submitted to our office with a determination of site eligibility and project effect. If eligible and adversely affected, a Memorandum of Agreement implementing appropriate mitigative measures will be required." (Wyoming SHPO, 2010). The applicant has agreed to condition the license, if issued, to include a stop-work provision.

The NRC staff has consulted with the Wyoming SHPO throughout the environmental review process.

Section 1.7.3.3 Interactions with Tribal Governments provides a listing of the nine tribes consulted to solicit their comments or concerns regarding cultural resources and the proposed Moore Ranch Project.

1.7.3 Coordination with Other Federal, Tribal, State, and Local Agencies

The NRC staff interacted with multiple federal, tribal, state, and local agencies and/or entities during preparation of this SEIS to gather information on potential issues, concerns, and environmental impacts related to the proposed ISR facility at the Moore Ranch site. The consultation and coordination process included, but was not limited to, discussions with the BLM, the Bureau of Indian Affairs (BIA), tribal governments, the WDEQ, the Wyoming State Engineer's Office (WSEO), and local organizations.

1.7.3.1 Coordination with the Bureau of Land Management

The BLM is responsible for administering the National System of Public Lands and the federal minerals underlying these lands. The BLM also manages split estate situations where federal minerals underlie a surface that is privately held or owned by state or local government. The proposed Moore Ranch Project contains no BLM surface-administered lands, but there are mineral rights administered by the BLM, which have been leased to the applicant.

While the BLM was not a cooperating agency for this SEIS, NRC staff coordinated with the BLM during its preparation. During the review, NRC staff met with personnel from BLM's Buffalo Field Office and the State Office in Cheyenne, Wyoming. BLM provided NRC staff with guidance documents; clarification regarding mineral leases administered on BLM lands; and expressed concerns related to water quality and hydrology, cumulative effects, and socioeconomic impacts.

1.7.3.2 Coordination with the Bureau of Indian Affairs

The U.S. Department of the Interior, Bureau of Indian Affairs' (BIA) mission is to enhance the quality of life, to promote economic opportunity, and to carry out the responsibility to protect and improve the trust assets of American Indians, Indian tribes, and Alaska Natives. BIA is responsible for the administration and management of 66 million acres of land held in trust by the United States for American Indian, Indian tribes, and Alaska Natives.

Introduction

The NRC staff met with staff from the BIA in Fort Washakie, Wyoming on January 15, 2009 (NRC, 2009c). The NRC staff briefed the BIA on proposed ISR facilities in Wyoming, and the involvement of BIA and Indian tribes in the environmental review process was discussed. The BIA stated that tribal governments should be consulted for any projects in the state. BIA also recommended that tribal elders be involved in cultural and historic surveys.

1.7.3.3 Interactions with Tribal Governments

In response to guidance from the Wyoming SHPO and to carry out Executive Order 13175, 'Consultations and Coordination with Indian Tribal Governments,' the NRC staff initiated discussions with tribes that possess heritage and cultural interest and ties to the Moore Ranch project area. Letters dated February 23, 2009 (NRC, 2009e), were sent to the following nine tribes to solicit their comments or concerns regarding cultural resources and the Moore Ranch Project:

- Eastern Shoshone
- Northern Arapaho
- Northern Cheyenne
- Blackfeet
- Three Affiliated Tribes
- Ft. Peck Assinboine/Sioux
- Oglala Sioux
- Crow
- Cheyenne River Sioux

To date, no responses from these Tribes have been received.

1.7.3.4 Coordination with the Wyoming Department of Environmental Quality

NRC staff met with the WDEQ in Cheyenne on January 12, 2009 (NRC, 2009c) to discuss the WDEQ's role in NRC's environmental review process for the Moore Ranch project (see NRC Trip Report in Appendix A). Issues that were brought up during the meeting included the WDEQ storm water program, air quality review and permitting, and noise quality. The WDEQ also clarified the injection well classification system. The WDEQ noted that groundwater quality should be restored to class of use conditions per statute and State regulations. Under NRC regulations, as a license condition, the ISR operator would be required to return the aquifer to baseline conditions or to the maximum contaminant levels provided in Table 5C of 10 CFR Part 40, Appendix A or to alternate concentration limits approved by NRC. The WDEQ requested early involvement in the NRC's review of applications for proposed ISR projects in the State and emphasized coordination with the BLM when ISR projects are located on BLM lands. NRC staff met with the WDEQ in both June and September 2009 to coordinate review of this SEIS and to ensure that NRC's assessment included issues of interest to the WDEQ.

NRC staff also met with the WDEQ staff in Sheridan and Lander, Wyoming on January 14, 2009 (NRC, 2009c). The WDEQ-Land Quality Division (LQD) explained the UIC Class III well application process, and noted that the WDEQ would require wellfield packages and groundwater restoration standards for future ISR operations. They expressed concern about potential excursions and unconfined aquifers. WDEQ-LQD staff also stated that groundwater parameters affected by ISR operations need to be restored to original background levels. They supported the use of solar evaporation ponds for wastewater disposal, but stated that ISR applicants, Native Americans, and the FWS have expressed concerns regarding the use of evaporation ponds. NRC has communicated with the WDEQ via teleconference and periodic meetings to discuss regulatory jurisdiction, the Wyoming permitting process, and the status of various issues.

1.7.3.5 Coordination with the Wyoming Game and Fish Department

The Wyoming Game and Fish Department (WGFD) is responsible for controlling, propagating, managing, protecting, and regulating all game and nongame fish and wildlife in Wyoming under Wyoming Statute (W.S.) 23-1-301-303 and 23-1-401. Regulatory authority given to WGFD allows for the establishment of hunting, fishing, and trapping seasons, as well as the enforcement of rules protecting nongame and state-listed species.

The proposed license area includes habitat for a variety of big game animals, raptors, migratory birds, and small mammals that may be affected by the proposed project. The WGFD has an interest in potential impacts to migratory behavior patterns, long-term population sustainability, and the effects of local hunting on big game; impacts to nesting raptors; and the loss of nesting habitat for the greater sage-grouse.

Based on a FWS recommendation, NRC staff sought information from the WGFD regarding sage grouse habitat within the proposed license area and appropriate mitigative measures to minimize potential impacts to the sage grouse. WGFD responded that there was no known sage grouse habitat on the proposed Moore Ranch site (WGFD, 2009). Discussions via conference calls have also been held to pinpoint specific project-related concerns regarding impacts to other fish and wildlife populations. Those concerns have been noted and addressed in project planning.

The WGFD also provided input on several terrestrial and aquatic habitats which have been considered in the discussion of the affected environment (Chapter 3) and in the impacts analysis (Chapter 4) of this draft SEIS. NRC staff has also informally consulted with WGFD.

1.7.3.6 Coordination with the Wyoming State Engineer's Office

NRC staff met with the WSEO on January 12, 2009 (NRC 2009c) to discuss well permitting. The WSEO was primarily concerned that proposed ISR facilities do not degrade the water quality, and that potential groundwater contamination be maintained onsite. They also expressed the need for applicants to ensure that there was close, professional supervision of well construction.

1.7.3.7 Coordination with the Wyoming Governor's Planning Office

NRC staff met with the Wyoming Governor's Planning Office on January 13, 2009 (NRC 2009c) and again on June 25, 2009. The Wyoming Governor's Planning Office briefed the NRC on the BLM Resource Management Plan for the Buffalo region. They stated that they are a cooperating agency with the BLM and are involved with anything related to natural resources, particularly BLM resource management plans, and with the Wyoming SHPO and WDEQ. They informed NRC of the statewide conservation and management efforts for sage grouse and noted that the governor has created a management plan for the protection of sage grouse. They emphasized that potential ISR facilities need to be geographically flexible to protect the core sage grouse areas.

Introduction

1.7.3.8 Coordination with the Wyoming Community Development Authority

NRC staff met with the Wyoming Community Development Authority on January 13, 2009 (NRC, 2009c) to discuss housing availability for employees of future potential ISR facilities. They noted that employees would typically look for housing in the surrounding communities and this might include hotels, apartments, or single-family homes.

1.7.3.9 Coordination with Localities

The NRC staff interacted with several local county and city entities in the vicinity of the proposed license area which has included teleconferences and face-to-face meetings. NRC met with several local county and city entities on January 13 and 15, 2009 (NRC 2009c) to discuss site-specific issues for the proposed Moore Ranch Project. Meetings were held with the following local entities: Douglas and Converse County Office, City of Casper Planning Office, City of Gillette and Campbell County Office, Converse Area New Development Organization, and the Town of Wright. Meetings with the local county and city entities focused on local economies, housing availability, and community services.

1.8 Structure of the SEIS

As noted in Section 1.4.1 of this document, the GEIS (NRC, 2009a) evaluated the broad impacts of ISR projects in a four-state region where such projects are common, but did not reach site-specific decisions for new ISR projects. In this SEIS, the NRC staff evaluated the extent to which information and conclusions in the GEIS could be incorporated by reference. The NRC staff also determined whether any new and significant information existed that would change the expected environmental impact beyond that discussed in the GEIS.

Chapter 2 of this SEIS describes the proposed action and reasonable alternatives considered for the proposed Moore Ranch Project, Chapter 3 describes the affected environment for the Moore Ranch site, and Chapter 4 evaluates the environmental impacts from implementing the proposed action and alternatives. Cumulative impacts are discussed in Chapter 5; Chapter 6 describes the environmental measurement and monitoring programs proposed for the Moore Ranch Project. A cost-benefit analysis is provided in Chapter 7, and a summary of environmental consequences from the proposed action and No-Action alternative is summarized in Chapter 8.

1.9 References

10 CFR Part 2. *Code of Federal Regulations*, Title 10, *Energy,* Part 2, "Rules of Practice for Domestic Licensing Proceedings and Issuance of Orders."

10 CFR Part 20. *Code of Federal Regulations*, Title 10, *Energy*, Part 20, "Standards for Protection Against Radiation."

10 CFR Part 40 Appendix A. *Code of Federal Regulations*, Title 10, *Energy*, Part 40 Appendix A, "Criteria Relating to the Operation of Uranium Mills and to the Disposition of Tailings or Wastes Produced by the Extraction or Concentration of Source Material from Ores Processed Primarily from their Source Material Content."

10 CFR Part 51. *Code of Federal Regulations*, Title 10, *Energy*, Part 51, "Environmental Protection Regulations for Domestic Licensing and Related Regulatory Functions."

40 CFR Part 1508. *Code of Federal Regulations*, Title 40, *Protection of Environment*, Part 1508, "Terminology and Index."

Atomic Energy Act of 1954, as amended. 42 USC 2011 et seq.

BLM, 2008. "Final Environmental Impact Statement for the West Antelope II Coal Lease Application." WYW163340. 2008.

BLM, 2003. "Final Environmental Impact Statement and Proposed Amendment for the Powder River Basin Oil and Gas Project (WY–070–02–065)," January 2003.

Energy Metals Corporation (EMC), 2007. U.S. "Application for USNRC Source Material License, Moore Ranch Uranium Project, Campbell County, Wyoming, Environmental Report." Casper, Wyoming: Uranium 1 Americas Corporation. ADAMS Accession Nos. ML072851222, ML072851229, ML072851239, ML07285249, ML07285253, ML07285255. October 2, 2007.

Endangered Species Act of 1973. 16 USC 1531–1544.

Executive Order 13175. 2000. *Consultation and Coordination with Indian Tribal Governments.* 65 FR 67249. (November 9).

FWS (U.S. Fish and Wildlife Service), 2008. Response to Request for Additional Information Regarding Endangered or Threatened Species and Critical Habitat for the Proposed License Application for Energy Metals Corporation's Moore Ranch Uranium Recovery Project. ADAMS Accession No. ML081420589. May 7, 2008.

National Environmental Policy Act of 1969 (NEPA). 42 USC 4321 et seq.

National Historic Preservation Act of 1966, as amended (NHPA). 16 USC 470aa et seq.

NRC (U.S. Nuclear Regulatory Commission), 2009a. NUREG–1910, "Generic Environmental Impact Statement for *In-Situ* Leach Uranium Milling Facilities." Washington, DC. May.

NRC, 2009b. "Notice of Intent to Prepare a Supplemental Environmental Impact Statement for Uranium One Incorporated, Moore Ranch *In-Situ* Recovery Project." *Federal Register*, Vol. 74, No. 161, pp. 42,332–42,333. August 21, 2009.

NRC, 2009c. Memo to A. Kock, Branch Chief, from I. Yu, B. Shroff, and A. Bjornsen, Project Managers, Office of Federal and State Materials and Environmental Management Programs. Subject: Informal Meetings with Local, State, and Federal Agencies in Wyoming Regarding the Environmental Reviews Being Conducted on the Moore Ranch, Nichols Ranch, and Lost Creek *In-Situ* Leach Applications for Source Material Licenses (Docket Nos. 040-09073, 040-09067, 040-09068, Respectively). ADAMS Accession No. ML090500544. March 2, 2009.

NRC, 2009d. Letter to Richard Currit, Wyoming State Historic Preservation Office, Uranium One Inc. Moore Ranch *In-Situ* Uranium Recovery Project—Section 106 Consultation. ADAMS Accession No. ML092790445. October 22, 2009.

Introduction

NRC, 2009e. Request for Information Regarding Tribal Historic and Cultural Resources Potentially Affected by the Proposed License Application for Uranium One Inc.'s Moore Ranch Uranium Recovery Project in Campbell County, Wyoming. ADAMS Accession No. ML090440139. February 23, 2009.

NRC, 2008a. "Notice of License Application Request of Energy Metals Corporation, WY and Opportunity to Request a Hearing." *Federal Register*. Vol 73, No 17, pp. 4,642–4,646. January 25, 2008.

NRC, 2008b. Letter to Brian Kelly, U.S. Fish and Wildlife Service. Request for Information Regarding Endangered or Threatened Species and Critical Habitat for the Proposed License Application for Energy Metals Corporation's Moore Ranch Uranium Recovery Project. ADAMS Accession No. ML080950201. April 9, 2008.

NRC, 2008c. Letter to Mary Hopkins, Wyoming State Historic Preservation Office, Initiation of Section 106 Process for Energy Metals Corporation's Moore Ranch Uranium Recovery Project License Request. ADAMS Accession No. ML080950161. April 9, 2008.

NRC, 2007. Acceptance Review of License Amendment Request for Energy Metals Corp. Moore Ranch Uranium Enrichment Project, 20 December 2007.

NRC, 2003. NUREG–1748, "Environmental Review Guidance for Licensing Actions Associated With NMSS Programs—Final Report." Washington, DC: August 2003.

NRC, 1997. NUREG–1508, "Final Environmental Impact Statement to Construct and Operate the Crownpoint Uranium Solution Mine Project, Crownpoint, New Mexico," ADAMS Accession No. ML082170248. February.

NRC, 1980. NUREG–0706, "Final Generic Environmental Impact Statement on Uranium Milling Project M-25." ADAMS Accession Nos. ML032751663, ML0732751667, ML032751669. September.

NRC, 1982. NUREG–0889, "Draft Environmental Statement Related to the Operation of Sand Rock Mill Project." Washington, DC. March 1982.

Uranium Mill Tailings Radiation Control Act of 1978 (UMTRCA). 42 USC 7901 et seq.

Uranium One, 2010a. Email from J.F. Winter, Manager Environmental and Regulatory Affairs, Uranium One to B. Shroff, Project Manager, NRC. Subject: Re: Waste Issues. May 12, 2010. ADAMS Package No. ML101330411.

Wyoming Statute (W.S.) 23-1-301-303 and 23-1-401.

WGFD (Wyoming Game and Fish Department), 2009. Letter to NRC staff Regarding Request for Information Regarding Sage-Grouse Habitats for the Proposed License Application for Uranium One Inc.. ADAMS Accession No. ML092920276. September.

Wyoming SHPO (Wyoming State Historic Preservation Office), 2008. Letter to NRC staff Regarding Energy Metals Corporation, Initiation of Section 106 Process for the Moore Ranch Recovery License Request. ADAMS Accession No. ML081850356. June 5.

2 IN-SITU URANIUM RECOVERY AND ALTERNATIVES

Chapter 2 of the Generic Environmental Impact Statement for *In-Situ* Leach Uranium Milling Facilities (NUREG–1910, GEIS) provided information on uranium recovery using the *in-situ* leach (ISL) process (NRC, 2009). This chapter describes the application of those processes and the alternatives considered for the issuance of a U.S. Nuclear Regulatory Commission (NRC) license to Uranium One for the construction, operation, aquifer restoration, and decommissioning of the proposed Moore Ranch Project. This chapter describes the proposed action and alternatives, which include a consideration of the No-Action alternative, as required by the National Environmental Policy Act (NEPA). Under the No-Action alternative, Uranium One would not construct, operate, restore the aquifer, or decommission the proposed Moore Ranch Project. The No-Action alternative is included to provide a basis for comparing and evaluating the potential impact of the proposed action.

Section 2.1 describes the alternatives considered for detailed analysis in this SEIS, including the proposed action. Section 2.2 describes those alternatives that were considered but eliminated from detailed analysis. Section 2.3 sets forth the final NRC staff recommendation. Section 2.4 provides the references cited for this chapter.

2.1 Alternatives Considered for Detailed Analysis

The NRC staff used a variety of sources to determine the range of alternatives to consider for detailed analysis in this draft Supplemental Environmental Impact Statement (SEIS). Those sources included the license application, including the Environmental Report (ER) submitted by Energy Metals Corporation, a wholly-owned subsidiary of Uranium One; the scoping and draft comments on NUREG–1910, "Generic Environmental Impact Statement for *In-Situ* Leach Uranium Milling Facilities;" the information gathered during the NRC staff site visit in January 2009; and interdisciplinary discussions held between the NRC staff and various stakeholders.

2.1.1 The Proposed Action (Alternative 1)

Under the proposed action, Uranium One is seeking an NRC source material license for the construction, operation, aquifer restoration, and decommissioning of the *in-situ* recovery (ISR) facility at the proposed Moore Ranch ISR Project as described in the license application. The applicant's proposed disposal method is via a Class I injection well discussed in Section 2.1.1.1; however, alternative wastewater disposal options for the proposed action are discussed in Section 2.1.1.2.

2.1.1.1 Disposal Via Class I Injection Well

The proposed Moore Ranch ISR Project includes a central processing plant and two wellfields, which are described in the following sections. The general ISR process is described in Chapter 2 of the GEIS. The information contained in the following sections was obtained either from the application (EMC, 2007a, b) or from the GEIS (NRC, 2009) unless otherwise stated.

2.1.1.1.1 Site Description

The license area for the proposed Moore Ranch Uranium Project (Moore Ranch Project) is comprised of about 2,879 ha [7,110 ac] in the Powder River Basin in Campbell County,

In-Situ Uranium Recovery and Alternatives

Wyoming. The actual surface area that would be affected by the proposed ISL operation would be less than 61 ha [150 ac] and would consist of the main processing area, the wellfields (and their associated infrastructure including pipelines and trunklines), extraction support facilities such as warehouses and chemical storage areas, and access roads. The location of the proposed license area is comprised of the following, or portions of the following, townships: Township 42 North, Range 74 West, Sections 25, 26, 27, 28, 33, 34, 35, 36; Township 41 North, Range 75 West, Sections 1, 2, 3, 4, 9 and 10; and Township 42 North, Range 74 West, Sections 30 and 31. The proposed main processing area (referred to herein as the central plant) would be in the northeast quarter of Section 34 of Township 42 North, Range 74 West (Figure 2-1). The proposed project site is located between the towns of Wright and Edgerton, which are approximately 40 km [25 mi] northeast and 39 km [24 mi] southwest from the proposed Moore Ranch Project. No occupied housing units exist in the proposed license area; it is primarily used for grazing. Numerous wells used for coal bed methane (CBM) production also exist in the proposed license area, as described in Section 3.2.1.3 of this SEIS. Under the proposed action, NRC would issue a license for the construction, operation, aquifer restoration, and decommissioning of facilities for ISR uranium milling and processing at the Moore Ranch Project. During the construction phase of the action, buildings, access roads, wellfields, and pipelines would be constructed, as described in Section 2.1.1.1.2 of this SEIS. Most of the significant surface and subsurface disturbance would occur within the construction area for the central plant (Figure 2-1) and result in the disturbance of approximately a 2.4 ha [6-ac] area. Wellfield 1 would cover an area of approximately 9.3 ha [23 ac] and Wellfield 2 would be developed over an approximate 13.7 [34 ac] area (Figure 2-1); wellfields development would entail laying pipeline and trunkline; and about 2 km [1.25 mi] of new gravel road would be constructed.

In addition, under the proposed action, between two and four Underground Injection Control (UIC) Class I injection wells would also be drilled for disposal of liquid effluent generated from production bleed, restoration [reverse osmosis (RO) brine], and miscellaneous plant wastewater. The formations into which the wells would be drilled are thousands of feet below the proposed ore production zone at the Moore Ranch Project, and they are described in more detail in Chapter 3. Uranium One submitted a permit application to WDEQ for a UIC Class I injection permit for deep well disposal at the proposed Moore Ranch Site. Disposal of liquid byproduct material would be reviewed under the NRC 20.2002 criteria. The Wyoming Department of Environmental Quality (WDEQ) is currently reviewing this permit application.

The proposed operations phase at the Moore Ranch Project would last approximately 12 years; however; each wellfield would be operational for about 3.25 years (Griffin, 2009). The central plant would operate at a maximum flow rate of approximately 11,364 L/min [3,000 gal/min] and the plant is expected to produce 2 to 3 million pounds of uranium per year. After uranium recovery has ended, the groundwater in the wellfield would contain constituents that were mobilized by the extraction process (the lixiviant). The applicant would initiate aquifer restoration in each wellfield after the uranium recovery operations end (NRC, 2008). NRC licensees are required to return water quality parameters to the standards in 10 CFR Part 40, Appendix A, Criterion 5B(5). Details associated with aquifer restoration and the establishment of alternate concentration limits (ACL) are detailed in Appendix C.

2.1.1.1.2   Construction Activities

General construction activities associated with ISR operations include drilling wells, as described in Section 2.4 of the GEIS, and include activities such as clearing and grading

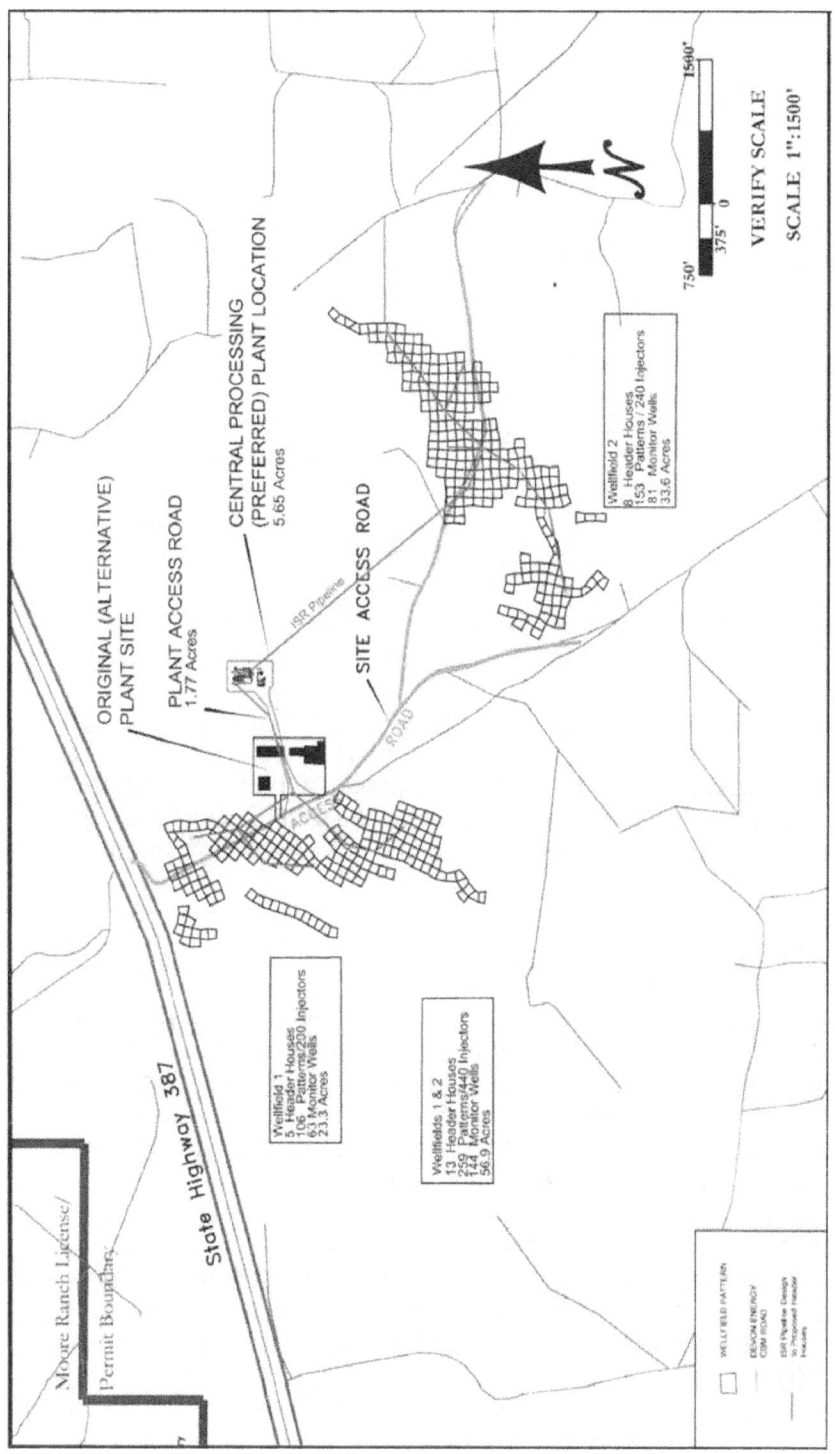

Figure 2-1. Site Layout at the Proposed Moore Ranch Project

In-Situ Uranium Recovery and Alternatives

associated with road construction and building foundations, building assembly, trenching, well drilling, and laying pipelines (NRC, 2009). The proposed facilities would consist of the central plant and associated infrastructure, such as the wellfields, pipelines, and roads. This section describes the physical plant, composed of the central plant including ion exchange (IX) column, wellfields, a construction and maintenance shop, warehouse, water treatment equipment, resin transfer facilities, lixiviant injection pumps, a small laboratory, and an employee break room. Figure 2-2 shows the different facilities that would comprise the proposed Moore Ranch Project.

2.1.1.1.2.1 Central Plant

The central plant at the proposed Moore Ranch Project would be constructed to provide chemical makeup of recovery solutions, recovery of uranium by IX, resin loading/unloading, elution and precipitation circuits, yellowcake drying capabilities, and groundwater restoration capabilities. The proposed Moore Ranch central plant facilities would be housed in a building with dimensions of approximately 107 × 30 m [350 × 100 ft] located within a 2.4 ha [6-ac] fenced area in the NE1/4, Section 34, T42N, and R75W. As stated previously, it would be designed and constructed to produce approximately 2–3 million pounds of uranium per year (EMC, 2007a). Uranium One has indicated that in the future they may process uranium-loaded resin from other potential Uranium One satellite projects in the area and would need to expand the central plant to accommodate an annual throughput of 4 million pounds of uranium per year. The NRC source material license, if granted, would have to be amended to permit the higher capacity and potentially the processing of off-site uranium-loaded resins. Section 2.1.1.1.3 of this SEIS describes the processing that would occur in the central plant. Figure 2-3 illustrates the central plant layout.

The applicant has stated that a concrete curb would be built around the central processing plant. It would be designed to contain the entire contents of the largest tank within the building in the event of a rupture. It would also contain the operating volumes of the two largest tanks, in the unlikely event that their simultaneous failures were to occur. Any spill of plant fluids would be contained by this curb, drained to the sump system, and pumped to the liquid byproduct material disposal system.

The design of the deep well pump houses and wellheads would be such that any release of liquids would be contained within the building or in a bermed containment area surrounding the facilities. Liquid inside the building would be contained and managed appropriately. Wells would be equipped with a high-level shutoff switch on the injection tubing to prevent operation of the pumps at pressures greater than the limiting surface injection pressure. In addition, the wells would be equipped with a low-pressure shutdown switch on the surface injection line that would deactivate the injection pump in the event of a surface leak. Lines leading to the deep well would be instrumented for leak detection and automatic shutoff.

Chemical storage facilities at the proposed Moore Ranch Project would include both hazardous and nonhazardous material storage areas. Bulk hazardous materials, which have the potential to impact radiological safety, would be stored outside and segregated from areas where licensed materials would be processed and stored. Figure 2-2 shows the proposed location of the supporting structures, including the warehouse/shop, an office, a chemical storage facility, and the location of stored carbon dioxide and oxygen tanks. The hazardous chemical storage would include bulk storage within the central plant, a chemical storage facility located about 30.5 m [100 ft] west of the central plant, and separate storage for carbon dioxide and oxygen tanks. Bulk sulfuric acid storage has been identified in a prior analysis because of its reactivity

In-Situ Uranium Recovery and Alternatives

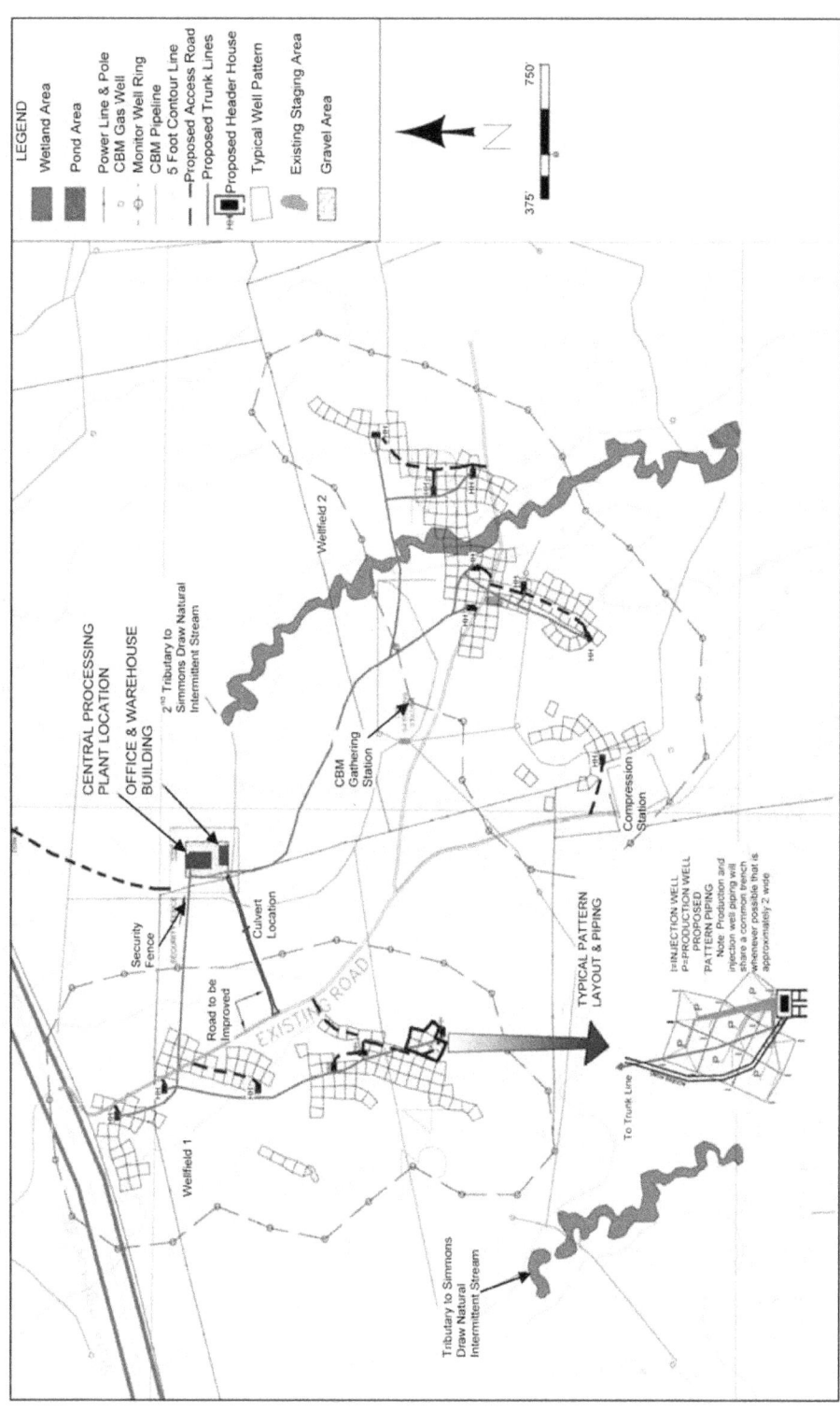

Figure 2-2. Wellfield Patterns and Infrastructure for the Proposed Moore Ranch Project (Uranium One, 2009a)

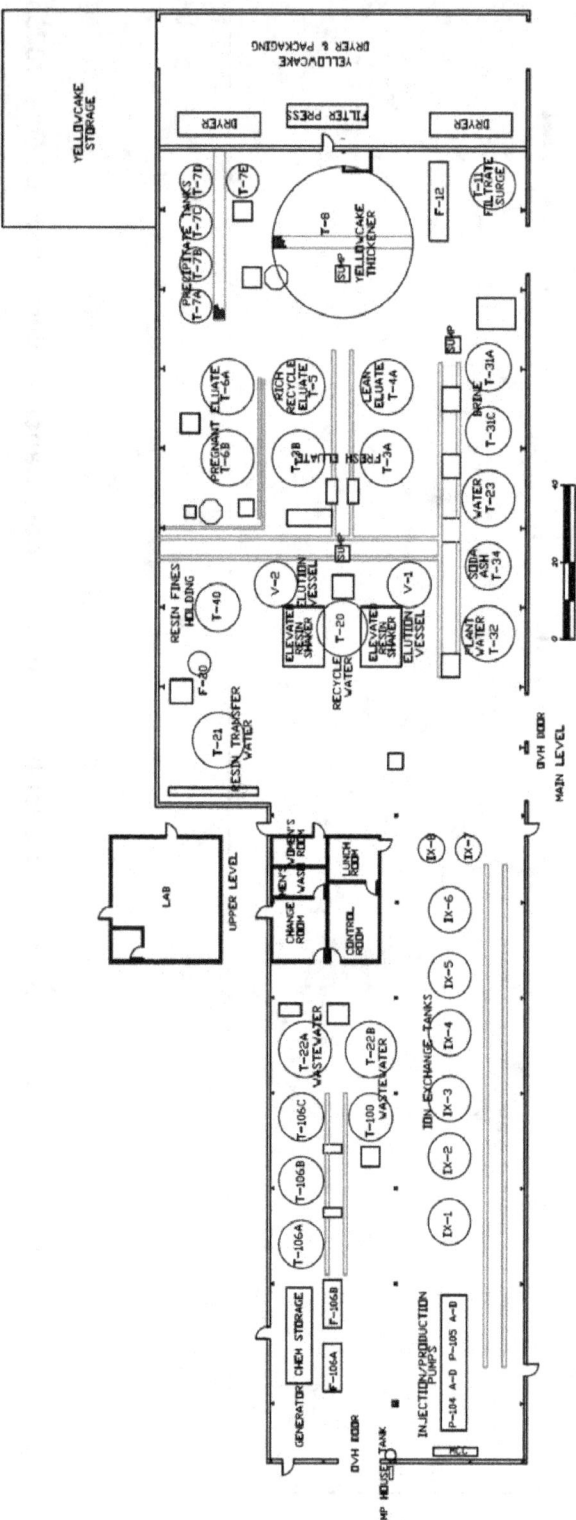

Figure 2-3. Layout of the Central Plant

and potential to impact chemical and radiological safety (Mackin, et al., 2001). A 22,727 L [6,000 gal] tank for sulfuric acid storage and a 22,727 L [6,000 gal] tank for storage of hydrogen peroxide have been proposed for use in the chemical storage area. These tanks would be located outside and separately from each other to minimize the potential for chemical reactions and vented into a water trap to limit the amount of vapors released into the atmosphere. The oxygen supplier would design and install an oxygen storage facility meet applicable industry standards. It would be delivered to the site by truck and stored on site under pressure in a cryogenic tank in liquid form (EMC, 2007a). Bulk storage of other nonhazardous chemicals, such as carbon dioxide, sodium carbonate, and sodium chloride would be inside the central plant to provide full containment of released materials.

The surface facilities would be designed and built using standard construction techniques, in accordance with appropriate building codes. Construction vehicles could include bulldozers, drilling rigs, water trucks, forklifts, pump hoist trucks, coil tubing trucks, pickup trucks, portable air compressors, and other support vehicles.

2.1.1.1.2.2 Access Roads

The primary method of transportation to the proposed Moore Ranch Project would be via U.S. or State highways. Access to the site from the east would be from State Highway (SR) 59 or SR 50 to SR 387; from the west, access would be from Interstate Highway (I)-25 to SR 259 to SR 387. Construction materials would be delivered to the Moore Ranch site via SR 87 (Figure 2-1). Access to the site from the highway is available through gravel and two-track roads established from CBM development and agricultural activity. A gravel access road located in T42N, R75W, Section 27 connects to the general location proposed for construction of the central plant and would require minor improvements and completion of a short spur road to accommodate truck and heavy equipment access during the construction and operation phases of the alternatives. The access road runs south through Section 34 and forks to the east through Section 35; it also continues south through the project boundary. This existing road would provide the primary access to all currently planned wellfields and facilities. Construction of the spur road would disturb approximately 1.2 ha [2 ac] of land. There are approximately 43 km [27 mi] of existing two track roads (neither paved nor graveled) used by CBM and the landowners. Secondary roads for wellfield header houses and facility access would fork off from the existing primary access road (Figure 2-2).

2.1.1.1.2.3 Wellfields

The underground infrastructure, consisting of wells and pipelines, would be established to inject, produce, and monitor groundwater and to transfer fluids between the wells and other production facilities. The proposed area to be developed in the wellfields is approximately 23 ha [57 ac]. The target mineralized zone for ISR at the proposed Moore Ranch Project is a sandstone formation located approximately 55 to 76 m [180 to 250 ft] below the surface (EMC, 2007b), referred to as the "70-Sand," which averages approximately 24 m [80 ft] thick in the proposed extraction areas. The geology and hydrology of this formation are described in greater detail in Sections 3.4.1 and 3.5.2 of this SEIS. The overall width of the mineralized area varies from 30 to 300 m [100 to 1,000 ft].

Two wellfields have been proposed for extraction at the Moore Ranch Project, as shown in Figure 2-2. Wellfield 1 would encompass approximately 9.3 ha [23 ac] in T42N, R47W in portions of Sections 27 and 34 and is located northwest of Wellfield 2. Wellfield 2 would encompass approximately 13.8 ha [34 ac] located in T42N, R47W, Section 34.

In-Situ Uranium Recovery and Alternatives

2.1.1.1.2.3.1 Injection and Production Wells

Injection wells would be used to introduce lixiviant into the uranium-bearing ore body; production wells would be used to extract uranium-rich solutions. Wells would be drilled using the rotary drilling technique and constructed so they could be used for either injection or recovery. By drilling dual-purposed injection and production wells, the applicant has the flexibility to change the wellfield flow patterns, as needed, to improve uranium recovery and restore groundwater more efficiently.

Prior to being placed into service, all wells would be developed to remove any drilling fluids, restore flow of formation water into the well, and establish stable formation water chemistry in the well (EMC, 2007a; Uranium One, 2008). In addition, the operator would use pressure packing field tests to ensure the mechanical integrity of all wells. These field tests involve sealing and over-pressuring the well to about 120 percent of maximum operating pressure. The well is then monitored for any pressure loss over time. Any well that does not maintain at least 90 percent of the pressure for 10 minutes would be taken out of service and either repaired and retested or plugged and abandoned (EMC, 2007a; Uranium One, 2008).

The injection/production pattern that would be used at the proposed Moore Ranch Project would be based on a conventional five-spot pattern, which would be modified to fit the characteristics of the ore body at the site. The standard production cell, referred to as a pattern, would contain four injection wells surrounding a centrally located production well (Uranium One, 2009a).

The initial number of wells to be installed at Wellfield 1 has been estimated as 245 injection wells, 160 production wells, and 63 monitoring wells. The initial number of wells to be installed at Wellfield 2 has been estimated as 227 injection wells, 195 production wells, and 81 monitoring wells. The proposed wellfield patterns are shown in Figure 2-2. The monitoring wells would be established in the overlying and underlying aquifers on 4-acre spacings to detect vertical excursions (Uranium One, 2009a). The wellfields would be brought into operation and restored based on a phased approach with separate schedules for each wellfield. Operations would begin in a portion of Wellfield 2, then move to the remaining portion of Wellfield 2 and part of Wellfield 1, then finish with Wellfield 1. As Wellfield 1 was operating, Wellfield 2 would begin restoration.

Injection and production wells would be connected to manifolds in a wellfield header house building. Header houses would be used to distribute injection fluid to injection wells and collect production solution from recovery wells. Each header house would be connected to two trunk lines, one for receiving injection fluid from the central plant and one for conveying recovery fluids to the central plant (Uranium One, 2008). The header house includes manifolds, valves, flow meters, pressure meters, booster pumps, and oxygen to incorporate into the lixiviant, as appropriate. Each header house would service approximately 40 to 60 wells (injection and production). Approximately five header houses have been proposed for Wellfield 1 and eight header houses have been proposed for Wellfield 2 (Uranium One, 2009a). The manifolds deliver the recovery solutions to the pipelines that transmit solutions to and from the IX facilities. Flow meters and control valves would be installed in the individual well lines to monitor and control flow rates and pressures. In addition to the injection and production wells, Class I disposal wells have been proposed for this project, as described in Section 2.1.1.1.6.

Within each wellfield, more water would be withdrawn than injected to create an overall hydraulic cone of depression. Under this pressure gradient, the natural groundwater movement would be from the surrounding area to the wellfield providing primary control of the production solution movement. The difference between the amount of water produced and injected is the wellfield "bleed." Fluids would be injected at a maximum rate of approximately 11,364 L/min [3,000 gal/min]. The average bleed rate at the proposed Moore Ranch Project has been estimated at 1 percent of the maximum injection rate or 114 L/min [30 gal/min]; it would be adjusted, as necessary, to maintain the wellfield cone of depression.

Designing, constructing, testing, and operating injection wells are regulated by the UIC program administered by the WDEQ who has primacy for the program, as delegated by the U.S. Environmental Protection Agency (EPA). The proposed program would require a UIC permit from the WDEQ to use Class III injection wells. Before ISR operations can begin, the portion of the aquifer designated for uranium recovery must be exempted as an underground source of drinking water, in accordance with the Safe Drinking Water Act (SDWA).

2.1.1.1.2.3.2 Monitoring Wells

Horizontal and vertical excursion monitoring wells would be installed at each wellfield, as dictated by geologic and hydrogeologic parameters. The horizontal monitoring wells screened in the uranium recovery zone would be located in a ring around the wellfields. Vertical monitoring wells would be placed in the first water-bearing sand underlying and overlaying the extraction zone. Typical locations of the monitoring well rings for proposed Wellfield 1 and Wellfield 2 are shown in Figure 2-2. The proposed well locations may be adjusted as the project progresses, to account for improved understanding of the geometry of the ore body and to adjust for surface topography variations.

2.1.1.1.2.3.3 Well Construction and Testing

At the proposed Moore Ranch Project, injection, production, and monitoring well casings would be constructed using plastic polyvinyl chloride (PVC) with a Standard Dimension Ratio of 17 and a nominal 5-in outside diameter. If a larger pump size was required, nominal casing diameters up to a 6-in outside diameter may be used. Casings in injection, production, and monitoring wells would use centralizers to ensure that the casing is centered in the drill hole. Each well would be cemented to strengthen the casing and plug the annulus of the hole to prevent vertical migration of solutions.

Each well would be tested for mechanical integrity before operation. As described in Section 2.3.1.1 of the GEIS, the purpose of this test is to verify that the well casing does not fail, causing water loss during injection or recovery operations. In a mechanical integrity test (MIT), the bottom and top of the casing are plugged (sealed) with a sealing device. The well is pressurized, and pressure gauges monitor pressure changes inside the casing. If the repaired well cannot be repaired after several tries, the well would be plugged and abandoned. Results of these MITs are maintained onsite and are available for NRC and WDEQ personnel inspection. Results of these MITs are also reported to the WDEQ on a quarterly basis.

During mud pit excavation associated with well construction, exploration drilling, and delineation drilling activities, topsoil would be separated from subsoil with a backhoe. The applicant has stated that when mud pit use was completed, subsoil would be replaced and topsoil applied. Mud pits typically remain open for less than 30 days. The WDEQ Land Quality Division (LQD) has guidelines on topsoil and subsoil management at uranium ISR facilities (WDEQ, 2000) that the applicant would follow in developing the wellfields at the proposed Moore Ranch Project.

In-Situ Uranium Recovery and Alternatives

2.1.1.1.2.3.4 Pipelines

The development of the wellfields at the proposed Moore Ranch Project would require the installation of underground piping, as described in this section. The locations of proposed trunk lines installed to support operations at the proposed Moore Ranch Project are shown in Figure 2-2. Individual well lines leading to the injection and production wells would travel to the local header house and trunk lines would lead in and out of the central plant through a pipe vault located on the northwest side of the central plant (Uranium One, 2009a).

In general, piping from the central plant, to and within the wellfield would be constructed of PVC or high-density polyethylene pipe with butt-welded joints or the equivalent. All pipelines would be pressure tested before final operation (EMC, 2007a). Wellfield piping would have an operating pressure of 150–300 psi. The network of process pipelines and cables would be buried to avoid freezing temperatures and to minimize the possibility of an accident. Burial trenches could be excavated as deep as 1.8 m [6 ft] below the ground surface to avoid any potential freezing issue. Trenches containing pipeline would typically be backfilled with native soil and graded to surrounding ground topography. The only exposed pipes would be at the central plant, at the wellheads, and in the wellfield header houses. Trunkline flows and manifold pressures would be monitored for process control (EMC, 2007a).

2.1.1.1.2.4 Other Structures and Systems

Liquid effluent wastes generated during uranium recovery operations would be disposed of in two to four Class I deep disposal wells. The applicant would have to obtain UIC permits for the construction and use of these deep disposal wells from the WDEQ, as previously described. These deep disposal wells would be completed in approved formations, and their exact locations would depend on field placement. The applicant estimates a maximum flow rate of 170 L/min [45 gal/m] during operations and an additional 380 L/min [100 gpm] from restoration of liquid effluent wastewater requiring deep well disposal.

Domestic liquid waste generated at the central processing facility would be disposed of in an on-site wastewater treatment (i.e., septic) system permitted by the County under the WDEQ-LQD Class V UIC Regulations.

2.1.1.1.2.5 Schedule

The applicant estimated that the construction phase of the proposed Moore Ranch Project would be approximately 9 months and would include the building of access roads, the central plant, and initiating development of Wellfield 2. As noted before the wellfields would be developed in phases along with the supporting infrastructure (i.e., header houses and pipelines). Wellfield 2 would be constructed first, followed by construction of Wellfield 1 within approximately 9 months (Griffin, 2009).

2.1.1.1.3 Operation Activities

As described in Section 2.4 of the GEIS, the ISL uranium recovery process involves two primary operations: uranium mobilization that occurs in underground aquifers when lixiviant is injected into the ore body and recovering the solutions when they are uranium laden (NRC, 2009). The uranium-laden solutions (referred to as pregnant lixiviant) would be pumped from the production wells to the central plant IX system.

Uranium recovery at the proposed Moore Ranch Project involves the following two processes. First is uranium mobilization, where solid uranium in subsurface ore bodies is dissolved and extracted from the ground. The second process is uranium processing, where the dissolved uranium is removed from and ultimately dried and packaged as yellowcake.

The next section describes the proposed operations at the Moore Ranch Project. Additionally, Section 2.4 of the GEIS provides general background information on the operations phase of an ISR facility (NRC, 2009).

2.1.1.1.3.1 Uranium Mobilization

Uranium mobilization at the proposed Moore Ranch Project would use the following steps: (1) injection of lixiviant, (2) oxidation and complexation of the uranium underground, (3) extraction or production of the pregnant lixiviant from the subsurface, and (4) excursion monitoring (EMC, 2007b). This process is described in the following sections.

2.1.1.1.3.2 Lixiviant Chemistry

The selected lixiviant must leach uranium from the host rock and keep it in solution during groundwater pumping from the host aquifer. The composition of the lixiviant is designed to reverse the natural geochemical conditions that led to the original uranium deposition. At the proposed Moore Ranch Project, the lixiviant for uranium recovery operations would be alkaline and would consist of varying concentrations and combinations of sodium carbonate/bicarbonate and oxygen added to the native groundwater to promote dissolution of uranium as a uranyl carbonate complex. The amenability of uranium deposits to ISR at the proposed Moore Ranch site has been demonstrated by nearby ISR operations in similar ore bodies in the Powder River Basin, including the Smith Ranch/Highland and the Christensen Ranch/Irigaray projects (EMC, 2007a). The lixiviant would be made up on a batch basis in the central plant and added continuously to the injection stream. Table 2.4-1 of the GEIS summarizes typical lixiviant chemistry (NRC, 2009). As noted in Section 2.4.1.1 of the GEIS, the principal geochemical reactions caused by the lixiviant are the oxidation and subsequent dissolution of uranium and other metals from the ore body and its subsequent extraction (NRC, 2009).

2.1.1.1.3.3 Lixiviant Injection and Production

At the proposed Moore Ranch Project, the lixiviant would be pumped down a total of approximately 472 injection wells in the two wellfields to the ore body, where it would oxidize and dissolve uranium from the formation. Solutions for milling would be injected at a maximum rate of approximately 11,364 L/min [3,000 gal/min] (EMC, 2007b). A water balance for the proposed project is shown in Figure 2-4. The liquid effluent generated at the central plant would be primarily production bleed containing liquid byproduct material that has been estimated at an average of 1 percent of the production flow, which would be 114 L/min [30 gal/min] at the maximum flow rate of 11,364 L/min [3,000 gal/min].

Downhole injection pressures would be maintained below the formation fracture pressure. A formation fracture pressure gradient of 1.0 psi for every 0.3 m [1 ft] of depth to the top of the

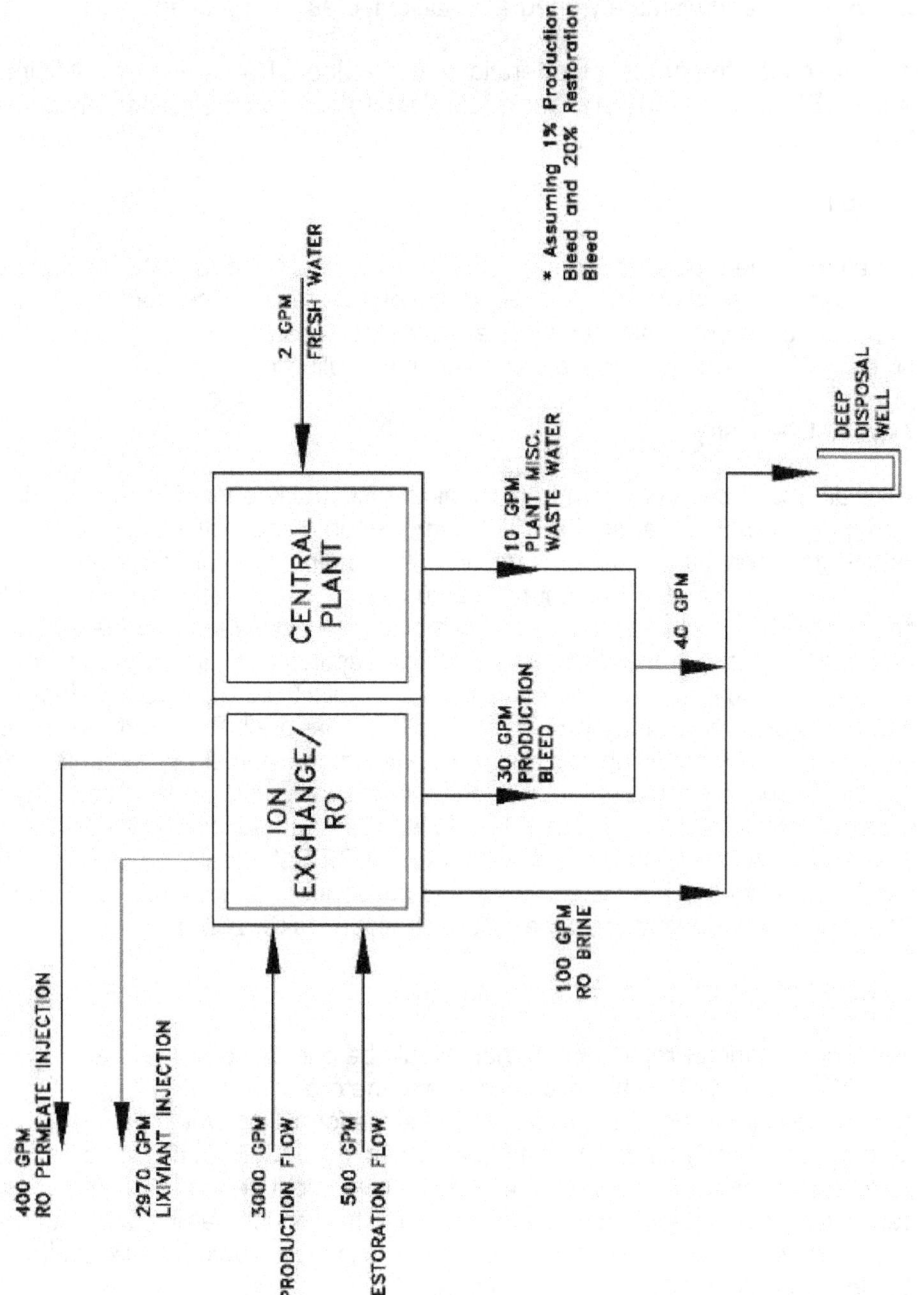

Figure 2-4. Water Balance for the Proposed Moore Ranch Project

screened interval was used to determine the injection pressures at the proposed Moore Ranch Project. The depth to the top of the anticipated screened interval ranges from approximately 49 m [160 ft] below land surface in Wellfield 2 to approximately 91.5 m [300 ft] in Wellfield 1. Therefore, injection pressures would range from 100 psi at the header houses located in shallower ore areas to no greater than 150 psi at header houses located in deeper ore areas (EMC, 2007b).

The uranium-bearing solution would migrate through the pore spaces in the sandstone and be recovered by a total of approximately 355 production wells, inclusive of both Wellfields 1 and 2. This uranium-rich, pregnant lixiviant would be pumped to the central plant IX facility, where it would be extracted through the processes described in Section 2.1.1.1.3.5 of this SEIS. Once the uranium has been extracted from the lixiviant (referred to as barren lixiviant), it would be recharged with carbonate/bicarbonate and oxidant and the solution reinjected into the ore body to continue extraction.

2.1.1.1.3.4 Excursion Monitoring

ISR operations can potentially affect the groundwater quality near a site when lixiviant moves from the production zone away from the injection wells, resulting in either a vertical or lateral excursion. Excursion monitoring is performed to monitor water flow to avoid a potential excursion. Uranium One proposes to install monitoring wells around the wellfields to monitor for horizontal excursions. Monitoring wells would also be installed in the overlying and underlying aquifers at a density of one well per every 4 acres of pattern area to monitor for vertical excursions (Uranium One, 2008). The final location of these wells would be determined when the final wellfield design and the wellfield package are submitted to WDEQ-LQD. During the safety review, the staff identified issues that could only be resolved after wellfield testing is complete. These issues will require that NRC staff review the hydrologic package for Wellfield 1 before ISR operations begin and for Wellfield 2, the NRC staff will review and approve the hydrologic package. The proposed monitoring program is described in more detail in Chapter 6 of this SEIS.

An excursion occurs when two or more excursion indicators in a monitoring well exceed their upper control limits (UCL) (NRC, 2003a). If an excursion is confirmed, the licensee notifies the NRC and takes several steps to confirm the excursion through additional sampling (NRC, 2003a). As described in NRC guidance (NRC, 2003a, Section 5.7.8.3), licensees typically retrieve horizontal excursions back into the production zone by adjusting the flow rates of the nearby injection and production wells to increase process bleed in the area of the excursion. Vertical excursions are more difficult to retrieve. If an excursion is suspected in groundwater monitoring wells, then the applicant is required to notify WDEQ and NRC within 24 hours, and the suspect wells would be put on excursion monitoring, meaning they would be monitored every seven days until the excursion indicators returned to nonexcursion levels. If an excursion cannot be recovered, the licensee may be required to stop lixiviant injection (NRC, 2003a). NRC license conditions require that licensees conduct biweekly sampling to detect excursions.

2.1.1.1.3.5 Uranium Processing

The uranium processing at the proposed Moore Ranch Project would use the following steps: (1) loading of uranium complexes onto IX resin; (2) elution (recovery) of uranium complexes from the resin; and (3) precipitation, drying and packaging of uranium (EMC, 2007b). This process is described in the following text.

In-Situ Uranium Recovery and Alternatives

2.1.1.1.3.5.1 Ion Exchange

The uranium-bearing solution, or pregnant lixiviant, pumped from the wellfield would be piped to the IX columns in the central plant to extract the uranium. The IX system would consist of eight fixed-bed IX vessels to be operated as three sets of two vessels in series with two vessels available for restoration. The IX system would be designed to process recovered solution at a rate of 11,364 L/min [3,000 gal/min] with each vessel sized for 14.2 m^3 [500 ft^3] of resin operated in a pressurized down flow mode. As the solution passed through the IX resin in the vessels, the uranyldicarbonate and uranyltricarbonate would, preferentially, be removed from the solution. The barren solutions leaving the IX units would be expected to contain less than 2 mg/l of uranium (EMC, 2007b).

After the barren lixiviant has left the IX system, carbon dioxide and/or carbonate/bicarbonate would be added, as necessary, to return the carbonate/bicarbonate concentration to the desired operating level. The solution would then be pumped back to the wellfield, with the oxidant (O_2 gas) added either as it left the central plant, or just before the solution was to be reinjected into the production zone.

2.1.1.1.3.5.2 Elution

A three-stage elution circuit has been proposed for use at the Moore Ranch Project, as illustrated in Figure 2-3. In a three-stage elution, the rich eluate would first pass through the elution vessels, which contain the IX resin. The rich eluate would strip approximately 84 percent of the uranyl carbonate ions from the resin and become pregnant eluate containing approximately 15,500 mg/l uranium. Then lean eluate would be contacted with resins and remove approximately 68 percent of the remaining uranyl carbonate to become rich eluate. Finally, fresh eluate would be passed through the resins in the elution vessels and remove approximately 35 percent of the remaining uranyl carbonate from the resins, resulting in lean eluate. At this stage, the resins would have a residual uranyl carbonate concentration of approximately 3.33 percent. The resins would be washed with freshwater or a sodium bicarbonate rinse, or both, and either transferred back to the appropriate vessel or to a resin transfer trailer, if the resins were to be shipped to an offsite extraction area. Each batch of eluate would be transferred from the respective eluate storage tank through the elution vessel at a rate of approximately 795.4 L/min [210 gal/min] (EMC, 2007a).

Approximately 125,000 L [33,000 gal] of eluate would contact 14 m^3 [500 ft^3] of resin. The first elution stage would generate approximately 1,500 ft^3 [11,220 gal] of pregnant eluate containing 10 to 20 g/L uranium. About 1,500 ft^3 [11,220 gal] of fresh eluate would be required per elution batch. The fresh eluate would be prepared by mixing the proper quantities of a saturated sodium chloride (salt) solution and saturated sodium carbonate (soda ash) solution and water to form a solution that would be about 9 percent NaCl and 2 percent Na_2CO_3. The saturated salt solution would be generated in a brine generator, and the saturated soda ash solution would be prepared by passing warm water [>105 °F] through a bed of soda ash. The eluate would be passed through a bank of 10 micron bag filters to remove entrained particulates prior to contacting the resin beds in the elution vessels (EMC, 2007a).

2.1.1.1.3.5.3 Precipitation, Drying, and Packaging

Approximately 795 L [210 gal] of sulfuric acid would be added to the pregnant eluate to break the uranyl carbonate complex, thus liberating carbon dioxide and freeing uranyl ions to form a uranyl sulfate ion complex. The acidic, uranium-rich fluid would then be pumped to the first of five agitated tanks arranged in series. The fluid would flow by gravity from one tank to the next. Hydrogen peroxide would be added to the first two tanks to form an insoluble uranyl peroxide compound. Sodium hydroxide would then be added to the solution in the third tank to raise the pH of the precipitate solution to near neutral for optimum crystal growth and settling. The uranium precipitate solution would then be pumped from the final precipitation tank to a 10.3 m [38-ft] diameter gravity thickener.

The yellowcake would be dried at approximately 120 °C [250 °F]. The drying and packaging area would include the vacuum dryer system and the packaging area immediately below the vacuum dryer. When the yellowcake was dry, it would flow through an enclosed chute and be deposited directly into a sealed hood on the drum for packaging. The off gases generated during the drying cycle would be filtered through a baghouse. Two rotary vacuum dryers would be located in a separate building containing the dryers and associated equipment but attached to the central plant.

The dried yellowcake would be removed from the rotary vacuum dryer by passing it through a rotary valve into 208-L [55-gal] steel drums. The dryer vacuum pump would be connected to the loading hood to minimize particulate emissions during drum loading. The dried yellowcake product would be stored in the steel drums in a restricted storage area, pending shipment by truck to other licensed facilities for further processing. An enclosed warehouse, adjacent to the yellowcake drying area, would be used to store yellowcake. The drummed yellowcake would be shipped by exclusive-use transport to a licensed conversion facility in Metropolis, Illinois.

2.1.1.1.3.6 Management of Production Bleed and Other Liquid Effluents

Uranium mobilization at the proposed Moore Ranch Project would produce excess water that must be properly managed. The production wells extract slightly more water than is reinjected into the host aquifer, which creates a net inward flow of groundwater into the wellfield. As mentioned earlier, during normal operations, production rates would be controlled by withdrawing a small portion of the barren solution from the IX circuit, which is then disposed of via the deep disposal wells. The estimated maximum flow rate of production bleed is 114 L/min [30 gal/min].

Other liquid effluents would be produced as part of the proposed Moore Ranch ISR Project. These include liquids from process drains, elution circuit bleed, and wash-down water. The maximum estimated flow rate of these other liquid effluents is 38 L/min [10 gal/min]. These waste streams would be handled in the same manner as the production bleed.

2.1.1.1.3.7 Schedule

The central plant at the proposed Moore Ranch Project would operate for approximately 12 years; however, the individual wellfields would operate for approximately 3.25 years (Griffin, 2009), although they would operate on separate schedules as described above in Section 2.1.1.1.2.3.1.

2.1.1.1.4 Aquifer Restoration Activities

Aquifer restoration within the wellfield ensures that the water quality and groundwater use in surrounding aquifers would not be adversely affected by the uranium recovery operation, as discussed in Section 2.5 of the GEIS (NRC, 2009). After uranium is recovered, the groundwater will contain constituents that were mobilized by the lixiviant. The process whereby groundwater

In-Situ Uranium Recovery and Alternatives

constituents are selected for monitoring throughout the life of the project is discussed in Section 6.3.1.2 of this SEIS. In compliance with 10 CFR Part, 40, Appendix A, Criterion 5B(5), groundwater quality in the exempted ore-bearing aquifer is required to be restored to (i) Commission-approved baseline; (ii) maximum concentration level of constituents (MCLs) listed in Table 5C, if the constituent is listed in Table 5C and if the baseline level of the constituent is below the value listed; or (iii) ACLs established by the Commission, if the baseline level of the constituents and the values listed in Table 5C are not reasonably achievable. The development of ACLs is described in Appendix C of this SEIS. These standards are implemented during aquifer restoration to ensure public health and safety. The applicant is required to provide financial sureties to cover planned and delayed restoration costs in compliance with 10 CFR Part 40, Appendix A, Criterion 9. NRC reviews the financial sureties annually.

Under the Federal UIC program, the exempted production aquifer will no longer be protected under the SDWA as an Underground Source of Drinking Water (USDW). In compliance with 40 CFR 146.4, the exempted aquifer does not currently serve as a source of drinking water and cannot now and will not in the future serve as a source of drinking water. Hence, groundwater in exempted aquifers cannot be considered as a source of drinking water after restoration.

The applicant needs to establish baseline water quality prior to the submission of a license application. The excursion parameters and UCLs are determined based on the baseline water quality sampled from monitoring wells placed in the ore-bearing, underlying, and overlying aquifers, when applicable. Therefore, the UCLs should be established prior to ISR operations. UCLs are used for control and management of excursions, if they occur, during ISR operations and restoration.

Aquifer restoration in each wellfield would begin as the uranium recovery operations end. By doing this, the period of groundwater contamination within the exempted aquifer is shortened. The preextraction class of use would be determined by the baseline water quality sampling program that would be performed for each wellfield compared to the use categories defined by the WDEQ, Water Quality Division. Restoration would be demonstrated to meet the requirements of the WDEQ, LQD Rules and Regulations (EMC, 2007b) and NRC requirements. Evaluation of the degree of groundwater restoration within the production zone would be based on the average baseline quality over the production zone. The applicant would collect baseline water quality data for each wellfield from the wells completed in the planned production zone. Restoration would be evaluated parameter-by-parameter, based on "Restoration Target Values" that would be established for a list of baseline water quality parameters.

Prior to initiation of operations and during the aquifer restoration phase, the applicant would have established a groundwater monitoring program as described in Chapter 3 to assess the impacts from operations on local groundwater and its subsequent restoration.

2.1.1.1.4.1 Groundwater Restoration Methods

The groundwater restoration program at ISR sites typically consists of two stages: restoration stage and stability monitoring. The restoration stage typically consists of three phases groundwater transfer, groundwater sweep; and groundwater treatment, as described in Sections 2.5.1, 2.5.2, and 2.5.3 of the GEIS (NRC, 2009). These phases are designed to optimize restoration equipment used in treating groundwater and to minimize the volume of groundwater consumed during the restoration stage. The following sections describe the groundwater restoration methods proposed for use at the proposed Moore Ranch Project.

Based on aquifer testing at Moore Ranch, Uranium One determined that efficient groundwater movement during the operations and restoration phases could be accomplished by "pulsing" the extraction wells by cycling them on and off. The pulsing would be achieved by either switching groups of extraction wells on and off or by alternating between injection and extraction cycles within individual well patterns. By doing this, those portions of the aquifer that may have been temporarily dewatered by an extraction well could be effectively resaturated (EMC, 2007a).

The sequence of groundwater restoration activities would be determined based on operating experience and wastewater system capacity. The specific mix of groundwater transfer and groundwater treatment would be determined as part of the groundwater restoration plan for each individual wellfield. A reductant, such as sulfide or sulfite, could also be added to the injection stream at any time during the restoration stage to lower the oxidation potential of the extraction zone.

The applicant at the proposed Moore Ranch Project would monitor the quality of groundwater in selected wells, as needed during restoration, to determine the efficiency of the operations and to decide if additional or alternate techniques would be necessary, as described in Section 4.5.2 of this SEIS. The evaluation of groundwater restoration within the production zone would be based on the average baseline quality over the production zone (EMC, 2007a). Online production wells used in restoration would be sampled for uranium concentration and for conductivity amongst other constituents to determine the progress of restoration on a pattern-by-pattern basis.

2.1.1.1.4.1.1 Groundwater Transfer

During the groundwater transfer phase of groundwater restoration at the proposed Moore Ranch Project, water could be transferred between a wellfield commencing restoration and another commencing extraction operations. Groundwater transfer could also occur within the same wellfield, if one area of the wellfield is in a more advanced state of restoration than another.

Baseline quality water from the wellfield commencing extraction would be pumped and injected into the wellfield in restoration. The higher total dissolved solids (TDS) water from the wellfield in restoration would be recovered and injected into the wellfield commencing extraction. The direct transfer of water would lower the TDS in the wellfield being restored by displacing affected groundwater with baseline quality water.

The goal of the groundwater transfer phase is to blend the water in the two wellfields until they become similar in conductivity. The water recovered from the restoration wellfield could be passed through ion-exchange columns or filtered during this phase if there were a significant amount of suspended solids that could block injection well screens.

For groundwater transfer between wellfields, a newly constructed wellfield must be ready to commence extraction. Therefore, this phase could be initiated any time during the restoration process. If a wellfield is not available to accept transferred water, groundwater sweep or some other activity could be the first phase of restoration. The groundwater transfer technique is advantageous to reduce the amount of disposed water.

In-Situ Uranium Recovery and Alternatives

2.1.1.1.4.1.2 Groundwater Treatment

During the groundwater treatment phase, water is pumped from the extraction zone to the surface for treatment. IX, RO, and electro dialysis reversal treatment equipment have been proposed for use during the groundwater treatment phase of the proposed Moore Ranch Project.

Groundwater recovered from the restoration field is passed through an IX unit prior to RO and electro dialysis reversal units. RO electro dialysis reversal units are used to separate clean water (permeate) from brine. Following treatment in the ion exchange unit, groundwater would either be reinjected into the wellfield or disposed. Prior to or following IX treatment, the groundwater could be passed through a decarbonation unit to remove residual carbon dioxide remaining in groundwater after uranium recovery.

All or some portion of the restoration recovery water could be sent to the RO unit to reduce the total dissolved solids in the affected groundwater, reduce the quantity of water that must be removed from the aquifer to meet restoration limits, concentrate the dissolved constituents into a smaller brine volume to facilitate waste disposal, and enhance the exchange of ions from the formation due to the large difference in ion concentration. The RO unit passes a high percentage of the water through the membranes, leaving 60 to 90 percent of the dissolved salts in the brine water or concentrate. The clean water, called permeate, is reinjected into the wellfield or stored for use in the extraction process. The permeate could also be decarbonated prior to reinjection into the wellfield. The brine water contains the majority of dissolved salts in the affected groundwater and would be disposed of via deep well injection. Make-up water, which could come from water produced from a wellfield that is in a more advanced state of restoration, water being exchanged with a new extraction unit, the purge of an operating wellfield or a combination of these sources, could be added prior to the RO or wellfield injection stream to control the amount of bleed in the restoration area.

At any time during the process, a chemical reductant, which would be used to create reducing conditions in the extraction zone, could be metered into the restoration wellfield injection stream. The concentration of reductant injected into the formation would be determined by how the extraction zone groundwater reacts with the reductant. The goal of reductant addition would be to decrease the concentrations of certain trace elements. Reductants are beneficial because several of the metals, which are solubilized during the leaching process, are known to form stable, insoluble compounds, primarily as sulfides. Dissolved metal compounds that are precipitated under reducing conditions include those of arsenic, molybdenum, selenium, uranium, and vanadium.

The groundwater restoration phase at the proposed Moore Ranch Project has been estimated to consume about 65.5 million and 94.2 million gallons of water at Wellfields 1 and 2. The numerical modeling to support this calculation is provided in the Uranium One technical report (EMC, 2007a). Upon completion of restoration activities, a minimum 12-month groundwater stability monitoring period would be implemented to demonstrate that the restoration goal has been adequately maintained, in accordance with WDEQ guidelines. Chapter 6 of this SEIS describes the restoration stability monitoring that would be conducted.

In-Situ Uranium Recovery and Alternatives

2.1.1.1.4.2 Schedule

The aquifer restoration phase of Wellfields 1 and 2 is estimated to last approximately 3.5 and 5.25 years (Griffin, 2009). The stabilization monitoring following completion of aquifer restoration is estimated to last approximately one year.

2.1.1.1.5 Decontamination, Decommissioning, and Reclamation Activities

The decommissioning of an ISR facility would be based on an NRC-approved decommissioning plan. Section 2.6 of the GEIS describes the general process for decontamination, decommissioning, and reclamation of an ISL facility (NRC, 2009). The applicant would submit a decommissioning plan to the NRC for review and approval at least 12 months before the planned commencement of final decommissioning. When approved, this plan would amend the license, initiate the decommissioning process, and provide NRC the detailed information required for NRC to evaluate the proposed decommissioning plan.

Prior to release of the property for unrestricted use, the applicant would conduct a comprehensive radiation survey to establish that any contamination is within the 10 CFR Part 40, Appendix A limits. The applicant would return all lands to their previous land use, unless an alternative was justified and approved by both the state and the landowner. For example, a rancher could decide to retain access roads. The goal of the decommissioning and reclamation process would be to return disturbed lands to production capacity of equal or better than existed prior to uranium recovery. The following sections describe the proposed decommissioning and surface reclamation plans for the proposed Moore Ranch Project. As part of this process, wells would be plugged and abandoned, disturbed lands would be reclaimed, contaminated equipment and materials would be removed, appropriate cleanup criteria for structures would be determined, items to be released for unrestricted use would be decontaminated to meet NRC requirements, and surveys would be performed to determine if there was residual contamination in soils and structures. The following sections described the general decommissioning activities that would occur at the proposed Moore Ranch Project.

2.1.1.1.5.1 Wellfield Decommissioning

Wellfield plugging and surface reclamation would be initiated when the regulatory agencies concur that the groundwater in a wellfield has been adequately restored and that the water quality is stable. All production, injection, and monitoring wells and drill holes would be abandoned and plugged, in accordance with WS–35–11–404 and Chapter VIII, Section 8 of the WDEQ-LQD Rules and Regulations, to prevent adverse impacts to groundwater quality or quantity. This process would involve removing pumps and tubing from the wells, plugging them with either cement or clay, cutting off the casing three feet below the ground surface, placing a steel plate on top of the casing that identifies the well and the date of plugging, and emplacing a cement plug at the top of the casing.

Reclamation in the wellfield production unit would involve removing surface and subsurface equipment. These would consist primarily of: injection and production feed lines; header houses; electrical and control distribution systems; well boxes and wellhead equipment; buried wellfield piping; recontouring, if necessary; and conducting a final background gamma survey over the wellfield to identify contaminated earthen materials requiring removal, final revegetation of the wellfield areas according to a revegetation plan; and surveying all piping, equipment, buildings, and wellhead machinery for contamination prior to release, in accordance with NRC decommissioning guidelines.

The applicant estimated that a significant portion of the equipment would meet radiological release limits, which would allow for disposal at an unrestricted landfill, as discussed in Section 2.1.1.1.6.3. Other materials would be decontaminated until they could be released. Equipment and materials that could not be decontaminated to meet release limits would be disposed of as byproduct material at a licensed facility.

2.1.1.1.5.2 Topsoil Handling and Replacement

Topsoil at Moore Ranch would be salvaged from building sites, permanent storage areas, main access roads, graveled wellfield access roads, and chemical storage sites, in accordance with WDEQ-LQD requirements. Conventional rubber-tired, scraper-type earth moving equipment is typically used to accomplish topsoil salvage operations. The exact location of topsoil salvage operations would be determined by wellfield pattern emplacement and designated access roads, which would be designated during final construction activities.

The topsoil thickness within the licensed area varies from nonexistent to several feet. However, typical topsoil stripping depths are expected to range from 0.7 to 1.5 cm [3 to 6 in]. Salvaged topsoil would be stored in designated stockpiles, generally located on the leeward side of hills, to minimize wind erosion and avoid drainage channels. The perimeter of large stockpiles would be bermed to control sediment runoff.

Surface soils would be cleaned up in accordance with the requirements in 10 CFR Part 40, Appendix A, considering ALARA goals for cleanup of soils. The methodologies for conducting postreclamation and decommissioning radiological surveys are discussed in Section 6.4 of the applicant technical report (EMC, 2007a). The cleanup goals for radium-226 and natural uranium are summarized in Table 2-1.

2.1.1.1.5.3 Final Contouring and Revegetation

The land surface would be recontoured, as necessary, to restore it to a surface configuration that would blend in with the natural terrain and would be consistent with the post-extraction land use. No major changes in the topography are anticipated for the proposed Moore Ranch Project.

Revegetation practices would be conducted in accordance with WDEQ-LQD regulations and the mine permit. During extraction operations the topsoil stockpiles and the disturbed wellfield areas would be seeded, as much as practicable, to establish a vegetative cover to minimize wind and water erosion. The WDEQ-LQD would approve the selected seed mix.

Table 2-1. Soil Cleanup Criteria and Goals*				
	Radium-226 (pCi/gm)		Natural Uranium (pCi/gm)	
Layer Depth	Limit	Goal	Limit	Goal
Surface (0–15 cm)	5	5	225	150
Subsurface (15 cm)	15	15	225	225

*Consistent with NRC requirements in 10 CFR Part 40, Appendix A, Criterion 6(6), the applicant proposes to apply a unity rule when more than one radionuclide is present so that the sum of the ratios of each radionuclide concentration to its limit does not exceed one.

In-Situ Uranium Recovery and Alternatives

The success of permanent revegetation in meeting land use and reclamation standards would be assessed prior to application for bond release by utilizing the Extended Reference Area method, as detailed in WDEQ-LQD Guideline No. 2–Vegetation (WDEQ, 1986). This method compares, on a statistical basis, the reclaimed area to the adjacent, undisturbed areas of the same vegetation type. The Extended Reference Area would be selected in consultation with the WQED-LQD, to ensure the representativeness of the undisturbed chosen area to the reclaimed area being assessed.

2.1.1.1.5.4 Procedures for Removing and Disposing of Structures and Equipment

Upon completion of the uranium recovery process, buildings, equipment, pipelines and associated materials would be removed as discussed in the following section.

2.1.1.1.5.4.1 Preliminary Radiological Surveys and Contamination Control

Prior to decommissioning the central plant and associated structures, a preliminary radiological survey would be conducted to characterize the levels of contamination on structures and equipment, to identify any potential hazards, and to support the development of procedures for dealing with such hazards prior to commencement of decommissioning activities. Based on the results of preliminary radiological surveys, gross decontamination techniques would be employed to remove loose contamination before decommissioning activities were initiated. This gross decontamination would generally consist of washing all accessible surfaces with high-pressure water. In areas where contamination is not readily removed by high-pressure water, a decontamination solution (e.g., dilute acid) may be used.

2.1.1.1.5.4.2 Removal of Process Buildings and Equipment

The majority of equipment in the process building would be reusable, as well as the building itself. Alternatives for the disposition of buildings and equipment would also be evaluated, including removal to a new location for future use, removal to another licensed facility for either permanent use or disposal, or decontamination to meet unrestricted release criteria. All potentially contaminated equipment and materials at the central plant such as tanks, filters, pumps, and piping, would be inventoried, listed, and designated by the applicant as discussed previously. For the proposed Moore Ranch Project, the process building would be decontaminated, dismantled, and released for use at another location. Materials that could not be decontaminated to meet release criteria would be sent to a permanent licensed disposal facility. Cement foundation pads and footings would be broken up and trucked to a solid waste disposal site or to a licensed disposal facility, if the residual materials could not meet release criteria.

2.1.1.1.5.4.3 Building Materials, Equipment, and Piping Released for Unrestricted Use

Salvageable building materials, equipment, pipe, and structures would be surveyed for alpha contamination, in accordance with NRC guidance and alpha contamination limits. Surface decontamination would be conducted, in accordance with ALARA, to reduce surface contamination levels as far below the limit as practical. Decontamination would focus, in particular, on inaccessible portions of equipment and structures in which radiological materials could accumulate, such as piping, traps, junctions, and access points. Nonsalvageable, contaminated equipment, materials, and dismantled structural sections would be sent to a licensed facility for disposal. In most cases, the byproduct material would be shipped as Low Specific Activity (LSA-I) material, UN2912, pursuant to 49 CFR 173.427.

2.1.1.1.5.5 Schedule

Wellfield and plant decommissioning is estimated to take approximately one year (Griffin, 2009).

2.1.1.1.6 Effluents and Waste Management

The operation of an ISR facility generates various types of effluents and waste. This section describes the types and volumes of effluents or wastes to be generated by operation of the proposed Moore Ranch Project. Also, the textbox below defines the different liquid and solid wastes that would be generated. The proposed disposal methods and locations for liquid and solid wastes are described in Section 3.13 and the impacts from generating and disposing these wastes are described in Section 4.14. Air quality and air emissions impacts are discussed in Section 3.7 and 4.7.

> The terms below define the various types of solid and liquid wastes generated at the Moore Ranch Project:
>
> **Liquid wastes**
>
> Liquid byproduct material (this term refers to all liquid wastes resulting from the proposed action except for sanitary wastewater and well development and testing wastewater)
>
> Sanitary Wastewater (ordinary sanitary (septic system) wastewater; this wastewater is nonhazardous, non-byproduct material wastewater)
>
> Well development and testing wastewaters (wastewater generated during well development and pumping tests; this water is nonhazardous, non-byproduct material wastewater and would not require treatment before disposal)
>
> **Solid wastes**
>
> Solid byproduct material (this term refers to all solid wastes resulting from the proposed action that exceed NRC limits in 10 CFR Part 20 for unrestricted release)
>
> Solid waste (nonhazardous, solid waste, including domestic/municipal wastes (trash), construction/ demolition debris, septic solids, and solid byproduct material resulting from the proposed action (e.g., equipment, soils) that has been determined to meet NRC criteria in 10 CFR Part 20 for unrestricted release)
>
> Hazardous waste (Resource Conservation and Recovery Act or state-defined hazardous waste that is non-byproduct material and includes universal hazardous wastes and used oil)

2.1.1.1.6.1 Gaseous and Airborne Particulate Emissions

During the construction, operations, aquifer restoration, and decommissioning phases of the proposed Moore Ranch Project, airborne emissions would be generated from fugitive dust; combustion engine exhaust, and radon gas emissions from lixiviant circulation. Uranium airborne particulate emissions from yellowcake drying would be zero to near zero because of the use of the rotary vacuum drying process, which is designed to capture virtually all escaping particulate matter. With the prevailing wind direction out of the south-southwest during the day time (EMC, 2007a), airborne emissions produced by the proposed Moore Ranch ISR Project would generally blow in the northeast direction.

Fugitive dust would be generated primarily during construction, transportation, and decommissioning activities by travel on unpaved roads and from land disturbance associated with the construction of buildings, wellfields, roads, and support facilities. The applicant estimated the total dust from vehicular traffic on gravel roads at 14 t/yr [15.5 T/yr], based on proposed activities and emissions factors provided by WDEQ (EMC, 2007b). Traffic-generating activities addressed in the calculations included employee commuting, wellfield construction,

In-Situ Uranium Recovery and Alternatives

operations and maintenance, delivery of supplies and materials, and yellowcake product shipments. The close proximity of the site to SR 387 limits the distance of unpaved access roads and, therefore, the associated fugitive dust that would be generated from proposed activities.

Combustion engine exhaust due to vehicular exhaust from workers commuting to the site; materials transport to the site; and diesel emissions from drilling rigs, diesel-powered water trucks, and other equipment used during the construction phase would also contribute to gaseous and particulate emissions. Emissions from diesel combustion engines in drilling rigs and construction equipment used predominantly during the construction and decommissioning phases were calculated by the NRC staff and discussed in detail in Appendix D. These calculations evaluated emissions of nitrogen oxides (NO_x), carbon monoxide (CO), sulfur oxides (SO_x), particulate matter (PM_{10}), formaldehyde, volatile organic compounds (VOC), and carbon dioxide (CO_2). Results show CO_2 and NO_x are the highest emissions of the pollutants evaluated. The calculated annual emissions for these pollutants during the construction phase bound emissions calculated for the decommissioning phase. Based on the applicant's proposed schedule for wellfield construction (one wellfield per year), and an NRC staff assumption that the applicant would drill two deep disposal wells in the first year, the calculated annual emissions of CO_2 and NO_x are 852 t/yr [940 T/yr] and 18.1 t/yr [20 T/yr]. Results of the NRC staff emissions calculations indicate the drilling of deep wells contributes a significant proportion to the total emissions during construction. Therefore, if the applicant chose to drill all four proposed deep wells in the first year, these emissions would increase to 1,400 t/yr [1,600 T/yr] CO_2 and 30 t/yr [33 T/yr] NO_x. Cumulative emissions approximations for CO_2 and NO_x for the lifecycle of the proposed facility (based on emissions from construction of all proposed wellfields, deep disposal wells, reclamation of these wellfields, and all surface facilities) are 2400 t/yr [2600 T/yr] and 55 t/yr [61T/yr], respectively. Results for all of the diesel engine emissions calculated are provided in Appendix D. Mobile road (vehicle) combustion emissions were not calculated because these engine emissions are controlled at the source by mandated emission controls, and the magnitude of proposed road vehicle activity is small relative to existing road traffic (Section 4.3).

The primary radioactive airborne effluent at the proposed Moore Ranch Project would be radon-222 gas. Radon-222 can be released in the wellfield when the pregnant lixiviant is brought to the surface from the ore zone aquifer. Radon-222 can be released during wellfield drilling, production, operation of the central plant, resin transfer operations, and aquifer restoration activities. The highest annual radon-222 releases would occur when multiple concurrent release activities occur during a single year. The applicant calculated the potential radon-222 emissions from the proposed Moore Ranch Project (EMC, 2007a) using methods documented in NRC Regulatory Guide 3.59. The NRC staff selected the highest annual radon-222 emissions from these results, approximately 15.0 TBq/yr [406 Ci/yr], as the sum of concurrent wellfield production releases for the proposed wellfields, releases from operation of the central plant, and releases from transferring resins from nearby satellite facilities (EMC, 2007a). This estimate accounts for the applicant's proposed use of pressurized downflow IX columns that the applicant estimates would limit radon-222 releases from IX operations to 10 percent of the radon-222 that would be available for release using IX columns open to atmospheric pressure (EMC, 2007a). Additional information on proposed radon-222 emissions and the evaluation of potential impacts are provided in Section 4.13.1.2.1.

The applicant has proposed installing separate ventilation systems for all indoor, nonsealed process tanks and vessels where radon-222 or process fumes would be expected. The system would consist of an air duct or piping system connected to the top of each of the process tanks.

In-Situ Uranium Recovery and Alternatives

Redundant exhaust fans would direct collected gases to discharge piping that would exhaust fumes to the outside atmosphere. The design of the fans would be such that the system would be capable of limiting employee exposures with the failure of any single fan. Discharge stacks would be located away from building ventilation intakes to prevent introducing exhausted radon-222 into the facility, as recommended in Regulatory Guide 8.31. Airflow through any openings in the vessels would be from the process area into the vessel and the ventilation system, controlling any releases that occur inside the vessel. Separate ventilation systems may be used, as needed, for the functional areas within the plant

The work area ventilation system would be designed to force air to circulate within the plant process areas. The ventilation system would exhaust air outside the building, drawing fresh air inside. During favorable weather conditions, open doorways and convection vents in the roof would provide work area ventilation. The design of the ventilation system would be adequate to ensure that radon-222 daughter concentrations in the facility are maintained below 25 percent of the derived air concentration specified in 10 CFR Part 20.

To ensure that the emission control system is performing within specified operating conditions, instrumentation would be installed that provides an audible alarm if the air pressure (i.e., vacuum level) falls below specified levels. Operation of this system would be checked and documented during dryer operations. In the event of system failure, the operator would perform and document checks of the differential pressure or vacuum every four hours. Additionally, during routine operations, data from the air pressure differential gauges for other emission control equipment would be observed and documented at least once per shift during dryer operations.

2.1.1.1.6.2 Liquid Effluents

During operations, the IX process would generate production bleed at an estimated rate of 114 L/min [30 gal/min]. During groundwater restoration, the discharge from IX and/or RO processes would increase by an estimated 380 L/min [100 gal/min]. Any wastewater generated during or after the operations phase would be classified as liquid byproduct material (NRC, 2000). This byproduct material is not regulated as a hazardous waste under the Resource Conservation and Recovery Act (RCRA). The anticipated water chemistry of the injected material is summarized in Table 2-2. The production bleed and groundwater restoration waters would be disposed of via deep well injection. Other liquid waste streams from the central plant would include plant wash-down water and bleed stream from the elution and precipitation circuits during operations, and would contribute an estimated 38 L/min [10 gal/min) to the total wastewater volume during operations.

Deep injection wells would be drilled to dispose of contaminated liquid effluents at the proposed project site. The applicant has submitted a permit application to WDEQ (currently under review) to construct from two to four UIC Class I disposal wells to inject liquid wastes into one of two geologic formations at a depth of: (i) the Lance Formation and Fox Hills Sandstone, or (ii) the Teckla Sandstone member of the Lewis Shale and the Teapot and Parkman Sandstone members of the Mesa Verde Formation. Both sets of geologic formations are located thousands of feet below the proposed uranium ore production zone. The WDEQ will evaluate the suitability of these formations for Class I deep well injection and to ensure human health and the environment is protected.

Table 2-2. Summary of Anticipated Liquid Byproduct Water Quality Parameters		
Chemical Species	Minimum (mg/L)	Maximum (mg/L)
pH	6	9
Ammonia as N	50	500
Sodium	150	3,000
Calcium	200	1,000
Potassium	10	1,000
Bicarbonate as HCO_3	1,500	4,000
Carbonate as CO_3	0	500
Sulfate	80	2,000
Chloride	200	4,000
Uranium as U_3O_8	1	15
Ra-226 (pCi/l)	300	3,000
TDS	4,000	15,000
Source: Uranium One, 2009c		

If injection is permitted in the Lance and Fox Hills sandstones, two UIC Class I disposal wells would be completed at a depth of between approximately 1,128 m to 2,286 m [3,700 ft to 7,500 ft], with injection rates of 114 L/min [30 gal/min] per well. The injection interval would be overlain by at least 61 m [200 ft] of mudstone aquitard in the upper part of the Lance Formation. Water analyses for the Lance Formation suggest that the concentration of total dissolved solids in the injection interval would be in the range of 1,000 to 3,000 parts per million (ppm).

Alternatively, injection may be permitted in the Teckla, Teapot, and Parkman sandstones, at a depth of approximately 2,413 m to 2,929 m [7,916 ft to 9,610 ft]. Depending on the permitted injection rate, up to four deep injection wells would be completed in these formations. The formations are overlain by the Lewis Shale, a regionally extensive aquitard that is about 122 m [400 ft] thick above the Teckla Sandstone. Total dissolved solids concentrations in the Teckla, Teapot, and Parkman sandstones are highly variable but are expected to be in the range of 7,000 to 15,000 ppm in the Moore Ranch area.

Domestic liquid wastes from the restrooms and lunchrooms would be disposed of in an approved septic system that meets WDEQ requirements for Class V underground injection wells. The septic system for the proposed Moore Ranch Project includes an approximately

In-Situ Uranium Recovery and Alternatives

3,788 L [1,000 gal] shop septic tank and a 7,575 L [2,000 gal] plant septic tank (Uranium One, 2009b).

A small amount of uncontaminated wastewater would result from well development and well testing. This water would not need treatment prior to discharge to the surface.

Stormwater runoff would also need to be managed at the proposed Moore Ranch Project. Facility drainage would be designed to route storm runoff water away from or around the plant, ancillary building and parking areas, and chemical storage areas. Federal and State agencies regulate the discharge of both stormwater runoff and the discharge of wastewater to surface waters through their permitting processes. The status of obtaining an industrial and construction National Pollutant Discharge Elimination System (NPDES) stormwater permit at the proposed Moore Ranch Project, as required under the Clean Water Act and WDEQ regulations, is summarized in Table 1-2. Uranium One would develop Best Management Practices (BMP) containing the procedures and engineering controls that would be implemented to manage stormwater runoff (EMC, 2007a).

2.1.1.1.6.3 Solid Wastes

All phases of the operational lifecycle of the proposed Moore Ranch Project could generate solid byproduct material and nonhazardous solid waste. Byproduct material could include spent resin, empty chemical containers, pipes and fittings, tank sediments, contaminated soil from leaks and spills, and contaminated construction and demolition debris. Nonhazardous solid waste includes septic solids municipal solid waste (general trash), and other solid wastes.

Solid byproduct material is material that does not meet the NRC criteria for unrestricted release (including any soils contaminated from the operations). This material would be disposed of at a licensed disposal site. The proposed Moore Ranch Project is estimated to annually produce 76 m^3 [100 yd^3] of byproduct material during the operational period (EMC, 2007b). These materials would be stored on site inside the restricted area until such time that a full shipment could be made to a licensed waste disposal site or mill tailings facility. Based on the use of covered roll-offs with a nominal capacity of 15 m^3 [20 yd^3], approximately five byproduct material shipments would occur per year.

The NRC staff calculated the amount of solid byproduct material that could be generated from decommissioning activities based primarily on information provided in the applicant's surety estimate (Uranium One, 2008) as 11,010 m^3 [14,390 yd^3] plus an additional [512 t [565 T] of concrete demolition material. This estimate includes materials resulting from removal of plant facilities and equipment, wellfield equipment and piping from the two proposed wellfields, and removal of any contaminated soils that do not meet NRC limits for unrestricted release. As mentioned earlier, the applicant does not presently have an agreement in place with a licensed site to accept its solid byproduct material for disposal. The applicant would be required to have a disposal byproduct material agreement in place prior to operations. The applicant's preferred destination for disposal of byproduct material is at the Pathfinder Mines Shirley Basin site in Mills, Wyoming. Chapter 3 describes the expected disposal site location, and Chapter 4 describes the impact of disposing solid byproduct material.

The Moore Ranch Project expects to produce approximately 1,530 m^3 [2,000 yd^3] per year of nonhazardous solid waste composed of municipal waste (facility trash), septic solids, and other solid wastes, such as uncontaminated equipment, hardware, and packing materials. Waste minimization and recycling processes would be used to reduce the quantity of nonhazardous solid waste generated. Since typical contract waste-haulage vehicles range in capacity from 15 to 30 m^3 [20 to 40 yd^3], a nominal capacity of 15 m^3 [20 yd^3] per vehicle was assumed, resulting in approximately 100 nonhazardous solid waste shipments per year, or an average of approximately two shipments per week, during the operational period (Uranium One, 2009b).

The NRC staff calculated the amount of nonhazardous solid waste that could be generated from decommissioning activities based primarily on information provided in the applicant's surety estimate (Uranium One, 2008) as 482 m^3 [630 yd^3] plus an additional 6,102 t [6,730 T] of concrete demolition material. This estimate includes materials resulting from removal of plant facilities and equipment and wellfield equipment from the two proposed wellfields that do not contain radioactive materials or that meet NRC limits for unrestricted release. Chapter 3 describes the expected disposal site location and capacity and Chapter 4 describes the impact of disposing nonhazardous solid waste from the proposed Moore Ranch Project.

As discussed in the prior section, domestic liquid wastes from the restrooms and lunchrooms would be disposed of in an approved septic system, which includes an approximately 3,788 L [1,000 gal] shop septic tank and a 7,575 L [2,000 gal] plant septic tank (Uranium One, 2009b). Solid materials collected in septic systems would be disposed of as solid waste, in accordance with WDEQ regulations.

Based on a preliminary screening of processes and materials to be used at the proposed Moore Ranch Project, the applicant expects that the facility would be classified as a Conditionally Exempt Small Quantity Generator (CESQG), under the RCRA and Wyoming regulations. This classification does not require an application to the WDEQ. A CESQG: (1) must determine if their waste is hazardous; (2) must not generate more than 100 kilograms per month of hazardous waste or, except with regard to spills, more than 1 kilogram of acutely hazardous waste; (3) may not accumulate more than 1,000 kilograms of hazardous waste on-site at any time; and (4) must treat or dispose of their hazardous waste either in a nonsite or off-site U.S. treatment storage or disposal (TSD) facility that meets specific requirements of 40 CFR 261.5. If the facility fails to meet these four criteria, it would lose CESQG status and be fully regulated as either a small-quantity generator (more than 100 but less than 1,000 kilograms of nonacute hazardous waste per calendar month) or a large-quantity generator (at least 1,000 kilograms nonacute hazardous waste per calendar month). Any hazardous wastes, such as organic solvents, paints, used oil and paint thinners, empty chemical containers, tank sediments/sludges, chemical wastes, and spent batteries would be disposed of in accordance with a management program that the facility will develop to meet applicable local, State, and Federal regulatory requirements.

2.1.1.1.7 Transportation

Primary transportation activities would involve truck shipping and worker commuting. A variety of truck shipments are planned to support proposed activities during all phases of the facility's lifecycle. This shipping activity involves construction equipment and materials, operational processing supplies, ion-exchange resins, yellowcake product, and waste materials.

During construction, Uranium One expects that 16 employees would commute daily. During the operational and restoration phases of the project, the applicant estimates that a maximum of 24 employees would commute (Uranium One, 2009a). Yellowcake production would be up to approximately 1,360,900 kg [3 million lbs]. With each outgoing truckload containing approximately 18,145 kg [40,000 lb], there would be 100 trips per year to the Honeywell Uranium Conversion Facility in Metropolis, Illinois, or the Cameco Corporation facility in Port

Hope, Ontario, Canada. This equates to an average of one shipment every 3.6 days (EMC, 2007a). Operational waste shipments of byproduct material and uncontaminated solid wastes (Section 2.1.1.1.6) would be 5 shipments per year and 100 shipments per year (Uranium One, 2009a). During the decommissioning phase, the applicant assumes 5 employees would commute daily.

The traffic generated during the decommissioning phase, related to shipment of waste materials offsite, is expected to represent most of the truck traffic during that period. The NRC staff estimated the annual and average daily number of shipments that would be expected from the proposed decommissioning activities based on the calculated amounts of decommissioning solid wastes discussed in Section 2.1.1.1.6.3 and the volume of material per shipment. Because the applicant proposes a 12 month duration for the decommissioning of each well field, the staff conservatively estimated the annual decommissioning waste generated by assuming the applicant completes decommissioning and reclamation of a single well field (Wellfield 2) and all the surface facilities in a single year. This resulted in approximately 850 shipments of waste material for one year. Approximately 41 percent of these shipments would go to a landfill and the remainder to a licensed byproduct facility. If the disposal facilities are assumed to accept shipments 5 days per week, and the shipments are assumed to occur throughout the year with each shipment resulting in 2 one-way truck trips, the annual average daily traffic contribution would be approximately 3 truck round trips per day and about 16 shipments per week or 32 one-way trips per week. This level of trucking activity for decommissioning waste shipments is lower than the applicant's estimate for trucking during the construction phase and higher than the applicant's estimate for truck traffic during the operations and aquifer restoration phases. Detailed traffic estimates for the proposed action and for the existing roads are provided in Section 4.3.

2.1.1.1.8 Financial Surety

As stated in Section 2.10 of the GEIS, NRC regulations [10 CFR Part 40, Appendix A, Criterion (9)] require that applicants cover the costs to conduct decommissioning, reclamation of disturbed areas, waste disposal, dismantling, disposal of all facilities including buildings and wellfields, and groundwater restoration. Uranium One would maintain financial surety arrangements to cover those costs for the proposed Moore Ranch ISR Project. The initial surety estimate would be based on the first year of operation. NRC and WDEQ require annual revisions to the surety estimate to reflect existing operations and planned construction or operation the following year. Once the NRC, WDEQ, and Uranium One have agreed to the estimate, Uranium One would submit a reclamation performance bond, irrevocable letter of credit, or other surety instrument to the NRC and WDEQ. The NRC reviews financial surety in detail as part of its review for the safety evaluation report (SER). For additional information on financial surety requirements, see 10 CFR Part 40, Appendix A and Section 2.10 of the GEIS.

2.1.1.2 Alternative Wastewater Disposal Options

Liquid wastes will be generated during the operations and aquifer restoration phases of the lifecycle for the proposed Moore Ranch ISR facility. These wastes are considered as byproduct materials and must be managed and disposed of in compliance with applicable State and

Federal regulations, as established by license and permit. The applicant indicated that the normal operational waste stream would be nonhazardous under RCRA. Predominantly, the liquid waste stream would consist of

1. Process bleed ranging from 1 to 3 percent of the total water extracted from the ore horizon

2. Effluents from the central processing plant, such as process drains, elution circuit bleed, and wash down water

3. Wellfield purge water

4. IX and RO reject brines produced during aquifer restoration.

Of these, the process bleed would be the largest component during operations. Assuming a total plant throughput of 11,364 L/min [3,000 gal/min], a 1 percent process bleed would produce about 114 L/min [30 gal/min] of liquid waste. The applicant estimates that operational liquid wastes could be as much as approximately 170 L/min [45 gal/min] that would ultimately need disposal. During the aquifer restoration phase of the facility, the majority of the liquid waste would be comprised of discharge from the IX and RO processes used to treat groundwater. The applicant estimates that the total would increase to about 380 L/min [100 gal/min] for disposal (Uranium One, 2008, 2009a).

In August 2009, the applicant submitted a permit application to WDEQ for liquid waste disposal via UIC Class I injection wells to the Lance Formation and Fox Hills Sandstones at a depth of approximately 1,128 m to 2,286 m [3,700 ft to 7,500 ft] (Uranium One, 2008, 2009a). Another option would be disposal into the Teckla, Teapot, and Parkman sandstones at depths from 2,413 to 2,929 m [7,916 to 9,610 ft], although this would require more disposal wells because of the lower anticipated permitted injection capacity. Disposal via Class I injection well is discussed in Section 2.1.1.1 of this SEIS. If the applicant failed to receive a UIC permit from the WDEQ, then they would have to get their NRC license amended to approve another disposal option before they initiated operations. While not proposed in the license application, the following is an expanded discussion of possible alternative wastewater disposal options that were mentioned in the GEIS. Table 2-4 compares the various options. The analysis of potential environmental impacts is discussed in Section 4.1.1.2 of the SEIS but is not included in the comparison of alternatives in Table 2-4. Table 2-4 considers the applicant's proposed wastewater disposal option to use a Class I UIC injection well.

Historically, ISR facilities have used several other methods to manage and dispose of liquid wastes. These include solar evaporation ponds, land application, and surface water discharge. The following sections consider these disposal options, as well as deep well injection through UIC Class V permitted wells (NRC, 2003a). Characteristics of each of these wastewater disposal options are summarized in Table 2-3.

Table 2-3. Comparison of Different Liquid Wastewater Disposal Options

Disposal Option	Land Size/Footprint	Relevant Regulations and Permits	Construction Requirements	Wastewater Storage Prior to Disposal	Wastewater Treatment
Class I Injection Well	0.1 ha [0.25 acres]	• 20 CFR Part 20, Subparts D, K • UIC Class I permit (WDEQ)	• Land clearing and excavation equipment for pad, mud pits. • Drilling rig	38,000 L [10,000 gal] storage tank	No additional treatment, but may add antifouling agent to reduce scaling in well.
Class V Injection Well	• 0.1 ha [0.25 acres] • Potentially additional land area required for radium-settling basins 0.1 to 1.5 ha [0.25 to 4 acres] and purge reservoirs 4 ha [10 acres] or more.	• 10 CFR Part 20, Subparts D, K, Appendix B • UIC Class V permit (WDEQ) • WYPDES permit (WDEQ)	• Land clearing and excavation equipment for pad, mud pits, radium-settling basins, treatment facilities. • Drilling rig	• Storage/surge tank(s) • Radium-settling basins, treatment facility if needed to reduce Ra, U, and other contaminant concentrations.	Decontamination through ion exchange (IX)/reverse osmosis (RO). Additional treatment to injection zone class of use/primary drinking water, whichever more stringent. May add antifouling agent to reduce scaling in well.
Evaporation Ponds	• Individual pond: 0.4 to 2.5 ha [1 to 6.25 acres], max 16.2 ha [40 acres]. • Pond System: about 40 ha [100 acres].	• 10 CFR Part 40, Appendix A • Wyoming State Engineer's Office • NESHAPS permit (40 CFR Part 61 Subpart W) • Contract for byproduct material disposal (liners, sludges).	• Land clearing and excavation equipment to prepare surface for pond(s). • Construction equipment to construct pond liner(s).	None	Decontamination through IX/RO. No additional treatment.
Land Application	40 ha [100 acres]	• 10 CFR Part 20, Subparts D, K, Appendix B • 10 CFR Part 40, Appendix A, Criterion 6(6) • Zero release WYPDES permit (WDEQ) • NESHAPS permit (40 CFR Part 61)	Land clearing and excavation equipment for roads, radium-settling basins, treatment facilities.	• Storage/surge tank(s) • Radium-settling basins, treatment facility if needed to reduce Ra, U, and other contaminant concentrations.	• Decontamination through IX/RO. • Radium-settling basins, treatment facility if needed to reduce Ra, U, and other contaminant concentrations.
Discharge to Surface Waters	• 0.1 ha [0.25 acres], depending on outfall. • Potentially additional land area required for radium-settling basins 0.1 to 1.6 ha [0.25 to 4 acres] and purge reservoirs 4 ha [10 acres] or more. • Potentially separate storage facilities (impoundments, tanks) to maintain separate waste streams.	• 10 CFR Part 20, Subparts D, K, Appendix B • Zero release WYPDES permit (WDEQ) • NESHAPS permit (40 CFR Part 61) • Zero release WYDES permit (40 CFR Part 440, Subpart C)	Land clearing and excavation equipment for roads, radium-settling basins, treatment facilities.	• Yes. Applicant may elect to maintain separate "process" and "mine" wastewater streams. • Radium-settling basins, treatment facility if needed to reduce Ra, U, and other contaminant concentrations.	Decontamination through IX/RO. Additional treatment class of use/primary drinking water, whichever more stringent.

Table 2-3. Comparison of Different Liquid Wastewater Disposal Options (continued)

Disposal Option	Decommissioning	Environmental Benefits	Climate Influences	Health & Safety Issues
Class I Injection Well	Plug and abandon well in accordance with WDEQ requirements.	• Isolation from accessible environment. Low exposure to individuals at surface. • Smallest footprint, no additional decommissioning wastes. • No added transportation impacts for wastes. • No additional waste streams created. • Minimal and temporary visual impacts from drilling.	Deeper drilling requires longer rig time, higher diesel emissions (approximately 20 × typical production well).	Potential pipeline leaks.
Class V Injection Well	• Radium-settling basin liners and sludges, treatment building debris to be disposed as byproduct material, additional transportation of wastes to licensed disposal facility. • Plug and abandon well in accordance with WDEQ requirements.	Wastewater treated to drinking water standards.	• Deeper drilling requires longer rig time, higher diesel emissions (approximately 20 × typical production well). • Additional equipment needed to construct wastewater storage and treatment facilities.	• Potential leaks from wastewater storage and treatment facilities. • Additional waste volume during decommissioning.
Evaporation Ponds	Pond liners and sludges to be disposed as byproduct material, additional transportation of wastes to licensed disposal facility.		Additional equipment needed to construct evaporation ponds.	• Potential leaks from evaporation ponds. • Additional waste volume during decommissioning.
Land Application	• Radium-settling basin liners and sludges, treatment building debris to be disposed as byproduct material, additional transportation of wastes to licensed disposal facility. • Application soils to be disposed as byproduct material if limits exceeded. • Additional transportation of wastes to licensed disposal facility.	• Wastewater treatment to reduce uranium, radium, and other constituents. • Limited construction needed for land application area.	Additional equipment needed to construct wastewater storage and treatment facilities.	• Potential leaks from wastewater storage and treatment facilities. • Additional waste volume during decommissioning.
Discharge to Surface Waters	Radium-settling basin liners and sludges, treatment building debris to be disposed as byproduct material, additional transportation of wastes to licensed disposal facility.	Wastewater treated to drinking water standards.	Additional equipment needed to construct wastewater storage and treatment facilities.	• Potential leaks from wastewater storage and treatment facilities. • Additional waste volume during decommissioning.

2.1.1.2.1 Evaporation Ponds

One commonly used method for disposal of liquid wastes is to pump the liquids to one or more ponds and allow for natural solar radiation to reduce the volume through evaporation. The waste streams are usually treated prior to being discharged into evaporation ponds, but radionuclides and other metals may still be present, which will be concentrated as the liquids evaporate. The basic design criteria for an evaporation pond system are contained in 10 CFR Part 40, Appendix A. The location of the ponds, design and construction of the necessary clay or geotextile liner systems and embankments for the ponds, as well as pond inspection and maintenance would be conducted in accordance with NRC regulations and guidance (NRC, 2003a, 2008), and established by NRC license conditions, as necessary. The siting and design of any impoundments would also take into account applicable EPA requirements in 40 CFR Part 264 (NRC, 2008). The Wyoming State Engineer's Office also has state permitting authority for new impoundments. An earlier study of potential locations for a tailings impoundment associated with a potential conventional uranium mill at the proposed Moore Ranch Project location identified nearby natural basins underlain by a mudstone (NRC, 1982). This low permeability layer would be a favorable condition to be considered in siting potential evaporation ponds for this waste disposal option. The effectiveness of this wastewater disposal option will depend on the evaporation rate compared to the rate at which liquid wastes are produced. The evaporation rate varies seasonally, depending on temperature and relative humidity; the rate tends to be highest during warm, dry conditions and is lower during cool, humid conditions. If the evaporation rate is low or the seasonal conditions favoring evaporation are short in duration, the operator can compensate to some extent by increasing the size, and therefore, the surface area of the evaporation ponds. Historically, the area of an individual evaporation pond at uranium ISR facilities has ranged from approximately 0.04 to 2.5 ha [0.1 to 6.2 ac] (NRC, 1997; 1998a,b; Cohen and Associates, 2008b), although these are for facilities that use a combination of waste disposal methods.

Regulatory requirements in 40 CFR Part 61, Subpart W limit maximum lined uranium mill tailings impoundments to 16.2 ha [40 acres], although these tailings ponds are intended for a somewhat different purpose. The total footprint of the evaporation pond system for all liquid waste streams has been estimated as high as 40 ha [100 acres] (NRC, 1997). The estimated average annual evaporation rate from free water surfaces in the vicinity of the proposed Moore Ranch Project is approximately 102 cm/yr [40 in/yr] (Wyoming State Climate Office, 2004). Using this estimate, the minimum total evaporation pond area needed to handle the anticipated wastewater volumes would be about 10 to 18 ha [24 to 44 acres]. Taking into account annual precipitation effectively reduces the evaporation rate, then the pond system would need to be about 25 percent larger. Also, additional area would be needed to build additional storage areas to facilitate wastewater transfer between ponds for maintenance or repair work. During the winter months in Wyoming, where temperatures would be anticipated to be below freezing, the ponds could ice over, reducing the evaporation effectively to zero. To maintain year-round liquid disposal capability at the proposed Moore Ranch ISR facility, the applicant would, therefore, need to have either sufficient storage capacity, or at least one other disposal option (e.g., deep well injection, land application) available.

To identify potential leaks from the evaporation pond system into the subsurface, the applicant would need to design, construct, and monitor a leak detection system and conduct routine inspections, typically on a daily, weekly, monthly, and quarterly basis, with special inspections as-needed (NRC, 2008). The applicant would also need to maintain sufficient freeboard (i.e., distance from the water level to top of the embankment) of about 1 to 2 m [3 to 6 ft], depending on the size of the individual pond, so that precipitation or wind-driven waves would not overtop the embankment (NRC, 2008). In addition, the applicant would need to maintain sufficient reserve capacity in the evaporation pond system to allow the entire contents of one or more pond(s) to be transferred to other ponds, in the event of a leak and subsequent corrective action and liner repair (NRC, 2009). As necessary, the applicant would implement measures such as perimeter fencing and netting to protect humans and wildlife. These requirements would be established as conditions in an NRC license and enforced through the NRC inspection program.

The applicant may need to obtain a National Emission Standards for Hazardous Air Pollutants (NESHAP) review by the WDEQ to demonstrate that radionuclides, such as radon, released to the air from this option met 40 CFR Part 61 requirements, in particular the provisions of Subpart W that incorporate the requirements of 40 CFR Part 192 (NRC, 2008; Cohen and Associates, 2008a). In developing the impoundment design, the applicant would also need to consider EPA surface impoundment regulations for surface impoundments in 40 CFR Part 264 (NRC, 2008; Cohen and Associates, 2008b).

Because ponds are open to the air, dust and dirt can be blown into a pond, and dissolved solids concentrations may increase through evaporation to the point where salts precipitate from the solution. The ponds may periodically need to be cleaned to maintain good repair and the necessary freeboard, and the accumulated salts and solids disposed as byproduct material at an NRC-licensed disposal facility. Similarly, when the operations and aquifer restoration phases end, the pond liners and any accumulated materials would also need to be disposed of as byproduct material. As an example of decommissioning waste volumes, the amount of byproduct material generated during decommissioning and reclamation of evaporation ponds at the Smith Ranch ISR facility in Converse County, Wyoming, was estimated in 2007 at 52 m^3 [68 yd^3] (NRC, 2009).

2.1.1.2.2   Land Application

Land application is a disposal technique that uses agricultural irrigation equipment to broadcast wastewater on a relatively large area of land for subsequent evaporation. Land application is authorized at several solution mines (NRC, 1995; 1998b). Water released in this fashion would require treatment to meet NRC release requirements in 10 CFR Part 20, Subparts D, K, and Appendix B, and WDEQ surface water discharge requirements imposed by a zero release Wyoming Pollution Discharge Elimination System (WYPDES) permit (NRC, 2003a). Water, soils, and vegetation would be monitored on a regular basis, as established by license condition, to ensure soil loadings and vegetation concentrations remain within permit limits (NRC, 1995, 2003a).

Liquid wastes pretreatment using IX columns, RO, and precipitation of barium/radium sulfate is typically incorporated into this process to decrease uranium and radium levels. This pretreatment is necessary to meet regulatory release limits and to minimize the potential buildup of radionuclides in surface soils and vegetation. Despite pretreatment, however, liquid waste disposal by land application typically requires large areas to remain below release requirements. For example, the Crow Butte facility near Crawford, Nebraska, has described about 40 ha [100 acres] as available for land application, if needed (NRC, 1998b), and the Highland Uranium Project in Converse County, Wyoming, identified two land application sites, each about 22 ha [54 acres] (NRC, 1995). Depending on how the applicant treated the waste water prior to land application, this disposal option might have additional land requirements related to constructing wastewater treatment facilities, radium-settling basins, and storage reservoirs (NRC, 1995). These facilities would add to the required footprint for this disposal option. For example,

In-Situ Uranium Recovery and Alternatives

radium-settling basins are typically on the order of 0.1 to 1.6 ha [0.25 to 4 acres] (NRC, 1995, 1997, 1998a); purge reservoirs for temporary storage of treated wastewater can be much larger, with a surface area on the order of 4 ha [10 acres] or more, depending on the permit terms (NRC, 1998a).

An additional consideration for this waste disposal option is radon released to the air. For example, calculations performed by NRC staff for land application over an area of 42 ha [104 acres], assuming average wastewater concentrations of 37 Bq/m3 [1 pCi/L] for radium and 1 mg/L [1 ppm] for uranium, indicated that potential doses were below regulatory limits (NRC, 1997). Similarly, representative calculations for 7 years of land application to an area of 18.5 ha [46 acres] with an assumed wastewater application rate of 1,514 L/min [400 gal/min], estimated a radon flux of 1.3 pCi/m^2-sec, not much over an assumed background of 1 pCi/m^2-sec (NRC, 2003a, Appendix D).

Areas used for land application would need to be included in decommissioning surveys at the end of the operations and aquifer restoration phases to ensure that soil concentration limits would not be exceeded, potentially adding to the total amount of material for disposal at a licensed facility (NRC, 2003a). In addition, any pond liners and precipitated solids accumulated in a radium-settling basin system would need to be disposed of as byproduct material. For example, the annual amount of radium-bearing sludges generated in a 1.6-ha [4-acre] radium-settling basin was estimated to be about 22.4 m^3/yr [29.3 ft^3/yr] (Powertech, 2009).

2.1.1.2.3   Surface Water Discharge

Another disposal method historically used at uranium ISR facilities is treatment of liquid effluent and discharge at the surface. Like land application, the water would need to be pretreated to meet NRC release requirements in 10 CFR Part 20, Subparts D, K, and Appendix B, the provisions of 10 CFR Part 40, Appendix A that require conformance with EPA regulations in 40 CFR Part 440, and WDEQ requirements imposed by a zero release WYPDES permit. The WYPDES permit would specify limits calculated to ensure the discharge does not cause a violation of water quality standards. WDEQ would not issue the permit if the discharge would either cause or contribute to the violation of water quality standards. Specific requirements for uranium ISL facilities are provided in EPA regulations at 40 CFR Part 440, Subpart C. Pretreatment of the liquid wastes using IX columns, RO, and barium/radium sulfate precipitation is typically incorporated into this process to decrease uranium and radium levels in the wastewater. Like the land application wastewater disposal option, this treatment might require additional land for the construction of radium-settling basins and storage reservoirs (NRC, 2003a).

The regulatory framework for wastewater disposal by surface discharge is complicated, and it requires the applicant to make the distinction between "process wastewater" generated during uranium recovery operations and "mine wastewater" generated during aquifer restoration (NRC, 2003a). The applicant would need to develop storage capabilities, depending on whether it intended to maintain separate wastewater streams or commingle (mix) "process" and "mine" wastewater prior to treatment to 10 CFR Part 20 standards. In addition, the applicant would need to address any radioactivity at the discharge point or from storage facilities (e.g., tanks, impoundments), radium-settling basins, and related sludges as part of decommissioning the facility (NRC, 2003a; Cohen and Associates, 2008b). In addition, the applicant would not be allowed to discharge "process" wastewater to navigable waters of the United States, in accordance with EPA regulations at 40 CFR 440.34 (NRC, 2003a).

2.1.1.2.4 Class V Injection Well

At the well, the techniques employed in disposing of liquid wastes through a UIC Class V deep injection well would be similar to those for deep injection of liquid wastes in a UIC Class I disposal well, as described previously in Section 2.1.1.1. The main difference would be the nature of the permit (WDEQ, 2001). For disposal through a UIC Class V well, WDEQ regulations assume that at least one underground source of drinking water would underlie the potential injection zone in the Lance Formation and Fox Hills sandstone, at a depth greater than 2,286 m [7,500 ft] at the proposed Moore Ranch ISR facility (Uranium One 2008, 2009a). Also, the waste stream to be injected could not be classified as hazardous. For this reason, the wastewater would need to be treated to meet NRC release standards in 10 CFR Part 20, Subparts D, K, and Appendix B, and to ensure that all toxic substances remained at concentrations less than the WDEQ class-of-use standards or any federal primary drinking water standards, whichever is more stringent (WDEQ, 2001). Similar to land application and surface discharge, wastewater would be pretreated using IX columns, RO, barium/radium sulfate precipitation, and potentially radium-settling basins to decrease the levels of uranium, radium, and other contaminants in the waste water. As a result, the applicant would need to address storage facilities (e.g., tanks, impoundments) or radium settling basins and sludges as part of the decommissioning of the facility (NRC, 2003a). In addition, a UIC Class V permit would require the applicant to implement a monitoring plan to ensure that wastes would be confined to the authorized injection zone (WDEQ, 2008).

2.1.2 No-Action (Alternative 2)

Under the No-Action alternative, the NRC would not approve the license application for the proposed Moore Ranch Project. The No-Action alternative would result in Uranium One not constructing, operating, restoring the aquifer, or decommissioning the proposed Moore Ranch Project. No facilities, road, or wellfields would be built; no pipeline would be laid, as described in Section 2.1.1.1.2. No uranium would be recovered from the subsurface ore body; therefore, injection, production, and monitoring wells would not be installed to operate the facility. No lixiviant would be introduced in the subsurface and no buildings would be constructed to process extracted uranium or store chemicals. Because no uranium would be recovered, neither aquifer restoration nor decommissioning activities would occur. No liquid or solid effluents would be generated. The No-Action alternative is included to provide a basis for comparing and evaluating the potential impacts of the other alternatives, including the proposed action.

2.2 Alternatives Eliminated from Detailed Analysis

As required by NRC regulations, the NRC staff considered other alternatives to the construction, operation, aquifer restoration, and decommissioning of the proposed Moore Ranch Project. The range of alternatives was determined by considering the purpose and need for the proposed action and the private party's objectives to extract uranium from a particular ore body. Reasonable alternatives considered in a site-specific environmental review depend on the proposed action and site conditions. This section describes alternatives to the proposed action that were considered but not carried forward for detailed analysis for reasons described in the following sections. Sections 2.2.1 and 2.2.2 describe different mining and associated milling alternatives for the proposed project site. Section 2.2.3 discusses an alternate geographic location, Section 2.2.4 discusses the use of different lixiviant chemistry, and Section 2.2.5 discusses the use of alternate treatment methods for process-related liquid waste streams (Uranium One, 2009b).

2.2.1 Conventional Mining and Milling

Uranium ore deposits at depth may be accessed either by open pit (surface) mining or by underground mining techniques. Open-pit mining is used to exploit shallow ore deposits, generally deposits less than 170 m [550 ft] below ground surface (EPA, 2008a). To gain access to the deposit, the topsoil is first removed and may be stockpiled for later site reclamation, while the remainder of the material overlying the deposit (i.e., the overburden) can be removed via mechanical shovels and scrapers, trucks or loaders, or by blasting (EPA, 1995; 2008a). The depth to which an ore body is surface mined depends on the ore grade, the nature of the overburden, and the ratio of the amount of overburden to be removed to extract one unit of ore (EPA, 1995).

Underground mining techniques vary depending on size, depth, orientation, ore body grade, surface strata stability, and economic factors (EPA, 1995, 2008a). In general, underground mining involves sinking a shaft near the ore body and then extending levels from the main shaft at different depths to access the ore. Ore and waste rock would need to be removed through shafts by elevators or by using trucks to carry these materials up inclines to the surface (EPA, 2008a).

In addition, once the open pit or underground workings are established, the mine may need to be dewatered to allow the extraction of the uranium ore. Dewatering can be accomplished either by pumping directly from the open pit or through pumping of interceptor wells to lower the water table (EPA, 1995). The mine water likely will require treatment prior to discharge, due to contamination from radioactive constituents, metals, and suspended and dissolved solids. Discharge of these mine waters may have subsequent impacts to surface water drainages and sediments, as well as to near-surface sources of groundwater (EPA, 1995).

Following the completion of mining, either by open pit or underground techniques, reclamation of the mine is needed. Stockpiled overburden can be reintroduced into the mine, either during extraction operations or following any topsoil reapplied, in an attempt to reestablish topography consistent with the surroundings. At the end of dewatering, the water table may rebound and fill portions of the open pit and underground workings. Historically, uranium mines have impacted local groundwater supplies, and the waste materials from the mines have contaminated lands surrounding the mines (EPA, 2008b).

Ore extracted from the open pit or underground mine would be processed in a conventional mill. As discussed in Appendix C of the GEIS (NRC, 2009), ore processing at a conventional mill involves a series of steps (handling and preparation, concentration, and product recovery). While the conventional milling techniques recover approximately 90 percent of the uranium content of the feed ore (NRC, 2009), the process does generate substantial wastes (known as tailings) since roughly 95 percent of the ore rock is disposed as waste (NRC, 2009). This process also can consume large amounts of water {e.g., approximately 534 liters per minute (Lpm); 141 gallons per minute [gpm]) for the proposed Pinon Ridge mill in Colorado (EFRC, 2009)}.

Tailings are disposed in extensive lined impoundments. NRC reviews the design and construction of the impoundments to ensure safe disposal of the tailings (NRC, 2009). Reclamation of the tailings pile generally involves evaporation of liquids in the tailings, settlement of the tailings over time, and covering the pile with a thick radon barrier and earthen material or rocks for erosion control. An area surrounding the reclaimed tailings piles would be fenced off in perpetuity, and the site transferred to either a State or Federal agency for

In-Situ Uranium Recovery and Alternatives

long-term care (EIA, 1995). The costs associated with final mill decommissioning and tailings reclamation can run into the tens of millions of dollars (EIA, 1995).

NRC evaluated the potential environmental impacts of conventional uranium milling operations in a programmatic context, including the management of mill tailings in the final generic environmental impact statement on uranium milling (NRC, 1980). This SEIS evaluated the nature and extent of conventional uranium milling to inform of the regulatory requirements for management and disposal of mill tailings and for mill decommissioning. The impacts from operating a conventional mill are significantly greater than for operating an ISR facility. For example, at the proposed Moore Ranch Project, approximately 61 ha [150 ac] would be used for uranium extraction operations (e.g., two wellfields, the central plant, pipeline infrastructure); however, for a conventional mill, more than twice that amount of land (150 ha [370 ac]) would be devoted to milling and allied activities during operations, and during mill construction a total of 300 ha [741 ac] could be impacted (NRC, 1980). Furthermore, the deposition of windblown tailings could further restrict land use near the tailings. Levels of contamination extended several hundred meters beyond the model site boundary evaluated in the GEIS for conventional milling. Therefore, conventional milling was eliminated from detailed analysis in the Moore Ranch SEIS.

2.2.2 Conventional Mining and Heap Leaching

Conoco, Inc. proposed to use conventional mining techniques to mine uranium in the same area as the proposed Moore Ranch Project and requested the issuance of an NRC Source and Byproduct Material License authorizing operation of the proposed Sand Rock Mill Project (NRC, 1982). Eleven different geographic locations at the site were evaluated to meet the performance objectives of reducing the length of slurry pipelines to minimize the potential for leaks and spills and to locate a prospective evaporation pond to minimize the potential impact on ecological resources. The development of a conventional mine would result in the disturbance of a larger land area with a greater risk to wildlife than an ISR facility, and the fugitive dust emissions would be greater with a conventional mine compared to an ISR facility and require more maintenance with an associated increase in occupational exposure compared to an ISR facility. In addition, evaporation ponds can produce large quantities of byproduct material in the form of pond liners and sludges. These materials lead to higher volumes of decommissioning wastes, requiring transportation to and disposal at an NRC-licensed facility. In addition, the effectiveness of evaporation ponds in Wyoming is subject to seasonal conditions such as winter freezing. For these reasons, the use of evaporation ponds associated with conventional mining at the proposed Moore Ranch Project was eliminated from detailed analysis.

Heap leaching is discussed in Appendix C of the GEIS (NRC, 2009). For low-grade ores, heap leaching is a viable alternative. Low-grade ore removed from open-pit or underground mining operations undergo further processing to remove and concentrate the uranium. Heap leaching is typically used when the ore body is small and situated far from the milling site. The low-grade ore is crushed to approximately 2.6 cm [1 in] in size and mounded above grade on a prepared pad. A sprinkler or drip system positioned over the top continually distributes leach solution over the mound. Depending on the lime content, an acid or alkaline solution can be used. The leach solution trickles through the ore and mobilizes the uranium, as well as other metals, into the solution. The solution is collected at the base of the mound by a manifold and processed to extract the uranium. The uranium recovery from heap leaching is expected to range from 50 to 80 percent, resulting in final tailings material of approximately 0.01 percent uranium content. Once heap leaching is complete, the depleted materials are AEA, as amended by UMTRCA,

In-Situ Uranium Recovery and Alternatives

byproduct material that must be placed in a conventional mill tailings impoundment, unless NRC grants a disposal exemption. While the impacts from heap leaching may be less than those from conventional milling, the impacts from the associated open-pit or underground mining would still be substantial. For these considerations, similar to those listed in Section 2.2.1, this alternative is not carried forward for detailed analysis.

2.2.3 Alternate Site Location

An alternate central plant location approximately 366 m [1,200 ft] west of the proposed location was considered by the applicant. Although this location would have been closer to Brown Road, it was determined to be less suitable because more changes to the existing topography would be required for the proposed plant layout than evaluated under the proposed action. At the alternate location, more cut and fill activity would have been required, thus resulting in more land disturbance and the potential to impact cultural resources.

2.2.4 Alternate Lixiviants

Alternate lixiviant chemistry was also considered for the operations phase of the proposed action, including acid leach solutions and ammonia-based lixiviants. Acid-based lixiviants such as sulfuric acid, dissolve heavy metals and other solids associated with uranium in the host rock and other chemical constituents that require additional remediation and have greater environmental impacts. At a small-scale research facility in Wyoming, test patterns were developed using acid-based lixiviants. During operations, two significant problems developed. The mineral gypsum precipitated on the well screens and in the aquifer, which plugged the wells and reduced the efficiency of the wellfield restoration. Aquifer restoration had limited success, because of the gradual dissolution of the precipitated gypsum, which resulted in increased salinity and sulfate levels in the affected groundwater (Uranium One, 2009b). Because it is technically more difficult to restore acid mine sites, the use of an acid-based lixiviant was eliminated from detailed analysis in the SEIS.

Ammonia-based lixiviants have been used at ISR operations in Wyoming. However, operational experience has shown that ammonia tends to adsorb onto clay minerals in the subsurface and then slowly desorbs from the clay during restoration, therefore requiring a much larger volume of groundwater be removed and processed during aquifer restoration (Mudd, 2001). Because of the greater consumptive use of groundwater to meet groundwater restoration requirements, the use of an ammonia-based lixiviant was eliminated from detailed analysis.

2.2.5 Alternate Wastewater Treatment Methods

A range of liquid treatment methods that considered the three primary waste streams generated at an ISR facility: plant eluant, wellfield purge water, and RO reject produced during wellfield restoration were considered for use at the proposed Moore Ranch Project. These methods included mechanical evaporation and chemical precipitation, as discussed in the following paragraphs.

Although mechanical evaporation could produce the smallest possible volume of brine for disposal, it would require larger storage tanks and more offsite shipment of materials than the proposed action, higher energy consumption approximately 16 times greater than the proposed action), a larger operations workforce, and from an environmental perspective mechanical evaporation would have a larger carbon footprint because of the greater power requirements (Uranium One, 2009b). From a safety perspective, mechanical evaporation would require

In-Situ Uranium Recovery and Alternatives

operating at high temperatures and pressures and would have high chemical requirements for solidification chemicals, thus increasing the potential for occupational exposure and accidents. Finally, the capital cost for mechanical evaporation would be approximately four times greater than those for deep disposal wells (Uranium One, 2009a). For these reasons, mechanical evaporation was eliminated from detailed analysis in the SEIS.

Chemical precipitation and RO to either pretreat the wastewater for more efficient operation of the RO system or for brine treatment were also considered. This practice would result in the formation of both brine residual and sludge. This method of treatment produces a higher volume of liquid residues and requires greater storage capacity than the proposed action (757,575 L [200,000 gal] brine storage tank compared to a 37,878 L [10,000 gal] storage tank for the proposed action). The brine would be concentrated waste that could potentially be characterized as either hazardous or mixed waste. The energy consumption for this treatment method would be approximately four times that of the proposed action, the labor would be approximately six times higher than that for the proposed action. This treatment method would involve the handling of the greatest amount of residues requiring onsite storage and transportation for offsite disposal, thus increasing the potential for occupational exposure and transportation accidents. For these reasons, chemical precipitation and RO were eliminated from detailed analysis in the SEIS.

2.2.6 Comparison of the Predicted Environmental Impacts

NUREG–1748 (NRC, 2003b) categorizes the significance of potential environmental impacts as follows:

SMALL: The environmental effects are not detectable or are so minor that they will neither destabilize nor noticeably alter any important attribute of the resource considered.

MODERATE: The environmental effects are sufficient to alter noticeably, but not destabilize important attributes of the resource considered.

LARGE: The environmental effects are clearly noticeable and are sufficient to destabilize important attributes of the resource considered.

In this section, for each of the three alternatives, the potential environmental impacts to each resource area are summarized for all four of the ISR phases: construction, operation, aquifer restoration, and decommissioning. The significance levels (SMALL, MODERATE, and LARGE) are specific to each resource and are defined in Chapter 4.

Chapter 4 of this SEIS presents a more detailed evaluation of the environmental impacts from the proposed action and the No-Action alternative. Table 2-4 compares the environmental impact by ISR phase of implementing the proposed action and the No-Action alternative and identifies the section of the SEIS where more detailed information can be found.

Table 2-4. Comparison of Predicted Environmental Impacts		
4.2 Land Use Impacts	Alternative 1—Proposed Action	Alternative 2—No-Action
Construction	SMALL 4.2.1.1	NONE 4.2.2
Operation	SMALL 4.2.1.2	NONE 4.2.2
Aquifer Restoration	SMALL 4.2.1.3	NONE 4.2.2
Decommissioning	SMALL 4.2.1.4	NONE 4.2.2
4.3 Transportation Impacts	Alternative 1—Proposed Action	Alternative 2—No-Action
Construction	SMALL 4.3.1.1	NONE 4.3.2.
Operation	SMALL 4.3.1.2	NONE 4.3.2
Aquifer Restoration	SMALL 4.3.1.3	NONE 4.3.2
Decommissioning	SMALL 4.3.1.4	NONE 4.3.2
4.4 Geology and Soils Impacts	Alternative 1—Proposed Action	Alternative 2—No-Action
Construction	SMALL 4.4.1.1	NONE 4.4.2
Operation	SMALL 4.4.1.2	NONE 4.4.2
Aquifer Restoration	SMALL 4.4.1.3	NONE 4.4.2
Decommissioning	SMALL 4.4.1.4	NONE 4.4.2
4.5.1 Water Resources Impacts (Surface Water)	Alternative 1—Proposed Action	Alternative 2—No-Action
Construction	SMALL 4.5.1.1.1	NONE 4.5.1.2
Operation	SMALL 4.5.1.1.2	NONE 4.5.1.2
Aquifer Restoration	SMALL 4.5.1.1.3	NONE 4.5.1.2
Decommissioning	SMALL 4.5.1.1.4	NONE 4.5.1.2
4.5.2 Water Resources Impacts (Groundwater)	Alternative 1—Proposed Action	Alternative 2—No-Action
Construction	SMALL 4.5.2.1.1	NONE 4.5.2.2
Operation	SMALL 4.5.2.1.2	NONE 4.5.2.2
Aquifer Restoration	SMALL 4.5.2.1.3	NONE 4.5.2.2
Decommissioning	SMALL 4.5.2.1.4	NONE 4.5.2.2

Table 2-4. Comparison of Predicted Environmental Impacts		
4.6 Ecological Resources Impacts (Vegetation)	Alternative 1—Proposed Action	Alternative 2—No-Action
Construction	SMALL 4.6.1.1.1	NONE 4.6.2
Operation	SMALL 4.6.1.2.1	NONE 4.6.2
Aquifer Restoration	SMALL 4.6.1.3	NONE 4.6.2
Decommissioning	SMALL 4.6.1.4	NONE 4.6.2
4.6 Ecological Resources Impacts (Wildlife)	Alternative 1—Proposed Action	Alternative 2—No-Action
Construction	SMALL 4.6.1.1.2	NONE 4.6.2
Operation	SMALL 4.6.1.2.2	NONE 4.6.2
Aquifer Restoration	SMALL 4.6.1.3	NONE 4.6.2
Decommissioning	SMALL 4.6.1.4	NONE 4.6.2
4.7 Meteorology, Climatology, and Air Quality Impacts	Alternative 1—Proposed Action	Alternative 2—No-Action
Construction	SMALL 4.7.1.1	NONE 4.7.3
Operation	SMALL 4.7.1.2	NONE 4.7.3
Aquifer Restoration	SMALL 4.7.1.3	NONE 4.7.3
Decommissioning	SMALL 4.7.1.4	NONE 4.7.3
4.8 Noise Impacts	Alternative 1—Proposed Action	Alternative 2—No-Action
Construction	SMALL 4.8.1.1	NONE 4.8.2
Operation	SMALL 4.8.1.2	NONE 4.8.2
Aquifer Restoration	SMALL 4.8.1.3	NONE 4.8.2
Decommissioning	SMALL 4.8.1.4	NONE 4.8.2
4.9 Impacts to Historical and Cultural Resources	Alternative 1—Proposed Action	Alternative 2—No-Action
Construction	SMALL 4.9.1.1	NONE 4.9.2
Operation	SMALL 4.9.1.2	NONE 4.9.2
Aquifer Restoration	SMALL 4.9.1.3	NONE 4.9.2
Decommissioning	SMALL 4.9.1.4	NONE 4.9.2

Table 2-4. Comparison of Predicted Environmental Impacts		
4.10 Visual and Scenic Resources Impacts	Alternative 1—Proposed Action	Alternative 2—No-Action
Construction	SMALL 4.10.1.1	NONE 4.10.2
Operation	SMALL 4.10.1.2	NONE 4.10.2
Aquifer Restoration	SMALL 4.10.1.3	NONE 4.10.2
Decommissioning	SMALL 4.10.1.4	NONE 4.10.2
4.11 Socioeconomic Impacts (Demographics)	Alternative 1—Proposed Action	Alternative 2—No-Action
Construction	SMALL 4.11.1.1.1	NONE 4.11.2
Operation	SMALL 4.11.1.2.1	NONE 4.11.2
Aquifer Restoration	SMALL 4.11.1.3	NONE 4.11.2
Decommissioning	SMALL 4.11.1.4	NONE 4.11.2
Socioeconomic Impacts (Income)	Alternative 1—Proposed Action	Alternative 2—No-Action
Construction	SMALL 4.11.1.1.2	NONE 4.11.2
Operation	SMALL 4.11.1.2.2	NONE 4.11.2
Aquifer Restoration	SMALL 4.11.1.3	NONE 4.11.2
Decommissioning	SMALL 4.11.1.4	NONE 4.11.2
Socioeconomic Impacts (Housing)	Alternative 1—Proposed Action	Alternative 2—No-Action
Construction	NONE 4.11.1.1.3	NONE 4.11.2
Operation	SMALL TO MODERATE 4.11.1.2.3	NONE 4.11.2
Aquifer Restoration	SMALL 4.11.1.3	NONE 4.11.2
Decommissioning	SMALL 4.11.1.4	NONE 4.11.2
Socioeconomic Impacts (Employment Structure)	Alternative 1—Proposed Action	Alternative 2—No-Action
Construction	SMALL 4.11.1.1.4	NONE 4.11.2
Operation	NONE 4.11.1.2.4	NONE 4.11.2
Aquifer Restoration	SMALL 4.11.1.3	NONE 4.11.2

Table 2-4. Comparison of Predicted Environmental Impacts		
Decommissioning	SMALL 4.11.1.4	SMALL 4.11.2
Socioeconomic Impacts (Local Finance)	Alternative 1—Proposed Action	Alternative 2—No-Action
Construction	SMALL 4.11.1.1.5	NONE 4.11.2
Operation	SMALL 4.11.1.2.5	NONE 4.11.2
Aquifer Restoration	SMALL 4.11.1.3	NONE 4.11.2
Decommissioning	SMALL 4.11.1.4	NONE 4.11.2
Socioeconomic Impacts (Education)	Alternative 1—Proposed Action	Alternative 2—No-Action
Construction	NONE 4.11.1.1.6	NONE 4.11.2
Operation	SMALL 4.11.1.2.6	NONE 4.11.2
Aquifer Restoration	SMALL 4.11.1.3	NONE 4.11.2
Decommissioning	SMALL 4.11.1.4	NONE 4.11.2
Socioeconomic Impacts (Health and Social Services)	Alternative 1—Proposed Action	Alternative 2—No-Action
Construction	NONE 4.11.1.1.7	NONE 4.11.2
Operation	SMALL 4.11.1.2.7	NONE 4.11.2
Aquifer Restoration	SMALL 4.11.1.3	NONE 4.11.2
Decommissioning	SMALL 4.11.1.4	NONE 4.11.2
4.12 Environmental Justice Impacts	Alternative 1—Proposed Action	Alternative 2—No-Action
Construction	NONE 4.12.2	NONE 4.12.3
Operation	NONE 4.12.2	NONE 4.12.3
Aquifer Restoration	NONE 4.12.2	NONE 4.12.3
Decommissioning	NONE 4.12.2	NONE 4.12.3

In-Situ Uranium Recovery and Alternatives

4.13 Public Occupational Health and Safety Impacts	Alternative 1—Proposed Action	Alternative 2—No-Action
Construction	SMALL 4.13.1.1	NONE 4.13.2
Operation	SMALL 4.13.1.2	NONE 4.13.2
Aquifer Restoration	SMALL 4.13.1.3	NONE 4.13.2
Decommissioning	SMALL 4.13.1.4	NONE 4.13.2
4.14 Waste Management Impacts	Alternative 1—Proposed Action	Alternative 2—No-Action
Construction	SMALL 4.14.1.1.1	NONE 4.14.2
Operation	SMALL 4.14.1.1.2	NONE 4.14.2
Aquifer Restoration	SMALL 4.14.1.1.3	NONE 4.14.2
Decommissioning	SMALL 4.14.1.1.4	NONE 4.14.2

2.3 Final Recommendation

After weighing the impacts of the proposed action and comparing the alternatives, the NRC staff, in accordance with 10 CFR 51.91(d), sets forth its NEPA recommendation regarding the proposed action. Unless safety issues mandate otherwise, the NRC staff's recommendation to the Commission related to the environmental aspects of the proposed action is that the source material license be issued as requested. This recommendation is based upon (1) the license application, including the ER submitted by Uranium One and the applicant's supplemental letters and responses to the NRC staff's requests for additional information; (2) consultation with Federal, State, Tribal, and local agencies; (3) the NRC staff independent review; (4) the NRC staff's consideration of comments received on the draft SEISs; and (5) the assessments summarized in this SEIS.

2.4 References

Cohen and Associates, 2008a. "Final Report History and Basis of NESHAPs and Subpart W." Vienna, Virginia: S. Cohen and Associates. September 25. <http://www.epa.gov/radiation/docs/neshaps/subpart-w/neshap-history.pdf> (25 May 2010).

Cohen and Associates, 2008b. "Final Report Review of Existing and Proposed Tailings Impoundment Technologies." Vienna, Virginia: S. Cohen and Associates. September 25. <http://www.epa.gov/radiation/docs/neshaps/subpart-w/tailings-impoundment-tech.pdf> (25 May 2010).

EMC (Energy Metals Corporation US), 2007a. "Application for USNRC Source Material License, Moore Ranch Uranium Project, Campbell County, Wyoming, Technical Report." Casper, Wyoming: Uranium 1 Americas Corporation. ADAMS Accession Nos. ML072851222,

ML072851258, ML072851259, ML072851260, ML072851268, ML072851350, ML072900446. October 2.

EMC, 2007b. "Application for USNRC Source Material License, Moore Ranch Uranium Project, Campbell County, Wyoming, Environmental Report." Casper, Wyoming: Uranium 1 Americas Corporation. ADAMS Accession Nos. ML072851222, ML072851229, ML072851239, ML07285249, ML07285253, ML07285255. October 2.

EIA (Energy Information Administration), 1995. "Decommissioning of U.S. Uranium Production Facilities." Office of Coal, Nuclear, Electric, and Alternate Fuels. DOE/EIA–0592. February.

EPA (U.S. Environmental Protection Agency), 2008a. "Technical Report on Technologically Enhanced Naturally Occurring Radioactive Materials from Uranium Mining: Mining and Reclamation Background." Volume 1. Office of Radiation and Indoor Air/Radiation Protection Division. EPA–402–R–08–005. April.

EPA, 2008b. "Health and Environmental Impacts of Uranium Contamination in the Navajo Nation: Five-Year Plan." Requested by House Committee on Oversight and Government Reform. June 9.

EPA, 1995. "Technical Resource Document: Extraction and Beneficiation of Ores and Minerals—Uranium." Volume 5. Office of Solid Waste / Special Waste Branch.

Griffin, M., 2009. <mike.griffin@uranium1.com. Phases of the Moore Ranch Project" September. [email communication]. ADAMS Accession No. ML092720144. (28 September 2009).

Mackin, P.C., D. Daruwalla, J. Winterle, M. Smith, and D.A. Pickett, 2001. NUREG/CR–6733, "A Baseline Risk-Informed Performance-Based Approach for *In-Situ* Leach Uranium Extraction Licensees." Washington, DC: NRC. September.

Mudd, G.M., 2001. "Critical Review of Acid *In-Situ* Leach Uranium Mining: 1-USA and Australia." Vol. 41. pp. 390–403.

NRC, 2009 NUREG–1910, "Generic Environmental Impact Statement for *In-Situ* Leach Uranium Milling Facilities." Washington, DC: NRC. May.

NRC, 2008. Regulatory Guide 3.11, "Design, Construction, and Inspection of Embankment Retention Systems at Uranium Recovery Facilities." Revision 3. Washington, DC: NRC. November 2008.

NRC, 2003a. NUREG–1569, "Standard Review Plan for *In-Situ* Leach Uranium Extraction License Applications—Final Report." Washington, DC: NRC. June 2003. Washington, DC: NRC. August.

NRC, 2003b. NUREG–1748, "Environmental Review Guidance for Licensing Actions Associated With NMSS Programs." Washington, DC: NRC. August.

NRC, 2000. "Staff Requirements–SECY–99–013 Recommendations on Ways to Improve the Efficiency of NRC Regulation at *In-Situ* Uranium Recovery Facilities." Washington, DC: July 26, 2000.

NRC, 1998a. "Environmental Assessment for Renewal of Source Material License No. SUA–1341, Cogema Mining, Inc. Irigaray and Christensen Ranch Projects, Campbell and Johnson Counties, Wyoming." Docket No. 40-8502. Washington, DC: NRC. June 1998a.

NRC, 1998b. "Environmental Assessment for Renewal of Source Material License No. SUA–1534 Crow Butte Resources Incorporated Crow Butte Uranium Project Dawes County, Nebraska." Docket No. 40-8943. Washington, DC: NRC. February.

NRC, 1997. NUREG–1508, "Final Environmental Impact Statement To Construct and Operate the Crownpoint Uranium Solution Mining Project, Crownpoint, New Mexico." Washington, DC: NRC. February.

NRC, 1995. "Environmental Assessment for Renewal of Source Materials License No. SUA–1511.Power Resources Incorporated Highland Uranium Project, Converse County, Wyoming."

NRC, 1982. NUREG–0889, "Draft Environmental Statement related to the Operation of Sand Rock Mill Project (Draft Report for Comment)." Washington, DC. March.

NRC, 1980. NUREG–0706, "Final Generic Environmental Impact Statement on Uranium Milling Project M-25." ADAMS Accession Nos. ML032751663, ML0732751667, ML032751669. September.

Powertech (USA), Inc., 2009 "Dewey-Burdock Project. Supplement to Application for NRC Uranium Recovery License Dated February 2009. Appendices." Greenwood Village, Colorado: Powertech (USA), Inc. August.

Uranium One (Uranium One Americas), 2009a. "Response to Request for Additional Information for the Moore Ranch In-Situ Uranium Recovery Project License Application (TAC JU011)." ADAMS Accession No. ML091900402. June 19.

Uranium One, 2009b. "Response to Request for Additional Information for the Moore Ranch In-Situ Uranium Recovery Project License Application (TAC JU011)." ADAMS Accession No. ML092450317. August 31.

Uranium One, 2009c. "Class I Permit Application and Comment Responses–Moore Ranch Uranium Project. Volumes I and II" Casper, Wyoming. Uranium One, Inc. August 17.

Uranium One, 2008. Response to Request for Additional Information for the Moore Ranch In-Situ Uranium Recovery Project License Application (TAC JU011)." ADAMS Accession No. ML082060527. July 11.

WDEQ (Wyoming Department of Environmental Quality), 2008. "UIC Program Policy: Permitting In-Situ Mine Waste Disposal Wells." 08-001.POL. Cheyenne, Wyoming: WDEQ, Water Quality Division. July 16.

WDEQ, 2001. "Water Quality Rules and Regulations, Chapter 16, Class V Injection Wells and Facilities Underground Injection Control Program." Cheyenne, Wyoming: WDEQ, Water Quality Division. July.

WDEQ, 2000. "Guideline No. 4–*In-Situ* Mining, Attachment III Topsoil and Subsoil Management and the Associated Erosion Control at Uranium *In-Situ* Leaching Operations." March.

WDEQ, 1986. "Guideline No. 2–Vegetation." March.

Wyoming State Climate Office, 2004. "Wyoming Climate Atlas." J. Curtis and K. Grimes, eds. Laramie Wyoming: Wyoming State Climate Office <http://www.wrds.uwyo.edu/sco/climateatlas/evaporation.html> (03 June 2010).

3 AFFECTED ENVIRONMENT

3.1 Introduction

The proposed Moore Ranch Project is located in the Powder River Basin in southwest Campbell County, Wyoming, in the Wyoming East Uranium Milling Region defined in the Generic Environmental Impact Statement for *In-Situ* Leach Uranium Milling Facilities (NUREG–1910, GEIS) (NRC, 2009a). The Powder River Basin is an energy-rich area that possesses some of the largest coal, coal bed methane (CBM) and natural gas deposits in the United States. The proposed project is located approximately 80 km [50 mi] southwest of Gillette, Wyoming, and approximately 80 km [50 mi] northeast of Casper, Wyoming (see Figure 1-1). The proposed license area encompasses approximately 2,879 ha [7,110 ac] of land; an estimated 61 ha [150 ac] of land surface could be directly disturbed by *in-situ* recovery (ISR) construction and operations.

This chapter describes the existing site conditions at the proposed Moore Ranch Project. For the purposes of this supplemental environmental impact statement (SEIS), the term license area refers to the 2,879 ha [7,110 ac] Moore Ranch Project plus an area extending 3.2 km [2 mi], as suggested in NUREG–1569 unless a different radius for a particular resource is specified (NRC, 2003). This section describes resource areas including land use, transportation, geology and soils, water resources, ecology, noise, air quality, historical and cultural resources, visual and scenic resources, socioeconomics, public and occupational health, and current waste management practices. Issues identified, based on agency and public concerns and regulatory and planning requirements, have been considered in the description of the affected environment. The information in this chapter forms the basis for assessing the potential impacts (see Chapter 4) of the proposed action and each alternative (Chapter 2).

3.2 Land Use

The proposed Moore Ranch Project is located in the northeast portion of Wyoming within the Powder River Basin in the following townships and ranges: Township 42 North, Range 75 West, Sections 26, 27, 33, 34, 35, 36, and portions of Sections 25 and 28; Township 41 North, Range 75 West, Sections 2, 3, 4, and portions of Sections 1, 9, and 10; and Township 42 North, Range 74 West, and portions of Sections 30 and 31.

Section 3.1.2.2 of the GEIS described the concept of split estate where the land surface and mineral rights can be owned by different entities, and in particular, where the U.S. Bureau of Land Management (BLM) owns the mineral rights (NRC, 2009a). This situation occurs at the proposed Moore Ranch Project. Of the 2,879 ha [7,110 ac] comprising the proposed Moore Ranch Project, over 85 percent of the surface rights are previously owned, and about 14 percent of the surface rights are owned by the State of Wyoming. About 59 percent of the mineral rights are owned by BLM but have been leased to Uranium One Americas (Uranium One), about 26 percent of the mineral rights are privately owned, and the State of Wyoming owns the mineral rights underlying their surface ownership. Figures 3-1 and 3-2 show the proposed Moore Ranch Project surface and mineral rights owners. The central plant would be located on privately owned land, and no mineral exploration (i.e., no wells to extract uranium) would occur at that location. At both Wellfields 1 and 2, the surface rights are privately owned; however, the mineral rights at Wellfield 1 have been leased from the BLM; and at Wellfield 2 the mineral rights are both publicly and privately owned. The Permit to Mine application submitted

Affected Environment

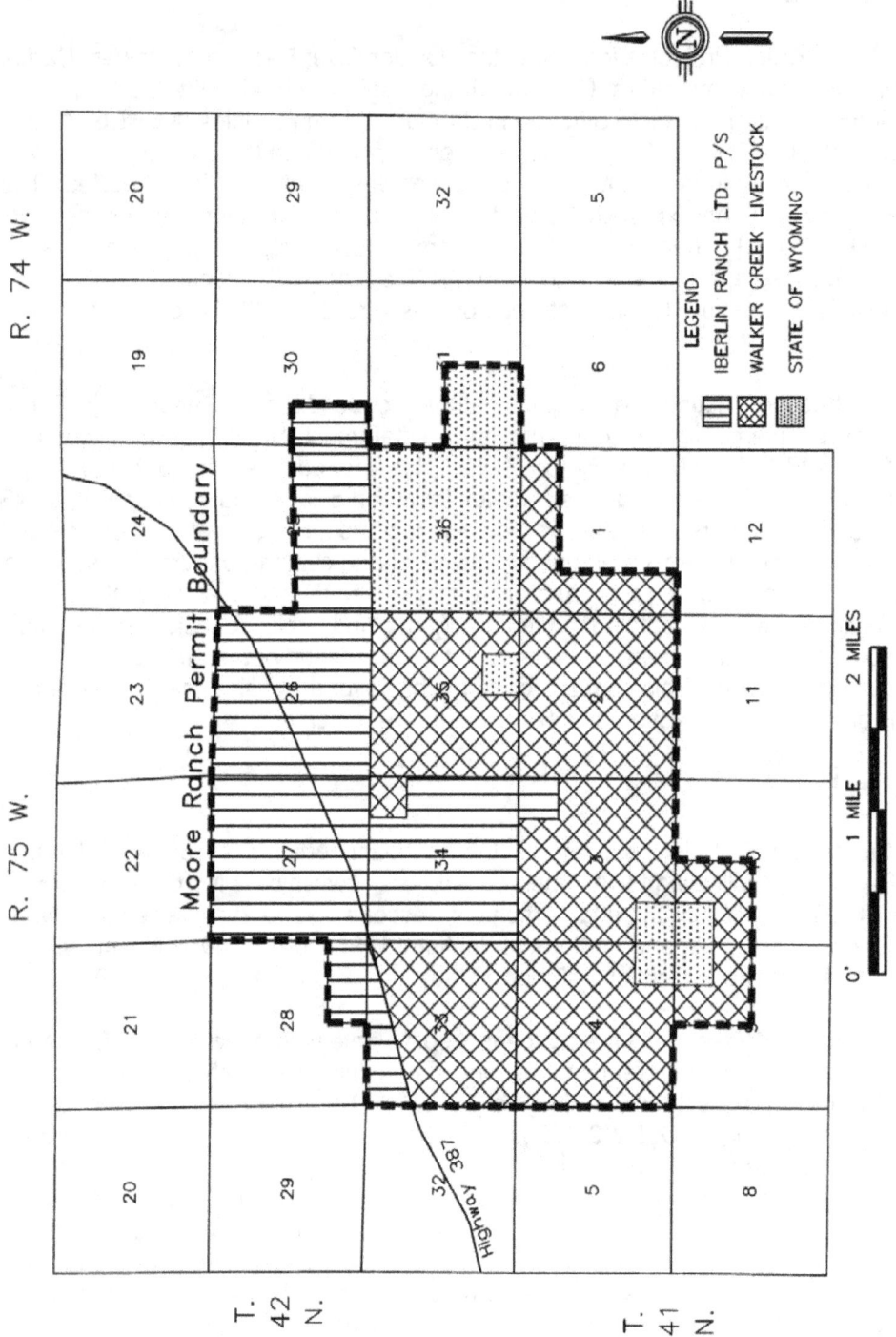

Figure 3-1. Surface Ownership at the Proposed Moore Ranch Project

Affected Environment

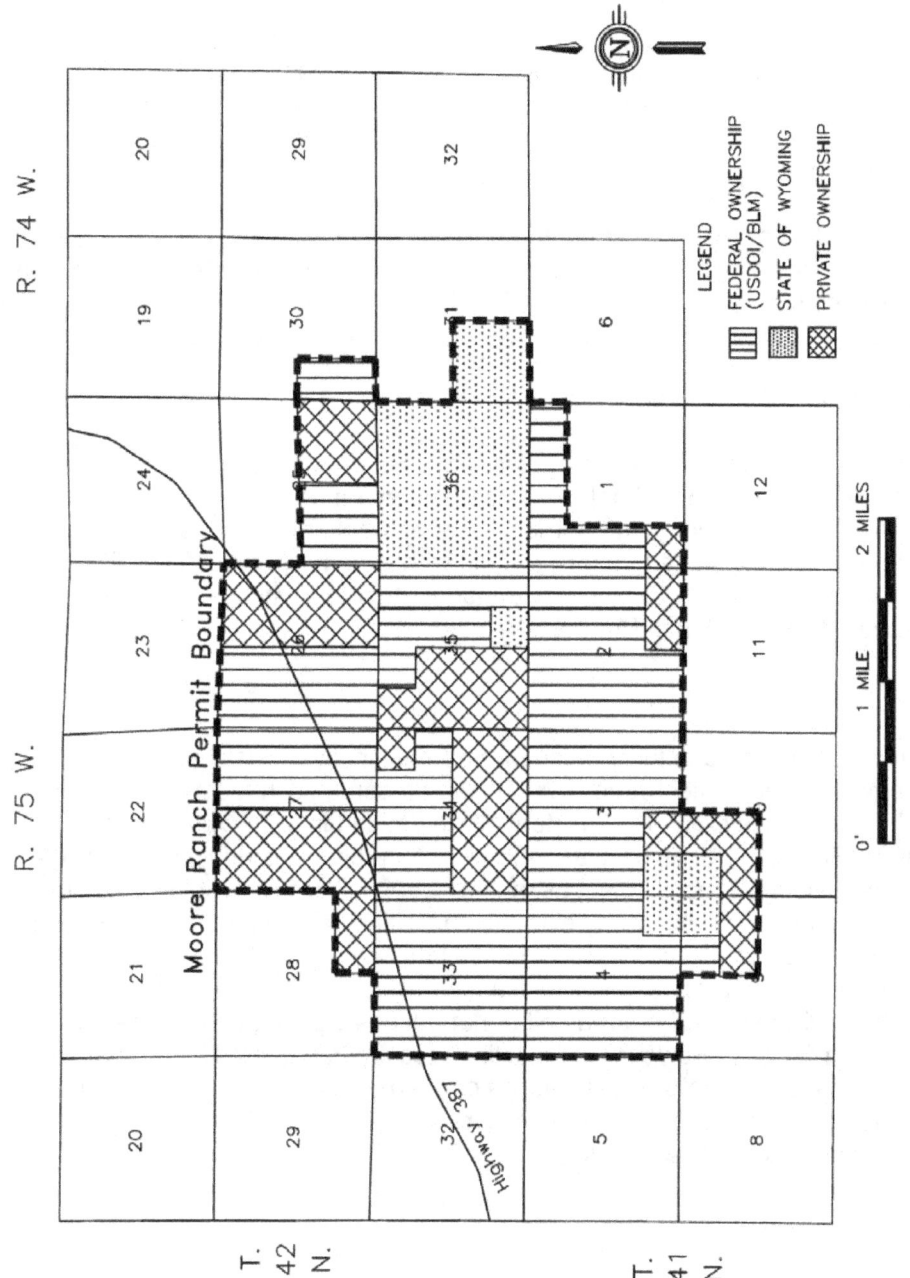

Figure 3-2. Mineral Rights Ownership at the Proposed Moore Ranch Site

Affected Environment

to Wyoming Department of Environmental Quality (WDEQ) shows the location of mineral leases in Wellfield 2.

3.2.1.1 Existing Uses

The proposed license area for the Moore Ranch Project is currently used for growing grass and grain for animal feed (pastureland), cattle grazing (rangeland), and for various types of CBM and oil and gas extraction, which are classified as a subcategory of rangeland. There are no other known land uses within the proposed Moore Ranch Project, or within a 3.2-km [2-mi] radius surrounding the property boundaries.

The proposed facility would be accessed from the east via State Highway (SR) 59 and SR 50, to SR 387 and from the west via Interstate (I)-25, to SR 259, to SR 387. The main access road connecting the proposed Moore Ranch Project with SR 387 is located in T42N, R75W, Section 27. Detailed discussion on transportation routes is provided in Section 3.3 of this SEIS.

3.2.1.2 Rangeland and Pastureland

Ranching is the predominant land use on and in the vicinity of the proposed Moore Ranch Project. Approximately 2,326 ha [5,748 ac] of land is used as rangeland supporting herds of cattle and sheep which graze among large herds of deer and antelope. Currently, an estimated 544 ha [1,344 ac] of land is being used as pastureland within the proposed Moore Ranch Project (EMC, 2007b). SR 387 is the only major transportation route that bisects the northern portions of the proposed Moore Ranch Project (see Figure 2-1).

3.2.1.3 Minerals and Energy

Uranium exploration has occurred in this portion of the Powder River Basin since the 1950s, and CBM development began in the 1980s. With the advancements in technology, development and production of CBM has substantially increased in the Powder River Basin since the mid-1990s (BLM, 2003). There were 465 wells located on or within a 3.2-km [2-mi] radius of the proposed Moore Ranch Project as of June 2009 for use as CBM or stock CBM wells (Uranium One, 2009b). These energy extraction facilities have attendant infrastructure systems, including pipelines, wellfields, and utility lines that occupy the land surface and the subsurface in the vicinity of the proposed Moore Ranch Project. As shown in Figure 3-3, there is approximately 15 km [10 mi] of either crude oil or natural gas pipeline that crosses the proposed Moore Ranch Project. However, no pipelines are located in areas where earthmoving activities would occur as part of the proposed action or alternatives. Approximately 64 km [40 mi] of either crude oil or natural gas pipeline occurs within a 3.2-km [2-mi] radius of the proposed Moore Ranch Project, as shown in Figure 3-3.

3.3 Transportation

As noted in the GEIS, the operation of ISR facilities has historically relied on roads for transportation of goods and personnel. Local roads are used to transport construction equipment and materials to support facility and wellfield construction activities (NRC, 2009a). The proposed Moore Ranch Project is located in an area served by two four-lane interstate highways. I-25, which is located approximately 48 km [30 mi] west of the proposed Moore

Ranch Project, extends north from Colorado, terminating where it merges with I-90 at Buffalo, Wyoming, about 120 mi to the northwest of the proposed Moore Ranch Project (Figure 1-1).

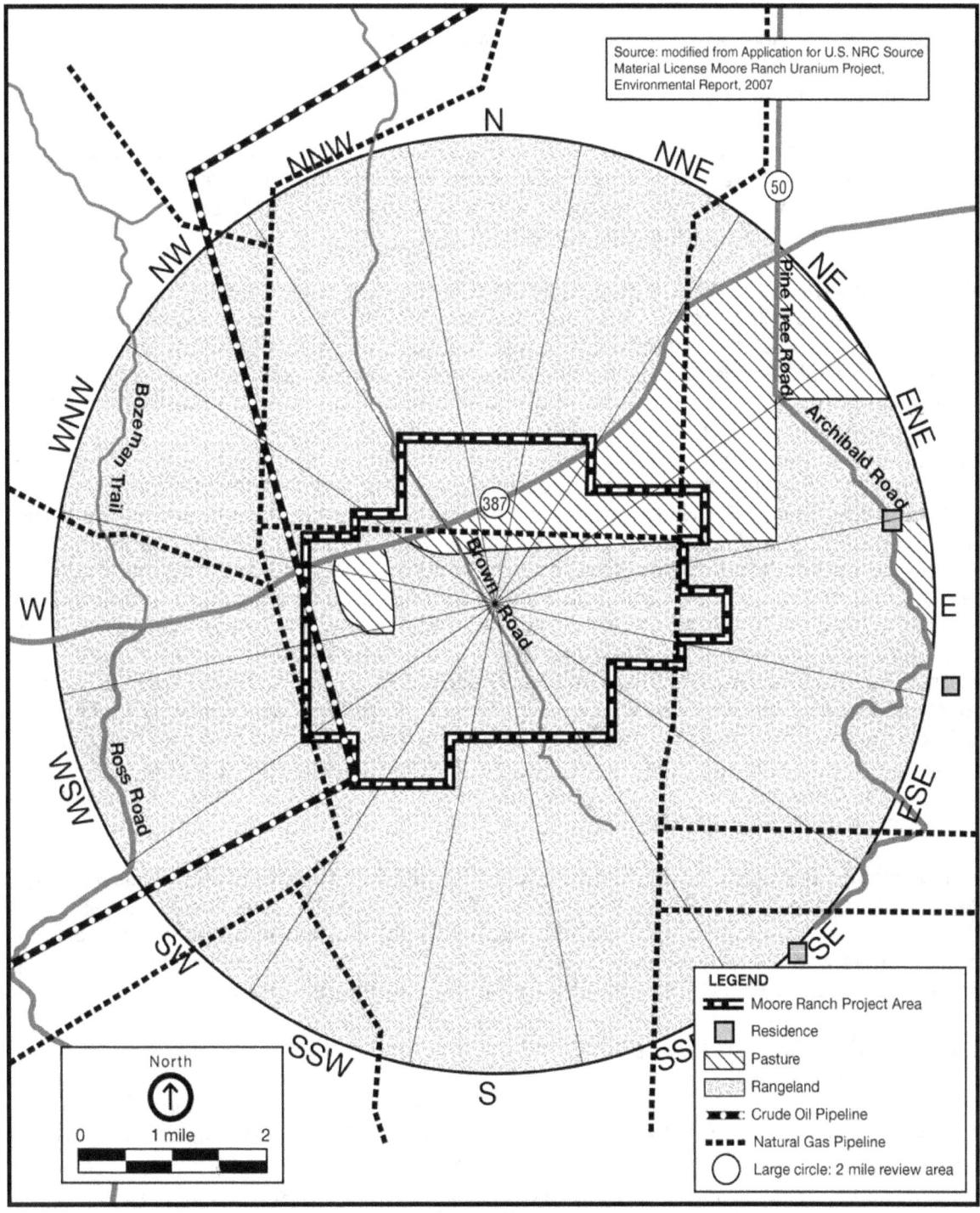

Figure 3-3. Land Use in the Vicinity of the Proposed Moore Ranch Project

Affected Environment

The primary transportation route to the proposed Moore Ranch Project from nearby communities would be via SR 387, a bidirectional (two-lane, opposing travel), asphalt-paved highway that connects the proposed license area to regional population and economic centers along I-25 to the west and SR 59 to the east (Uranium One, 2009b). The City of Gillette is located approximately 80 km [50 mi] northeast of the proposed Moore Ranch Project. SR 387 runs east-west from the town of Wright to I-25 at the town of Midwest, bisecting the northern portion of the proposed license area (Figure 1-1). SR 50 commences in Gillette and runs to the south and connects with SR 387 approximately 6.4 km [4 mi] east of the proposed Moore Ranch Project. SR 59 connects with SR 387 at Wright, located approximately 48 km [30 mi] east of the proposed Moore Ranch Project. Both SR 50 and SR 59 are also bidirectional, asphalt-paved highways in good to average condition. The lanes on SR 50, 59, and 387 are 3.6 m [12 ft] wide, and the total width of paved roadway ranges from 8 to 12 m [26 to 40 ft], based on the varying width of the paved shoulder (Uranium One, 2009b).

All state highways adjacent to the proposed Moore Ranch Project are access controlled and maintained year round by the Wyoming Department of Transportation. Highway maintenance includes snow removal, debris removal, and road repairs (Uranium One, 2009b). Onsite road maintenance would include periodic grading of the primary access roads, snow plowing, applying water or other agents for controlling fugitive dust emissions, and regular inspections to ensure the adequacy of erosion control measures (Uranium One, 2009b).

Approximately 7.2 km [4.5 mi] of SR 387 crosses the northern portion of the proposed Moore Ranch Project (Figure 2-1). Numerous county roads provide access to public and private lands, many of which consist of maintained gravel surfaces. Unimproved or minimally improved private roads are common. Brown Road, an existing gravel road, accesses the general location selected for construction of the central plant and is currently used for agricultural and oil and gas activities in the area. The proposed location of the central plant would be approximately 0.8 km [0.5 mi] from the intersection of SR 387 and Brown Road. Brown Road may require minor improvements to accommodate trucks and heavy equipment access during the construction and operation phases of the proposed Moore Ranch Project. In addition, approximately 1.2 km [0.8 mi] of gravel roads would be constructed to connect the central plant to Brown Road, to connect Wellfield 1 to Brown Road, and to connect Wellfield 2 to an existing access road (Figure 2-2). Other roads enter the proposed Moore Ranch Project, but none provide access to residences or other public destinations.

Annual average daily traffic counts for trucks using SR 387 in the vicinity of the proposed Moore Ranch Project ranged from 220 to 410 trucks in 2006. The figure for all vehicle types was 970 to 3,130 per day (Uranium One, 2009b). For SR 50, the annual average daily traffic count for all vehicles was 550 in 1999 (BLM, 2003). No traffic count data are available for Brown Road. Table 3-1 provides traffic count data for the surrounding state routes.

3.4   Geology and Soils

The proposed Moore Ranch Project is located in the Pumpkin Buttes Uranium District in the Powder River Basin within the Wyoming East Uranium Milling Region evaluated in the GEIS (NRC, 2009a). Section 3.3 of the GEIS provides a general description of the geology and soils within this area. Section 3.4.1 of this SEIS provides a site-specific discussion of the geology and soils in the vicinity of the proposed Moore Ranch Project.

Affected Environment

Table 3-1. Annual Average Daily Traffic Counts						
Route Name	Description	All Vehicles				Trucks	
		1998	1999	2005	2006	2005	2006
SR 59	Gillette South of Urban Limits	18,690	17,760				
SR 59	Johnson-Campbell County Line	1,110	1,210				
SR 59	Wright	2,150	2,250	3,630	3,930	690	750
SR 59	Converse-Campbell County Line	1,350	1,450				
SR 387	Johnson-Campbell County Line	1,110	1,210				
SR 387	Between SR 50 and SR 59			970–3,130	970–3,130	210–410	220–410
Source: Uranium One, 2009b						

The Powder River Basin is a large structural and topographic depression that parallels the Rocky Mountains. The basin is bounded to the north by the Miles City Arch in southeastern Montana, to the south by the Hartville Uplift and the Laramie Range, to the east by the Black Hills, and to the west by the Big Horn Mountains and the Casper Arch. As indicated in the GEIS, the basin was formed during the Laramide Orogeny (mountain-building era) approximate 50 to 65 million years ago (NRC, 2009a). Rapidly subsiding portions of the basin received thick clastic wedges (i.e., made of fragments of other rocks) of predominantly arkosic sediment (i.e., sediments containing a significant fraction of feldspar), while large more slowly subsiding portions of the basin received a greater proportion of paludal (marsh) and lacustrine (lake) sediments.

The sedimentary rock sequence in the basin ranges in age from recent (Holocene) to early Paleozoic (Cambrian–500 million to 600 million years ago) and overlies a basement complex of Precambrian (more than a billion years old) igneous and metamorphic rocks. As noted in the GEIS, erosion has removed the upper part of the sedimentary sequence in the Powder River Basin, leaving only the Tertiary-aged White River, Wasatch, and Fort Union Formations. The White River Formation is of the Oligocene age and is the shallowest Tertiary unit in the Powder River Basin. It is underlain by the Wasatch, which is of the Eocene Age. The Paleocene age Fort Union Formation directly underlies the Wasatch Formation, which directly overlies the Cretaceous Lance Formation. Figure 3.3-5 of the GEIS provides a stratigraphic section of Tertiary-aged formations in the Wyoming East Uranium Milling Region (NRC, 2009a). The Tertiary-aged Wasatch Formation hosts the uranium deposits proposed for mining at the proposed Moore Ranch Project.

Affected Environment

3.4.1 Geology

The early Eocene Wasatch Formation unconformably overlies the Fort Union Formation around the margins of the Powder River Basin. However, within the basin center and in the vicinity of the proposed Moore Ranch Project, the two formations are conformable. The relative amount of coarse, permeable clastics increases near the top of the Fort Union Formation, and the overlying Wasatch Formation contains numerous sandstone beds that can be correlatable over wide areas. Except in isolated areas of the Powder River Basin, the Wasatch-Fort Union contact is arbitrarily set at the top of either thick coals or a thick sequence of clays and silts. The applicant considers the top of the Roland coal as the boundary of the Wasatch Formation at the proposed Moore Ranch Project.

The Wasatch Formation reaches a maximum thickness of about 488 m [1,600 ft] and outcrops at the surface in the proposed license area. The Wasatch Formation is composed of interbedded sandstones, siltstones, clays, and coals and was deposited in a fluvial (river) environment. These sandstone horizons are the host rock for several uranium deposits in the southern Powder River Basin. Within the proposed Moore Ranch Project, mineralization occurs in a 15 to 30-m [50- to 100-ft] thick sandstone lens, which extends over an area of several townships. This formation dips gently to the northwest from 1 degree to 2½ degrees in the southern part of the Powder River Basin (EMC, 2007b).

Locally, remnants of the overlying White River Formation are known to occur on top of the Pumpkin Buttes. A basal conglomerate forms the resistant cap rock on top of the buttes. This formation is not known to contain significant uranium resources in this area.

Detailed stratigraphic analysis of a portion of the Wasatch Formation was performed because the target ore zone occurs within the sands in this formation. The site-specific stratigraphy has been characterized based on subsurface data collected from thousands of well borings in and around the proposed license area in the 1970s and 1980s. This data is associated with the Conoco, Inc. application to the U.S. Nuclear Regulatory Commission (NRC) to construct and operate a uranium mill associated with an open-pit mine in the same area being evaluated for the proposed Moore Ranch Project (NRC, 1982). This data was supplemented with data collected from additional applicant well borings drilled in late 2006 and early 2008. The underlying Fort Union Formation was not studied in detail because it would not be influenced by the project. The top of the Roland Coal, which separates the Wasatch Formation, from the underlying Fort Union Formation is approximately 335 m [1,100 ft] thick across the proposed Moore Ranch Project.

The applicant adopted the Conoco, Inc. nomenclature for the hydrostratigraphic units of interest within the proposed Moore Ranch Project. Sands occurring stratigraphically above the Roland Coal are numbered, increasing toward the surface. Figure 3-4 illustrates the stratigraphic sequence at the proposed Moore Ranch Project. The applicant generated 13 geologic cross sections to characterize the vertical and lateral stratigraphy at the site. Figure 3-5 illustrates an isopach map of the 70 Sand showing the areal distribution and thickness of the unit containing the ore zone. Figure 3-6 illustrates the stratigraphy in Wellfield 1, and Figure 3-7 illustrates the stratigraphy in Wellfield 2. The applicant technical report provides figures and maps showing the areal distribution and thickness of overlying and underlying sand and shale sequences (EMC, 2007b).

Affected Environment

The 40 and 50 Sands that occur immediately above the Roland Coal are regionally extensive and considered significant aquifers (EMC, 2007b). The approximate thickness of the 40 and

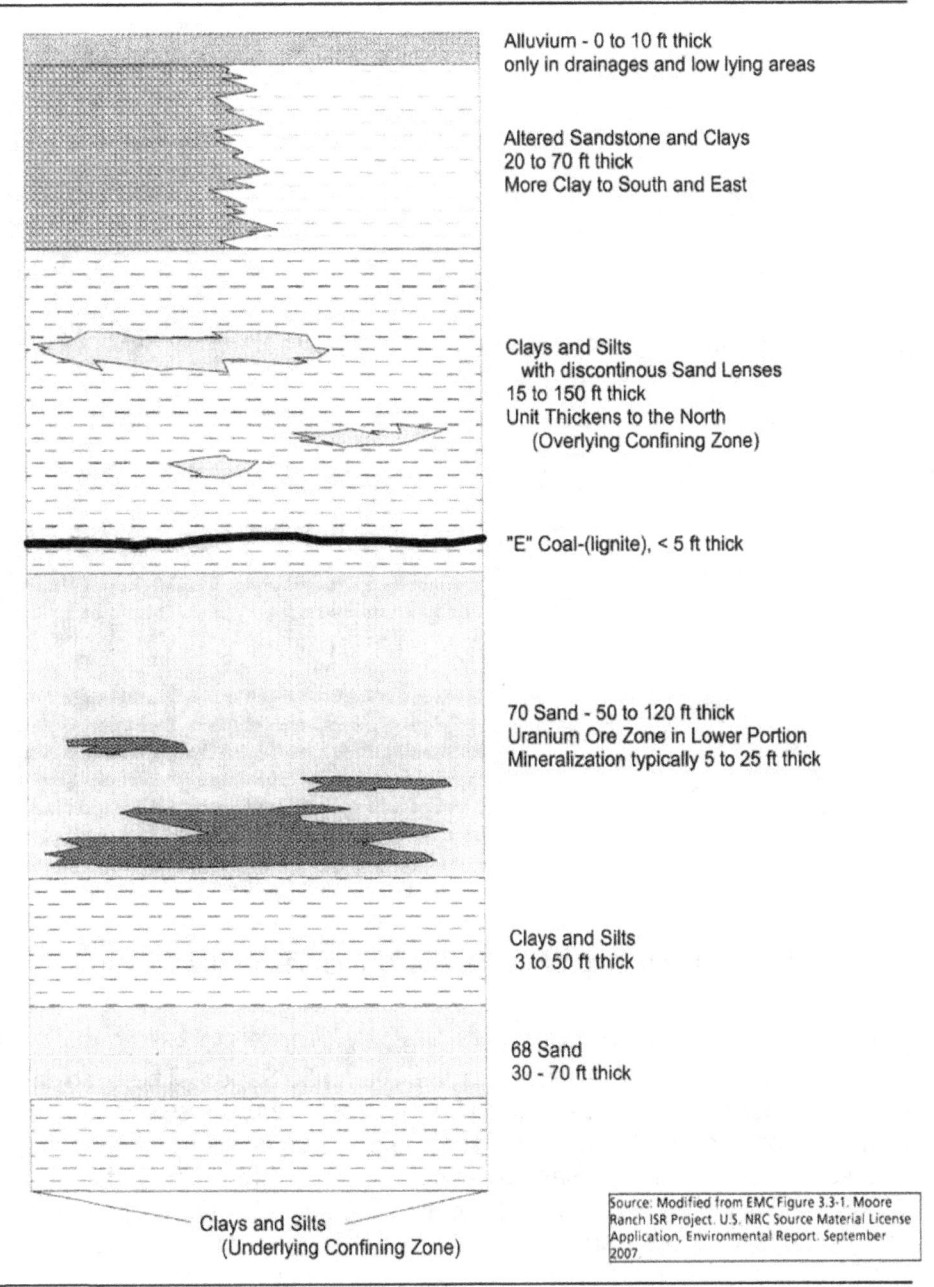

Figure 3-4. Generalized Stratigraphy Sequence Showing the 70 Sand Ore Production in the Wasatch Formation

Affected Environment

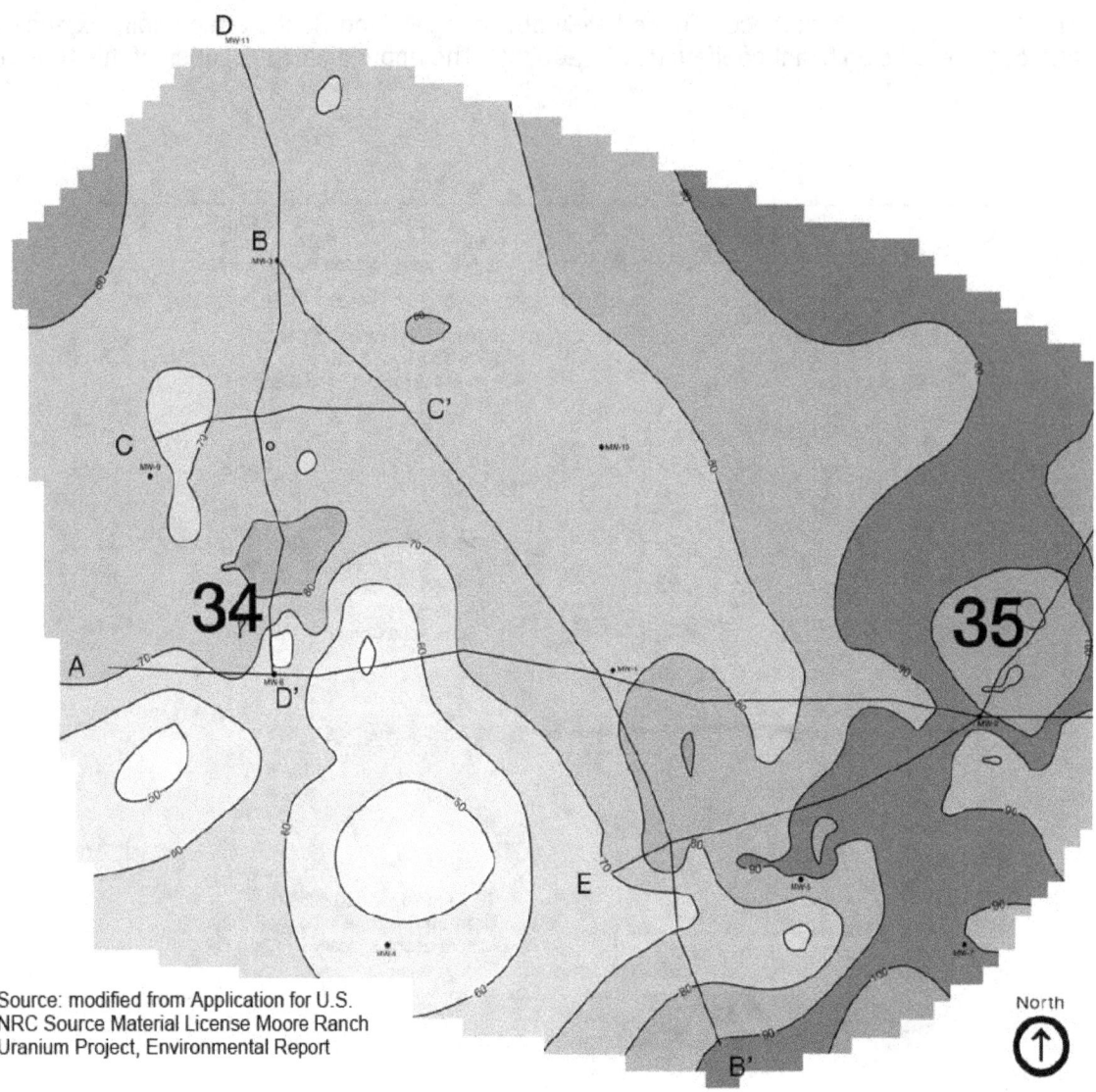

Figure 3-5. Isopach Map Showing the Thickness of the 70 Sand

50 Sands within the proposed Moore Ranch Project are 24 and 27 m [80 and 90 ft]. A 1.5 to 12 m [5 to 40 ft] thick shale or mudstone separates the 40 and 50 Sands. The overlying 58 Sand varies in thickness from 1.5 to 24 m (5 to 80 ft) across the proposed project area. The overlying 60 Sand is approximately 30 m [100 ft] thick, is continuous throughout the proposed project area, and separated from the 58 Sand by about 1.5 to 21 m [5 to 70 ft] of shale and mudstone. The 68 Sand is the first sand underlying the 70 Sand, which contains the economic ore deposits in the proposed Moore Ranch Project. The 68 Sand is separated from the 60 Sand by 0 to 8 m [0 to 25 ft] of shale or mudstone. The 68 Sand ranges from to 12 to 30 m [40 to 100 ft] across the proposed license area and coalesces with the 60 Sand on the west side of the proposed license area. The 70 Sand, the proposed ore production zone, coalesces with the 68 Sand in Wellfield 2. The 70 Sand is laterally extensive and ranges from 12 m to 37 m [40 to 120 ft] thick across the proposed Moore Ranch Project. The dip is generally less than one

Affected Environment

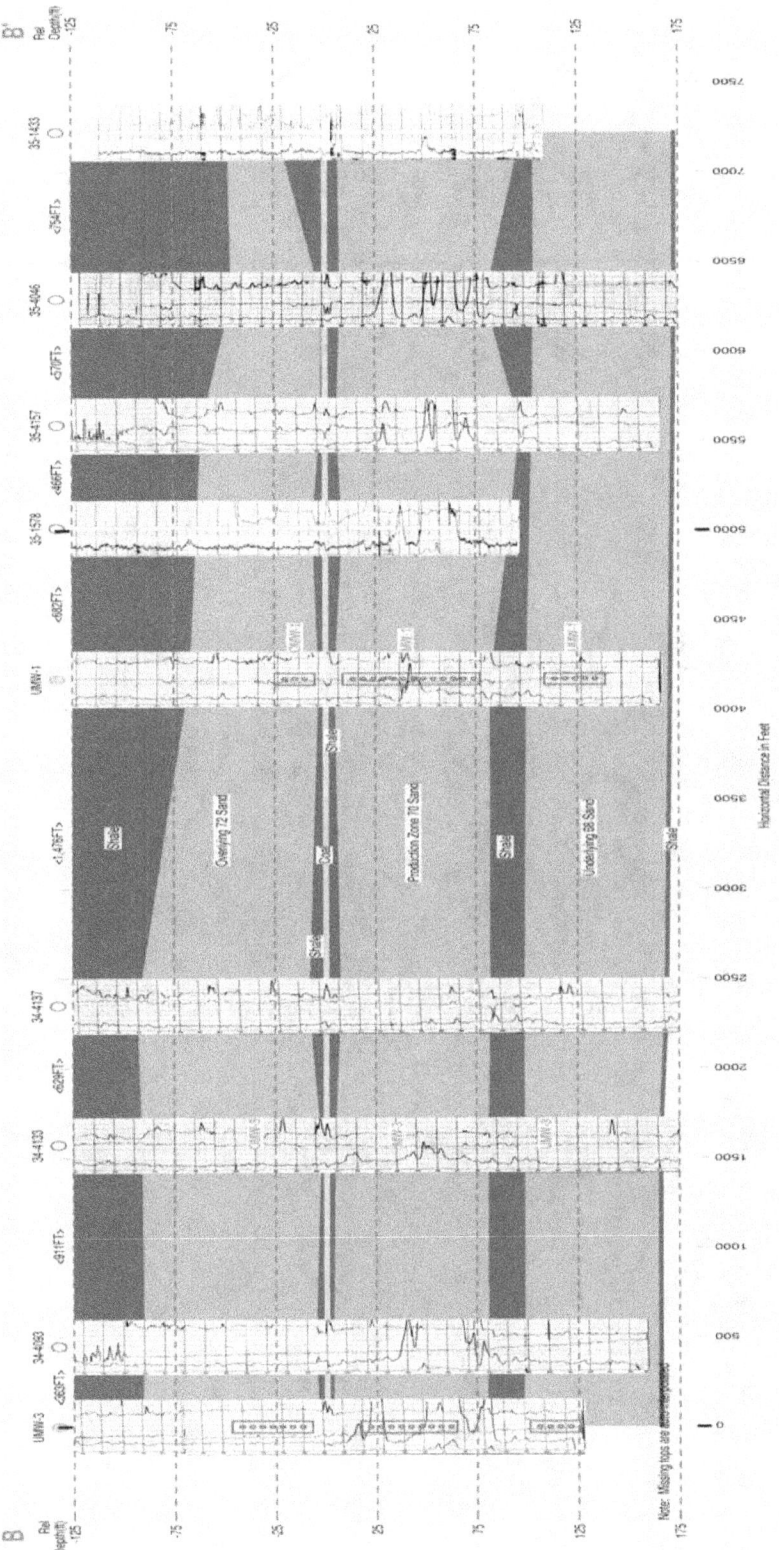

Figure 3-6. West-East Stratigraphic Cross Section of Wellfield 1

Affected Environment

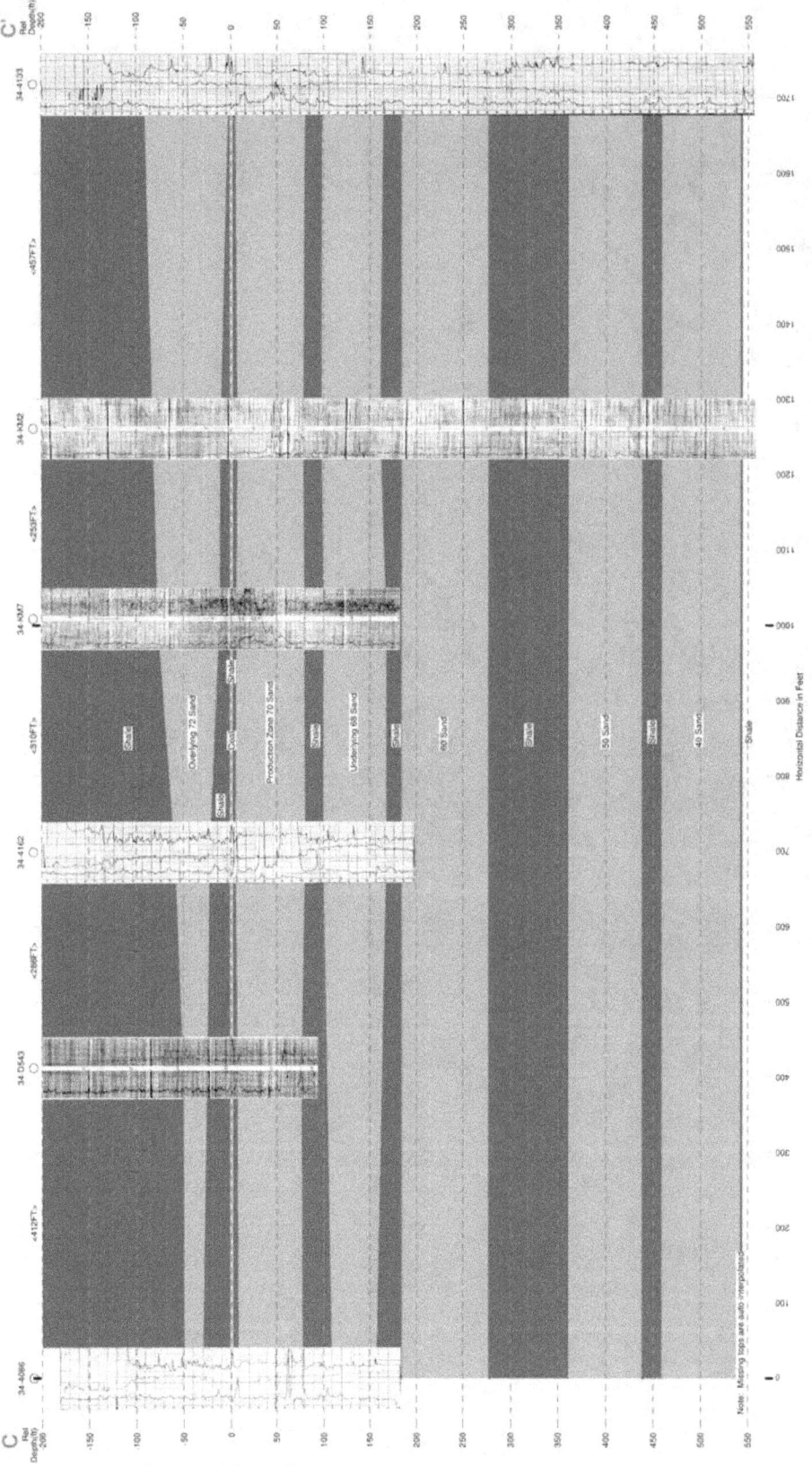

Figure 3-7. West-East Stratigraphic Cross Section of Wellfield 2

Affected Environment

degree toward the northwest. A coal layer, referred to as the E Coal, that ranges in thickness from 0.3 to 0.9 m
[1 to 3 ft] typically occurs a few feet above the top of the 70 Sand. The 72 Sand overlies the 70 Sand and is the shallowest sand occurring across the proposed project area. The 70 Sand is separated from the overlying 72 Sand by a shale sequence ranging in thickness from a few feet to 49 m [160 ft] in some areas.

The applicant has proposed two different injection zones for use in waste disposal at the proposed Moore Ranch Project. The applicant has submitted an Underground Injection Control (UIC) permit application to WDEQ that evaluates injection into the Teapot-Teckla-Parkman Sandstones, at a depth ranging from 2,413 to 2,929 m [7,916 to 9,610 ft] below ground surface and the Lance Formation and Fox Hills Sandstone, at a depth ranging from 1,128 to 2,286 m (3,700 to 7,500 ft) below ground surface. Both of these formations are thousands of feet deeper than the 70 Sand ore production zone.

3.4.2 Soils

The applicant performed a soil survey of the proposed Moore Ranch Project in 2007 to define the existing topsoil resource and determine the extent, availability, and suitability of soils material for use in reclamation. A site-specific map was generated to show the areal distribution of different soil types, and soil map units and soil series descriptions were included in the applicant environmental report (EMC, 2007b). The general topography of the proposed Moore Ranch Project includes rolling hills and ridges, as well as drainages. Soils occurring in this area were generally fine-textured throughout with patches of sandy loam on upland areas. Fine-textured soils occur near or in drainages. The proposed project area contains deep soils on lower toe slopes and flat areas near drainages; shallow and moderately deep soils occur on upland ridges and shoulder slopes. Soils on the proposed Moore Ranch Project are typical for semiarid grasslands and shrublands in the Western United States. Most soils have some suitable topsoil. The primary limiting factor is texture.

The 2007 soil fieldwork characterized the soils within the proposed license area with respect to topsoil salvage depths and related physical and chemical properties. Based on data from samples collected from within the proposed license area, from field observations and knowledge of the soils in southern Campbell County, an approximate salvage depth for each map series was identified. These salvage depths ranged from 0.24 m to 1.5 m [0.8 to 5 ft]. An average salvage topsoil depth over the proposed Moore Ranch Project was estimated as 1.1 m [3.6 ft].

The potential for wind and water erosion of soil within the proposed Moore Ranch Project varies from slight to severe and is mainly a factor of the texture and organic content of the surface soil. Because the surface soils throughout the proposed Moore Ranch Project have a fine loamy and sandy texture, the soils are more susceptible to erosion from wind than water.

No prime farmland has been documented at the proposed Moore Ranch Project, based on a reconnaissance survey by the National Resources Conservation Service.

Affected Environment

3.5 Water Resources

3.5.1 Surface Waters and Wetlands

As noted in Section 3.3.4.1 of the GEIS, the Wyoming East Uranium Milling Region encompasses 10 primary watersheds (NRC, 2009a). The Antelope Creek Watershed drains the location for the proposed Moore Ranch Project. Surface water features, both in the vicinity of and within the proposed license area, include intermittent streams that flow to the southeast ultimately to the Cheyenne River. Water bodies within the Wyoming East Uranium Milling Region are mainly classified as Class 3B surface waters, according to the state classification of designated uses. The designated uses for Class 3B surface waters are recreation, aquatic life other than fish, wildlife, agriculture, industry, and scenic value.

3.5.1.1 Drainage Basins

The proposed Moore Ranch Project area lies within the Ninemile Creek drainage basin, which covers an area of 163 km^2 [63 mi^2]. Ninemile Creek is a tributary to Antelope Creek, which is a tributary of the South Cheyenne River, which ultimately flows to the Missouri River. Seven subwatersheds occur within the proposed license area and are associated with Ninemile Creek, Simmons Draw, Pine Tree Draw and their tributaries (Figure 3-8). Each of these subwatersheds drains to the southeast; Simmons Draw and Pine Tree Draw flow into Ninemile Creek. As shown in Figure 3-8, Wash #1 is an intermittent tributary to Simmons Draw and flows to the west of Wellfield 1. Upper Wash #2 is another intermittent stream to Simmons Draw, and it bisects the central portion of Wellfield 2.

3.5.1.2 Surface Water Features

The arid conditions in eastern Wyoming limit the formation of year-round surface water and wetland features. Regional annual rainfall averages approximately 35.5 cm [14 in] per year, while annual lake evaporation may reach 101 cm [40 in] per year. Surface waters, particularly in the upper headwaters of watersheds, are seasonal in nature, responding to springtime snow melt. In some instances, surface waters may manifest intermittent flow conditions in response to extreme rainfall events. Otherwise, rainfall is normally absorbed into the soil.

Despite the arid conditions and headwater setting, linear wetland features and nine small, artificial ponds persist and are scattered across the proposed Moore Ranch Project within low lying drainages in response to the CBM operations that occur throughout the area. CBM-produced water in the vicinity of the proposed Moore Ranch Project is estimated to contribute 9 to 52 percent of surface flows and could result in perennial flows in formerly intermittent channels (Uranium One, 2009a). Approximately 31 CBM wells occur within the proposed license area, with another 101 located within a 3.2-km [2-mi] radius of the boundary of the proposed Moore Ranch Project (Figure 3-9). These operations discharge extracted groundwater onto the surface and are responsible for sustaining the existing surface water features (wetlands and ponds).

The CBM discharges are monitored through three Wyoming Pollutant Discharge Elimination Systems (WYPDES permits) issued to operators located either within or adjacent to the proposed license area. Surface water flow is discussed in Section 3.5.1.3 of this SEIS.

Affected Environment

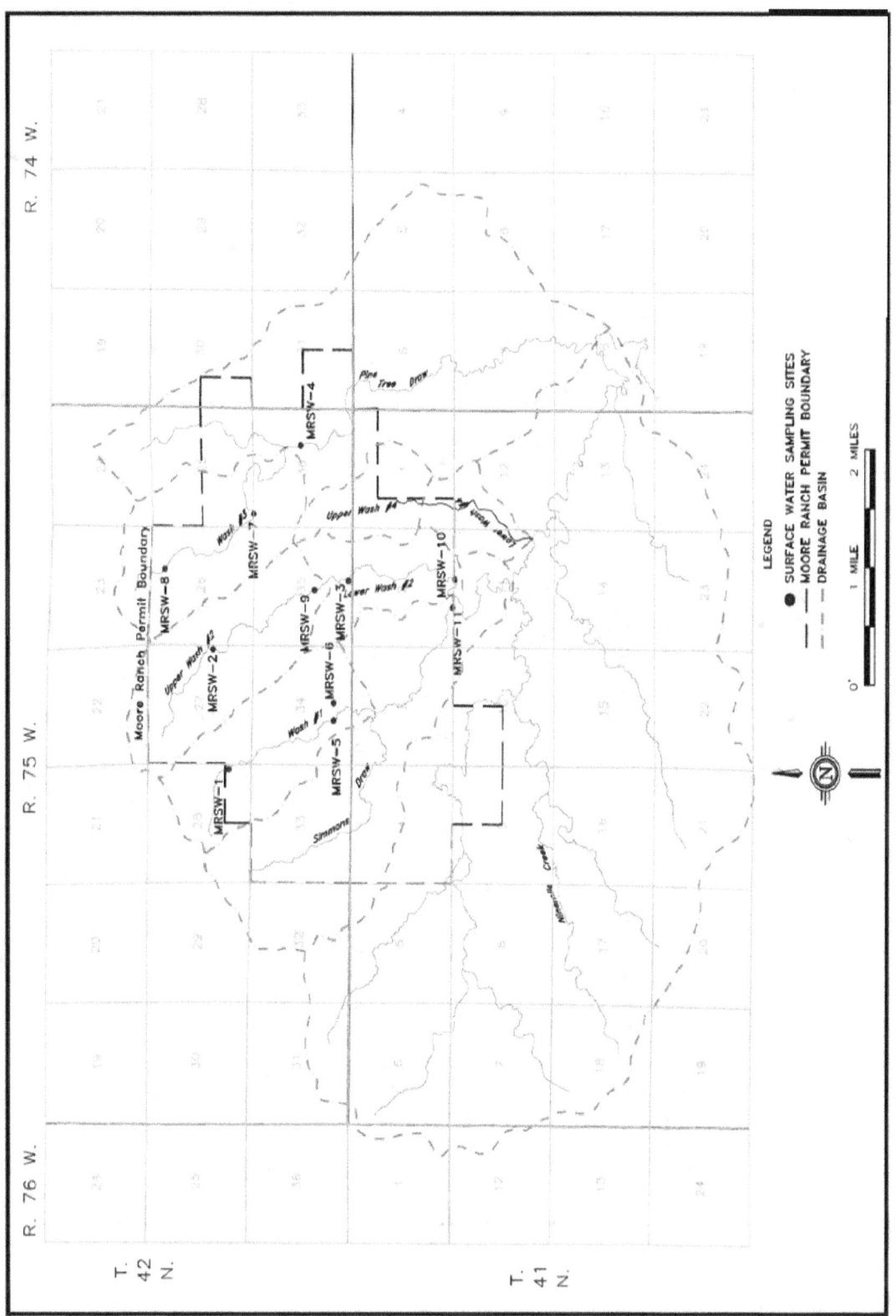

Figure 3-8. Drainage Basin and Sub-Watersheds at the Proposed Moore Ranch Project

Affected Environment

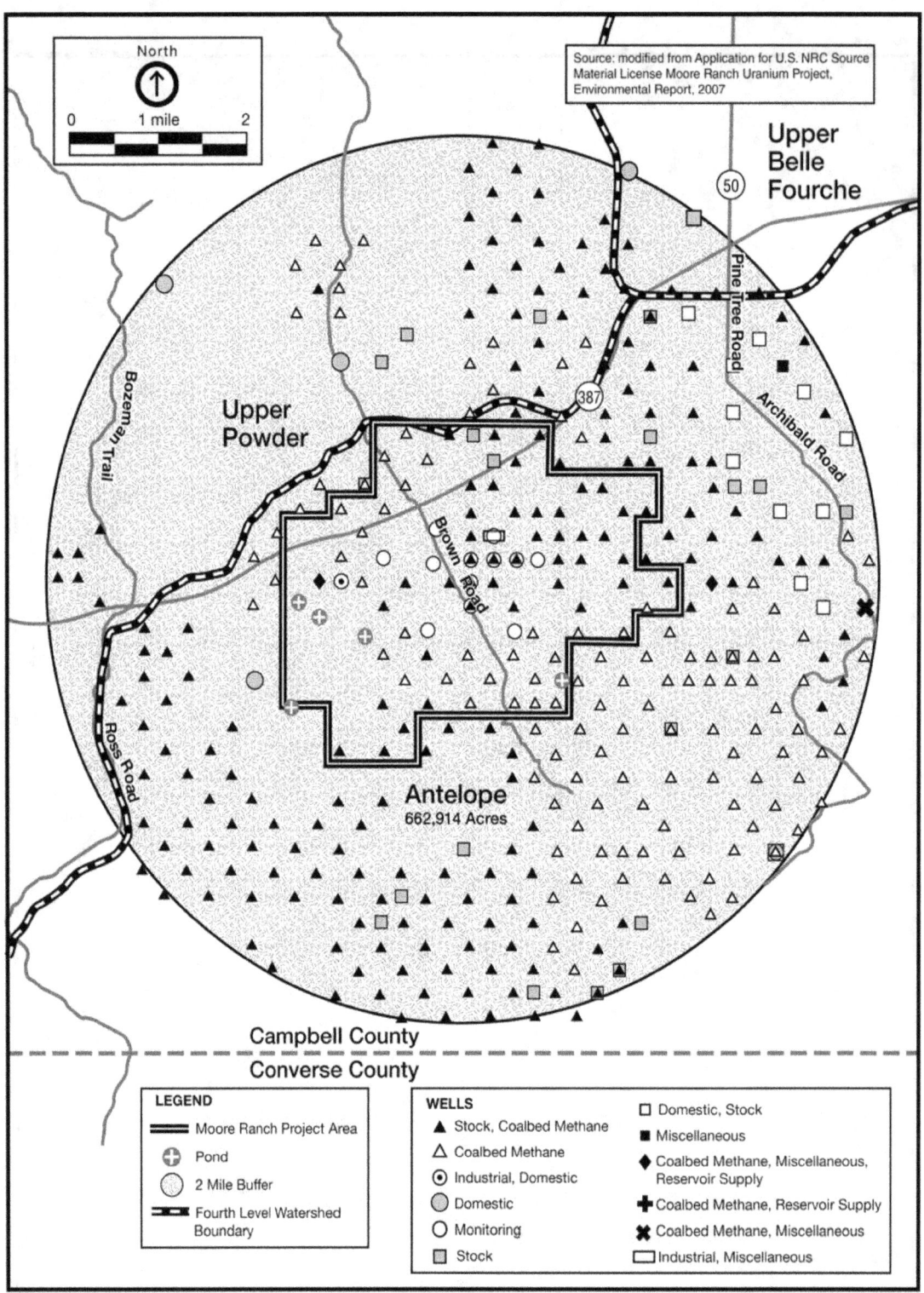

Figure 3-9. CBM Production Near the Proposed Moore Ranch Project

Affected Environment

3.5.1.2.1 Intermittent Streams

Ninemile Creek, Simmons Draw, and Pine Tree Draw are the dominant streams within the proposed license area. Each of these intermittent streams collect surface water runoff from the numerous drainages or "washes" carved into the landscape (Figure 3-8). Because these channels remain dry most of the year, the channels contain upland vegetation growth. Hydrophytes (plants adapted to saturated soil conditions) persist yearlong only in short reaches where near-surface soil saturation extends well into the summer months from discharge from CBM and livestock wells. As previously noted, none of these intermittent streams drain the basins that encompass the proposed wellfields, except for Upper Wash #1 and Upper Wash #2.

The stream channels on the proposed Moore Ranch Project are briefly described in the following text. For purposes of this document, unnamed channels are given designations such as "Tributary A" or "Wash #1." Subtitle designations such as "Upper" and "Lower" relate to subwatersheds, based on water quality sampling stations discussed later in Section 3.5.1.4. A map of the stream channels and contributing watersheds is provided in Figure 3-8.

Ninemile Creek flows through approximately 2.4 km [1.5 mi] of the southwest corner of the proposed license area and drains a total area of 16,316 ha [63 mi^2]. The elevation difference from the headwaters to the mouth of Ninemile Creek is 186 m [610 ft] over an approximate channel length of 32 km [20 mi], with an average gradient of 0.6 to 0.7 percent. Simmons Draw, a tributary to Ninemile Creek, flows to the southeast through the western boundary of the proposed license area, approximately 13.8 km [6.8 mi] at a gradient of 0.7 percent, with a drainage area of 21 km^2 [8.1 mi^2]. The total basin elevation difference is 79 m [260 ft]. Simmons Draw has two main tributaries: Simmons Draw Tributary A (Wash #1) and Simmons Draw Tributary B (Wash #2). Wash #1 has a length of approximately 4.5 km [2.8 mi] with a 1.4 percent gradient; Tributary B (subdivided into Upper Wash #2 and Lower Wash #2) has drainage areas of 4.9 and 2.5 km^2 [1.9 and 0.95 mi^2], with channel lengths of 0.74 and 2.1 km [0.46 and 1.3 mi] and average gradients of 0.012 and 0.007 ft/ft. Each of these tributaries are intermittent with fragmented wetlands and ponds based primarily on discharges from CBM and livestock wells.

Pine Tree Draw has a drainage area of 124 ha [8.2 mi^2] and drains the eastern side of the proposed license area. The total basin elevation difference is 110 m [360 ft] over a channel length of approximately 12.2 km [7.6 mi], resulting in a gradient near 0.9 percent. Pine Tree Draw Tributary A has a drainage area of 4.6 km^2 [1.8 mi^2], a channel length of 5.1 km [3.2 mi], and an average gradient of 0.014 ft/ft.

Pine Tree Spring, a relatively short tributary to Pine Tree Draw, drains the far eastern side of the proposed license area and has a channel length of approximately 1.1 km [0.7 mi].

3.5.1.2.2 Ponds

Nine small, disconnected artificial ponds (reservoirs) are scattered across the proposed license area and occur within the channels of Ninemile Creek, Simmons Draw, Pine Tree Draw, and their principal tributaries. These reservoirs have been permitted through the Wyoming State Engineers Office (WSEO) within the proposed license area because they could be impacted by CBM-produced water discharge. The ponds are classified as palustrine unconsolidated bottom, in accordance with the Cowardin classification system (Cowardin, et al., 1978), and are generally less than 0.4 ha [1 ac] in size. These surface water features result from accumulation

Affected Environment

behind structures (dams and dikes), in excavated pits, or from the discharge of pumped groundwater from CBM operations, windmills, or livestock watering tanks.

3.5.1.3 Surface Water Flow

The CBM-produced water in the vicinity of the proposed Moore Ranch Project is estimated to contribute 9 to 52 percent of the surface water flows and could result in perennial flows in formerly intermittent channels (Uranium One, 2009a). The CBM discharges are monitored through three WYPDES permits issued to CBM operators located either within or adjacent to the proposed license area. Thirty outfalls are monitored under the e WYPDES permits; 7 outfalls are located upstream of the proposed Moore Ranch Project, and the remaining 22 outfalls are located on the proposed Moore Ranch Project. Eight of these locations are in the vicinity of Wellfield 1 and Wellfield 2.

The average historic discharge rate of the CBM unit with the most permitted outfalls on the proposed Moore Ranch Project was 106,061 L/day [28,800 gal/day] over a period of eight years (2000 to 2008) compared to a maximum permitted limit of 2.6×10^6 L/day [680,000 gal/day]. Flow from this CBM unit is anticipated to be less than 22,727 L/day [6,000 gal/day] by the year 2013 (EMC, 2007b). The average discharge from outfalls located in the vicinity (which were not dry) was approximately 57,197 L/day [15,100 gal/day].

Peak flood flows were also calculated for each of the drainage basins on the proposed Moore Ranch Project, as part of the Draft EIS for the Sand Rock Mill Project, docket No. 40-8743 (NRC, 1982). The Draft EIS calculations were reviewed to determine the validity of the analytical methods and estimate surface water runoff. The applicant used different methods to estimate peak flood discharges, as described in their technical report (EMC, 2007b). Based on this analysis, it was determined that Wellfield 1 and the central plant were located higher than any region that could potentially be flooded. However, Wellfield 2 could potentially be flooded by a 100-year flood event. Therefore, the applicant proposed to minimize damage to infrastructure in a potential flooding event by avoiding installation in main channels of drainages; sizing culverts properly; and implementing best management practices for embankments, culverts, and drainage crossings.

3.5.1.4 Surface Water Quality

The proposed Moore Ranch Project lies entirely within the Antelope Creek drainage basin as shown in Figure 3-8. The EPA listed Antelope Creek and its tributaries as unimpaired surface waters. The WDEQ classifies Antelope Creek as a Class 3B surface water. This classification of waters includes intermittent and ephemeral streams that are able to support aquatic communities but are not known to support fish populations or be used as a drinking water supply. Class 3B waters may also support recreation, agriculture, industry, and provide scenic value.

All surface water sample locations within the proposed Moore Ranch Project are characterized as existing stock ponds or areas in drainages where ponding occurs. Water ponded at all surface water locations are typically fed by springtime snow melt runoff or high intensity rain events in the summer.

As noted above, 31 CBM outfalls occur in the proposed license area under 3 WYPDES permits. These permits monitor maximum flow, pH, specific conductance, chlorides, total recoverable arsenic and barium, and dissolved iron. Other chemical species are also monitored, including total radium-226.

Three sets of surface water samples collected during fall 2006, early spring 2007, and late spring 2007 were analyzed from nine locations within the proposed Moore Ranch Project, as shown on Figure 3-8. No surface water samples were collected from locations MRSW-10 and MRSW-11, which were both dry during the above sampling events. Table 3-2 summarizes the sample results.

The sample results indicate a seasonal variability in surface water quality largely influenced by the CBM operations in the area. The surface water in ponds typically exhibit saline characteristics of CBM surface discharge (higher conductivity, total dissolved solids (TDSs), and bicarbonate readings) in the summer and fall when there is less precipitation. The surface water sample results indicate that surface waters are basic, with numerous samples exceeding the CBM-permitted pH limit of 9.0. The average of all pH readings during all sampling periods was 9.08, which is above the Wyoming Class I (domestic use), Class II (agricultural use), and Class III (livestock use) standards. Although sampling for lead appears to exceed the 0.015 mg/L Class I standard, the minimal detection limit in the laboratory for lead was set at 0.05 mg/L for the fall 2006 samples. Therefore, the actual lead concentration fell below the 0.05 mg/L detection limit. Subsequent sampling indicated lead concentrations below the lead Class I standard.

As expected, the water samples taken during the fall months at CBM discharge locations commonly exhibited significantly higher values for bicarbonate, carbonate, chloride, conductivity, fluoride, TDS, gross alpha, gross beta, nitrogen, arsenic, potassium, magnesium, and sodium compared to samples taken during the spring months, indicating that surface water quality improves during the springtime as a result of diluted surface water from snow melt or large precipitation events, or both.

3.5.1.5 Wetlands

Wetlands are areas that are inundated or saturated by surface or groundwater at a frequency and duration to support, and under normal circumstances do support, a prevalence of vegetation typically adapted to life in saturated soil conditions (USACE, 1987).

A wetland delineation was performed, as part of the baseline assessment for the proposed Moore Ranch Project, using the methodologies outlined in the Regional Supplement to the U.S. Army Corps of Engineers (USACE) Wetland Delineation Manual: Great Plains Region (2006) to support reclamation planning and wellfield infrastructure (EMC, 2007b). The wetland survey methodology is described in Energy Metal Corporation (EMC) (2007b). Identification of potential wetlands was based on visual assessment of vegetation, hydrology indicators, and intrusive soil sampling to determine the presence of wetland criteria indicators (EMC, 2007b).

The proposed license area was found to contain nine wetlands classified as palustrine emergent based on the Cowardin, et al. (1979) wetland classification system (EMC, 2007b). Emergent wetlands are located in channels and total 12.6 ha [31.2 ac]. Similarly, nine ponds classified as palustrine open water were delineated and total 1.7 ha [4.1 ac]. These wetlands and ponds are shown in Figure 3-10.

Affected Environment

Table 3-2. Surface Water Quality at the Proposed Moore Ranch Project

Sample Parameters	MRSW-1[+]	MRSW-2	MRSW-3	MRSW-4	MRSW-5	MRSW-6[#]	MRSW-7[#]	MRSW-8	MRSW-9[#]
Bicarbonate as HCO3, mg/l	782	763	161	199	1064	457	665	402	99
Carbonate as CO3, mg/L	37	36	7	21	63	61	17	580	14
Chloride, mg/L	5	5	5	11	8	3	6	8	2
Conductivity, umhos/cm	1305	1170	694	1087	1843	839	979	1528	204
Fluoride, mg/L	0.5	0.6	0.5	0.5	0.7	0.5	0.5	1.1	0.2
pH, s.u.	8.99	8.96	9.10	9.37	8.80	9	9	9.40	9
Solids, Total Dissolved TDS @ 180C, mg/L	801	729	327	711	1159	540	646	1017	122
Sulfate, mg/L	14	4	219	350.33	62	6	13	8	4
Gross Alpha, pci/L	3.9	2.25	10.3	4.05	6.7	1.1	5.4	3.35	1
Gross Beta, pci/L	16.05	11.85	11.6	9.75	21.85	6.9	13.1	15.5	2
Lead 210 pci/L*	57.3*	<1.0	<1.0	1.0	4.0	<1.0	<1.0	<1.0	4.8
Polonium 210 pci/L	<1.0	<1.0	<1.0	<1.0	<1.0	<1.0	<1.0	<1.0	<1.0
Radium 226, pci/L	<0.2	<0.2	<0.2	0.2	1.3	0.9	<0.2	<0.2	<0.2
Radium 228, pci/L	<1.0	<1.0	1.3	1.0	<1.0	<1.0	<1.0	<1.0	<1.0
Thorium 230, pci/L	<0.2	<0.2	<0.2	<0.2	<0.2	<0.2	<0.2	<0.2	<0.2
Nitrogen, Ammonia as N, mg/L	0.12	<0.05	0.08	0.27	0.20	0.14	0.09	0.33	<0.05
Nitrogen, Nitrate+Nitrite as N, mg/L	0.3	<0.1	<0.1	0.1	0.4	<0.1	<0.1	<0.1	<0.1
Aluminum, mg/L	0.6	0.1	<0.1	0.1	<0.1	0.7	0.3	0.1	0.2
Arsenic mg/L	0.003	0.002	0.002	0.006	0.005	0.004	0.004	0.011	0.002
Barium, mg/L	0.4	0.5	0.1	0.1	0.2	0.3	0.4	0.3	<0.1

Affected Environment

Table 3-2. Surface Water Quality at the Proposed Moore Ranch Project									
Sample Parameters	MRSW-1[+]	MRSW-2	MRSW-3	MRSW-4	MRSW-5	MRSW-6[#]	MRSW-7[#]	MRSW-8	MRSW-9[#]
Boron, mg/L	<0.1	<0.1	<0.1	0.1	0.1	<0.1	<1.0	0.1	<0.1
Cadmium, mg/L	<0.005	<0.005	<0.005	<0.005	<0.005	<0.005	<0.005	<0.005	<0.005
Calcium, mg/L	21	17	50	26	32	18	21	10	14
Chromium, mg/L	<0.05	<0.05	<0.05	<0.05	<0.05	<0.05	<0.05	<0.05	<0.05
Copper, mg/L	<0.01	0.02	<0.01	0.01	<0.01	<0.01	<0.01	<0.01	<0.01
Iron, mg/L	0.25	0.11	0.08	0.13	0.35	0.33	0.65	0.32	0.11
Lead, mg/L	<0.05	0.019	<0.05	0.050	<0.05	0.001	<0.001	<0.05	<0.001
Magnesium, mg/L	35	30	16	22	54	13	14	26	5
Manganese, mg/L	<0.01	0.02	<0.01	0.02	0.03	0.02	0.02	0.01	<0.01
Mercury, mg/L	<0.001	<0.001	<0.001	<0.001	<0.001	<0.001	<0.001	<0.001	<0.001
Molybdenum mg/l	<0.1	<0.1	<0.1	<0.1	<0.1	<0.1	<0.1	<0.1	<0.1
Nickel, mg/L	<0.05	<0.05	<0.05	<0.05	<0.05	<0.05	<0.05	<0.05	<0.05
Potassium, mg/L	12	10	7	8	16	7	9	12	5
Selenium, mg/L	<0.0002	<0.0002	0.001	0.001	0.002	<0.002	<0.001	0.0013	<0.001
Silica, mg/L	5.1	2.6	4.8	6.8	7.5	7.6	8.0	5.6	5.2
Sodium, mg/L	244	238	84	189	348	155	218	369	22
Uranium, mg/L	0.0022	0.000467	0.0097	0.0044	0.0022	0.0003	0.0005	0.0020	0.0017
Vanadium, mg/L	<0.1	<0.1	<0.1	<0.1	<0.1	<0.1	<0.1	<0.1	<0.1
Zinc, mg/L	<0.01	0.015	<0.01	0.010	0.01	0.01	<0.01	<0.01	<0.01
Iron, TOTAL mg/L	0.65	0.157	0.22	0.16	0.45	0.62	0.69	0.374	0.14
Manganese, TOTAL mg/L	0.02	0.013	0.015	0.05	0.04	0.03	0.03	0.01	<0.01
Lead 210, suspended pci/L	<1.0	<1.0	<1.0	<1.0	<1.0	<1.0	<1.0	<1.0	<1.0

Affected Environment

Table 3-2. Surface Water Quality at the Proposed Moore Ranch Project									
Sample Parameters	MRSW-1[+]	MRSW-2	MRSW-3	MRSW-4	MRSW-5	MRSW-6[#]	MRSW-7[#]	MRSW-8	MRSW-9[#]
Polonium 210 suspended pci/L	<1.0	<1.0	<1.0	<1.0	<1.0	<1.0	<1.0	<1.0	<1.0
Radium 226, suspended pci/L	<0.2	<0.2	<0.2	<0.2	0.97	0.3	<0.2	<0.2	<0.2
Thorium 230 suspended pci/L	<0.2	<0.2	<0.2	<0.2	<0.2	<0.2	<0.2	<0.2	<0.2
Uranium suspended, pci/L	<0.0003	<0.0003	<0.0003	<0.0003	<0.0003	<0.0003	<0.0003	<0.003	<0.0003

Source: EMC, 2007b
Refer to Figure 3-8 for sample locations.
+ The average of three samples collected from 2006 to 2007 except as noted.
# The average of two samples collected during either 2006 or 2007.
* Average contains an anomalous value considered analytical error.
No samples collected from MRSW-10 and MRSW-11; location was always dry.

Wetlands comprise narrow, linear emergent systems within drainages and stream channels, as a direct result of CBM and livestock well discharges. Several CBM outfalls also jointly serve as livestock watering holes, comprising open water pools located along or within drainages. Some of the outfalls and accompanying watering tanks were observed releasing water and influencing the presence of wetland parameters. In those drainages where water is released, the wetland characteristics are actively present. In those drainages where there is a gradual decrease in the volume of CBM water being discharged via the outfalls described in Section 3.5.1.3 of this SEIS, the wetland parameters are receding, particularly wetland hydrology and hydrophytic vegetation, and upland vegetation is encroaching into the streambeds.

The wetlands delineated on the proposed Moore Ranch Project include the following systems: a single thread confined to the Ninemile Creek channel at the southern end of the proposed license area; three systems found within the Simmons Draw channel; one wetland within Simmons Draw Tributary A and Simmons Draw Tributary B; and three wetlands within the Pine Tree Draw drainage basin, as shown in Figure 3-10.

On May 10, 2010, the USACE concurred with the applicant's methods used to identify wetlands within the proposed project area and deemed that the methods are consistent with the USACE Wetland Delineation Manual and the Great Plains Region supplement (USACE, 2010). This verification of wetland delineation is valid for 5 years (until May 10, 2015). Additionally, the USACE determined that authorization is not required for any construction activities within Wellfield 1. Installation of wells and associated pipelines within wetland areas at Wellfield 2 are authorized by Nationwide Permit (NP) 12 (USACE, 2010). WDEQ-specific conditions have been incorporated as regional conditions of NP 12.

Affected Environment

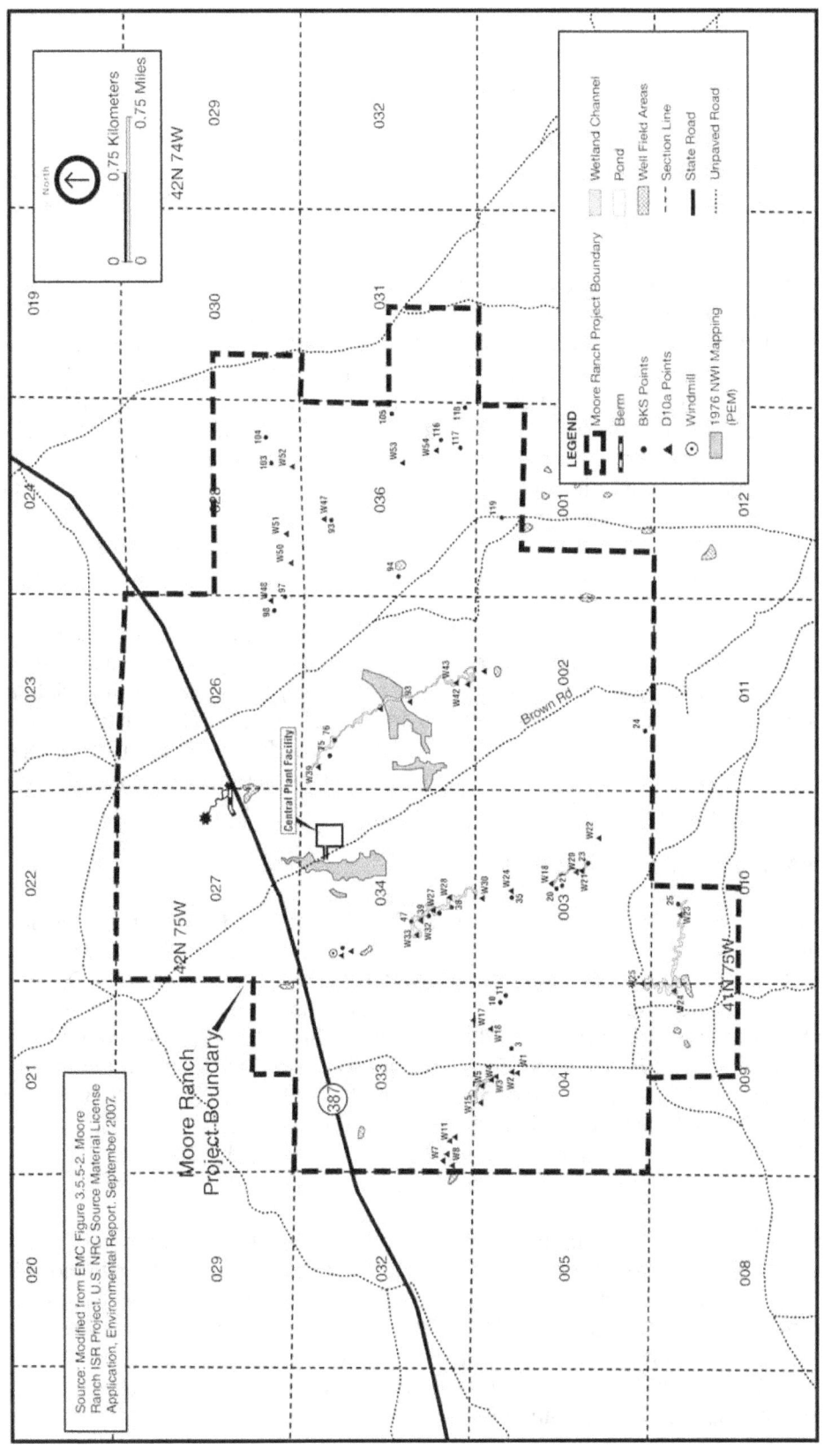

Figure 3-10. Wetlands Located on the Proposed Moore Ranch Project

Affected Environment

3.5.2 Groundwater

3.5.2.1 Regional Groundwater Resources

As noted in Section 3.3.4.3 of the GEIS, the Powder River Basin where the proposed Moore Ranch Project is located is part of the Wyoming East Uranium Milling Region (NRC, 2009a). In this region, uranium bearing aquifers are part of the Northern Great Plains regional aquifer system, which extends over one-third of Wyoming.

The Northern Great Plains aquifer system consists of five major aquifers, which from shallowest to deepest are designated as the Lower Tertiary, Upper Cretaceous, Lower Cretaceous, Upper Paleozoic, and Lower Paleozoic aquifers. The shallowest Lower Tertiary aquifers are located in sandstone beds within the Wasatch and Fort Union Formations, which are up to 1,400 m [4,600 ft] thick. These aquifers act as important regional water supplies for drinking water and livestock. Below them are the Upper Cretaceous aquifers, which are found in sandstone beds in the Lance, Hell Creek, and Fox Hills sandstones. These formations when combined are up to 1,070 m [3,850 ft] thick. The Fox Hills Sandstone is a significant water source. The next Lower Cretaceous aquifers are located beneath a regional thick sequence of shales known as the Pierre, Lewis, and Steele shales. Water yielding aquifers in the Lower Cretaceous are widespread and include the Muddy Sandstone and the Inyan Kara in the Powder River Basin.

These Lower Cretaceous aquifers contain little freshwater. The Upper Paleozoic aquifers are the Madison and the Tensleep Limestone in the western portion of the Powder River Basin and the Minnelusa Formation in the eastern portion. They are deeply buried and contain little to no freshwater. The Lower Paleozoic aquifers are the Winnipeg sandstone, Red River limestone, and Stonewall limestone formations. They are not typically used for water supplies because they are very deep and slightly saline to moderately saline in the southern extent and contain freshwater only in a small area in north-central Wyoming.

3.5.2.2 Local Groundwater Resources

The uranium-bearing aquifer at the proposed Moore Ranch Project is located in the Wasatch Formation, which is part of the shallow Lower Tertiary aquifer system. The Wasatch formation is described as an arkosic fine- to coarse-grained sandstone with siltstone, claystone, and coals. The contact between the underlying Fort Union Formation and the Wasatch Formation is gradational in the vicinity of Moore Ranch and is generally arbitrarily set at the top of the thicker coals or thick sequence of clays and silts. The applicant has identified the boundary between the two formations to be the top of the Roland Coal. The Wasatch Formation total thickness ranges from 244 to 335 m [800 to 1,100 ft] in the proposed project area. In the southern portion of the Powder River Basin, the Wasatch Formation generally dips to the northwest at 1.0 to 2.5 degrees.

There are commonly multiple water-bearing sands within the Wasatch Formation. Due to their higher permeability, these water-bearing sands provide the primary sources for groundwater withdrawal. Groundwater within the Wasatch aquifers is typically under confined (artesian) conditions, although locally unconfined conditions exist. Well yields from the Wasatch Formation in the southern part of the Powder River Basin where the site is located are reported to be as high as 1,900 Lpm [500 gpm]. The overall flow of groundwater in the shallow aquifers in the vicinity of the proposed Moore Ranch Project is toward the Powder River Basin to the north-northwest.

Affected Environment

As previously discussed in Section 3.4.1, the applicant has adopted the nomenclature used by Conoco, Inc., for the hydrostratigraphic units of interest within the proposed Moore Ranch Project. Sands above the Roland Coal are numbered, increasing upward. The 40 and 50 Sands lie immediately above the Roland Coal and are regionally extensive sands that are considered significant aquifers. The approximate thickness of the 40 and 50 Sands in the proposed license area are 24 to 27 m [80 and 90 ft]. The 58 Sand varies in thickness from 1.5 to 24 m [5 to 80 ft]. The 60 Sand is approximately 30 m [100 ft] thick and is continuous throughout the proposed project area. It is separated from the 58 Sand by about 1.5 to 21 m [5 to 70 ft] of shale and mudstone. The 68 Sand is the first sand underlying the 70 Sand, which contains the economic ore deposits in the area. The 68 Sand ranges from 12 to 30 m [40 to 100 ft] across the proposed license area and coalesces with the 60 Sand on the west side of the proposed license area. The 70 Sand is the proposed ore production zone and coalesces with the 68 Sand in one of the proposed wellfields. The 72 Sand overlies the 70 Sand and is the shallowest sand over the majority of the proposed license area. The 70 Sand is separated from the overlying 72 Sand by a continuous shale layer ranging in thickness from 1 to 49 m [3.3 to 160 ft] in some areas. Over small portions of the proposed license area, the 80 Sand overlies the 72 Sand.

3.5.2.3 Uranium-Bearing Aquifers

The 70 Sand is the proposed production aquifer located 30.5 to 91.4 m [100 to 300 ft] below ground surface in the proposed project area. The 70 Sand is laterally extensive and ranges from 12.2 to 36.6 m [40 to 120 ft] thick. The 70 Sand dips to the northwest at about 1 degree. It outcrops approximately 1.6 km [1 mi] south of the proposed project area. The 70 Sand is not completely saturated over its thickness in most of the proposed license area. Since the water levels in the 70 Sand are below the overlying shale, it is defined as an unconfined aquifer. Water produced from wells in unconfined aquifers comes from physical drainage of water from the formation pores, not from compression of the sediments and expansion of water due to pressure decreases, as in confined aquifers. The natural groundwater flow is estimated to be to the northwest in the 70 Sand at about 2.4 m/yr [7.8 ft/yr], based on the reported gradient of 0.004 ft/ft and hydraulic conductivity of 5.36 ft/day.

3.5.2.3.1 Hydrogeologic Characteristics

The hydraulic properties of the 70 Sand production aquifer have been evaluated through a series of pumping tests. Aquifer testing was performed between 1978 and 1980 while Conoco was investigating the Moore Ranch site as a possible surface mine site. The applicant conducted additional pumping tests in 2007 and 2008. Analysis of data from the 2008 test estimated the transmissivity and hydraulic conductivity of the 70 Sand to be 37.6 m^2/day [405 ft^2/day] and 1.63 m/day [5.36 ft/day], respectively. Estimates of specific yield for the unconfined aquifer ranged from 0.011 to 0.039.

3.5.2.3.2 Level of Confinement

The 70 Sand is separated from the overlying 72 Sand by a continuous shale and coal seam across the proposed license area. Water levels in the 72 Sand are much higher than the 70 Sand. The 70 Sand is also not completely saturated. These two features demonstrate that the aquifers are not hydraulically connected. All of the pumping tests conducted in the 70 Sand to date have demonstrated no response in the 72 Sand, which supports the lack of a hydraulic interconnection. Because the 70 Sand is not completely saturated, the groundwater in the 72 Sand is likely perched on the shale separating the two aquifers.

Affected Environment

The 70 Sand is separated from the underlying 68 Sand by shale over much of the proposed project area. Pumping tests conducted to date have identified no hydraulic interconnection between these sands in proposed Wellfield 1. In portions of proposed Wellfield 2, however, boring data indicates that the shale is missing or less than 5 ft thick. In this area, the 68 Sand coalesces with the 70 Sand. Pumping tests in Wellfield 2 where the shale is absent have shown a hydraulic connection between the 68 and 70 Sand. Water levels in the 68 and 70 Sands are also similar, supporting a potential hydraulic connection. In the area in Wellfield 2 where the 68 and the 70 Sands coalesce, the applicant considers the 60 Sand to be the underlying aquifer.

3.5.2.3.3 Groundwater Quality

Baseline groundwater quality programs have characterized the quality of groundwater within the shallow Wasatch aquifers within the proposed Moore Ranch Project area (Table 3-3). Groundwater quality in the 72 Sand aquifer and production zone 70 Sand exceed the WDEQ Class I standards for TDS and sulfate. The radionuclides radium-226 and uranium are elevated above EPA maximum contaminant levels (MCL) for drinking water in the majority of samples collected from the production zone 70 Sand aquifer and the underlying 68 Sand aquifer. The

Table 3-3. Average Preoperational Baseline Groundwater Quality for the "72 Sand" Overlying Aquifer, the "70 Sand" Extraction zone Aquifer, the "68 Sand," and the "60 Sand" Underlying Aquifer

Water Quality Parameter	Average			
	"72 Sand"	"70 Sand"	"68 Sand"	"60 Sand"
Bicarbonates as HCO_3 (mg/L)	208.2	277.2	148.6	225.8
Carbonates as CO_3 (mg/L)	1.6	1.0	10.4	4.4
Chloride (mg/L)	4.1	2.3	2.0	2.6
Conductivity (umhos/cm)	1051.6	1034.2	753.3	621.0
Fluoride (mg/L)	0.2	0.2	0.2	0.2
pH (s.u.)	8.07	7.58	8.99	8.77
TDS (mg/L)	770.6	712.5	416.5	414.4
Sulfate (mg/L)	401.0	330.3	162.0	128.4
Gross Alpha (pCi/L)	5.7	259.1	78.24	74.7
Gross Beta (pCi/L)	14.4	80.6	40.09	16.4
Lead-210 (pCi/L)	2.0	9.2	6.81	1.37
Polonium-210 (pCi/L)	1.2	5.6	3.55	1.17
Radium-226 (pCi/L)	1.1	95.6	21.1	0.71
Radium-228 (pCi/L)	1.9	1.7	1.2	1.32
Thorium-230 (pCi/L)	0.3	0.3	0.2	0.05
Nitrogen, Ammonia as N (mg/L)	0.214	0.1	0.5	Nondetect
Nitrogen, Nitrate + Nitrite as N (mg/L)	0.4	0.2	0.3	Nondetect

Table 3-3. Average Preoperational Baseline Groundwater Quality for the "72 Sand" Overlying Aquifer, the "70 Sand" Extraction zone Aquifer, the "68 Sand," and the "60 Sand" Underlying Aquifer

Water Quality Parameter	Average			
	"72 Sand"	"70 Sand"	"68 Sand"	"60 Sand"
Aluminum (mg/L)	0.1	0.1	0.1	Nondetect
Arsenic (mg/L)	0.001	0.002	0.002	0.001
Barium (mg/L)	0.1	0.1	0.1	Nondetect
Boron (mg/L)	0.1	0.1	0.1	Nondetect
Cadmium (mg/L)	0.005	0.005	0.005	Nondetect
Calcium (mg/L)	137.4	135.2	54.4	58.8
Chromium (mg/L)	0.05	0.05	0.05	Nondetect
Copper (mg/L)	0.01	0.01	0.011	Nondetect
Iron (mg/L)	0.082	0.151	0.052	Nondetect
Lead (mg/L)	0.001	0.002	0.005	Nondetect
Magnesium (mg/L)	37.3	31.8	7.4	6.1
Manganese (mg/L)	0.064	0.033	0.016	0.014
Mercury (mg/L)	0.001	0.001	0.001	Nondetect
Molybdenum (mg/L)	0.1	0.1	0.1	Nondetect
Nickel (mg/L)	0.05	0.05	0.05	Nondetect
Potassium (mg/L)	16.6	10.2	14.7	8.3
Selenium (mg/L)	0.001	0.025	0.135	0.083
Silica (mg/L)	10.9	13.0	11.2	Nondetect
Sodium (mg/L)	32.3	33.3	63.4	70.2
Uranium (mg/L)	0.001	0.161	0.050	0.0532
Vanadium (mg/L)	0.1	0.1	0.1	Nondetect
Zinc (mg/L)	0.01	0.01	0.01	Nondetect

Source: EMC, 2007b, Table 2.7.3-21.

average radium 226-228 concentration in the production zone is an order of magnitude greater than the EPA MCL. The 68 Sand aquifer also exceeds the EPA MCL for selenium. In Wellfield 2, the 60 Sand aquifer exceeds the EPA MCL for selenium and uranium. Elevated concentrations of these radionuclides is consistent with the presence of uranium ore-bodies.

Using WDEQ standards, Uranium One classified the class of use for each shallow aquifer on a well-by-well basis in the proposed license area. WDEQ Class I is drinking water, Class II is agricultural use, Class III is for livestock use, and Class VI is water that is unsuitable for any of these uses. The single well in the perched 80 Sand aquifer was classified as Class VI. One well in the 72 Sand aquifer was classified as Class I, another as Class II, and two others as

Affected Environment

Class III. In the 70 Sand production zone aquifer, all eight wells were Class VI and one well outside the ore zone was Class I. All four wells in the 68 Sand and three wells in the 60 Sand were found to be Class VI.

For ISR operations to be conducted in a proposed ore-bearing aquifer within the permit boundaries of the proposed ISR Site, the aquifer must be declared as an exempted aquifer, in compliance with 40 CFR Part 146. The applicant is required to obtain a Class III UIC exemption permit from the State. The State requests an aquifer exemption from the EPA for the proposed Class III UIC permit. The applicant must have both the UIC permit and the exemption before operations may begin. The water quality of the 70 Sand production zone aquifer in the project area is Class VI under WDEQ standards, which means the groundwater cannot be used for drinking, livestock, or agricultural use as a consequence of its uranium and radium-226 concentrations. It would, therefore, be a candidate for an exempt aquifer declaration. The 68 Sand would also be a candidate, given its water quality is also Class VI.

3.5.2.3.4 Current Groundwater Uses

According to a search of the WSEO database, there are 559 wells with groundwater rights located within the 3.2-km [2-mi] radius of the proposed Moore Ranch Project as of June 2009. Groundwater rights for wells are granted on a well-by-well basis through the WSEO (Uranium One, 2009b). Domestic and stock wells have a limit of 94.7 L/min [25 gal/min] per well. There are no minimum water levels entitled with the groundwater rights. The vast majority of water rights in and near the proposed Moore Ranch Project area are for CBM activities in the Fort Union Formation at depths exceeding 244 m [800 ft].

Of the wells identified in the search, 465 are CBM or stock–CBM wells. All of these CBM and stock–CBM wells that have completion records, are greater than 213 m [700 ft] deep. Of those with no completion records it is unlikely they are completed in shallower Wasatch Formation sands, as the target for CBM production is the Fort Union Formation, which is located at depths exceeding 244 m [800 ft] in the proposed license area. Given the depth of these wells, it is unlikely they will be impacted by operations in the 70 Sand production zone.

Within the 3.2 km [2 mi] radius, there are three domestic water wells ranging from 41.7 to 134 m [137 to 440 ft] deep. Two are located east of the proposed Moore Ranch Project area near the limit of the 3.2-km [2-mi] radius. One well is located in the license area and is permitted as an industrial, domestic well by Rio Algom Mining Corporation. While these wells are permitted for domestic use, there are no currently occupied residences within the proposed license area and 3.2-km [2-mi] radius; therefore, these wells are not being primarily used for human consumption. Given both the distance from the proposed operations and the well's upgradient locations, they would unlikely be impacted by the proposed operations.

Also within the 3.2-km [2-mi] radius there are 27 permitted stock wells, of which 3 are located in the proposed project area. Of these wells, 25 are completed at depths greater than 213 m [700 ft] and two are completed at less than 213 m [700 ft]. At least four other unpermitted stock wells are known to be in the proposed project area, for which no completion information is available, but they are estimated to be in the 68, 70, and 72 Sands. An inoperable windmill with an unpermitted well is also located in the proposed project area. Some of these wells that are located in the shallow 68, 70, and 72 Sands may be impacted by operations within the 70 Sand production zone.

Affected Environment

There are no irrigation water wells within the 3.2 km [2 mi] radius. The deepest water well that has groundwater rights within the 3.2-km [2 mi] radius is permitted as a CBM well and is 430 m [1,410 ft] deep. It is not likely to be impacted by ISR operations. The remaining deep wells in the proposed project area are oil and gas wells.

If the domestic well is completed in the exempted portion of the ore-bearing aquifer, the well cannot be used as a source of drinking water, in compliance with 40 CFR Part 146. In this case, the domestic well is required to be properly plugged and abandoned prior to commencement of ISR operations. If industrial or livestock wells are completed in the exempted portion of the aquifer, these wells would be required to be properly plugged and abandoned to avoid potential negative impacts on targeted bleed rates during ISR operations and also to minimize potential adverse impacts on the environment. Upon completion of ISR operations, the applicant is required to return groundwater quality in the exempted portions of the production aquifer to restoration standards, in compliance with 10 CFR Part 40, Appendix A, Criterion 5B(5).

3.5.2.4 Surrounding Aquifers

In addition to the sands of the Wasatch discussed previously, the underlying Fort Union Formation and Fox Hills Sandstones include potentially important aquifers. However, because of the relatively shallow depth of 30.5 to 91.4 m [100 to 300 ft] for the 70 Sand production zone in the overlying Wasatch, these deeper aquifers that are separated by thick sequences of shale are not likely to be impacted by ISR operations in the production zone.

The shallowest deep aquifer is the Lance Formation at depths of 1,128–1,738 m [3,700–5,700 ft] below ground surface. The TDS content of groundwater in the Lance Formation is on average 1,200 ppm (EMC 2007a; Section 3.4.3.3). Because of its low water yields and large depths, the Lance Formation is not expected to be a drinking water supply at or in the vicinity of the permit area. If the Lance Formation is considered as a candidate deep disposal aquifer for the proposed Moore Ranch Project, the State of Wyoming would evaluate the feasibility of using the formation for deep well disposal via a Class I injection well. The State and EPA would only grant such a permit to the applicant if it can demonstrate that liquid effluent could be safely isolated in a deep aquifer.

3.6 Ecology

The Wyoming East Uranium Milling Region, as described in the GEIS, encompasses the Wyoming Basin, Northern Great Plains, Southern Rockies, and Western High Plains. The proposed Moore Ranch ISR Project is located within the Powder River Basin of the Northwestern Great Plains ecoregion. Section 3.3.5.1 of the GEIS provides the following description of this region:

> The Northwestern Great Plains encompass the Missouri Plateau section of the Great Plains. This area includes semiarid rolling plains of shale and sandstone derived soils punctuated by occasional buttes and badlands. For the most part, it has not been influenced by continental glaciation. Cattle grazing and agriculture with spring wheat and alfalfa farming are common land uses. Agriculture is affected by erratic precipitation and limited opportunities for irrigation. In Wyoming, mining for coal and coal-bed methane production is prevalent, with a large increase in the number of coal-bed methane wells drilled in recent years. Native grasslands and some woodlands persist, especially in areas of steep or broken topography (Chapman, et al., 2004).

Affected Environment

Section 3.3.5.1 of the GEIS provides the following description of the Powder River Basin:

> The Powder River Basin ecoregion of the Northwestern Great Plains covers rolling prairie and dissected river breaks surrounding the Powder, Cheyenne, and Upper North Platte Rivers. The Powder River Basin has less precipitation and less available water than the neighboring regions. Vegetation within this region is composed of sagebrush and mixed-grass prairie dominated by blue grama (*Bouteloua gracilis*), western wheatgrass (*Elymus smithii*), prairie junegrass (*Koeleria macrantha*), Sandberg Bluegrass (*Poa secunda*), needle-and-thread grass (*Stipa comata*), rabbitbrush (*Chrysothamnus*), fringed sage (*Artemisia frigida*), and other forbs, shrubs, and grasses (Chapman et al., 2004).

The applicant conducted a number of ecological studies at the proposed Moore Ranch Project to accomplish the objectives specified in NUREG–1569, "Standard Review Plan for *In-Situ* Leach Uranium Extraction License Applications," and to meet the applicable State of Wyoming requirements. These studies include vegetation and wetland surveys conducted in the spring/summer of 2007 and wildlife surveys conducted from fall 2006 through summer 2007.

3.6.1 Terrestrial Ecology

The proposed project area is comprised primarily of grassland, with areas of sagebrush in the southwest corner. Interspersed among those major plant communities are less abundant seeded grasslands (improved pastures) habitats and intermittent streams, as described next. No perennial streams or other permanent water bodies exist within the proposed Moore Ranch Project (EMC, 2007a).

3.6.1.1 Vegetation

The applicant conducted baseline vegetation and wetland surveys during the spring/summer of 2007, in accordance with applicable State and Federal guidelines. The applicant's environmental report (EMC,2007a) provides a detailed description of the survey results. The spatial distribution of the vegetation types within the proposed Moore Ranch Project are shown in Figure 3-11.

The proposed license area for the Moore Ranch Project is approximately 2,879 ha [7,110 ac] and consists primarily of four vegetation communities: Meadow Grassland, Upland Grassland, Agricultural Grassland, and Big Sagebrush Shrublands, as shown in Figure 3-11. Approximately 61 ha [150 ac] or about 2 percent of the proposed license area would be disturbed by the proposed action. Each vegetation community was investigated to establish a baseline, in support of the NRC license application. No threatened or endangered plant species were encountered within the proposed license area. Two State-listed species of concern, or State designated weeds, Canada thistle (*Cirsium arvense*) and field bindweed (*Convolvulus arvensis*), were identified in the proposed project area and should be managed.

Affected Environment

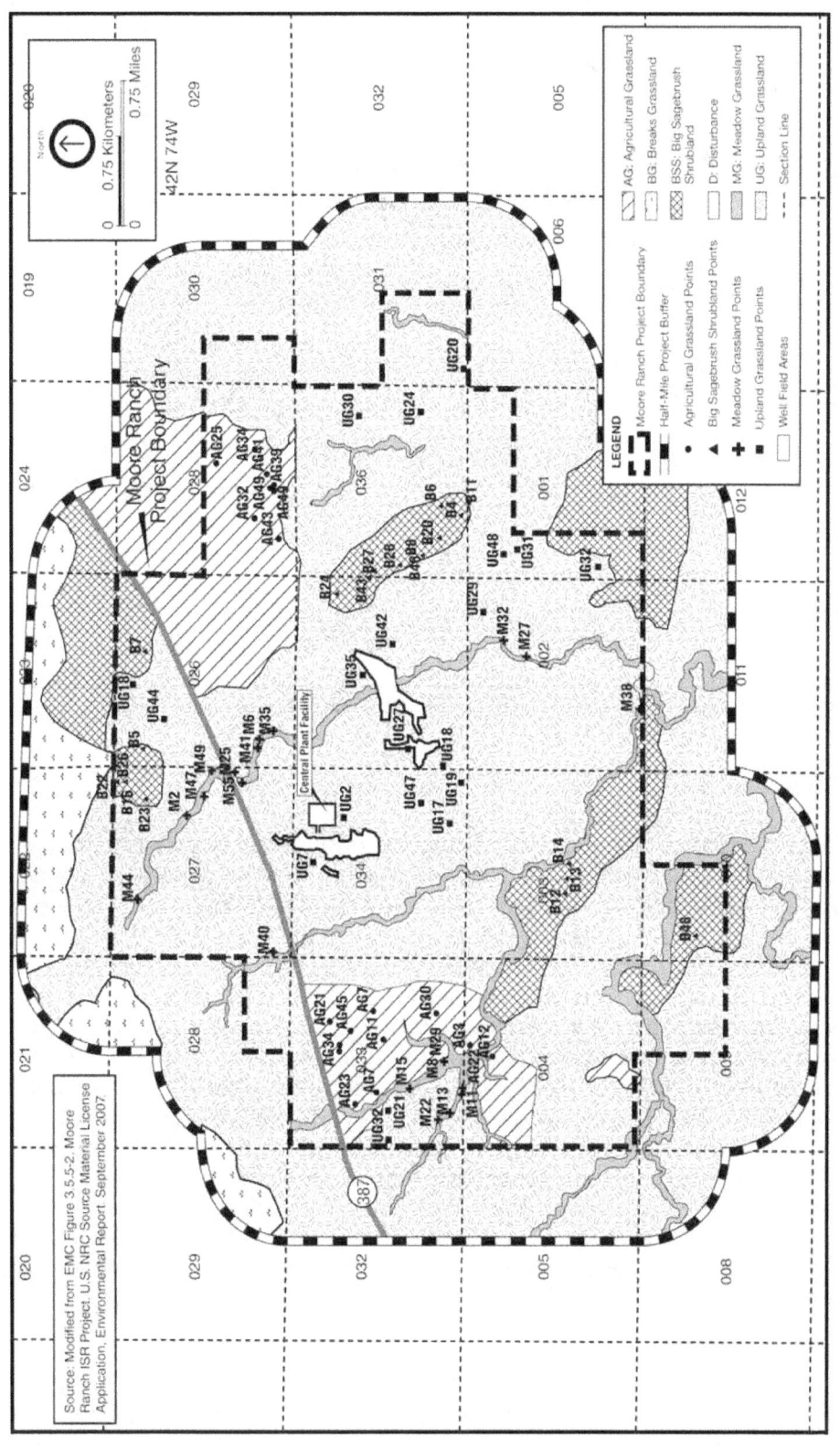

Figure 3-11. Vegetation Communities on the Proposed Moore Ranch Project

Affected Environment

Table 3-4 summarizes the area of each vegetation community within the proposed license area. The applicant's environmental report summarizes the vegetation community mapping units for the proposed Moore Ranch Project (EMC, 2007a). Upland grassland is the predominant vegetation type within the proposed license area, as summarized in Table 3-4.

Table 3-4. Areal Distribution of Vegetation Communities within the Proposed License Area			
Mapping Unit	**Proposed License Area (in hectares)**	**Proposed License Area (in Acres)**	**Percent of Area**
Meadow Grassland	130.9	323.32	5
Upland Grassland	2,027	5,006.69	70
Agricultural Grassland	377	931.19	13
Big Sagebrush Shrubland	286.43	707.48	10
Disturbance	53.5	132.15	2
TOTAL	2,875	7,100.83	100
Source: EMC, 2007a			

3.6.1.2 Wildlife

Baseline wildlife information for the proposed Moore Ranch Project is available from previous data collection efforts conducted for CBM plans-of-development that generally covered all but the extreme southeastern and western sections of the proposed Moore Ranch Project area and the perimeter. These annual surveys were conducted from the years 2003 through 2006 and included numerous wildlife species, habitat features such as bald eagle nesting and winter roost sites, sage-grouse leks, raptor nests, and surveys for avian species of concern. In addition, the applicant conducted a site-specific survey from October 2006 through June 2007. Detailed results of these investigations are documented in the applicant's environmental report (EMC, 2007a). Because much of the proposed license area has been included in annual wildlife monitoring since 2003, the Wyoming Game and Fish Department (WGFD) delineated the study area for raptors and other migratory birds to those portions of the proposed Moore Ranch Project and a 1.6-km [1-mi] perimeter not encompassed by the previous overlapping studies.

Site-specific wildlife surveys of the proposed Moore Ranch Project targeted bald eagle winter roost sites, sage-grouse leks, nesting raptors (including eagles), mountain plovers, and other avian species of concern.

3.6.1.2.1 Big Game

Pronghorn antelope (*Antilocapra americana*) and mule deer (*Odocoileus hemionus*) are the only two big game species that regularly occur on the proposed license area. No crucial big game habitat or migration corridors occur in or within several kilometers of the area (WGFD, 2009).

Pronghorn antelope are more abundant than mule deer in the proposed Moore Ranch Project area, but neither species is prevalent because it is not their preferred habitat. The WGFD classified the proposed license area as a yearlong pronghorn antelope range, meaning that a portion of a population of animals makes general use of this habitat on a year-round basis. The proposed Moore Ranch Project spans two WGFD pronghorn antelope herd units bisected by highway SR 387. The WGFD estimated the 2006 post-season pronghorn antelope populations in those two hunt areas to be approximately 36,500 and 32,300, which is above the WGFD population objectives to manage and regulate big game herds (WGFD, 2006a,b).

Mule deer use nearly all habitats but prefer sagebrush-grassland and are not abundant in the proposed license area. Monitoring data indicate that mule deer are not very migratory in the vicinity of the proposed Moore Ranch Project. The majority of the proposed license area has been classified by WGFD as a yearlong mule deer range, except portions south of the highway that are considered inadequate to support mule deer. The WGFD estimated the 2006 post-season mule deer population to be approximately 12,350 and 9,700 animals, compared to herd objectives of 11,000 and 9,100 (WGFD, 2006a,b).

3.6.1.2.2   Avian Species

This section of the SEIS describes bird species that have been identified at the proposed Moore Ranch Project, based on the surveys described previously.

Upland Game Birds

The mourning dove (*Zenaida macroura*) is the only upland game bird known to regularly occur in the vicinity of the proposed Moore Ranch Project, and it is a relatively common breeder in Campbell County. Most sightings at the proposed Moore Ranch Project occur during migration near sites with water sources or trees, though they were occasionally recorded in upland grassland habitats.

The greater sage-grouse (*Centrocercus urophasianus*) is listed as a Federal candidate species and an avian species of special concern in Wyoming, and it is discussed in more detail in Section 3.6.3 (75 FR13090; WGFD, 2005a). Sage-grouse are found in sagebrush shrubland habitats, and sagebrush is essential during all seasons and for every phase of their life cycle. Sage-grouse are rare in the vicinity of the proposed Moore Ranch Project because of the limited habitat to support their existence. No large expanses of contiguous sagebrush occur within several kilometers of the proposed Moore Ranch Project. Consequently, few sage-grouse have been documented in the area; and no grouse leks have ever been discovered either on or near the proposed Moore Ranch Project. The nearest known sage-grouse lek is located approximately 4.0 km [2.5 mi] to the northwest of the proposed Moore Ranch Project area (BLM, 2009).

Raptors

Suitable habitat for several raptor species occurs in the proposed Moore Ranch Project area and within a 1.6-km [1-mi] perimeter of the site. These raptor species include golden eagle (*Aquila chrysaetos*), ferruginous hawk (*Buteo regalis*), red-tailed hawk (*Buteo jamaicensis*), Swainson's hawk (*Buteo swainsoni*), northern harrier (*Circus cyaneus*), American kestrel (*Falco sparverius*), prairie falcon (*Falco mexicanus*), great horned owl (*Bubo virginianus*), burrowing owl (*Athene cunicularia*), and short-eared owl (*Asio flammeus*) (EMC, 2007a). Nests have been

Affected Environment

observed for the ferruginous hawk, red-tailed hawk, great horned owl, and Swainson's hawk but not for the other raptors, based on BLM data (BLM, 2007a).

Thirty-six raptor nest sites have been identified within the vicinity of the proposed Moore Ranch Project since 2003 (EMC, 2007a). Nineteen nest sites were located within the proposed Moore Ranch Project, and the remaining 17 were located around the perimeter of the site (EMC, 2007a).

Waterfowl and Shorebirds

The proposed Moore Ranch Project has extremely limited and marginal habitat for waterfowl and shorebirds because of the lack of surface water. Natural aquatic habitats are mainly present during spring migration and consist of small, isolated pools that can be completely dry during summer. Recent CBM development in the proposed Moore Ranch Project area has increased the number of localized water sources with limited depth, geographic area, and duration. Several common species of waterfowl and shorebirds, including the mallard (*Anas platyrhynchos*) and killdeer (*Charadrius vociferus*), have been infrequently observed in the proposed Moore Ranch Project area (EMC, 2007a).

3.6.1.2.3 Other Mammals, Reptiles, and Amphibians

A variety of small- and medium-sized mammal species occur in the vicinity of the proposed Moore Ranch Project. These include predators and furbearers, such as the coyote (*Canis latrans*), red fox (*Vulpes vulpes*), swift fox (*Vulpes velox*), bobcat (*Lynx rufus*), striped skunk (*Mephitis mephitis*), weasel, badger (*Taxidea taxus*), muskrat (*Ondatra zibethicus*), and raccoon (*Procyon lotor*). Prey species include various rodents, such as mice, rats, voles, gophers, ground squirrels, chipmunks, and lagomorphs (jackrabbits and cottontails). These prey species are cyclically common and widespread throughout the region and are important for raptors and other predators. No occupied black-tailed prairie dog (*Cynomys ludovicianus*) colonies have been documented on or in the vicinity of the proposed Moore Ranch Project, based on repeated surveys (EMC, 2007a).

Few reptiles and amphibians have been recorded during recent wildlife surveys because of the lack of suitable habitat (EMC, 2007a). The common bullsnake (*Pituophis melanoleucas sayi*) was the only herpetological species recorded on the proposed Moore Ranch Project in the baseline studies conducted in 2006 and 2007.

3.6.2 Aquatic Ecology

Aquatic habitat on and near the proposed Moore Ranch Project is limited by the intermittent nature of surface waters in the proposed license area. The lack of deep-water habitat and extensive and persistent water sources precludes the presence of fish and limits the abundance and diversity of other aquatic species. No perennial drainages are present in the proposed license area.

Affected Environment

3.6.3 Protected Species

Threatened and Endangered Species

Table 3-5 lists species that are Federally-listed under the Endangered Species Act (ESA) of 1973, State-listed under the Final Comprehensive Wildlife Conservation Strategy for Wyoming, or BLM-listed as sensitive species and that occur in Campbell and Johnson Counties.

No threatened and endangered species occur within the proposed Moore Ranch Project study area. Based on consultation with the U.S. Fish and Wildlife Service (FWS) (2008a), federally listed threatened and endangered species (or their designated habitat) that could potentially occur in the proposed project area are discussed in the following paragraphs.

The Ute ladies'-tresses orchid (*Spiranthes diluvialis*) is Federally-listed as threatened. The species is a perennial, terrestrial orchid that occurs in Nebraska, Wyoming, Colorado, Utah, Idaho, Montana, and Washington. Within Wyoming, it inhabits moist meadows with moderately dense but short vegetative cover. The species is found at elevations of 1,280 to 2,130 m [4,200 to 7,000 ft], though no known populations occur in Wyoming above 1,680 m [5,500 ft] (Fertig, 2000). Generally, this orchid is found in low densities of four to eight flowering plants per square meter (Fertig, 2000). The species is likely to inhabit silt, sand, or gravely soils in areas with ample sunlight (Fertig, 2000). It is characterized by 12- to 50-cm [4.7- to 20-in] stems with linear basal leaves up to 28 cm [11 in] long and spikes of small white to ivory flowers that bloom between early August and early September (Fertig, 2000). Urbanization, livestock grazing, pesticide use, competition with noxious weeds, and loss of pollinators threaten this species' survival (Fertig, 2000). This species was not identified during the applicant vegetation inventories and it is not known to occur on or in the vicinity of the proposed project site.

The black-footed ferret (*Mustela nigripes*) is listed as an endangered species that inhabits prairie dog colonies. A black-footed ferret survey was not required, since black-footed ferrets live exclusively in prairie dog colonies, which are not present on or within 1.6 km [1 mi] of the proposed Moore Ranch Project area.

The black-footed ferret is a small mammal in the weasel family with a natural to buff-colored body and black face, feet, and tail. Adults are 46 to 61 cm [18 to 24 in] long and weigh 0.7 to 1.1 kg [1.5 to 2.5 lb], with males generally larger than females (FWS, 2008b). Generally, black-footed ferret occurrence coincides with prairie dog habitat (black-tailed [*Cynomys ludovicianus*], Gunnison's [*C. gunnisoni*], and white-tailed [*C. leucurus*]) because prairie dog is the main prey of the ferret, and the ferret also uses prairie dog burrows for shelter (FWS, 2008b). Black-footed ferrets are more likely to occur in black-tailed prairie dog habitat than in other prairie dog species' habitat; historically, it is estimated that 85 percent of all black-tailed ferrets occurred in black-tailed prairie dog habitat, 8 percent in Gunnison's prairie dog habitat, and 7 percent in white-tailed prairie dog habitat (FWS, 2008b).

The bald eagle (*Haliaeetus leucocephalus*) was delisted from threatened status in 2007, but it is still protected under the Bald and Golden Eagle Protection Act and the Migratory Bird Treaty Act. Potential habitat for bald eagle nesting and roosting activities is quite limited within the proposed license area because of the lack of trees. Nor does the proposed Moore Ranch Project contain unique or sizeable, concentrated prey sources (e.g., fisheries, waterfowl wintering areas) that would be expected to attract bald eagles. There have been no bald eagle sightings during either site-specific winter roost surveys or other baseline surveys completed by the applicant in 2006 and 2007 nor have they been observed in annual surveys conducted since

Affected Environment

2003 (EMC, 2007a). In addition to the wildlife inventories conducted by Uranium One, a BLM environmental assessment for the Yates Petroleum Corporation All Day Plan of Development (BLM, 2008) identifies information concerning bald eagle roosts within the Hank Unit of the proposed Nichols Ranch ISR Project, located about 32 km [20 mi] northwest of the proposed Moore Ranch Project. The environmental assessment documented observation of 7 bald eagles on December 3, 2007, 5 bald eagles on December 16, 2008, 1 bald eagle on January 12, 2009, and 13 bald eagles on February 11, 2009 (BLM, 2008).

The blowout penstemon (*Penstemon haydenii*) is Federally-listed as endangered. The species is a perennial herb that is endemic to the Nebraska Sandhills in north-central Nebraska and to the northeastern region of the Great Divide Basin in Carbon County, Wyoming (Fertig, 2008). The species is found exclusively in sparsely vegetated, early successional sand dunes or blowout areas at elevations of 1,790 to 2,270 m [5,860 to 7,440 ft] (Fertig, 2008). The proposed Moore Ranch ISR Project does not have sand dune habitat and is outside of the elevation range in which this species is typically found. This species was not identified during vegetation inventories and it is not known to occur on or in the vicinity of the proposed site.

The swift fox (*Vulpes velox*) is a State of Wyoming species of concern and a BLM-designated sensitive species. The species was removed from the ESA Candidate List in 2002 due to successful conservation measures and reintroduction efforts in western states. The species is native to the Great Plains region, and in Wyoming, the swift fox inhabits flat terrain east of the Continental Divide with shortgrass or mixed-grass prairie and is often associated with prairie dog colonies (WGFD, 2005b). Swift foxes are nocturnal and use underground dens year round. Threats to the species' continued survival include loss of prairie habitat, trapping and hunting, and predator control campaigns (WGFD, 2005b). This species was not identified during the applicant wildlife inventories and it is not known to occur on or in the vicinity of the proposed project site.

The Greater sage-grouse (*Centrocercus urophasianus*) is Federally-listed as a candidate species, a State of Wyoming species of concern, and a BLM-designated sensitive species. On March 5, 2010, the FWS published a finding in the *Federal Register* stating that listing of the species was warranted but precluded by higher priority listing actions (75 FR 13909). The Wyoming Governor issued an Executive Order in August 2008 that sets out 12 provisions for oil and gas resources management within core and noncore population areas to protect the species on the State level (State of Wyoming, 2008). WGFD published Recommendations for Development of Oil and Gas Resources Within Important Wildlife Habitats, and the Wyoming BLM issued an instructional memorandum on March 5, 2010, which supplements the BLM 2004 National Sage-Grouse Habitat Conservation Strategy, to be consistent with the Governor's Executive Order (WGFD, 2010; BLM, 2004; BLM, 2010). This guidance was again updated in April 2010 (WGFD, 2010). The species inhabits open sagebrush plains in the western United States and is found at elevations of 1,200 to 2,700 m [4,000 to 9,000 ft], corresponding with the occurrence of sagebrush habitat (69 FR 933). The greater sage-grouse is a mottled brown, black, and white ground-dwelling bird that can be up to 0.6 m [2 ft] tall and 76 cm [30 in] in length (69 FR 933). Breeding habitat, referred to as leks, and stands of sagebrush surrounding leks are used in early spring and are particularly important habitat because birds often return to the same leks and nesting areas each year. Leks are generally more sparsely vegetated areas such as ridgelines or disturbed areas adjacent to stands of sagebrush habitat. Threats to this species' survival include loss of habitat, agricultural practices, livestock grazing, hunting, and land disturbances from energy/mineral development and the oil and gas industry (Sage-Grouse

Working Group, 2006). This species was not identified during the applicant wildlife inventories and few have ever been documented on or in the vicinity of the proposed site.

Species of Concern

The Wyoming Field Office of the USFWS uses the list, Migratory Bird Species of Management Concern in Wyoming (also known as Migratory Birds of High Federal Interest) for conducting reviews related to noncoal surface disturbance projects (FWS, 2002). This list (Table 3-5) is based on the Wyoming Bird Conservation Plan (Nicholoff, 2003). Seventy seven avian species of concern are identified on this list; 22 species are identified as being species in need of

Table 3-5. Migratory Bird Species of Management Concern Observed on the Proposed Moore Ranch Project

Species	Primary Nesting Habitat(s)	Status/Occurrence in Project Region*	Occurrence Within Proposed License Area
Species Of Level I Concern–Conservation Action Needed			
McCown's longspur *Calcarius mccownii*	Short-grass prairie, shrub-steppe	Breeder	Observed, presumed breeder
Ferruginous hawk *Buteo regalis*	Shrub-steppe grasslands	Breeder	Observed, breeder
Species Of Level II Concern–Continued Monitoring Recommended			
Lark Bunting *Calamospiza melanocorys*	Short-grass prairie, shrub steppe	Breeder	Observed, presumed breeder
Chestnut-collared Longspur *Calcarius ornatus*	Short-grass prairie	Potential breeder	Observed, likely breeder
Vesper Sparrow *Pooecetes gramineus*	Shrub-steppe	Breeder	Observed, presumed breeder

Reference: EMC, 2007a
*Wyoming lat/long encompassing Moore Ranch Project

conservation action (Level I); and the remaining 55 species are classified as Level II concern, for which continued careful monitoring is recommended.

Surveys for avian species of concern, including mountain plovers (*Charadrius montanus*), sage-grouse, and bald eagle, were conducted annually from 2003 through 2007 at the proposed Moore Ranch Project. Most surveys have occurred in the spring and summer to document migrating and breeding birds; winter surveys were conducted for bald eagle roost sites. The study area for previous surveys included most of the proposed project area and a 0.8-km [0.5-mi] perimeter {1.6-km [1-mi] for bald eagles}. The applicant surveyed the entire proposed Moore Ranch Project from fall 2006 through early summer 2007.

Table 3-6 lists the five avian species of concern that were observed in the proposed Moore Ranch Project area during the applicant 2006–2007 baseline studies, including their primary nesting habitats and historical occurrence in the general area. For these five species, BLM Wyoming has enacted the Sensitive Species Policy and List to focus species management efforts within BLM lands and ensure that actions authorized, funded, or carried out by BLM do not contribute to the need for any species to become listed under the ESA.

Affected Environment

Table 3-6. Federally- and State-Listed Species

Scientific Name	Common Name	Federal Status*	State Status†	County of Occurrence‡
Amphibians				
Ambystoma tigrinum	tiger salamander	–	SGCN	CAM; JOH
Bufo cognatus	Great Plains toad	–	SGCN	CAM
Rana pipiens	northern leopard frog	–	SGCN; BLM-SS	CAM; JOH
Rana pretiosa	spotted frog	–	BLM-SS	CAM; JOH
Rana sylvatica	wood frog	–	SGCN	JOH
Birds				
Accipiter gentilis	northern goshawk	–	SGCN; BLM-SS	JOH
Aegolius funereus	boreal owl	–	SGCN	JOH
Ammodramus bairdii	Baird's sparrow	–	BLM-SS	CAM; JOH
Ammondramus savannarum	grasshopper sparrow	–	SGCN	CAM; JOH
Amphispiza belli	sage sparrow	–	SGCN; BLM-SS	CAM; JOH
Asio flammeus	short-eared owl	–	SGCN	CAM; JOH
Athene cunicularia	burrowing owl	–	SGCN; BLM-SS	CAM; JOH
Buteo regalis	ferruginous hawk	–	SGCN; BLM-SS	CAM; JOH
Calcarius mccownii	McCown's longspur	–	SGCN	CAM; JOH
Calcarius ornatus	chestnut-collared longspur	–	SGCN	CAM
Centrocercus urophasianus	greater sage-grouse	Candidate	SGCN; BLM-SS	CAM; JOH
Charadrius montanus	mountain plover	–	SGCN	CAM; JOH
Coccyzus americanus	yellow-billed cuckoo	–	SGCN; BLM-SS	JOH
Cygnus buccinator	trumpeter swan	–	BLM-SS	CAM; JOH
Dolichonyx oryzivorus	boblink	–	SGCN	CAM
Egretta thalus	snowy egret	–	SGCN	JOH
Falco peregrinus anatum	American peregrine falcon	DL	SGCN; BLM-SS	CAM; JOH
Gaviea immer	common loon	–	SGCN	JOH
Haliaeetus leucocephalus	bald eagle	DL	SGCN	CAM; JOH
Lanius ludovicianus	loggerhead shrike	–	BML-SS	CAM; JOH
Numenius americanus	long-billed curlew	–	SGCN; BLM-SS	CAM
Nycticorax nycticorax	black-crowned night-heron	–	SGCN	CAM; JOH
Oreoscoptes montanus	sage thrasher	–	BLM-SS; SGCN	CAM; JOH
Plegadis chihi	white-faced ibis	–	BLM-SS	CAM; JOH
Rallus limicola	Virginia rail	–	SGCN	JOH
Sitta pygmaea	pygmy nuthatch	–	SGCN	CAM; JOH
Spizella breweri	Brewer's sparrow	–	BLM-SS; SGCN	CAM; JOH
Fish				
Hiodon alosoides	goldeye	–	SGCN	JOH
Hybognathus argyritis	western silvery minnow	–	SGCN	CAM; JOH
Macrhybopsis gelida	sturgeon chub	–	SGCN	CAM; JOH

Table 3-6. Federally- and State-Listed Species				
Scientific Name	Common Name	Federal Status*	State Status†	County of Occurrence‡
Oncorhynchus clarki bouvieri	Yellowstone cutthroat trout	–	BLM-SS	CAM; JOH
Scaphirhynchus platorynchus	shovelnose sturgeon	–	SGCN	CAM; JOH
Stizostedion canadense	sauger	–	SGCN	CAM; JOH
Mammals				
Corynorhinus townsendii	Townsend's big-eared bat	–	BLM-SS; SGCN	CAM; JOH
Cynomys leucurus	white-tailed prairie dog	–	SGCN	JOH
Cynomys ludovicianus	black-tailed prairie dog	–	SGCN	JOH
Euderma maculatum	spotted bat	–	BLM-SS	CAM; JOH
Lasionycteris noctivagans	silver-haired bat	–	SGCN	CAM; JOH
Lasiurus cinereus	hoary bat	–	SGCN	CAM; JOH
Lontra canadensis	river otter	–	SGCN	JOH
Martes pennanti	fisher	–	SGCN	JOH
Microtus richardsoni	water vole	–	SGCN	JOH
Mustela nigripes	black-footed ferret	E	SGCN	CAM; JOH
Mustela nivalis	least weasel	–	SGCN	JOH
Myotis ciliolabrum	western small-footed myotis	–	SGCN	JOH
Myotis evotis	long-eared myotis	–	BLM-SS; SGCN	CAM; JOH
Myotis thysanodes	fringed myotis	–	BLM-SS; SGCN	JOH
Myotis volans	long-legged myotis	–	SGCN	JOH
Perognathus fasciatus	olive-backed pocket mouse	–	SGCN	CAM; JOH
Sorex haydeni	Hayden's shrew	–	SGCN	JOH
Sorex nanus	dwarf shrew	–	SGCN	CAM; JOH
Vulpes velox	swift fox	–	BLM-SS; SGCN	CAM; JOH
Reptiles				
Coluber constrictor flaviventris	eastern yellowbelly racer	–	SGCN	CAM; JOH
Plants				
Anemone narcissiflora ssp. zephyra	zephyr windflower		PSC	JOH
Arnica lonchophylla	northern arnica	–	PSC	JOH
Cymopterus williamsii	Williams' waferparsnip	–	BLM-SS; PSC	JOH
Cypripedium montanum	mountain lady-slipper	–	PSC	JOH
Draba fladnizensis var. pattersonii	white artiv whitlow grass	–	PSC	JOH
Festuca hallii	Hall's fescue	–	PSC	JOH
Juncus triglumis var. triglumis	three-flower rush	–	PSC	JOH
Papaver kluanense	alpine poppy	–	PSC	JOH
Parnassia kotzebuei	Kotzebuei's grass-of-parnassus	–		JOH
Pedicularis contorta var. ctenophore	coil-breaked lousewort	–	PSC	JOH
Penstemon haydenii	blowout penstemon	E	–	CAM; JOH
Physaria lanata	woolly twinpod	–	PSC	CAM; JOH

Affected Environment

Table 3-6. Federally- and State-Listed Species				
Scientific Name	Common Name	Federal Status*	State Status†	County of Occurrence‡
Polygala verticillata	whorled milkwort	–	PSC	CAM
Polygonum spergulariiforme	fall knotweed	–	PSC	JOH
Potamogeton amplifolius	large-leaved pondweed	–	PSC	JOH
Psilocarphus brevissimus	dward woolly-heads	–	PSC	CAM
Puccinellia cusickii	Cusick's alkali-grass	–	PSC	JOH
Pyrrocoma clementis var. *villosa*	hairy tranquil goldenweed	–	HCP	JOH
Rubus acaulis	northern blackberry	–	PSC	JOH
Schoenoplectus heterochaetus	slender bulrush	–	PSC	CAM
Sesuvium verrucosum	sea purslane	–	PSC	CAM
Spiranthes diluvialis	ute ladies'-tresses	T	–	CAM; JOH
Sporobolus compositus	longleaf dropseed	–	PSC	CAM
Triodanis leptocarpa	slim-pod Venus' looking-glass	–	PSC	CAM
*DL = delisted; E = endangered; T = threatened; – = not listed. †BLM-SS = BLM Wyoming-designated Sensitive Species; PSC = plant species of concern, as designated by the WYNDD; SGCN = species of greatest conservation need, as designated by the WGFD ‡CAM = Campbell County, Wyoming; JOH = Johnson County, Wyoming Sources: BLM, 2002; FWS, 2008b; U.S. Department of Agriculture, 2009; WGFD, 2005b; Wyoming Natural Diversity Database (WYNDD), 2009a,b				

3.7 Meteorology, Climatology, and Air Quality

3.7.1 Meteorology and Climatology

The majority of Wyoming is dominated by mountain ranges and rangelands of the Rocky Mountains and high altitude prairies. The closest mountain ranges to the proposed Moore Ranch Project are the Bighorn Mountains, the Black Hills, and the northern Laramie Range located approximately 80, 137, and 80 km [50, 85, and 50 mi] from the proposed Moore Ranch Project. Because of these distances, the site does not experience significant wind channeling or shielding from any of these three mountain ranges (Uranium One, 2008). The average elevation over the eastern and southern prairie region, also known as the High Plains region, is over 1,828 m [6,000 ft] above mean sea level (AMSL). The Rocky Mountains are perpendicular to the prevailing westerly winds and provide an effective barrier to the significant Pacific-generated weather systems. Much of the moisture produced from these systems is dropped along the western slopes of the Rocky Mountains, thereby leaving the eastern portion of the state in a semiarid condition.

The proposed Moore Ranch Project is located within the Powder River Basin in northeastern Wyoming. The physical setting of the Powder River Basin is characterized by semiarid plains with low hills and buttes, little vegetation, and few substantial topographical features. The basin stretches approximately 190 km [120 mi] east to west and 320 km [200 mi] north to south in southeast Montana and northeast Wyoming. The region has extensive natural resources, such as coal, CBM, and uranium. It is both a topographic drainage and geologic structural basin. The proposed Moore Ranch Project area is located at 43°34'12.83" N latitude, 105°50'49.72" W longitude in the south-central portion of the Powder River Basin. The elevation of the proposed project area is approximately 1,670 m [5,500 ft] AMSL. This region of Wyoming experiences

Affected Environment

diverse weather patterns that fluctuate throughout the year, largely because of its proximity to the Rocky Mountain system and its relatively high elevation. The area is characterized by long winters, generally from December to April, which can bring frequent snow storms. Summer can be hot in the Powder River Basin due to the lack of moisture; however, the summer season tends to be short, with occasional hail, thunder, or snow storms. The Powder River Basin is

treated as a single air quality control area by State and Federal regulators because of the uniformities in geography and climate.

Because of the extensive surface coal mining that has developed over the last 30 years, the Powder River Basin airshed is heavily monitored. Coal production in the Powder River Basin grew from a few million tons in 1973 to over 400 million tons in 2006. A parallel growth in ambient air quality monitoring throughout the Powder River Basin accompanied the growth in coal production through the enactment of the Clean Air Act and the Surface Mining Control and Reclamation Act of the 1970s. There are more than 100 particulate monitoring samplers and more than 20 meteorological monitoring towers in the Powder River Basin to support air quality permitting, compliance, and research objectives (Uranium One, 2008).

Since no onsite meteorological data are available for the proposed Moore Ranch Project, data from the Antelope Mine, located approximately 40 km [25 mi] southeast of the proposed Moore Ranch Project were used to describe the expected meteorological conditions in the proposed project area. The Antelope mine location has similar topographic features as the Moore Ranch Project area, characterized by mildly rolling hills covered with grass and sparse shrubs. No mountain ranges channel or shield the wind between these two locations nor are there bodies of water that would alter the general meteorological conditions at either the proposed Moore Ranch Project or the Antelope mine (Uranium One, 2008). As a result of the NRC's safety review, Uranium One will be required by a license condition to collect onsite meteorological data.

3.7.1.1 Temperature

Temperatures fluctuate greatly throughout the year in the Powder River Basin. Located in a semiarid climate, summer temperatures at the proposed project site can be quite warm, while winters are commonly quite cold. The annual average temperature in the project area region is 7 °C [46 °F]. The average maximum daily temperature is 32 °C [90 °F], with July yielding the warmest average temperatures. The average minimum daily temperature is −12 °C [10 °F], with January being the coldest month on average. Large, diurnal temperature variations occur in the region due to its high altitude and low humidity. Spring and summer daily variations of 8 to 14 °C [15 to 25 °F] are common, with maximum temperature variations of 17 to 21 °C [30 to 40 °F] observed during extremely dry periods. Less daily variation is observed during the cooler portions of the year; fall and winter have fluctuations of 5 to 7 °C [10 to 15 °F] (EMC, 2007a).

3.7.1.2 Wind

Wyoming is quite windy, and frequently winter winds reach 48 to 64 kph [30 to 40 mph], with gusts up to 80 to 97 kph [50 to 60 mph]. Prevailing wind directions vary from west-southwest through west to northwest. In many localities, winds are so strong and constant that trees (when present) show a definite lean towards the east or southeast. Average wind speeds within the project area vary from 24 to 27 kph [15 to 17 mph] from the west-northwest throughout the year.

Affected Environment

Wind data for the proposed project area were obtained from Glenrock Coal Company, approximately 70 km [45 mi] south of the project area; and Antelope Coal Company, approximately 60 km [35 mi] east of the project area. The average annual wind speed is approximately 20.6 kph [14.8 mph] at Glenrock Coal Company and approximately 17.9 kph [11.1 mph] at Antelope Coal Company. Maximum hourly averages of greater than 80 kph (50 mph) have been recorded at both mine sites. Seasonal wind roses for the Antelope Coal Company site are shown in Figure 3-12. As noted in Section 3.7.1, data from the Antelope Coal Company are considered to be most representative of the proposed Moore Ranch Project.

3.7.1.3 Precipitation

The proposed project area receives relatively little rainfall. The mean annual precipitation within the area is approximately 35 cm [13.7 in]. May has been the wettest, and January has been the precipitation gauges capture only a small proportion of snowfall under windy conditions. Severe storms generated from severe weather conditions that could bring wind, rain, snow, or hail from any given direction are rare because the surrounding mountains effectively block or weaken storms (EMC, 2007a).

Table 3-7 summarizes average temperature, precipitation, and snow fall trends taken from a National Climate Data Center weather station located in the town of Midwest, approximately 32 km [20 mi] southwest of the proposed project area. Table 3-7 reflects the large temperature fluctuations between seasons, as well as the relatively small amount of precipitation that occurs at the proposed project area.

3.7.1.4 Evaporation

As discussed in Section 3.3.6.1 of the GEIS, the annual evaporation rates in the Wyoming East Uranium Milling Region range from about 102 to 127 cm [40 to 50 in] (NWS, 1982 in NRC, 2009a). The low humidity, sunshine, and high winds contribute to a high rate of evaporation. driest month on average. The actual annual moisture may be somewhat higher because

3.7.1.5 Climate Change and Greenhouse Gases

On a larger scale, climate change is a subject of national and international interest. The recent compilation of the state of knowledge in this area by the U.S. Global Change Research Program (GCRP), a Federal Advisory Committee (GCRP, 2009), was considered in preparation of this SEIS. Average U.S. temperatures have risen more than 1.1 °C [2 °F] over the past 50 years and are projected to rise more in the future. During the period from 1993–2008, the average temperature in the Great Plains increased by approximately 0.83 °C [1.5 °F] from the 1961 to 1979 baseline (GCRP, 2009). The projected change in temperature over the period from 2000 to 2020, which encompasses the period the proposed Moore Ranch Project would be licensed, ranges from a decrease of approximately 0.28 °C [0.5 °F] to an increase of approximately 1.1 °C [2 °F]. Although the GCRP did not incrementally forecast a change in precipitation by decade, they did project a change in spring precipitation from the baseline period (1961 to 1979) to the next century (2080 to 2099). For the region in Wyoming where the proposed Moore Ranch Project is located, the GCRP forecasted a 10 to 15 percent increase in spring precipitation (GCRP, 2009).

The EPA determined that potential changes in climate caused by greenhouse gas (GHG) emissions endanger public health and welfare based on a body of scientific evidence assessed by the U.S. Global Climate Research Program, the Intergovernmental Panel on Climate Change, and the National Research Council (74 FR 66496). The Administrator issued an endangerment finding based on the Technical Support Document compiled by the above-referenced scientific organizations which indicates that, while ambient concentrations of GHG emissions do not cause direct adverse health effects (such as respiratory or toxic effects), public

Table 3-7. Climate Data for Midwest, Wyoming, Climate Station		
Temperature (°C/°F)	Mean-Annual	7.5/45.5
	Low-Monthly Mean	−5.7/21.7
	High-Monthly Mean	21.5/70.7
Precipitation (cm/in)	Mean-Annual	35.0/13.7
	Low-Monthly Mean	1.4/0.5
	High-Monthly Mean	6.5/2.5
Snowfall (cm/in)	Mean-Annual	135/53
	Low-Monthly Mean	0/0
	High-Monthly Mean	22.6/8.8
Source: National Climatic Data Center (NCDC), 2009. "Climatography of the United States No. 20: Monthly Station Climate Summaries, 1971–200." Asheville, North Carolina: National Oceanic and Atmospheric Administration (2004 in NRC).		

health risks and impacts can result indirectly from changes in climate. Based on EPA's determination, NRC recognizes that GHGs may have an effect on climate change. In CLI-09-21, the Commission provided guidance to the NRC staff to consider carbon dioxide and other GHG emissions in its National Environmental Policy Act (NEPA) reviews. GHG emissions were considered as an element of the existing air quality assessment. Relevant GHG emissions are discussed in both Chapters 4 and 5 of this SEIS.

Affected Environment

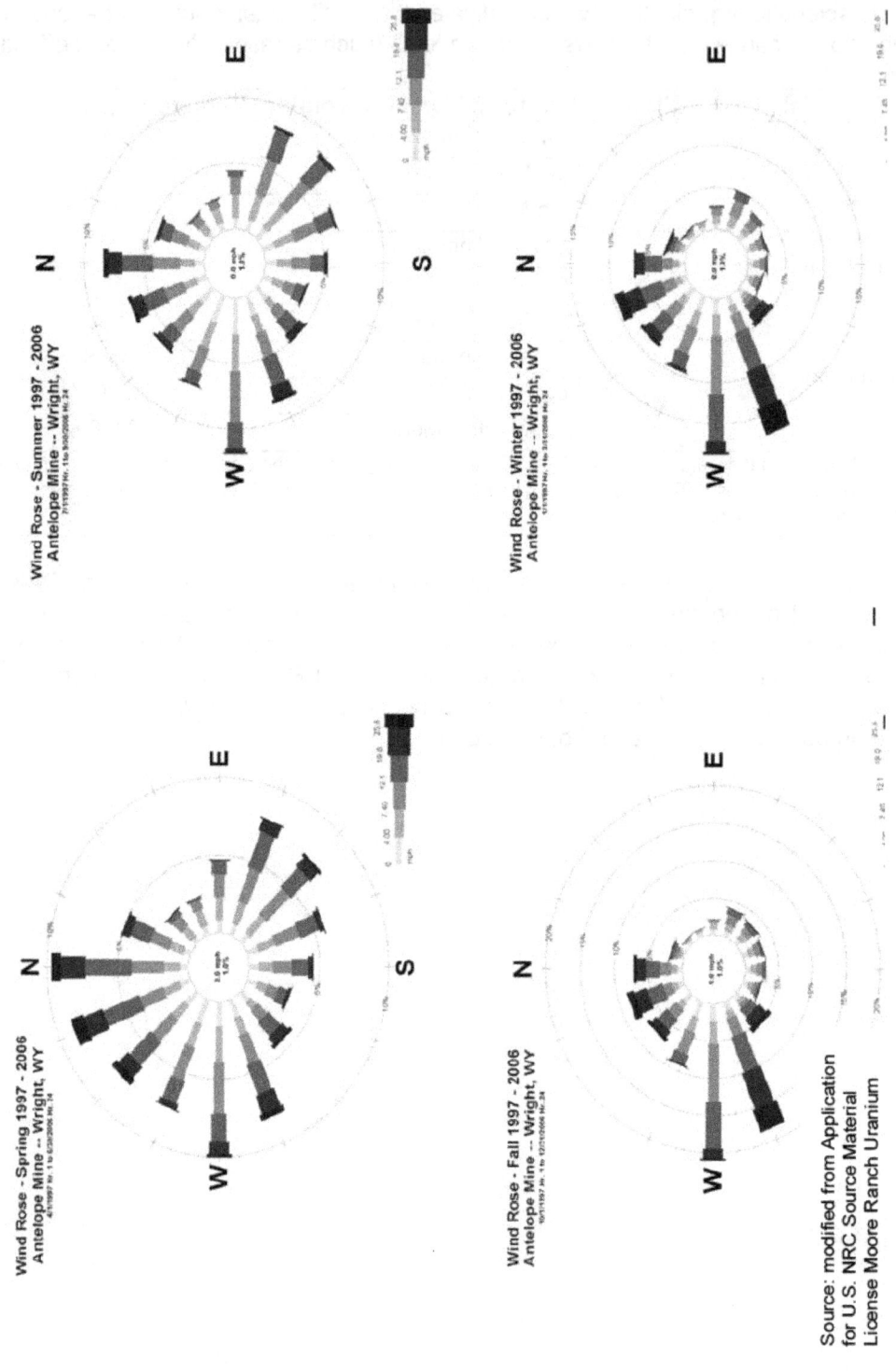

Figure 3-12. Seasonal Wind Roses for the Antelope Coal Company

Affected Environment

3.7.2 Air Quality

The proposed Moore Ranch Project is located in an attainment area for all the primary pollutants. The terrain within the area, combined with windy conditions, provides good conditions for dispersion of air pollutants. The closest resident is approximately 4.5 km [2.8 mi] from the center of the proposed Moore Ranch Project.

The WDEQ adopted the EPA National Ambient Air Quality Standards (NAAQS), as summarized in Table 3.2-8 of the GEIS (NRC, 2009a). The proposed Moore Ranch Project is located in Campbell County, Wyoming, which is an attainment area for all the primary pollutants. The dominant emissions from activities at the proposed Moore Ranch Project would be carbon monoxide (CO) and particulate matter (PM). CO is an odorless and colorless pollutant. In general, CO at ISR projects would be generated primarily by engine combustions (including all vehicles as well as stationary motors such as generators). Over 90 percent of CO comes from motor vehicles.

As discussed in Section 3.3.6.2 of the GEIS, the EPA has established air quality standards to promote and sustain healthy living conditions. These standards, known as the NAAQS, address CO, lead (Pb), nitrogen dioxide (NO_2), particulate matter (PM_{10} and $PM_{2.5}$), ozone (O_3), and sulfur dioxide (SO_2). EPA revised the NAAQS standards after the preparation of the GEIS. This includes a new rolling 3-month average standard for lead at 0.15 µg/m^3 and a new one hour nitrogen dioxide standard at 100 parts per billion. EPA revisions to SO_2 and O_3 standards are under consideration but are not finalized (EPA, 2010). States may develop standards that are stricter or that supplement the NAAQS. Wyoming has a more restrictive standard for SO_2 (annual at 60 µg/m^3 and 24 hours at 260 µg/m^3) and supplemental standards for particulate matter (annual PM_{10} at 50 µg/m^3 and 24-hour $PM_{2.5}$ at 65 µg/m^3) (WDEQ, 2008).

Particulate matter refers to particles found in the air. Some particles are large enough to be seen as dust, soot, or smoke, while others are too small to be visible. As noted previously, the NAAQS for PM_{10} and $PM_{2.5}$, limit the allowable concentration of particulate matter particles smaller than 10 and 2.5 micrometers. Emissions from highway and nonroad construction vehicles compose approximately 28 percent of total PM_{10} and $PM_{2.5}$ emissions. The large sources of PM include fugitive dust from paved and unpaved roads, agricultural and forestry activities, wind erosion, wildfires, and managed burning.

The WDEQ Air Quality Division analyzes measurements from 26 stations located throughout Wyoming to ensure ambient air quality is maintained, in accordance with NAAQS. Annually, the results are synthesized into the Wyoming Ambient Air Monitoring Annual Network Plan (WDEQ, 2009). The baseline air quality conditions of the proposed Moore Ranch Project were determined by evaluating data from four monitoring stations in the region, to provide a reasonable representation of the air pollutant levels that could be expected to occur at the site. Monitoring data were reviewed for the Wamsutter, Casper, Lander, and Murphy Ridge monitoring locations. Furthermore, the GEIS reported that all areas within the Wyoming East Uranium Milling Region were classified as being in attainment for NAAQS (NRC, 2009a).

WDEQ monitors air quality and annually reports the results to EPA. The 2007 monitoring results are consistent with the areas attainment status (WDEQ, 2009). Construction activities at two locations resulted in a couple of anomalous PM_{10} readings; however, these were attributable to localized, temporary construction activities, and, therefore, not representative. Table 3-8 presents the air quality monitoring data for all of the monitoring stations.

Affected Environment

As discussed in Section 3.3.6.2 of the GEIS, Prevention of Significant Deterioration (PSD) requirements identify maximum allowable increases in concentrations for particulate matter, SO_2, and NO_2 for areas designated as attainment. There are several different classes of PSD areas, with Class I areas having the most stringent requirements. No Class I areas are present in the Wyoming East Uranium Milling Region (NRC, 2009b). GEIS Table 3.4-9 identifies the Class I areas in Wyoming, South Dakota, Montana, and Nebraska. GEIS Figures 3.2-16 and 3.4-20 maps the locations of Class I areas. Wind Cave National Park, the closest Class I area to the proposed action, is located about 188 km [117 mi] to the east of the Moore Ranch site. Cloud Peak Wilderness Area, the closest Class II area to the proposed action, is located about 124 km [77 mi] to the northwest of the Moore Ranch site.

3.8 Noise

The proposed Moore Ranch Project area is located in rural Campbell County, Wyoming. The known land uses within both the proposed Moore Ranch Project and within a 3.2 km [2 mi] radius of the proposed project boundary are grazing, wildlife habitat, and CBM recovery operations, none of which generate significant noise. Traffic along the road leading to the site would generate some noise. Sound levels from CBM operations would be expected to be

Table 3-8. Existing Conditions—2007 Ambient Air Quality Monitoring Data					
Monitoring Stations	**Wamsutter**	**Casper**	**Lander**	**Murphy Ridge**	
Distance to Site	290 km (180 mi)	97 km (60 mi)	258 km (160 mi)	484 km (300 mi)	
Pollutant					**Standards (Averaging Time)**
Carbon Monoxide	N/A	N/A	N/A	0.7 ppm	9 ppm (8-hour)
	N/A	N/A	N/A	0.9 ppm	35 ppm (1-hour)
Lead	1.5 µg/m^3	N/A	N/A	N/A	1.5 µg/m^3 (Quarterly Average)
Nitrogen Dioxide	0.007 µg/m^3	N/A	N/A	0.003 µg/m^3	100 µg/m^3 (Annual Arithmetic Mean)
Particulate Matter (PM$_{10}$)	227.0 µg/m^3 (Note: 2006 was 73.0 µg/m^3)	30 µg/m^3	40 µg/m^3	64 µg/m^3	150 µg/m^3 (24-hour)
Particulate Matter (PM$_{2.5}$)	N/A	N/A	26.0 µg/m^3	N/A	15 µg/m^3 (Annual) (Arithmetic Mean)
	N/A	N/A	7.6 µg/m^3	N/A	35 µg/m^3 (24-hour)
Ozone	0.064 ppm	N/A	N/A	0.068 ppm	0.08 ppm (8-hour)
Sulfur Dioxide	0.001 ppm	N/A	N/A	0.001 ppm	0.03 ppm (Annual)
	0.010 ppm	N/A	N/A	0.002 ppm	0.14 ppm (24-hour)
Source: WDEQ, 2009					

Affected Environment

unnoticeable from distances of 490 m [1,600 ft] and beyond (BLM, 2003). The closest residence to the proposed Moore Ranch Project is located approximately 4.5 km [2.8 mi] from the center of the proposed license area. No people reside in the proposed license area (EMC, 2007a).

The Federal Highway Administration (FHWA) and the Wyoming Department of Transportation (WYDOT) have noise impact assessment procedures and criteria to help protect the public health and welfare from excessive vehicular traffic noise. FHWA established Noise Abatement

Criteria described in Table 3-9 according to land use, recognizing that different areas are sensitive to noise in different ways. A person is considered to be impacted by noise according to WYDOT procedures when existing or expected future sound levels approach [within 1 decibels (dBA)], are or exceed the Noise Abatement Criteria, or when expected future sound levels exceed existing sound levels by a substantial amount (15 dBA). These criteria were used to assess impacts at the proposed Moore Ranch Project. Cattle grazing, the primary land use within the proposed project area generates minor noise. However, SR 387, which crosses through the northern portion of the proposed project area and Brown Road, which accesses the site, are line sources of noise. Vehicular traffic sound a distance of 15 m [50 ft] from the receptor has been estimated at 54 to 62 dBA for passenger cars and 58 to 70 dBA for heavy trucks (NRC, 2009a). Because noise from line sources such as roads is reduced by approximately 3 dBA per doubling of distance (NRC, 2009a), the maximum truck sound level of 70 dBA on the shoulder of either SR 387 or Brown Road would diminish to the level of a Category "A" Activity, shown in Table 3-9, approximately 480 m [1,575 ft] from the source, excluding the noise dampening characteristics of topographic interference and vegetation.

Table 3-9. Noise Abatement Criteria: 1-Hour, A-Weighted Sound Levels in Decibels (dBA)		
Activity Category	$L_{eq}(h)$*	Description of Activity Category
A	57 (Exterior)	Lands on which serenity and quiet are of extraordinary significance and serve an important public need and where the preservation of those qualities is essential if the area is to continue to serve its intended purposes.
B	67 (Exterior)	Picnic areas, recreation areas, playgrounds, active sports areas, parks, residences, motels, hotels, schools, churches, libraries, and hospitals.
C	72 (Exterior)	Developed lands, properties, or activities not included in Categories A or B above.
D	--	Undeveloped lands
E	52 (Interior)	Residences, motels, hotels, public meeting rooms, schools, churches, libraries, hospitals, and auditoriums.
*$L_{eq}(h)$ is an energy-averaged, one-hour, A-weighted sound level in decibels (dBA). Source: 23 CFR Part 772		

It was assumed that sound levels beyond a distance of 480 m [1,575 ft] from SR 387 and Brown Road would approximate 40 dBA, to conservatively overestimate a baseline that is consistent

Affected Environment

with the GEIS statement that existing ambient noise levels in this region would be 22 to 38 dBA (NRC, 2009a). Figure 3.2-17 of the GEIS provides examples of sound levels for common activities (NRC, 2009a).

With regard to wildlife located on the site, field observations suggest that noise from oil and gas and CBM operations could affect greater sage-grouse lek activity (Braun, 1998; Wisdom, et al., 2002). The construction and operation of ISR facilities would involve similar activities. However, sage-grouse leks have not been discovered on or near the proposed project area, based on a 2007 survey (EMC, 2007a).

3.9 Historical and Cultural Resources

Section 3.3.8 of the GEIS provides a general overview of historical and cultural resources for the Wyoming East Uranium Milling Region in which the proposed Moore Ranch Project is located (NRC, 2009a). This section discusses the cultural background and historic and cultural resources identified at the proposed Moore Ranch Project and in the surrounding area. No structures or buildings were evaluated for the proposed Moore Ranch Project site since potential buildings and structures were previously investigated. Brunette (2007) noted the occurrence of "active and abandoned ranch headquarters/ranching related buildings [and] earthen dikes/stock ponds" in and around the proposed Moore Ranch Project. None of these sites are located in the immediate vicinity of the central plant and proposed wellfields. Site 48CA146 includes features and remains associated with an abandoned ranch located about 0.8 km [0.5 mi] south of the proposed wellfield areas. Site 48CA3400, the remains of an historic homestead, is located approximately 1.5 mi southwest of the Moore Ranch Project area (Brunette, 2007). Site 48CA6173 also contains the remains of an historic homestead. Site 48CA6173 is located southwest of the proposed project area (Brunette, 2007).

The National Historic Preservation Act (NHPA) requires federal agencies to consider the effects of their undertakings on historic properties. Historic properties are defined as resources that are eligible for listing on the *National Register of Historic Places* (NRHP). The criteria for eligibility are listed in Title 36, "Parks, Forests, and Public Property," Part 60, Section 4, "Criteria for Evaluation," of the *Code of Federal Regulations* (36 CFR Part 60.4) and include (1) association with significant events in history; (2) association with the lives of persons significant in the past; (3) embodies distinctive characteristics of type, period, or construction, and (4) or sites or places that have yielded or are likely to yield important information (ACHP, 2010). The historic preservation review process (Section 106 of the NHPA) is outlined in regulations issued by the Advisory Council on Historic Preservation in Title 36, Parks, Forests, and Public Property, Part 800, Protection of Historic Properties (36 CFR Part 800). The NRC has coordinated its Section 106 review for the proposed Moore Ranch Project through NEPA per 36 CFR 800.8(c).

The issuance of a materials license is a federal action that could possibly affect either known or undiscovered historic properties located on or near the proposed Moore Ranch Project. In accordance with the provisions of the NHPA, the NRC is required to make a reasonable effort to identify historic properties in the area of potential effect (APE). The APE for this review is area that may be impacted by construction, operation, aquifer restoration, and decommissioning activities associated with the proposed action. If no historic properties are present or affected, the NRC is required to notify the State Historic Preservation Office before proceeding. If it is determined that historic properties are present, the NRC is required to assess and resolve possible adverse effects of the undertaking.

Affected Environment

Cultural resources identification and assessment also considers the ARPA [16 United States Code (USC) 469-469c-e] as amended, which covers permitting of archaeological investigations on public land such as that managed by the BLM. Finally, State of Wyoming laws dealing with protection of archaeological resources also are considered. These various laws and regulations were discussed in Appendix B of the GEIS.

The NRC initiated consultation with the Wyoming State Historic Preservation Office (SHPO), under Section 106 of the NHPA (NRC, 2008). A response to an NRC letter was received from the Wyoming SHPO on June 5, 2008. By letter dated November 3, 2009, the Wyoming SHPO concurred with the determination that the sites located in the proposed project area were ineligible for listing on the NRHP (Wyoming SHPO, 2010).

3.9.1 Cultural History

The archaeological cultural sequence for the proposed Moore Ranch Project is divided between the prehistoric periods (Paleoindian, Archaic, and Late Prehistoric) and the recent protohistoric/historic era. The former encompasses about 11,000 years between 12,000 B.P. (before present) and 250 B.P. (about A.D. 1700). The protohistoric/historic era ranges from A.D. 1700 to A.D. 1959.

3.9.1.1 Prehistoric Era

As mentioned previously, the prehistoric periods are divided into Paleoindian, Archaic, and Late Prehistoric. The hallmark artifact forms for the Paleoindian period (12000 to 8500 B.P.) in the region include, from oldest to youngest, Clovis, Folsom/Goshen, Agate Basin, Hell Gap, Eden, Scottsbluff, and Cody. Paleoindian sites in the region, yielding both Pleistocene megafauna and Paleoindian artifacts, include the James Allen site in southwestern Wyoming; Hell Gap and Agate Basin in eastern Wyoming, located east and southeast of the project; and Medicine Lodge Creek in central Wyoming. The Paleoindian period comes to a close in the terminal Pleistocene/early Holocene era. The Pleistocene megafauna (e.g., mammoth, muskox) are replaced by modern antelope, bison, deer, and elk. These smaller grazers were better adapted to the change from savannah to grassland communities that resulted from the onset of warmer and drier conditions in the Holocene. The Archaic period (8500 to 1500 B.P.) in eastern and northeastern Wyoming is broken into three subperiods: Early (8500 to 5000 B.P.), Middle (5000 to 3000 B.P.), and Late (3,000 to 1,500 B.P).

In general, the regional Early Archaic sites are marked by the presence of various side- and corner-notched projectile points and side-notched knives. The subperiod is known for semisubterranean houses that are usually marked by the presence of one or more hearths, firepits, storage pits, and milling basins. The latter is of particular interest because features clearly indicate that floral species were playing an important role in subsistence strategies. The Middle Archaic site assemblages reflect a relatively broad spectrum of gathering and hunting responses, with an emphasis on bison procurement. By the Late Archaic times, communal bison kills occur and recorded examples contain diagnostic Yonkee points (large, corner-notched projectile points), which are the preferred method of felling the bison through the subperiod. Late Archaic faunal assemblages demonstrate the presence of smaller game animals and midsize ungulates (deer and antelope).

The Late Prehistoric period (1500 to 300 B.P.) heralds the acceptance of new technologies, such as smaller projectile points adapted to use with arrows. Prior to the Late Prehistoric period, the points were hafted on spears. Also introduced at this time is earthenware technology, which improves food preparation techniques. Stewing, braising, and boiling were now possible, which significantly broadened the number of floral and faunal species that could be used. At some time between 1000 and 600 B.P., there is considerable movement of people into Wyoming from several directions. Kiowa-Apache and Shoshone-Comanche move into the region first, probably in response to several factors including population pressures from eastern sedentary groups who had partially adapted to horticultural regimes. Between about 600 B.P. (A.D. 1300) and A.D. 1700, the Crow, Cheyenne, and Arapaho all move into Wyoming to pursue their bison-oriented lifestyles.

3.9.1.2 Protohistoric/Historic Era

The Protohistoric period dates between about A.D. 1700 and 1840. This period includes the time when European goods and the domesticated horse are introduced into the region. There is no appreciable European presence in the region, with the exception of French fur traders moving up and down the Missouri River. Across the northern High Plains, there was active trading in European material goods, including metal knives, pots, and glass beads. However, Native American goods in similar styles also continued to be produced. The Native American tribes continued to pursue Native traditions into the 1900s in the region, although the majority of the tribal members were relocated to the Wind River Reservation.

The Historic era is subdivided into seven periods: Early Historic (A.D. 1801–1842), Preterritorial (A.D. 1843–1867), Territorial (A.D. 1868–1889), Expansion (A.D. 1890–1919), Depression (A.D. 1920–1939), World War II (A.D. 1940–1946), and Post-World War II (A.D. 1947–1959). Various themes have been identified that crosscut the periods. The proposed project area was historically used for cattle ranching, with limited oil and gas exploration in the nearby vicinity. There is no indication from the proposed project sites identified to date that there were earlier historic occupations of the area. Thus, at best, historic occupations are limited to the expansion and post-expansion periods.

3.9.2 Historic and Cultural Resources Identified and Places of Cultural Significance

3.9.2.1 Previous Cultural Resources Investigations

Seven cultural resource investigations have been conducted on the proposed license area dating from 1981 to the present. These investigations have been conducted in support of energy extraction activities, including the data used to support the analysis in the NRC draft EIS for the Sand Rock Mill Project (NRC, 1982). Four historic investigations overlap the areas that could potentially be directly disturbed by the proposed action.

3.9.2.1.1 Archaeology–Identification and Evaluation

The proposed Moore Ranch Project site has been subjected to three Class III surveys; two of the surveys were completed for earlier projects. In 1981, the Office of the Wyoming State Archaeologist (project #WY 56-81) completed a survey in support of the Conoco, Inc. license application to the NRC for the Sand Rock Mill (Brunette, 2007). The second survey was conducted for permitting CBM wells (Brunette, 2007).

Systematic cultural resource investigations for the proposed Moore Ranch Project were conducted in 2007 (Brunette, 2007), and the results of the site file research and the archaeological survey of the project are filed under BLM Cultural Resource Use Permit No. 320-WY-SR05. Brunette (2007) requested or conducted site file searches in August and December 2006 and April 2007 prior to fieldwork. Archaeological surveys were conducted in phases between September 2006 and July 2007 using standard BLM-mandated survey approaches and following the general guidance provided in the State Protocol for the execution of Class III surveys.

Brunette systematically surveyed a total of 492 ha [1,215 ac] using 30-m [100-ft] transects. The Brunette 2006 and 2007 surveys resulted in the relocation of Sites 48CA962 through 48CA965, 48CA967, 48CA970; the identification of seven new sites; and the recording of 25 isolated resources. Per the State Protocol between BLM and the Wyoming SHPO, isolated finds are ineligible to the NRHP, and no further archaeological consideration was recommended (Brunette, 2007). Table 3-10 summarizes newly identified and relocated archaeological sites at the proposed Moore Ranch Project.

Table 3-10. Newly Identified and Relocated Archaeological Sites by Location and Site Characteristics		
48CA962	Multicomponent lithic and historic scatter	Not eligible. Site boundaries for site 48CA6695 have been expanded to include 48CA962.
48CA963	Multicomponent lithic and historic scatter	Not eligible
48CA964	Multicomponent lithic and historic scatter with historic feature	Unevaluated for NRHP eligibility pending further evaluative testing. The site would not be impacted by proposed action.
48CA965	Prehistoric: lithic scatter	Not eligible
48CA967	Prehistoric: lithic scatter	Not eligible
46CA970	Multicomponent lithic and historic debris scatters	Not eligible, site no longer extant
48CA6691	Prehistoric: lithic scatter with hearth	Not eligible.
48CA6692	Prehistoric: lithic scatter with hearth and fire-cracked rock (FCR) concentration	Not eligible.
48CA6693	Multicomponent lithic and historic debris scatters with hearth	Not eligible.
48CA6694	Multicomponent lithic and historic debris scatters with shallow historic depression	Unevaluated for NRHP eligibility pending further evaluative testing. The site is outside areas proposed for development. The site would not be impacted by proposed actions as currently planned.

Affected Environment

Table 3-10. Newly Identified and Relocated Archaeological Sites by Location and Site Characteristics

48CA6695	Prehistoric: lithic and FCR scatter	Not eligible.
48CA6696	Prehistoric: lithic, FCR, and groundstone scatter.	Unevaluated for NRHP eligibility pending further evaluative testing. The site is outside areas proposed for development. The site would not be impacted by proposed actions as currently planned.
48CA6697	Historic: dump	Not eligible.

Source: Brunette 2007, Wyoming SHPO, 2009; 2010

The seven newly identified cultural resources include Sites 48CA6691 through 48CA6697. Of this grouping, Brunette (2007) recommended Sites 48CA6694 and 48CA6696 as eligible to the NRHP. Brunette (2007) also relocated six previously recorded sites from the 1981 survey, including area sites 48CA962 through 48CA965, 48CA967, and 48CA970. Site 48CA970 no longer exists and requires no additional consideration (Wyoming SHPO, 2009). Of the newly identified and relocated sites, three were recommended eligible to the NRHP (48CA964, 48CA6694, and 48CA6696). However, these sites have not been formally investigated to determine their NRHP eligibility. The Wyoming SHPO stated that, until further testing is conducted, the NRHP eligibility status of these sites is unevaluated (Wyoming SHPO, 2009). These sites are discussed in detail in Section 3.9.3.

3.9.2.1.2   Ethnology–Identification and Evaluation

Consultation with the tribes that have heritage interest in the proposed Moore Ranch Project is ongoing. Section 106 tribal consultation letters were sent to the following tribes on December 24, 2008: Blackfeet, Cheyenne River Sioux, Crow, Eastern Shoshone, Ft. Peck Assiniboine/Sioux, Northern Arapaho, Northern Cheyenne, Oglala Sioux, and Three Affiliated Tribes. No response has been received indicating that traditional cultural properties or landscapes of importance occur within the proposed license area.

3.9.3   Historic Properties Listed in the National Register of Historic Places

No cultural resources on the proposed Moore Ranch Project are currently listed in the *National Register of Historic Places* (NRHP). Three sites on the proposed Moore Ranch Project were recommended eligible to the NRHP: 48CA964, 48CA6694, and 48CA6696. Two of the sites are multicomponent; Site 48CA6696 has only prehistoric artifacts. Site 48CA964 was originally recommended eligible to the NRHP by archaeologists from the Office of the Wyoming State Archaeologist (project #WY 56-81). The site is a multicomponent, prehistoric/historic site. The site has two distinct areas separated by an intermittent drainage. The historic artifacts date from the 19^h century and include solder dot cans, an oval iron ring, and a tin stove. The prehistoric component consisted of a corner-notched projectile point and chipped stone flakes. The 1981 flake types are not noted in the Brunette (2007:21) summary of the site; however, the occurrence of two chert, tertiary flakes was noted during the 2007 relocation of the resource.

Site 48CA6694 is also a multicomponent prehistoric/historic site located near the central plant and proposed wellfields. Historic components recovered include architectural remains of a structure and associated 20th century refuse. Prehistoric artifacts recovered include fire-cracked rock (FCR), a groundstone mano, an Eden projectile point, bifaces, unifaces, and flakes. The Eden point is representative of the late Paleoindian Cody Complex. Brunette (2007:12) notes that two of the bifaces may be Paleoindian point fragments as well. No subsurface investigations were conducted at the site, so it is unknown if the FCR is associated with subsurface features. This site was recommended to be eligible to the NRHP because of its diagnostic, prehistoric artifacts and the possibility that data from the site could be used to address research questions concerning settlement patterns, subsistence strategies, seasonal migration rounds, landscape evolution, and climatic reconstruction (Brunette, 2007).

Site 48CA6696 is located to the east of the proposed Moore Ranch Project. The site is prehistoric and consists of a surface scatter of artifacts including groundstone, FCR, chipped stone tools, and tools. The tool assemblage lacks temporal diagnostics but does include both formed and expedient tools. The flakes recovered suggest that late stage reduction and tool maintenance may have occurred at the site. Brunette (2007) recommended the site to be eligible to the NRHP because dates obtained from it could be used to address research questions concerning settlement, subsistence, and landscape use strategies. However, these sites have not been formally investigated to determine their NRHP eligibility. The Wyoming SHPO stated that until further testing is conducted, the NRHP eligibility status of these sites is unevaluated (Wyoming SHPO, 2009).

3.9.4   Tribal Consultation and Places of Cultural Significance

Consultation with Native American tribes was initiated in 2008 (see Section 3.9.2.1.2). No places of cultural significance have been identified by Native American tribes or others in the proposed project area. Consultation is ongoing and will continue throughout this review.

3.10   Visual and Scenic Resources

The proposed Moore Ranch Project is located on private land; therefore, no public agency protects scenic quality. However, it is located in prairie landscape of the Powder River Basin in the vicinity of public lands that are administered by the Buffalo Field Office of the BLM. The BLM evaluates the scenic quality of the land it administers through a Visual Resource Inventory (BLM, 2007b) to ensure that the scenic (visual) value is preserved. As part of this inventory, the BLM completes a scenic quality evaluation, a sensitivity level analysis, and a delineation of distance zones in order to group areas into one of four visual resource management (VRM) classes. Class I is the most protected of visual and scenic resources, and Class IV is the least restrictive.

The portion of the Powder River Basin in which the proposed Moore Ranch Project is located is characterized as basin and range country with prominent buttes and ridges interspersed by rolling grasslands. Semi-permanent streams are fed by intermittent drainages, which seasonally drain the adjacent uplands. Past changes to land surfaces include those associated with human habitation; the development of stock ponds and reservoirs; access roads; and the introduction of gas, oil, and other energy development infrastructure. The proposed license area is comprised of about 2,879 ha [7,110 ac] of privately owned land. The surface area affected by the proposed operation would be about 61 ha [150 ac] and would consist of the central plant, wellfields, and support facilities such as warehouses and chemical storage facilities.

Affected Environment

The BLM has established VRM classifications and has resource management plans for all of the Wyoming East Uranium Milling Region, which includes the proposed Moore Ranch Project (NRC, 2009a). The VRM classifications for the region are shown in Figure 3.3-17 of the GEIS (NRC, 2009a). In the past, the landscape has been extensively modified in urban areas and in several rural areas by oil, natural gas, coal production, and power generation. The bulk of the Wyoming East Uranium Milling Region is categorized as VRM Class III (along highways) and Class IV (open grassland, oil and natural gas, urban areas). The BLM resource management plans for this region do not identify any VRM Class I resources.

The area considered for visual resources associated with the proposed Moore Ranch Project includes the proposed project site, access roads, and a 3.2 km [2 mi] buffer area outside of the proposed license area. Beyond this distance, any changes to the landscape would be in the background distance zone, which would either be unobtrusive or imperceptible to viewers. Areas and associated viewer types considered to be potentially sensitive to visual changes include park, recreation, and wilderness study areas; major travel routes; and residential areas.

No parks, recreation areas, wilderness study areas, or residential areas occur within the proposed Moore Ranch Project. As shown in Figure 2-1, SR 387 traverses the northern section of the proposed project site. In addition to the highway, the proposed project area is currently used for pastureland, rangeland, and for various types of CBM and gas extraction (see Section 3.2, Land Use). These energy extraction facilities have attendant infrastructure systems including pipelines, wellfields, and utility lines that occupy land surface areas in the vicinity of the proposed project area.

The BLM has inventoried the landscape within the proposed Moore Ranch Project and the surrounding 3.2-km [2-mi] area and rated the areas as VRM Class IV. The management objective of VRM Class IV is to provide for management activities that require major modification of the existing character of the landscape. The level of change to the characteristic landscape can be high, and the proposed action is compatible with these objectives.

3.11   Socioeconomics

This section of the SEIS describes current socioeconomic factors that have the potential to be directly or indirectly affected by the construction and operation of a new uranium recovery facility at the proposed Moore Ranch Project site. The proposed Moore Ranch Project is located in the Wyoming East Uranium Milling Region, which is described in Section 3.3.10 of the GEIS (NRC, 2009a). The proposed ISR facility and the people and communities that would support it can be described as a dynamic socioeconomic system. The communities provide the people, goods, and services required to construct and operate the facility. Construction and operations, in turn, create the demand for people, goods, and services and pays for them in the form of wages, salaries, and benefits, and payments for goods and services. Income from wages and salaries and payments for goods and services is then spent on other goods and services within the community, thus creating additional opportunities for employment and income.

The proposed Moore Ranch Project is located in a rural portion of Campbell County between the small towns of Midwest and Wright. The city of Gillette is located approximately 80 km [50 mi] to the northeast of the proposed project site, and it is home to over half of the Campbell County population (approximately 20,000 people). The city of Casper is located approximately

Affected Environment

85 km [53 mi] southwest of Moore Ranch. Casper is located in Natrona County and has a relatively large population of approximately 50,000 people. The city of Douglas, located approximately 96 km [60 mi] southwest of the proposed project area, and the town of Glenrock, located approximately 77 km [48 mi] south of the proposed project area may also provide workers and housing for the proposed ISR construction and operations (NRC 2009b; USCB, 2009). The socioeconomics region of influence (ROI) is defined by the area where employees and their families would reside, spend their income, and use their benefits, thereby affecting the economic conditions of the region.

Most of the construction and operations workers for the proposed ISR facility would likely come from several surrounding communities in Campbell County. Additional workers would also come from communities in Converse, Johnson, and Natrona counties. Given that most employees would reside near the ISR facility, the most significant impacts of plant construction and operations are likely to occur in Campbell County. The focus of the analysis in this SEIS is, therefore, on the impacts of the proposed ISR facility in the ROI, Campbell County.

The following subsections describe the demographics, income, housing, employment structure, local finance, and education and public services in the ROI surrounding the proposed ISR facility at the proposed Moore Ranch site.

3.11.1 Demographics

Campbell County is currently home to an estimated population of approximately 40,000 residents (USCB, 2010). The population of Campbell County is mostly comprised of White nonHispanics, with Hispanic, American Indian, and other races each comprising less than 5 percent of the population. Table 3-11 shows population projections and growth rates from 1980 to 2050 in Campbell County. The population in Campbell County has grown and is projected to continue to grow at a declining rate through 2050.

Table 3-11. Population and Percent Growth in Campbell County, Wyoming, from 1980 to 2050		
Year	Population	Percent Growth*
1980	24,367	—
1990	29,370	20.5
2000	33,698	14.7
2008	41,473	23.1
2010	43,440	28.9
2020	52,130	20.0
2030	59,990	15.1
2040	68,403	14.0
2050	76,678	12.1
— = No data available. *Percent growth rate is calculated over the previous decade. Sources: Population data for 1980–2000 [U.S. Census Bureau (USCB), 2010]; 2008 estimate (USCB, 2010); projected population data for 2010–2030 Sources: USCB, 2010; Wyoming Department of Administration and Information, Economic Analysis Division, 2008 <http://eadiv.state.wy.us] July 2008>; population projections for 2040 and 2050 (calculated)].		

Affected Environment

The 2000 demographic profile of the population in Campbell County is presented in Table 3-12. Persons self-designated as minority individuals comprise about 6.0 percent of the total population in 2000. The minority population is composed largely of Hispanic or Latino residents According to the U.S. Census Bureau (USCB) 2006–2008 American Community Survey 3-Year Estimates, minority populations were estimated to have increased by approximately 1,300 persons and comprised 8.3 percent of the county population (see Table 3-13). Most of

Table 3-12. Demographic Profile of the Population in Campbell County in 2000		
	Campbell County	Percent
Total Population	33,698	—
Race (Not-Hispanic or Latino)		
White	31,701	94.1
Black or African American	47	0.1
American Indian and Alaska Native	280	0.8
Asian	100	0.3
Native Hawaiian and Other Pacific Islander	28	0.1
Some other race	11	0.0
Two or more races	340	1.0
Ethnicity		
Hispanic or Latino	1,191	3.5
Minority Population (including Hispanic or Latino ethnicity)		
Total minority population	1,997	5.9
Source: US Census Bureau 2010, Census 2000 Summary File 1 (SF 1) 100-Percent Data, Table P4, Hispanic or Latino, and Not Hispanic or Latino by Race for Campbell County, Wyoming <http://factfinder.census.gov>; USBC, 2010.		

Table 3-13. Demographic Profile of the Population in Campbell County, 2006–2008 Three-Year Estimate		
	Campbell County	Percent
Total Population	40,121	—
Race (Not-Hispanic or Latino)		
White	36,805	91.7
Black or African American	189	0.5
American Indian and Alaska Native	380	0.9
Asian	204	0.5
Native Hawaiian and Other Pacific Islander	0	0.0
Some other race	82	0.2
Two or more races	481	1.2
Ethnicity		
Hispanic or Latino	1,980	4.9
Minority Population (including Hispanic or Latino ethnicity)		
Total minority population	3,316	8.3
Source: US Census Bureau, 2010; ACS Demographic and Housing Estimates, 2006–2008; American Community Survey 3-Year Estimates; Hispanic or Latino and Race for Campbell County, Wyoming http://factfinder.census.gov USBC, 2010		

this increase was due to an estimated influx of Hispanic or Latinos (approximately 800 persons), an increase in population of over 66 percent from 2000. The next largest increase in minority population was Black or African American, an increase of approximately 140 persons from 2000.

3.11.2 Income

Estimated income information for the ROI is presented in Table 3-14. According to the USCB 2006–2008 American Community Survey 3-year estimates, median household and per capita income in Campbell County were both above the Wyoming average. An estimated 5.1 percent of the population and 4.2 percent of families in Campbell County were living below the official poverty level (USCB, 2010).

According to the USCB 2006–2008 American Community Survey 3-year estimates, the annual unemployment average for Campbell County was 3.1 percent, which was slightly lower than the annual unemployment average of 3.5 percent for Wyoming (USCB, 2010); however, those rates doubled by the first quarter of 2009 to 4.1 percent.

Table 3-14. Income Information for the Region of Influence 2006–2008 American Community Survey 3-Year Estimates

	Campbell County	Wyoming
Median household income (dollars)*	76,666	53,096
Per capita income (dollars)*	31,122	27,873
Percent of families below the poverty level	4.2	5.5
Percent of persons below the poverty level	5.1	8.9
Source: U S Census Bureau (2010), 2006–2008 American Community Survey 3-Year Estimates, Economic Characteristics for Campbell County and Wyoming <http://factfinder.census.gov> USCB, 2010. *In 2008 inflation-adjusted dollars.		

3.11.3 Housing

Table 3-15 lists the total number of occupied housing units, vacancy rates, and median value in Campbell County. According to the 2000 Census, there were over 13,000 housing units in the ROI, of which approximately 12,000 were occupied. The median value of owner-occupied units was $102,900.

By 2008, the total number of housing units in Campbell County had grown by almost 1,700 units to 14,959, while the total number of occupied units also grew by 1,700 units to 13,907. As a result, the number of available vacant housing units decreased slightly by almost 30 units to 1,052, or 7 percent of all housing units (USCB, 2010).

3.11.4 Employment Structure

Between 2000 and 2008, the civilian labor force in Campbell County increased by approximately 31 percent to 24,566 (USCB, 2010). The largest source of employment in Campbell County is the mining industry, which accounts for 27 percent of all jobs but 40 percent of all earnings in the county. Government-related jobs are the second largest employers in Campbell, providing 13 percent of the total job force, and retail trade accounts for 10 percent of the employment.

3.11.5 Local Finance

Campbell County taxes commercial personal property. The County determines assessed valuation of commercial property at 11.5 percent of the market value (Wyoming Department of Revenue, 2001).

Affected Environment

Wyoming has a 5 percent sales tax and allows counties to increase sales tax up to 4 percent above the state rate. Campbell County has an additional 0.25 percent sales and use tax for a total of 5.25 percent (Liu, 2008). The additional tax added by the county comes back to

Table 3-15. Housing in Campbell County, Wyoming	
2000	
Total	13,288
Occupied housing units	12,207
Vacant units	1,081
Vacancy rate (percent)	8.1
Median value (dollars)	102,900
2006–2008; 3-year Estimate	
Total	14,959
Occupied housing units	13,907
Vacant units	1,052
Vacancy rate (percent)	7.0
Median value (dollars)	200,200
Source: US Census Bureau, 2010, 2006–2008 American Community Survey 3-Year Estimates, Housing Characteristics for Campbell County for years 2000 and 2006–2008 <http://factfinder.census.gov> USCB, 2010.	

the county and only receives a portion of the 5 percent state tax. The average property tax rate in Campbell County is 6.25 percent.

3.11.6   Education

The Campbell County School district currently enrolls 7,500 students. Campbell County School District #1, including the Gillette area, had a student-to-teacher ratio of 12.98 to 1 in 2007 (Wyoming Department of Education, 2007). By 2009, the student-to-teacher ratio was 19.2 to 1 (Campbell County School District, 2009).

3.11.7   Health and Social Services

The primary care facility in Campbell County is the Campbell County Memorial Hospital, which is located in Gillette. The hospital also has two branch clinics located in Gillette and the town of Wright (Wyoming Hospital Association, 2009). The closest medical center offering full service emergency services is the Wyoming Medical Center in Casper, located approximately 87 km (54 mi) southwest of the proposed Moore Ranch Project. There are a variety of utility service providers in the area.

3.12   Public and Occupational Health and Safety

The purpose of this section is to summarize the natural background radiation levels in and around the proposed Moore Ranch Project area. Descriptions of these levels are known as "preoperational" or "baseline" radiological conditions, and they would be used for evaluating potential radiological impacts associated with ISR operations. Also included in this chapter of the document are descriptions of applicable safety criteria and radiation dose limits that have been established for public protection and occupational health and safety.

Radiation dose is a measure of the amount of ionizing energy that is deposited in the body. Ionizing radiation is a natural component of the environment and ecosystem, and members of the public are exposed to natural radiation continuously. Radiation doses to the general public occur from radioactive materials found in the earth soils, rocks, and minerals. Radon-222 is a radioactive gas that escapes into ambient air from the decay of uranium (and its progeny, radium-226) found in most soils and rocks. Naturally-occurring low levels of uranium and radium are also found in drinking water and foods. Cosmic radiation from outer space is another natural source of radiation. In addition to natural sources of radiation, there are also artificial or manmade sources that contribute to the dose received by the general public. Medical diagnostic procedures using radioisotopes and x-rays are a primary manmade radiation source. The National Council for Radiation Protection (NCRP) in its Report No. 160, estimates the annual average dose to the public from all natural background radiation sources (terrestrial and cosmic) is 3.1 millisieverts ([mSv] [310 millirem (mrem)]. The annual average dose to the public from all sources (natural and manmade) is 6.2 mSv [620 mrem] (NCRP, 2009).

3.12.1 Background Radiological Conditions

In accordance with NRC regulations contained in 10 CFR Part 40, Appendix A, Criterion 7, a preoperational monitoring program was developed and implemented to establish baseline conditions at the proposed project site. Results of the baseline radiological environmental monitoring provide data on background levels that can be used for evaluating future impacts from routine facility operations or from accidental or unplanned releases. The scope of the baseline program conducted for the proposed Moore Ranch Project is generally consistent with the NRC guidelines in Regulatory Guide 4.14 (NRC, 1980). As a result of the NRC's safety review, Uranium One will be required by a license condition to collect additional baseline data.

Following the guidance of Regulatory Guide 4.14 (NRC, 1980), some of the specific sampling methods included

- An integrated gamma scan survey using gamma sensitive NaI(Tl) detectors using global positioning systems (GPS) for mapping the ambient gamma radiation levels across the proposed site;

- Soil samples, including surface soil (top 5 cm depth), 15-cm depth samples and 1-m depth samples. All samples were analyzed for radium-226. Selected samples were also analyzed for uranium, thorium-230, and lead-210;

- Sediment samples from primary stream drainage areas and surface water impoundments;

- Ambient gamma and radon monitoring, using thermoluminescent dosimeters for total ambient gamma and alpha track etch dosimeters for radon;

- Airborne particulate sampling, collected weekly with quarterly composite (by location) analysis. Samples were analyzed for uranium, thorium-230, radium-226, and lead-210;

- Groundwater and surface water sampling with analysis for gross alpha and gross beta, uranium, thorium-230, radium-228, radium-226, polonium-210, and lead-210; and

Affected Environment

- Vegetation (short grasses and clover) samples with analysis for uranium, thorium-230, radium-226, polonium-210, and lead-210.

The intent of the overland gamma survey was to characterize and quantify natural background or preoperational radiation level and radionuclide concentrations in soils throughout the proposed license area. As shown in Section 6.1.2 of the environmental report, the average results for measured gamma radiation are within the range of gamma radiation levels typically measured in this region of Wyoming. The applicant identified elevated areas as likely attributable to their physical features, such as hilltops and exposed rock, which are known to demonstrate elevated levels of natural background radioactivity (EPA, 2006). Similar variability in surface or near-surface measurements taken at other Wyoming sites have been attributed to natural radioactivity potentially influenced by weathering factors, such as erosion and/or deposition (Whicker, et al., 2008).

Surface and subsurface soil samples were analyzed for radium-226, uranium, thorium-230, and lead-210. As presented in Section 6.1.3 of the environmental report and Addendum 6.2A, surface soil samples results were consistent with typical background U.S. ranges (EMC, 2007a). The average radium-226 concentration for surface samples from the proposed Moore Ranch Project was 1.2 pCi/g, with a maximum value of 4.8 pCi/g. The average radium-226 concentration for subsurface samples was 2.5 pCi/g, with a maximum value of 9.2 pCi/g. Sediment samples collected from streambeds were analyzed for radium-226, uranium, thorium-230, and lead-210. As presented in Section 6.1.4 and Addendum 6.2A of the applicant environmental report, results for the majority of the sediment samples were consistent with typical U.S. background ranges (EMC, 2007a). The average radium-226 concentration was 1.2 pCi/g, with a maximum value of 3.1pCi/g. The uranium average was 1.9 pCi/g, with a maximum value of 9.6 pCi/g. The lead-210 average was 3.3 pCi/g, with a maximum value of 11 pCi/g. The thorium-230 average was 1 pCi/g, with a maximum value of 3.2 pCi/g. Generally, all average values are consistent with the typical range of background concentrations [0.5 to 2 pCi/g] for these radionuclides (EMC, 2007a). Similar results were reported for pond sediment samples.

The applicant placed radon samplers, along with passive gamma detectors, in 10 downwind and upwind locations and obtained baseline measurements. Twelve months of sampling results are presented in Section 6.1.5 of the applicant environmental report and Addendum 6.2A (EMC, 2007a). Reported average radon-222 results for all sampling locations range from 0.1 to 1.7 picocuries/L in air and are consistent with typical background levels. Gamma measurements collected at these same sampling locations range from 0.41 to 0.78 mSv [41 to 78.5 mrem] per quarter and are consistent with typical U.S. background levels (NCRP, 2009).

The applicant also collected air particulate samples at 4 locations over 12 consecutive months. Air samples were collected on a weekly basis to prevent dust loading the filters; composited for a given quarter; and analyzed for radium-226, uranium, thorium-230, and lead-210. Air samplers were located at the nearest residence, an upwind (background) location, and selected downwind locations within the proposed license area, based on NRC regulatory Guide 4.14 criteria. Results were reported in Section 6.1.6 of the applicant environmental report and Addendum 6.2A and include the following (EMC, 2007a):

- Uranium: Concentrations ranged from zero (with a detection level of 1×10^{-16} microcuries per milliliter) to 7.22×10^{-16} µCi/mL.

Affected Environment

- <u>Thorium-230 (Th-230)</u>: Concentrations ranged from zero (with a detection level of 1×10^{-16} µCi/mL) to 2.14×10^{-15} µCi/mL.

- <u>Radium-226 (Ra-226)</u>: Concentrations ranged from zero (with a detection level of 1×10^{-16} µCi/mL) to 8.64×10^{-16} µCi/mL.

- <u>Lead-210 (Pb-210)</u>: Concentrations ranged from zero (with a detection level of 2×10^{-15} µCi/mL) to 3.59×10^{-14} µCi/mL.

These values are within levels measured at other locations across the region and the United States (NCRP, 2009).

The applicant collected groundwater samples at 11 locations over 12 consecutive months. Water samples were collected on a quarterly frequency and analyzed for gross alpha, gross beta, radium-226, radium-228, uranium, thorium-230, polonium-210, and lead-210. Monitoring wells are located within the proposed production and project areas, based on NRC Regulatory Guide 4.14 criteria. The monitoring results are reported in Section 6.1.8 of the applicant environmental report and Addendum 6.2A (EMC, 2007a). Except for a limited number of elevated values for radium and uranium, as may be expected for an environment with such a number of elevated uranium deposits and where there was historic drilling and exploration activity, the results were consistent with typical groundwater background levels.

Surface water samples were collected at eleven locations over two consecutive quarters. There was insufficient water to sample during the last two quarters of the year. Water samples were collected on a quarterly frequency and analyzed for gross alpha, gross beta, radium-226, radium-228, uranium, thorium-230, polonium-210, and lead-210. Sampling locations were located within the proposed production and project areas, following NRC Regulatory Guide 4.14. The sampling results are reported in Section 6.1.9 of the applicant environmental report (EMC, 2007a). Except for lead-210 and uranium, sample concentration results were either below limits of detection or considered consistent with the range of values for typical background surface water measurements. Two samples from within the same drainage area had lead-210 results above the proposed EPA drinking water standard of 1 pCi/L in the Fall of 2006. One of the samples is considered an analytical error (EMC, 2007a). Follow-up samples taken 5 months later from the same sample points were below analytical reporting limits for lead-210. Most sample results identified dissolved uranium above analytical reporting limits, with some results approaching the EPA drinking water standard of 30 µg/L (approximately 20 pCi/L). These elevated uranium results appear to be the background levels for dissolved uranium in surface water for the area.

Section 6.1.10 of the applicant environmental report presents results for three vegetation samples collected within the proposed license area (EMC, 2007a). Vegetation types sampled included sage brush and grasses. Samples were analyzed for radium-226, uranium, thorium-230, lead-210, and polonium-210. All results are consistent with typical vegetation background levels.

The results of the sampling and analysis as summarized in Section 6.1 of the applicant's environmental report, provide data suitable for describing the natural, preoperational background radiation levels for the area surrounding the proposed facility (EMC, 2007a).

Affected Environment

3.12.2 Public Health and Safety

NRC has the statutory responsibility, under the Atomic Energy Act (AEA), as amended by the Uranium Mill Tailings Radiation Control Act, to protect the public health and safety and the environment. The NRC regulations in 10 CFR Part 20 specify annual dose limits to members of the public of 1 mSv [100 mrem] total effective dose equivalent and 0.02 mSv per hour [2 mrem per hour] from any external radiation sources. This public dose limit from NRC licensed activities is a fraction of the background radiation dose, as discussed in Section 3.12.1.

A review of the area within 80 km (50 mi) around the proposed facility indicated that there is one current and several potential uranium mining facilities:

- Smith Ranch-Highland—An operational *in-situ* uranium facility located approximately 58 km [36 mi] south of Moore Ranch

- Christensen Ranch—Irigaray Located approximately 31 and 42 km [19 and 26 mi] northwest of Moore Ranch. The Christensen Ranch site was recently granted an NRC license amendment to restart *in-situ* recovery operations.

- North Butte Project—Located approximately 26 km [16 mi] north of Moore Ranch. This is a satellite facility for the Smith Ranch-Highland facility. It is not currently constructed or operational.

- Ruth Project—Located approximately 21 km [13 mi] west of Moore Ranch. This is a satellite facility for the Smith Ranch-Highland facility. It is not currently constructed or operational.

- Nichols Ranch-Hank Unit—Located approximately 32 km [20 mi] northwest of Moore Ranch. This is a proposed *in-situ* uranium facility that is currently undergoing licensing activities.

Because of their relative distances, none of these projects are expected to cause an appreciable contribution to the background radiation exposures to individuals in the area. Other than CBM, there are no major sources of nonradioactive, chemical releases into the atmosphere or water-receiving bodies in the immediate area surrounding the proposed site.

3.12.3 Occupational Health and Safety

NRC regulates occupational health and safety risks to workers from exposure to radiation, mainly through its Radiation Protection Standards contained in 10 CFR Part 20. In addition to annual radiation dose limits, these regulations incorporate the principal of maintaining doses "as low as reasonably achievable," (ALARA) taking into consideration the purpose of the licensed activity and its benefits, technology for reducing doses, and the associated health and safety benefits. To comply with these standards, radiation safety measures are implemented for protecting workers at uranium ISR facilities, ensuring radiation exposures and resulting doses are less than the occupational limits as well as ALARA.

Also of concern, with respect to occupational health and safety, are industrial hazards and exposure to nonradioactive pollutants, which for an ISR operation can include normal industrial airborne pollutants associated with service equipment (e.g., vehicles), fugitive dust emissions

Affected Environment

from access roads and wellfield activities, and various chemicals used in the *in-situ* extraction process. Industrial safety aspects associated with the use of hazardous chemicals at the proposed Moore Ranch Project would be regulated by the Occupational Safety and Health Administration (OSHA). The type of chemicals and permitted levels are discussed in Section 4.13.1.

As an industry, *in-situ* uranium recovery represents a lower level of health and safety risks to its workers, compared with conventional mining, considering the less intrusive mining methods and reduced exposure to hazards common with open-pit or shaft mining (IAEA, 2001, EMC, 2007a).

3.13 Waste Management

Chapter 2 of this SEIS described the types and volumes of liquid and solid waste that would be generated by the operation of the proposed Moore Ranch Project. The disposal options being considered include the use of a sanitary landfill for disposal of nonradioactive solid wastes, a licensed waste disposal site or mill tailings facility for byproduct material, deep disposal wells for liquid effluents, and an onsite septic system for sanitary waste. No mixed waste would be generated from implementing the alternatives. The applicant expects that the proposed Moore Ranch Project would be classified as a Conditionally Exempt Small Quantity Generator of hazardous waste under the Resource Conservation and Recovery Act. The WDEQ will make that determination. Section 2.1.1.1.6 of this SEIS discusses the expected annual waste volumes that would be generated. This section describes the disposition of the wastes that would be generated by the proposed Moore Ranch Project.

3.13.1 Liquid Waste Disposal

Liquid wastes generated from operation of the proposed Moore Ranch Project would include sanitary waste water, waste water generated from well development and well testing, and liquid effluent generated by the ISR process [liquid byproduct material]. Domestic waste water from restrooms and lunchrooms would be disposed of in a WDEQ-approved septic system. Except for well development and well test waters (which would be uncontaminated and could be discharged to the surface), all remaining liquid effluent generated from production bleed and plant wash-down water would be byproduct material to be disposed of via deep well injection, as described under the proposed action in Section 2.1.1.1.6.2 of this SEIS.

3.13.2 Solid Waste Disposal

Solid byproduct material (including radioactively contaminated soils or other media) that does not meet NRC unrestricted release criteria must be disposed of at a facility permitted to receive byproduct material. As discussed in Section 2.1.1.1.6.3, the proposed action would annually generate approximately 76 m3 [100 yd^3] of solid byproduct material (that does not meet NRC criteria for unrestricted release). Because the applicant is proposing to construct more than one wellfield, the cumulative estimate for byproduct material from decommissioning the plant facilities and all wellfields (over a planned 5-year period) is 3,295 m^3 [4,310 yd^3], plus 423 t [470 T] of concrete. The applicant has indicated that its preferred destination for byproduct material generated by operations of the proposed Moore Ranch Project would be the Pathfinder Mines Corporation Shirley Basin site, located approximately 213 km [132 mi] from the proposed Moore Ranch Project. The Pathfinder Mines Corporation site is limited, under an agreement with the WDEQ, to receive a total of 37,490 m^3 [49,000 yd^3] of waste. A formal disposal agreement between the applicant and Pathfinder Mines Corporation is being negotiated but has not yet been finalized (Uranium One, 2009a, 2010). This agreement must be in place before the applicant begins its proposed ISR operations. The applicant has identified an alternate location for disposal of this waste as the Energy Solutions disposal site in Clive, Utah. The applicant is negotiating a draft disposal agreement with this site, as a contingency measure (Uranium One, 2010).

As discussed in Section 2.1.1.1.6.3, solid wastes are materials that are not hazardous and are either nonradioactive or comply with NRC unrestricted release limits. Solid wastes generated by the proposed Moore Ranch Project would include general facility trash, septic system solids, construction/demolition debris, and any solid byproduct material (such as piping, valves, instrumentation, or equipment) that has been decontaminated to meet NRC criteria for unrestricted release. The proposed operations activities during operations would annually generate approximately 1,530 m^3 [2,000 yd^3], and decommissioning activities would generate about 45,500 m^3 (59,500 yd^3) of solid waste (Uranium One, 2010). The applicant has proposed disposing of solid wastes at the City of Casper landfill in Casper, Wyoming, approximately 97 km [60 mi] from the proposed Moore Ranch Project site. The applicant would transport the wastes either directly to the Casper facility or to a permitted transfer station located at the Midwest/Edgerton landfill, located approximately 39 km [24 mi] from the proposed Moore Ranch Project site. The Casper landfill has a permitted capacity of 317,000,000 m^3 [414,000,000 yd^3] of compacted solid waste and a life expectancy of 1,400 years (Uranium One, 2010).

As discussed in Section 2.1.1.1.6.3, the applicant likely anticipates being classified as a Conditionally Exempt Small Quantity Generator of hazardous wastes, generating less than 100 kilograms per month of these wastes. The City of Casper operates a hazardous waste collection program for conditionally exempt small quantity generators at its landfill special waste and diversion facility. The applicant proposes to transport the small quantities of hazardous wastes it generates to this facility (Uranium One, 2010).

3.14 References

23 CFR Part 772. *Code of Federal Regulations*, Title 23, *Highways*, Part 772, "Procedures for Abatement of Highway Traffic Noise and Construction Noise."

36 CFR Part 800. *Code of Federal Regulations*, Title 36, Parks, *Forests, and Public Property*, Part 800, "Protection of Historic Properties."

69 FR 933, U.S. Fish and Wildlife Service, 2004. "Endangered and Threatened Wildlife and Plants; 90-day Finding for a Petition to List the Eastern Subspecies of the Greater Sage-Grouse as Endangered." Federal Register: Vol. 69, No. 4. pp. 933–936. January 7.

74 FR 66496, EPA (U.S. Environmental Protection Agency), 2009. "Endangerment and Cause or Contribute Findings for Greenhouse Gases under Section 202(a) of the Clean Air Act." Federal Register: Vol. 74, No. 239. pp. 66,496-66,546. December 15.

75 FR 13909, FWS (U.S. Fish and Wildlife Service), 2010. "Endangered and Threatened Wildlife and Plants; 12-Month Findings for Petitions to List the Greater Sage-Grouse (Centrocercus urophasianus) as Threatened or Endangered." *Federal Register.* Vol. 75, No. 55. pp. 13,909–13,959. March 23.

ACHP (Advisory Council on Historic Preservation), 2010. "National Register Evaluation Criteria." <http://www.achp.gov/nrcriteria.html> (12 July 2010).

Basin Electric, 2009. "Dry Fork Station – Basin Electric Power Cooperative." <http://basinelectric.com/Projects/Dry_Fork_Station/index.html> (06 November 2009).

BLM (U.S. Bureau of Land Management, 2010. Instruction Memorandum No. 2010-071. Subject: Gunnison and Greater Sage-grouse Management Considerations for Energy Development (Supplement to National Sage-Grouse Habitat Conservation Strategy). Washington, DC: U.S. Department of Interior Bureau of Land Management.

BLM, 2009. "Sage-Grouse Management." <http://www.blm.gov/wy/st/en/field_offices/Buffalo/wildlife/data.html#SG> (14 April 2009).

BLM, 2008. "Bureau of Land Management Buffalo Field Office Environmental Assessment for Yates Petroleum Corporation All Day POD Plan of Development WY–070–08–026." Cheyenne, Wyoming: BLM.

BLM, 2007a. "Raptor Nest Sites 2007 Located Within the Wyoming BLM Buffalo Field Office Jurisdiction." GIS data files from Bureau of Land Management Buffalo field office. Accessed from Wyoming Geolibrary.

BLM, 2007b. "Visual Resource Management." Manual 8400. <http://www.blm.gov/nstc/VRM/8410.html>

BLM, 2004. National Sage-Grouse Habitat Conservation Strategy. U.S. Department of Interior Bureau of Land Management. <http://www.blm.gov/pgdata/etc/medialib/blm/wo/Planning_and_Renewable_Resources/fish__wildlife_and.Par.9151.File.dat/Sage-Grouse_Strategy.pdf>.

BLM, 2003. "Final Environmental Impact Statement and Proposed Amendment for the Powder River Basin Oil and Gas Project (WY–070–02–065)." January.

BLM, 2002. "BLM Wyoming Sensitive Species Policy and List." <http://www.blm.gov/pgdata/etc/medialib/blm/wy/wildlife.Par.9226.File.dat/02species.pdf> (28 September 2009).

Braun, C., 1998. "Sage-Grouse Declines in Western North America: What are the Problems?" Proceedings of the Western Association of Fish and Wildlife Agencies 78: pp. 139–156. 1998.

Brunette, J.A., 2007. "Class III Cultural Resources Inventory for the Energy Metals Corporation, Moore Ranch In-Situ Uranium Project." (nonpublic due to sensitive information on cultural resources).

Campbell County School District, 2009. Gillette, WY. <http://www.campbellcountyschools.net>.

Chapman, S.S., S.A. Bryce, J.M. Omernik, D.G. Despain, J. ZumBerge, and M. Conrad, 2004. "Ecoregions of Wyoming" (color poster with map, descriptive text, summary tables, and photographs). Reston, Virginia: U.S. Geological Survey (map scale 1:1,400,000).

Cowardin, et al., 1979. "Classification of Wetlands and Deepwater Habitats of the United States." FWS/OBS–79/31.

Department of Administration and Information, Economic Analysis Division, State of Wyoming. <http://eadiv.state.wy.us/s&utax/Report_FY08.pdf>

Affected Environment

EMC (Energy Metals Corporation U.S.), 2007a. "Application for USNRC Source Material License, Moore Ranch Uranium Project, Campbell County, Wyoming, Environmental Report." Casper, Wyoming: Uranium One Americas Corporation. ADAMS Accession Nos. ML072851222, ML072851229, ML072851239, ML07285249, ML07285253, ML07285255.

EMC, 2007b "Application for USNRC Source Material License, Moore Ranch Uranium Project, Campbell County, Wyoming, Technical Report." Casper, Wyoming: Uranium One Americas Corporation. ADAMS Accession Nos. ML072851222, ML072851258, ML072851259, ML072851260, ML072851268, ML072851350, ML072900446.

Endangered Species Act of 1973. 16 USC 1531–1544.

EPA (U.S. Environmental Protection Agency), 2010. "National Ambient Air Quality Standards (NAAQS)." < http://epa.gov/air/criteria.html> (06 May 2010).

EPA, 2006. "Assessment of Variations in Radiation Exposure in the United States (Revision 1)." Contract Number EP–D–05–02.

Fertig, W., 2008. "State Species Abstract: *Penstemon Haydenii*, Blowout Penstemon." Wyoming Natural Diversity Database, University of Wyoming. <http://www.uwyo.edu/wynddsupport/docs/Reports/SpeciesAbstracts/Penstemon_haydenii.pdf> (28 September 2009).

Fertig, W., 2000. "Status Review of the Ute ladies tresses (*Spiranthes diluvalis*) in Wyoming." Wyoming Natural Diversity Database, University of Wyoming. Prepared for the Wyoming Cooperative Fish and Wildlife Research Unit, U.S. Fish and Wildlife Service, and Wyoming Game and Fish Department. <http://uwyo.edu/wynddsupport/docs/Reports/WYNDDReports/U00FER01WYUS.pdf> (28 September 2009).

FWS (U.S. Fish and Wildlife Service), 2009c. "Sage-Grouse Fact Sheet." http://www.fws.gov/mountain-prairie/species/birds/sagegrouse/sagegrousefactsheet.pdf > (29 September 2009).

FWS, 2008a. Letter from U.S. Fish and Wildlife Service Ecological Services to G.Suber, U.S. Nuclear Regulatory Commission. May 7.

FWS, 2008b. "Black-Footed Ferret (*Mustela nigripes*)—Five-Year Review: Summary and Status Evaluation." South Dakota Field Office. Pierre, South Dakota. <http://ecos.fws.gov/docs/five year review/doc2364.pdf> (29 September 2009).

FWS, 2002. "Migratory Bird Species of Management Concern in Wyoming." Cheyenne, Wyoming: U.S. Fish and Wildlife Service.

GCRP (U.S. Global Climate Change Research Program), 2009. "Global Climate Change Impacts in the United States." Cambridge, England: Cambridge University Press.

Liu, W., 2008. "Wyoming Sales, Use, and Lodging Tax Revenue Report.". Department of Administration and Information, Economic Analysis Division, State of Wyoming <http://eadiv.state.wy.us/s&utax/Report_FY08.pdf>.

Affected Environment

NCDC (National Climatic Data Center), 2009. "Climatography of the United States No. 20: Monthly Station Climate Summaries, 1971–200." Asheville, North Carolina: NCDC.

NCRP (National Council on Radiation Protection and Measurements), 2009. "Ionizing Radiation Exposure of the Population of the United States." Report No.160. March 3, 2009. Bethesda, Maryland.

NHPA (National Historic Preservation Act of 1969), as amended. 16 USC 470aa et seq.

Nicholoff, S.H., compiler., 2003. *Wyoming Bird Conservation Plan.* Version 2.0. Wyoming Partners in Flight. Wyoming Game and Fish Department, Lander, WY. May 1, 2003.http://www.blm.gov/wildlife/plan/WY/Wyoming%20Bird%20Conservation%20Plan.htm (6 October 2009). ADAMS Accession No. ML092940037.

NRC (U.S. Nuclear Regulatory Commission), 2009a. NUREG–1910, "Generic Environmental Impact Statement for *In-Situ* Leach Uranium Milling Facilities." Washington, DC: NRC. May.

NRC, 2009b. Memo to A. Kock, Branch Chief, from I. Yu, B. Shroff, and A. Bjornsen, Project Managers, Office of Federal and State Materials and Environmental Management Programs. Subject: Informal Meetings with Local, State, and Federal Agencies in Wyoming Regarding the Environmental Reviews Being Conducted on the Moore Ranch, Nichols Ranch, and Lost Creek *In-Situ* Leach Applications for Source Material Licenses (Docket Nos. 040-09073, 040-09067, 040-09068, Respectively). ADAMS Accession No. ML090500544. March 2.

NRC, 2008. Letter to Mary Hopkins, Wyoming State Historic Preservation Office, Initiation of Section 106 Process for Energy Metals corporation's Moore Ranch Uranium Recovery Project License Request. ADAMS Accession No. ML080950161. April 9.

NRC, 2003. NUREG–1569, "*In-Situ* Leach Uranium Extraction License Applications– Final Report." Washington, DC: NRC. June.

NRC, 1982. NUREG–0889, "Draft Environmental Statement Related to the Operation of Sand Rock Mill Project (Draft Report for Comment)." Washington, DC: NRC. March.

NRC, 1980. Regulatory Guide 4.14, "Radiological Effluent and Environmental Monitoring at Uranium Mills." Rev. 1. Washington, DC: NRC. April.

Sage-Grouse Working Group (Northeast Wyoming Sage-Grouse Working Group), 2006. "Northeast Wyoming Sage-Grouse Conservation Plan. August 15. <http://gs/state/wy.us/ wildlife/wildlife_management/sagegrouse/Northeast/NEConsvPlan.pdf> (29 September 2009).

State of Wyoming, 2008. *Executive Order 2008-2.* Governor Dave Freudenthal. August 1. <http://gf.state.wy.us/wildlife/wildlife_management/sagegrouse/SageGrouseExec Order2008-2[1].pdf>.

Uranium One (Uranium One Americas), 2010. "Responses to NRC Waste Management Questions." ADAMS Accession No. ML101330379. May 12.

Uranium One (Uranium One Americas), 2009a. "Responses to Request for Additional Information for the Moore Ranch *In-Situ* Uranium Recovery Project License Application (TAC JU011)." ADAMS Accession No. ML092450317. August 31.

Affected Environment

Uranium One, 2009b. "Response to Request for Additional Information for the Moore Ranch *In-Situ* Uranium Recovery Project License Application (TAC JU011)." ADAMS Accession No. ML091900402. June 19.

Uranium One, 2008a. "Response to Request for Additional Information for the Moore Ranch *In-Situ* Uranium Recovery Project License Application (TAC JU011)." ADAMS Accession No. ML082060527. July 11.

USACE (U.S. Army Corps of Engineers), 2010. "Subject: Response to a Preconstruction Notification (PCN)." Letter to J. Winter from M.A. Bilodeau, Program Manager, Department of the Army Corps of Engineers. May 10.

USACE, 2006. "Interim Regional Supplement to the Corps of Engineers Wetland Delineation Manual: Great Plains Region." August.

USACE, 1987. "Corps of Engineers Wetlands Delineation Manual." Technical Report Y–87–1. Department of the Army.

USCB (U.S. Census Bureau), 2010. "Annual Population Estimates of the Resident Population for Counties: April 1, 2000, to July 1, 2009, for Wyoming." <http://www.census.gov/popest/counties/CO-EST2009-01.html> (20 May 2010).

USCB, 2009. "State and County QuickFacts for Campbell, Johnson, and Natrona Counties Wyoming." <http://quickfacts.census.gov> (10 September 2009).

USDA (U.S. Department of Agriculture), 2009. "PLANTS Database." Distribution of State-Listed Plants in Campbell and Johnson Counties, Wyoming. <http://plants.usda.gov/index.html> (28 September 2009).

Wyoming Department of Administration and Information, Department of Economic Analysis, 2008. "Population of Wyoming, Counties, Cities, and Towns: 2000 to 2013." July. <http://eadiy.state.wy.us/pop/wyc&sc30.htm>. (24 May 2010).

WDEQ (Wyoming Department of Environmental Quality), 2009. "Wyoming Ambient Air Monitoring Annual Network Plan 2009." <http://deq.state.wy.us/aqd/downloads/AirMonitor/Network%20Plan_2009.pdf> (14 September 2009).

WDEQ, 2008. "Chapter 2, Ambient Standards." <http://deq.state.wy.us/aqd/standards.asp> (6 May 2010).

WGFD (Wyoming Game and Fish Department), 2010. "Recommendations for Development of Oil and Gas Resources Within Important Wildlife Habitats." Version 6.0. Cheyenne, Wyoming: WGFD. April.

WGFD, 2009. "Big Game Crucial Range Maps, Cheyenne, Wyoming." <http://gf.state.wy.us/habitat/index.asp> April.

WGFD, 2006a. "2006 Big Game Job Completion Reports—Job Completion Report, Sheridan Wyoming Region." <http://gf.state.wy.us/wildlife/jobcompetionreports/index.asp> pp. 2–21, 124–133.

WGFD, 2006b. "2006 Big Game Job Completion Reports—Job Completion Report, Casper Wyoming Region." <http://gf.stae.wy.us/wildlife/jobcompletionreports/index.asp> pp. 34–44, 82–91.

WGFD, 2005a. "Avian Species of Special Concern, Native Species Status List." Buffalo, Wyoming: WGFD. January.

WGFD, 2005b. "Swift Fox (Vulpes velox)" in Final Comprehensive Wildlife Conservation Strategy." <http://gf.state.wy.us/wildlife/CompConvStrategy/Species/Mammals/PDFS/Swift%20Fox.pdf> (29 September 2009).

Whicker, et al. "Radiological Site Characterizations, 2008. 27 Gamma Surveys, Gamma/Ra-226 Correlations, and Related Spatial Analysis Techniques." *Health Physics 95.* Supplement 5. p. 18

Winter, J., 2010. Uranium One, Email Communication to Behram Shroff, U.S. Nuclear Regulatory Commission, May 12, 2010. ADAMS Accession No. ML101330441.

Wisdom, M.J., et al., 2002. "Performance of Greater Sage Grouse Models for Conservation Assessment in the Interior Columbia Basin, USA." *Conservation Biology.* Vol. 16. pp. 1,232–1,242.

Wyoming Department of Education, 2007. "2007 Teacher, Pupil and School Counts." <http://www.k12.wy.us/statistics/stat2/2007_teacher_pupil_school_counts.pdf>

Wyoming Department of Employment, 2009. "Wyoming Labor Market Information, Local Area Unemployment Statistics." <http://doe.state.wy.us/lmi/laus.htm>

Wyoming Department of Employment, Research and Planning, 2009. Wyoming Unemployment Rate Increases to 5.0 percent in May 2009. <http://wydoe.state.wy.us/LMI/news.htm>

Wyoming Department of Revenue (WDOR), 2001. "State of Wyoming Property Tax System." June 8, 2001. <http://revenue.state.wy.us/PortalVBVS/uploads/propertytaxsystem.pdf>

Wyoming Department of Revenue and Excise Tax, 2001. Memorandum To all owners and lessees of industrial plants and mines in Wyoming <http://revenue.state.wy.us/PortalVBVS/uploads/Memorandum_3WyomingResidentgeneral_primercontractors.pdf>

Wyoming Economic Analysis Division, 2009a. "Economic Summary: 1Q09." <http://eadiv.state.wy.us/wef/Economic_Summary1Q09.pdf>

Wyoming Economic Analysis Division, 2009b. "Wyoming Cost of Living for the Fourth Quarter 2008." <http://eadiv.state.wy.us/wcli/NewsRelease-4Q08.pdf>

Wyoming Hospital Association, 2009. "Campbell County Memorial Hospital." <http://www.ccmh.net> (06 November 2009).

Wyoming SHPO (Wyoming State Historic Preservation Office), 2010. "Subject: Status of Site 48CA962." E-mail to J. Davis from R.L. Currit, Senior Archaeologist, Wyoming State Historic Preservation Office. June 15. ADAMS Accession No. ML101660667.

Affected Environment

Wyoming SHPO, 2009. "Subject: Uranium One, Inc. More Ranch *In-Situ* Uranium Recovery Project Cultural Resources Inventory (SHPO File# 0608RLC007)." Letter to A. Kock from R.L. Currit, Senior Archaeologist, Wyoming State Historic Preservation Office. November 3. ADAMS Accession No. ML093170805.

WYNDD (Wyoming Natural Diversity Database), 2009a. "Plant Species of Concern." University of Wyoming. <http://www.uwyo.edu/wynddsupport/docs/SOC_PLANTS/2007_Plant_SOC.pdf> (28 September 2009).

WYNDD, 2009b. "Animal Species of Concern." University of Wyoming. <http://www.uwyo.edu/wynddsupport/docs/SOC_Animals/2003_Animals_SOC.pdf> (28 September 2009).

4 ENVIRONMENTAL IMPACTS AND MITIGATIVE ACTIONS

4.1 Introduction

The Generic Environmental Impact Statement for *In-Situ* Leach Uranium Milling Facilities (NUREG-1910, GEIS) evaluated the potential environmental impact of implementing *in-situ* recovery (ISR) operations in four distinct geographic regions, including the Wyoming East Uranium Milling Region, where the proposed Moore Ranch Project is located. This chapter evaluates the potential environmental impacts from implementing the proposed action and alternative wastewater disposal options and the No-Action alternative at the Moore Ranch site. Other reasonable alternatives considered for the proposed Moore Ranch Project included an alternative site location, alternate lixiviants, conventional mining and milling, and conventional mining and heap leach processing, all of which were eliminated from detailed analysis as described in Section 2.2.

This chapter analyzes the four lifecycle phases of *in-situ recovery* (ISR) uranium extraction (construction, operations, aquifer restoration, and decommissioning/reclamation) at the proposed Moore Ranch Project consistent with the analytical approach used in the GEIS (NRC, 2009a). The results of the GEIS impact analyses for the Wyoming East Uranium Milling Region, as summarized in Table 1-1 of this supplemental environmental impact statement (SEIS), were used to focus the site-specific environmental review at the proposed Moore Ranch Project. If the GEIS concluded that that there could be a wide range of impacts on a particular resource area (e.g., the impacts could range from SMALL to LARGE, for example) then that resource area was evaluated in greater detail within this site-specific SEIS.

Sections 4.2 through 4.14 evaluate the impact from both the proposed action (which includes construction, operation, aquifer restoration, and decommissioning/reclamation using a Class I injection well for management of process-related liquid waste streams) and the No-Action alternative, which means no ISR facility would be built and operated at the proposed Moore Ranch Project. The No Action alternative is assessed to provide a baseline to compare the potential impacts from the proposed action.

NRC established a standard of significance for assessing environmental impacts in the conduct of environmental reviews based on the Council of Environmental Quality regulations, as discussed in NRC (2003a) and summarized as follows:

> **SMALL:** The environmental effects are not detectable or are so minor that they would neither destabilize nor noticeably alter any important attribute of the resource.
>
> **MODERATE:** The environmental effects are sufficient to noticeably alter, but not destabilize, important attributes of the resource.
>
> **LARGE:** The environmental effects are clearly noticeable and are sufficient to destabilize important attributes of the resource.

4.2 Land Use Impacts

A potential environmental impact on land use at the proposed Moore Ranch Project could occur during all phases of the ISR facility lifecycle. The impact could be from land disturbance from

Environmental Impacts and Mitigative Actions

construction and decommissioning, grazing and access restrictions, and competing access for mineral rights. Potential impacts on land use could be greater in areas with higher percentages of private land ownership and Native American land ownership or in areas with a complex patchwork of land ownership. At the end of operations, all lands would be returned to their preextraction land use of livestock grazing and wildlife habitat, unless an alternative use is justified and approved by the state and the landowner [i.e., the rancher desires to retain roads or buildings (EMC, 2007a)].

Detailed discussion of the potential environmental impacts on land use from construction, operation, aquifer restoration, and decommissioning for the proposed Moore Ranch Project are provided in the following sections.

4.2.1 Proposed Action (Alternative 1)

4.2.1.1 Construction Impacts

Section 4.3.1.1 of the GEIS described land use impacts that could occur during construction from land disturbances and access restrictions that could limit other mineral extraction, grazing, or recreational activities. The GEIS concluded that land disturbances during construction would be temporary and limited to small areas within permitted boundaries, and that well sites, staging areas, and trenches would be reseeded and restored. The GEIS further noted that changes to land use access including grazing restrictions and impacts on recreational activities, would be limited because of the small size of the restricted area, the temporary nature of restrictions, and the availability of other land for these activities. As summarized in Table 1-1, the GEIS determined that potential construction impacts on land use in the Wyoming East Uranium Milling Region could range from SMALL to LARGE, depending on the factors described previously (NRC, 2009a). The impacts that contributed to a greater than SMALL conclusion in the GEIS analysis considered potential alterations to ecological, historical, and cultural resources where the impact could range from SMALL to LARGE. For this SEIS, potential impacts to ecological, historical, and cultural resources are discussed in Sections 4.6 and 4.9, respectively. Section 4.4 of this SEIS evaluates potential impacts to soil from surface disturbances. Therefore, the discussion below assesses land use impacts at the proposed Moore Ranch Project considering the proposed land disturbances and associated access restrictions that could limit other mineral extraction activities, grazing activities, or recreational activities.

Construction phase activities including drilling, trenching, excavating, grading, and surface facility construction would have the largest direct land use impact. As described in Section 2.1.1.1.2 of this SEIS, construction related activities would disturb and fence approximately 61 ha [150 ac] of the proposed Moore Ranch Project. Constructing the central plant would disturb approximately 2.4 ha [6 ac] and developing the wellfields would disturb approximately 23 ha [57 ac]. Other land disturbance would be associated with developing the infrastructure that includes laying pipeline and constructing access roads. The first phase of construction, which would include the first of the two wellfields and the central plant and ancillary facilities, is estimated to last for approximately 9 months (Griffin, 2009). Livestock that currently reside within the areas proposed for development would be moved and livestock access to certain areas would be limited by fences, which could alter current rangeland leases within the affected area and could affect rangeland use. Coal bed methane (CBM) and other exploratory drilling would be restricted from the 23 ha [57 ac] being developed for the wellfields. The applicant has stated that close communication between themselves and CBM operators during the laying of pipeline would help to further limit the potential land use impact. Because the types of land use activities are similar to those evaluated in the GEIS, and the area of land surface disturbance for the proposed action is small and at the low end of the range {50 to 750 ha [120 to 1,860 ac] was considered in the GEIS} of surface disturbance considered in the GEIS, the U.S. Nuclear Regulatory Commission (NRC) staff conclude that land use impacts would be SMALL. Furthermore, the staff has not identified any new and significant information during its independent review that would change the expected environmental impact beyond that discussed in the GEIS.

4.2.1.2 Operation Impacts

As discussed in the GEIS (Section 4.3.1.2), land use impacts from operational activities would be similar to impacts during the construction phase from access restrictions because the infrastructure would be in place. No additional land disturbances would occur from operational activities. Because impacts from access restrictions and land disturbances would either be similar to, or less than, those for construction, the GEIS concluded that overall potential land use impacts from operational activities at an ISR facility would be SMALL.

During the operations phase of the proposed Moore Ranch Project, land use would be restricted, as described in Section 4.2.1.1 of this SEIS. Livestock grazing and natural resources extraction and drilling would continue to be restricted from the wellfields and the central plant during the operations phase, which the applicant has estimated would last approximately 12 years (Griffin, 2009). Because the types of land use activities are similar to those evaluated in the GEIS, and the amount of land surface disturbance for the proposed action is small and at the low end of the range of surface disturbance considered in the GEIS, the NRC staff conclude that land use impacts would be SMALL. Furthermore, the staff has not identified any new and significant information during its independent review that would change the expected environmental impact beyond those discussed in the GEIS.

4.2.1.3 Aquifer Restoration Impacts

Section 4.3.1.3 of the GEIS describes aquifer restoration impacts to land use (NRC, 2009a). Since aquifer restoration uses the same infrastructure that existed during operations, the land use impacts from aquifer restoration would either be similar to, or less than, those from operations. As aquifer restoration proceeds and wellfields are closed, operational activities would diminish. Therefore, the GEIS concluded aquifer restoration impacts to land use would be SMALL.

The GEIS concluded that land use impacts during aquifer restoration would be similar to those during the operations phase because no additional land disturbance would occur. Wellfield access during the aquifer restoration phase, estimated to last for approximately 3.5 to 5.25 years for Wellfields 1 and 2, respectively (Griffin, 2009) would be restricted from other uses such as livestock grazing, and CBM and other exploratory drilling, as described in Section 4.2.1.1 of this SEIS. Because the types of land use activities are similar to those evaluated in the GEIS, and the amount of land surface disturbance for the proposed action is small and at the low end of the range of surface disturbance considered in the GEIS as discussed in Section 4.2.1.1, the NRC staff conclude that land use impacts from aquifer restoration activities would be SMALL. Furthermore, the staff has not identified any new and significant information during its independent review that would change the expected environmental impact beyond that discussed in the GEIS.

Environmental Impacts and Mitigative Actions

4.2.1.4 Decommissioning Impacts

The land use impact from decommissioning is discussed in Section 4.3.1.4 of the GEIS, which concluded that the impacts from decommissioning an ISR facility would be comparable to that described for the construction phase, with a temporary increase in land-disturbing activities for dismantling; removing; and disposing of facilities, equipment, piping, and excavated contaminated soils. Access restrictions would remain in place until decommissioning and reclamation were completed, although a licensee could decommission and reclaim the site in stages. Reclamation of land to preexisting conditions and uses would help to mitigate long-term potential impacts. The GEIS concluded that impacts to land use during decommissioning could range from SMALL to MODERATE and would be SMALL when decommissioning and reclamation were completed (NRC, 2009a).

At the proposed Moore Ranch Project, the impact from dismantling and decontaminating the central plant, roads, and support facilities would be consistent with the conclusions reached in the GEIS. The 61 ha [150 ac] potentially disturbed as part of the proposed action (less than 2 percent of the proposed license area) would be returned to its preextraction condition and available for other uses such as livestock grazing, CBM, and other exploratory drilling. Topsoil removed and stored as part of the proposed action would be replaced and areas reseeded. The areas most directly impacted would include the central plant, wellfields and their infrastructure (i.e., pipeline and header houses), and access roads constructed for the proposed action. As decommissioning and reclamation proceeded, the area of disturbed and fenced land would decrease.

During decommissioning, the applicant would perform surface reclamation to return disturbed land equal to preextraction use. The applicant would perform the following activities. All contaminated equipment and materials and structures (including piping) would be removed from the site to a licensed facility for disposal or reuse. Equipment decontaminated to levels consistent with NRC requirements would be released for unrestricted use. All production, injection, and monitoring wells and drillholes would be abandoned, in accordance with applicable WDEQ-LQD rules and regulations (EMC, 2007a). Well casing would be cut off at least 0.9 m [3 ft] below the ground surface. Final surface reclamation of each wellfield production unit would be completed after approval of groundwater restoration stability and the completion of well-abandonment activities. Surface preparation would be accomplished as needed to blend any disturbed areas into the contour of the surrounding landscape (EMC, 2007a). Permanent vegetation would be established on disturbed areas (EMC, 2007a). At the completion of decommissioning activities, because the reclaimed land would be released for other uses and no longer restricted, the land use impacts for disturbed areas would be MODERATE until the reestablishment of vegetation in seeded areas. Once vegetation is established in reclaimed areas, the NRC staff conclude the land would be returned to a condition that would support a variety of land uses and, therefore, land use impacts would be SMALL.

4.2.2 No-Action (Alternative 2)

Under the No-Action alternative, the proposed Moore Ranch Project would not be licensed and the land would be available for other uses such as grazing, CBM, oil, and gas production. No construction activities would occur; therefore, the 61 ha [150 ac] of land surface potentially disturbed during the proposed action would not be disrupted; no access restrictions would be

Environmental Impacts and Mitigative Actions

in-place to restrict wildlife usage. No wells would be drilled, no pipeline would be laid, and no access roads would be constructed.

There would be no operations impacts to land use because no ISR facility would have been constructed, and no subsurface injection of lixiviant would occur. The current land uses of natural resources extraction and grazing lands would continue with no access restrictions within the proposed license area. Operations impacts to current land uses from the continued CBM and oil and gas extraction activities within the study area could occur from accidental breaks or failures in equipment and infrastructure systems; however, the occurrence of such accidents is beyond the scope of this SEIS. There would be no impact from operations activities associated with the proposed Moore Ranch Project.

Under the No-Action alternative, there would be no impact from aquifer restoration activities such as the injection, production, or monitoring of subsurface fluids from the proposed Moore Ranch Project because no wells would have been drilled, nor wellfields developed. Aquifer restoration activities that could involve the pumping of wells would not occur, and there would be no impact to the current land uses.

There would be no impact to land use from decommissioning activities because the proposed Moore Ranch Project would not be developed. No buildings requiring decontamination and decommissioning would exist, no topsoil would need to be reclaimed, no land surfaces would need to be revegetated.

4.3 Transportation Impacts

Potential environmental impacts from transportation to and from an ISR facility could occur during all phases of the facility lifecycle. Impacts would result from workers commuting to and from the site and from the shipment of materials and chemicals used in the ISR process. Impacts could also occur from fugitive dust emissions, noise, incidental wildlife or livestock kills, increased traffic on local roads, and from accident occurrence. Fugitive dust impacts are evaluated as air quality impacts in Section 4.7, noise impacts are evaluated in Section 4.8, and livestock kills are evaluated as potential ecological impacts in Section 4.6.1.1.2.

Detailed discussion of the potential environmental impacts from transportation to and from the proposed Moore Ranch Project during construction, operation, aquifer restoration, and decommissioning is provided in the following sections.

4.3.1 Proposed Action (Alternative 1)

4.3.1.1 Construction Impacts

Section 4.3.2.1 of the GEIS concluded that ISR construction activities would generate low volumes of additional traffic (relative to local traffic counts) and would not significantly increase traffic or accidents on many of the roads in the region. Roads that have low traffic counts could be moderately impacted by additional workers commuting during periods of peak employment. Additionally, the GEIS concluded that, depending on site specific conditions, there could be a moderate impact from fugitive dust, noise, and incidental wildlife or livestock kills on, or near, site access roads. For these reasons, the GEIS determined that the construction phase of ISR projects could result in transportation impacts that ranged from SMALL to MODERATE

Environmental Impacts and Mitigative Actions

(NRC, 2009a). This section of the SEIS discusses the potential impact on the local transportation system from implementing the proposed action at the Moore Ranch Project.

The primary access road to the proposed Moore Ranch Project is Brown Road, a gravel road that intersects State Highway (SR) 387 roughly 6.4 km [4 mi] west of SR 50 and 48 km [30 mi] east of I-25 (Figure 1-1). As discussed in Section 3.3, SR 387 is the primary transportation route to the proposed license area and it connects the site to regional population and economic centers along I-25 to the west and SR 59 to the east. These roads were evaluated for potential traffic impacts in the GEIS. The applicant estimated the maximum anticipated increase in vehicle traffic on SR 387 by phase of the proposed Moore Ranch Project. For this analysis, the applicant obtained vehicle traffic counts, from the Wyoming Department of Transportation for trucks and automobiles. The daily truck traffic estimate for the proposed action included trucks that haul heavy equipment (cranes, bulldozers, graders, track hoes, trenchers, front-end loaders) to the construction site. The applicant estimated the average daily increase in auto traffic based on workforce levels for each phase of the ISR project and conservatively assumed that there would be one employee per vehicle for each vehicle trip in the auto traffic projections. The proposed commuter traffic was bounded by the number of workers assumed in the GEIS, however, the number of truck shipments during the construction period for the proposed action was well above the number of shipments assumed in the GEIS. As shown in Table 4-1, during the construction phase of the proposed Moore Ranch Project, when the proposed traffic is considered in the context of existing traffic counts, the applicant estimates a maximum 9.1 percent increase in daily truck traffic and a 4.8 percent increase in automobile traffic along sections of SR 387 (Uranium One, 2009a). This magnitude of change in existing traffic is considered by staff to be unnoticeable and, therefore, impacts to traffic would be consistent with the SMALL traffic-related impacts evaluated in the GEIS.

Brown Road, the spur road to the central plant, and access roads to the wellfields would be periodically graded and cleared of snow, as necessary, to ensure site access was maintained.

Table 4-1. Estimated Annual Average Daily Traffic on State Highway 387 for the Construction Phase of the Proposed Moore Ranch Project								
Section Description	Mile Route Signs	2007*			Projected Volume	Percent Increase	Projected Volume	Percent Increase
	Begin† (mile marker)	All Vehicles	Trucks	Percent Trucks	Trucks	Trucks	Auto traffic	Auto traffic
Johnson-Campbell County Line	118.726	1,500	370	24.7	390	5.4	1,162	2.8
JCT 300 (WY50 and Pinetree JCT)	131.793	890	220	24.7	240	9.1	702	4.8
JCT County Roads North & South	137.12	900	220	24.4	240	9.1	712	4.7
JCT Local Roads North & South	149.24	2,000	410	20.5	430	4.9	1,622	2.0
Wright	150.63	3,390	480	14.2	500	4.2	2,942	1.1

*Year 2007 used as base for projected traffic volume
†Begin=Mile Marker Start of Section
Construction phase traffic: Assuming a construction workforce of 10 plus 6 wellfields staff traveling one round trip per day to the site (32 auto traffic trips per day), and 10 truck round trips (20 trips) per day for equipment being hauled to the site.

Environmental Impacts and Mitigative Actions

The applicant proposes to spray road surfaces with water for dust suppression and conduct regular inspections for erosion and sediment control. Figure 2-2 shows the location of the proposed spur road, wellfield roads, and other access roads that would be required to support the proposed Moore Ranch Project. Because of the limited distance between Highway 387 and the central plant, the NRC staff considered the MODERATE potential impacts the GEIS concluded could result during construction (e.g., road dust, livestock and wildlife kills) would be less likely at the proposed Moore Ranch Project and, therefore, the impact would be SMALL. Section 4.7 contains additional analysis of potential impact from fugitive dust emissions, and impacts to wildlife are addressed in Section 4.6.1.1.2.

The applicant estimated that approximately 50 percent of the construction workforce (8 workers) would be based in Campbell County (EMC, 2007b). Shorter commuting distances would reduce road surface wear and the likelihood of traffic accidents. Traffic interactions between commuters and tractor trailers would be minimized because heavy equipment would be transported primarily during off-peak hours.

Because of the small increase in anticipated traffic, transportation impacts during the construction phase of the proposed Moore Ranch Project were estimated to be SMALL. Furthermore, the staff has not identified any new and significant information during its independent review that would change the expected environmental impact beyond that discussed in the GEIS.

4.3.1.2 Operation Impacts

As discussed in Section 4.3.2.2 of the GEIS, the low level of facility-related traffic would not noticeably increase traffic or the occurrence of accidents on most roads, although local, less-traveled roads could be moderately impacted during periods of peak employment. There could be impacts from fugitive dust emissions, noise, and possible incidental wildlife or livestock kills either on or near site access roads as described in Section 4.3.1.1 for the construction phase of ISR facilities.

The GEIS also assessed the potential for and consequence from accidents involving the transportation of hazardous chemicals and radioactive materials. The GEIS recognized the potential for high consequences from a severe accident involving transportation of hazardous chemicals in a populated area. The probability of such accidents occurring was determined to be low because of the small number of shipments, comprehensive regulatory controls, and the applicant's use of best management practices (BMP). For radioactive material shipments [yellowcake product, ion-exchange (IX) resins, waste materials], compliance with transportation regulations was expected to limit radiological risk for normal operations. The GEIS concluded there would be a low radiological risk in the unlikely event of an accident. The use of emergency response protocols would help to mitigate the consequences of severe accidents that involved the release of uranium. The GEIS concluded that the potential environmental impact from transportation during operations could range from SMALL to MODERATE (NRC, 2009a).

The operations phase of the proposed Moore Ranch Project would last approximately 12 years and involve a peak workforce of approximately 24 employees, which equates to a maximum average of 48 auto trips per day, conservatively assuming one employee per vehicle per one-way vehicle trip (Uranium One, 2009a). Truck traffic during this phase of the proposed Moore Ranch Project would include yellowcake shipments, byproduct material shipment, solid

Environmental Impacts and Mitigative Actions

waste shipment, and regular operation deliveries. The highest levels of project-related automobile traffic would be from the operations workforce commuting to and from the site. The proposed workforce and operational truck traffic was bounded by the traffic evaluated in the GEIS. The maximum anticipated increase in vehicle traffic on SR 387 was estimated for the operations phase of the proposed Moore Ranch Project and is summarized in Table 4-2. The maximum expected increase in truck and automobile usage of SR 387 was estimated at 0.5 and 7.2 percent during the operations phase. This magnitude of change in existing traffic is considered by NRC staff to be unlikely to be noticed and therefore impacts to traffic would be SMALL, consistent with traffic-related impacts evaluated in the GEIS.

Onsite road maintenance during the operations phase would consist of the applicant performing periodic grading of the primary access roads, snow plowing, applying water or other agents to control fugitive dust emissions, and regular inspections to ensure the adequacy of erosion control measures. The GEIS concluded there could be MODERATE transportation impacts based on the distance traveled, since traveling longer distances on dirt roads could result in greater fugitive dust emissions and there would be greater potential for incidental livestock and wildlife kills. Because of the short distance traversed by the access road between Highway 387 and the central plant at the proposed Moore Ranch Project both fugitive dust emissions and the potential for incidental livestock and wildlife kills would be reduced. Therefore, the impact would be SMALL. Section 4.7 provides additional analysis of potential impacts from fugitive dust emissions and potential impacts to wildlife are addressed in Section 4.6.1.1.2.

Section 4.2.2.2 of the GEIS evaluated yellowcake transportation, assuming shipment volumes ranging from 34 to 145 yellowcake shipments per year, which could result in a risk of 0.04 and 0.003 latent cancer fatalities if an accident were to occur, given the larger number of shipments (NRC, 2009a). Considering the annual maximum production rate of yellowcake at the proposed

Table 4-2. Estimated Annual Average Daily Traffic on State Highway 387 for the Operations Phase of the Proposed Moore Ranch Project									
Section Description	Mile Route Signs	2007*			Projected Volume	Percent Increase	Projected Volume	Percent Increase
	Begin† (mile marker	All Vehicles	Trucks	Percent Trucks	Trucks	Trucks	Auto traffic	Auto traffic
Johnson-Campbell County Line	118.726	1500	370	24.7	371	0.3	1178	4.2
JCT 300 (WY50 & Pinetree JCT)	131.793	890	220	24.7	221	0.5	718	7.2
JCT County Roads North & South	137.12	900	220	24.4	221	0.5	728	7.1
JCT Local Roads North & South	149.24	2000	410	20.5	411	0.2	1638	3.0
Wright	150.63	3390	480	14.2	481	0.2	2958	1.6

*Year 2007 used as base for projected traffic volume
†Begin=Mile Marker Start of Section
Operational phase traffic: Assuming a site workforce of 24 (48 auto traffic trips per day), and 1.12 average total truck trips per day (yellowcake shipments = 100 round trips per year or 200 truck trips/year = .548 trips/day; nonradioactive waste shipments = 100 round trips/year or 200 trucks trips/year = 0.548 trips/day;) byproduct material shipments = 5 round trips/year or 10 trips/year = 0.027 trips/day.

4-8

Environmental Impacts and Mitigative Actions

Moore Ranch Project of 4 million pounds and an approximate capacity of 40,000 pounds for each yellowcake shipment, (Uranium One, 2009a), a maximum of 100 shipments per year is estimated for the proposed action, or an average of one shipment every 3.5 days. Therefore, the shipment of yellowcake at the proposed Moore Ranch Project is bounded by the GEIS analyses, and the number of shipments would not significantly affect the project-related traffic relative to the expected commuting workforce. Therefore, the environmental impact from transportation during the operations phase of the proposed Moore Ranch Project would be SMALL.

The GEIS reported that accidents involving yellowcake releases result in up to 30 percent of shipment contents being released, which is less than the fraction used in the previous calculation (NRC, 2009a). To minimize the risk of an accident involving resin or yellowcake transport, all such materials would be transported in accordance with the U.S. Department of Transportation (USDOT) and NRC regulations, handled as low-specific activity materials, and shipped using exclusive-use-only vehicles. Only properly licensed and trained drivers would transport low-specific activity materials.

Should a transportation accident occur, the staff conclude the consequences of such accidents would be limited because the applicant would develop an emergency response plan for yellowcake and other transportation accidents that could occur during either shipment to or from the proposed Moore Ranch Project, and would ensure their personnel received proper emergency response training. Emergency response protocols would include communication equipment and emergency spill kits on each vehicle and emergency response kits at shipping and receiving facilities. Yellowcake shipments would be made in accordance with USDOT and NRC regulations (EMC, 2007a). Section 5.2 of the Moore Ranch Environmental Report (EMC, 2007a) and Section 7.5.5 of the technical report (EMC, 2007b) provides additional details on the applicant's emergency response plan.

The applicant estimated that approximately four bulk chemical, fuel, and supply deliveries would be made per day throughout the operations phase of the proposed Moore Ranch Project (Uranium One, 2009a). This number of shipments is greater than the daily number chemical supply shipments considered in the GEIS, however, the incoming process chemicals are commonly used in industrial applications and their transport would be made in accordance with the applicable USDOT hazardous materials shipping provisions. The applicant's plans to use an alternative processing chemical to anhydrous ammonia further limits the chemical transportation hazards relative to what was considered in the GEIS. As a result, the staff concludes the chemical shipments can be executed safely and potential environmental impacts of these shipments would be SMALL.

Similarly, the transportation of byproduct material (including contaminated equipment and soils) would pose a small potential for an environmental impact in the event of an accident. This conclusion is based on the low number of annual shipments and the potential risk relative to concentrated yellowcake product shipments considered previously and in the GEIS. Trip frequency and the associated risk of an accident would be curtailed by storing byproduct material in a restricted area within the central plant until a full shipment could be transported for disposal. The applicant has estimated an annual byproduct material production rate of 76.5 m^3 [100 yd^3]. Based on the use of roll-off containers with a nominal capacity of 15.3 m^3 [20 yd^3], there would be 5 shipments annually to a licensed disposal facility. Shipments of nonhazardous solid waste either to the Midwest-Edgerton landfill or to a future transfer facility expected to operate at that location would be required approximately twice a week (Uranium One, 2010).

Environmental Impacts and Mitigative Actions

Based on the small increase in anticipated traffic during the operations phase, the low and manageable risks associated with yellowcake, chemical, and waste transportation, the impacts from the operation phase of the proposed Moore Ranch Project would be SMALL. Furthermore, the staff has not identified any new and significant information during its independent review that would change the expected environmental impacts beyond those discussed in the GEIS.

4.3.1.3 Aquifer Restoration Impacts

Section 4.3.2.3 of the GEIS concluded that the magnitude of transportation activities during aquifer restoration would be lower than that for the construction and operations phases of an ISR facility. Aquifer restoration-related transportation activities would primarily be limited to supply shipments, waste shipments, onsite transportation, and employee commuting. The GEIS concluded that transportation impacts from the aquifer restoration phase of an ISR facility could range from SMALL to MODERATE, if the roads traveled had less traffic (NRC, 2009a).

At the proposed Moore Ranch Project, the applicant estimated that the transportation impacts during aquifer restoration would be the same as that described for the operations phase but less than what would occur during the construction phase. Table 4-3 summarizes the maximum anticipated increase in vehicle traffic on SR 387 during aquifer restoration. The applicant estimated the expected increase in truck and automobile usage of SR 387 at 0.5 and 7.2 percent, respectively, (Uranium One, 2009a), comparable to the operations phase impacts. Therefore, transportation impacts during the aquifer restoration phase would be SMALL. Furthermore, the staff has not identified any new and significant information during its independent review that would change the expected environmental impact beyond that discussed in the GEIS.

Table 4-3. Estimated Annual Average Daily Traffic on State Highway 387 for the Aquifer Restoration Phase of the Proposed Moore Ranch Project									
Section Description	Mile Route Signs Begin** (mile marker)	2007			Projected Volume	Percent Increase	Projected Volume	Percent Increase
		All Vehicles	Trucks	Percent Trucks	Trucks	Trucks	Auto traffic	Auto traffic
Johnson-Campbell County Line	118.726	1,500	370	24.7	371	0.3	1,178	4.2
JCT 300 (WY50 & Pinetree JCT)	131.793	890	220	24.7	221	0.5	718	7.2
JCT County Roads North & South	137.12	900	220	24.4	221	0.5	728	7.1
JCT Local Roads North & South	149.24	2,000	410	20.5	411	0.2	1,638	3.0
Wright	150.63	3,390	480	14.2	481	0.2	2,958	1.6

*Year 2007 used as base for projected traffic volume
†Begin = Mile Marker Start of Section
Aquifer restoration phase traffic: Assuming a site workforce of 24 (48 auto traffic trips per day) and a maximum of 1.12 truck trips per day (using operation phase truck traffic data).

4.3.1.4 Decommissioning Impacts

Section 4.3.2.4 of the GEIS concluded that transportation activities occurring during decommissioning would be similar to those that occurred during the construction and operation phases of an ISR facility, except that the magnitude of transportation activities (e.g., number and types of waste and supply shipments, no yellowcake shipments) could be lower than that during operations (NRC, 2009a). Therefore, the potential transportation impacts would be smaller. The accident risk from transportation during decommissioning would be bounded by the yellowcake transportation risk during operations. The GEIS concluded that the potential environmental impact from transportation activities during the decommissioning phase would be SMALL because fewer transportation activities would occur.

The site-specific analysis at the proposed Moore Ranch Project was in agreement with the GEIS conclusions. The maximum anticipated increase in vehicle traffic on SR 387 was estimated for the decommissioning phase, as summarized in Table 4-4. During the decommissioning phase, the maximum expected increase in automobile traffic on SR 387 was estimated at 1.5 percent compared to approximately 9.1 and 7.1 percent during the construction and operation phases, respectively, of ISR operations. The increase in truck traffic would be 2.7 percent. Based on the foregoing analysis, while the details of site-specific waste volumes were different than the GEIS assumptions (i.e., less byproduct material generated from decommissioning during the proposed action, but more nonhazardous solid waste generated) the overall magnitude of the impact from annual transportation during the decommissioning phase did not increase to change the proposed action impact conclusion relative to the GEIS conclusions. Therefore, transportation impacts during the decommissioning phase of the proposed Moore Ranch Project would be SMALL.

Table 4-4. Estimated Annual Average Daily Traffic on State Highway 387 for the Decommissioning Phase of the Proposed Moore Ranch Project								
Section Description	Mile Route Signs Begin† (mile marker	2007*			Projected Volume	Percent Increase	Projected Volume	Percent Increase
		All Vehicles	Trucks	Percent Trucks	Trucks	Trucks	Auto traffic	Auto traffic
Johnson-Campbell County Line	118.726	1,500	370	24.7	376	1.6	1,140	0.9
JCT 300 (WY50 & Pinetree JCT)	131.793	890	220	24.7	226	2.7	680	1.5
JCT County Roads North & South	137.12	900	220	24.4	226	2.7	690	1.5
JCT Local Roads North & South	149.24	2,000	410	20.5	416	1.5	1,600	0.6
Wright	150.63	3,390	480	14.2	486	1.2	2,920	0.3
*Year 2007 used as base for projected traffic volume
†Begin = Mile Marker Start of Section
Decommissioning phase traffic: Assuming a site workforce of 5 (10 auto traffic trips per day) and 3 truck round trip per day (6 truck trips per day) for equipment and waste shipments.								

Environmental Impacts and Mitigative Actions

4.3.2 No-Action (Alternative 2)

Under the No-Action alternative, traffic volumes and patterns would remain the same as described in Section 3.3 of this SEIS. There would be no transportation of materials to and from the site to support licensed activities. There would be no transportation of either radioactive or solid waste attributable to the proposed action because the facility would neither be licensed nor constructed and operated. The current transportation activities to support ongoing CBM and oil and gas exploration and production activities would continue. Existing land use activities would persist. Two companies have active CBM claims within the proposed license area.

4.4 Geology and Soils Impacts

Potential environmental impacts to geology and soils could occur during all phases of the Moore Ranch ISR facility lifecycle. However, these impacts would largely occur during the construction phase of the proposed project.

4.4.1 Proposed Action (Alternative 1)

4.4.1.1 Construction Impacts

Section 4.3.3.1 of the GEIS indicated that during construction of ISR facilities, the principal impacts on geology and soils would result from earthmoving activities associated with constructing surface facilities, access roads, wellfields, and pipelines (NRC, 2009a). Earth-moving activities that could impact soils would include clearing of ground or top soil and preparing surfaces for the processing plant, satellite facilities, pump houses, access roads, drilling sites, and associated structures. Similarly, excavating and backfilling trenches for pipelines and cables could impact soils in the proposed project area.

The GEIS concluded that construction impacts on geology and soils would depend on local topography, surface bedrock geology, and soil characteristics. The earth-moving activities are normally limited to only a small portion of the proposed project. Consequently, earth-moving activities would result in SMALL and temporary (months) disturbance of soils-impacts that are commonly mitigated using accepted BMP. Construction activities would also increase the potential for erosion from both wind and water from the removal of vegetation and the physical disturbance from vehicle and heavy equipment traffic. However, these activities would result in SMALL impacts if equipment operators adopt construction BMP to either prevent or substantially reduce erosion (NRC, 2009a).

The GEIS also indicated that ISR extraction activities would not result in the removal of any rock matrix or structure and that subsidence would not occur from the collapse of overlying rock strata in the extraction zone, which could occur in underground mining operations. No other geologic impacts would be anticipated to occur with the ISR extraction method.

At the proposed Moore Ranch Project, a maximum of 61 ha [150 ac] out of the 2,879 ha [7,110 ac] license area, or approximately 2 percent of the total license area, would be disturbed by the proposed action. This amount of surface disturbance is at the low end of the range evaluated in the GEIS (NRC, 2009a). Soil disturbance would primarily result from the same types of practices that were evaluated in the GEIS, and would be limited to the central plant area, the wellfields, access roads, and from developing drill sites and laying pipeline. These disturbances would be temporary for the duration of the proposed action because the applicant

Environmental Impacts and Mitigative Actions

would restore affected areas and reclaim the soils as described in Section 2.1.1.1.5 of this SEIS.

As described in Section 2.1.1.1.2.3.3 and 2.1.1.5.2, the applicant has proposed to remove and stockpile topsoil from areas that would potentially be disturbed by the proposed action (i.e., the central plant and wellfields header houses) prior to the construction of these facilities. Conventional rubber-tired, scraper-type earth-moving equipment would typically be used to accomplish such topsoil salvage operations. Stockpiles would generally be located on the leeward side of hills to minimize wind erosion, drainage channels, or to avoid other locations that could result in a loss of material. The applicant may surround large stockpiles by a berm and seed them with wheatgrass to control and minimize sediment runoff. The staff conclude the applicants proposed measures to remove and stockpile topsoil are adequate to preserve topsoil until the site areas are reclaimed and therefore impacts to topsoil would be SMALL.

The interpretation of soil mapping conducted across the proposed Moore Ranch Project indicates the potential for water erosion (see Section 3.4.2 of the SEIS) varies from slight to severe, and the potential for wind erosion varies from moderate to severe (EMC, 2007a). Soils across the proposed Moore Ranch Project are prone to wind erosion because of their fine loamy and sandy texture and the semiarid climate. Within the 2.4-ha [6-ac] central plant fenced area, the underlying soils have a slight potential for water erosion and a severe potential for wind erosion. The soils within the proposed wellfields have a moderate to severe risk from both wind and water erosion.

The applicant has proposed mitigation measures to limit soil erosion, including reestablishing temporary or permanent native vegetation as soon as possible after disturbance and implementing BMPs to retain sediment within the disturbed areas [e.g., silt fencing or retention ponds (Uranium One, 2009b).] Roads would be constructed to minimize erosion by surfacing with a gravel road base, constructing stream crossings at right angles using adequate embankment protection and culvert installation, and by providing adequate road drainage with runoff control structures. The NRC staff conclude the applicant's proposed measures to limit soil erosion are adequate to limit potential soil erosion impacts and therefore the potential impacts to soil would be SMALL.

During wellfield construction, drilling activities and the installation of piping would impact soils because, as described in Section 2.1.1.1.2.3.3, the applicant is planning to use mud pits during drilling activities and trenches during wellfield construction. Approximately 970 wells would be drilled in the development of Wellfields 1 and 2. Excavating mud pits would result in surface soil disturbance and entail first removing topsoil and placing it in a separate location. The subsoil would then be removed and deposited next to the mud pit. The applicant stated that when mud pit use was completed, subsoil would be replaced and topsoil applied. Mud pits typically remain open for less than 30 days. Pipeline trench excavation would follow a sequence similar to mud pit excavation, with topsoil stored separately from subsoil and with topsoil deposited on the soil after the pipeline ditch has been backfilled. The applicant would mitigate potential soil compaction by placing multiple pipelines and/or utilities in the same trench when possible. The WDEQ-LQD has guidelines on topsoil and subsoil management at uranium ISR facilities (WDEQ, 2000). Following review of the applicant's proposed use of mud pits and pipeline trenching, the staff conclude the potential impacts to soils would be SMALL based on the limited size of the wellfield areas, the proposed topsoil removal and stockpiling activities, and the short duration of mud pit usage and pipeline trenching that was proposed by the applicant.

Environmental Impacts and Mitigative Actions

In summary, based on (i) the aforementioned similarities between the proposed action and the analysis in the GEIS, (ii) the limited construction area and associated amount of surface disturbance expected from the proposed action, (iii) the applicant's proposed erosion control measures, and (iv) the short duration for use of the mud pits and pipeline trenching activities, the NRC staff conclude the potential environmental impact to geology and soils from construction activities at the proposed Moore Ranch Project would be SMALL. Furthermore, the staff has not identified any new and significant information during its independent review that would change the expected environmental impact beyond that discussed in the GEIS.

While the NRC staff conclude impacts to soils from construction are expected to be SMALL, the staff recognize that alternative methods are available that the applicant could choose to implement to further limit potential impacts from the use of mud pits during well drilling activities. Alternatives or mitigating measures to using mud pits during well drilling operations include, for example, lining the mud pits with an impermeable membrane, disposing of potentially contaminated drilling mud and other fluids off-site, and using portable tanks or tubs to contain drilling mud and other fluids.

4.4.1.2 Operation Impacts

As discussed in Section 4.3.3.2 of the GEIS, during ISR operations, a nonuranium-bearing (barren) solution or lixiviant is injected via wells into the mineralized zone. The lixiviant moves through the pores in the host rock, dissolving uranium and other metals. Production wells withdraw the resulting "pregnant" lixiviant containing uranium and other dissolved metals, and pump it to a processing facility for further uranium recovery and purification.

The removal of uranium from the target sandstones during ISR operations would result in a permanent change to the composition of uranium-bearing rock formations. However, the uranium mobilization and recovery process in the target sandstones does not result in the removal of rock matrix or structure and; therefore, no significant matrix compression or ground subsidence is expected. Therefore, the GEIS concluded the impacts on geology from ground subsidence at ISR projects would be SMALL (NRC, 2009a).

Section 4.3.3.2 of the GEIS also indicated soils could be impacted from ISR operations if pipelines were to leak during transfer of barren and pregnant uranium-bearing lixiviant to and from the processing facility in aboveground and underground pipelines. If a pipe ruptured or failed, lixiviant could be released and (1) pond on the surface, (2) run off into surface water bodies, (3) infiltrate and adsorb in overlying-soil and rock, or (4) infiltrate and percolate to groundwater. In the case of spills from pipeline leaks and ruptures, licensees would have established immediate spill response procedures through onsite, standard operation procedures established before operations (e.g., NRC, 2003b, Section 5.7). As part of the monitoring requirements at ISR facilities, licensees must report certain spills to the NRC within 24 hours. Licensees located in the State of Wyoming must also comply with applicable WDEQ spill response and reporting requirements.

Based on these considerations, the GEIS (Section 4.3.3.2) concluded that impacts to soils from spills during operation could range from SMALL to LARGE, depending on the volume of soil affected by the spill. Because of the requirement for immediate responses at ISR facilities, spill recovery actions, and routine monitoring programs, impacts from spills would be temporary, and the overall long-term impact to soils would be SMALL.

Environmental Impacts and Mitigative Actions

At the proposed Moore Ranch Project, the response to surface releases would be discussed in a site-specific Spill Contingency Plan (EMC, 2007a). As described in Section 2.1.1.1.2.1 of this SEIS, the plant would be designed to capture and drain spilled liquids from potential process tank failures to a sump for transferring spilled solutions to either appropriate tankage or to the waste disposal system. As previously noted, the potential also exists for a release from piping that transfers fluids between the central plant and the wellfield. Piping system break, leak, or separation releases would generally be small because of the applicant's proposed engineering controls to detect pressure changes in the piping system that would alert the plant operator. The program to monitor wellfield and pipeline flow and pressure is discussed in Chapter 6 of this SEIS. If a release were to occur, the applicant would be required to remediate the release, remove contaminated soils, and dispose of contaminated materials at a licensed disposal facility.

Based on these considerations, the potential environmental impacts to soils from spills during operation of the proposed Moore Ranch Project could range from SMALL to LARGE, depending on whether a release occurred and the volume of soil potentially affected. The central plant and shop buildings would be self-contained, and all exterior chemical and fuel tanks would have secondary containment (Uranium One, 2009b). Based on the proposed design of the central plant, which includes containment, the proposed monitoring capabilities and plans for immediate spill response, and proposed spill recovery actions, the impact from a potential spill would be temporary, and the long-term impact to soils would be SMALL, consistent with GEIS conclusions. Furthermore, the staff has not identified any new and significant information during its independent review that would change the expected environmental impact beyond those discussed in the GEIS.

Regarding the potential impacts to the rock matrix, the processes described in the GEIS also apply to the planned uranium recovery operations at the proposed Moore Ranch Project. The proposed uranium recovery activities would not remove rock matrix; therefore, no significant matrix compression or ground subsidence would be expected because the volume of fluid (bleed) withdrawn from the formation would typically be one percent or less. No subsidence would occur because no collapse of overlying rock strata in the mining zone would result from the proposed uranium recovery operations. No other geologic impacts are anticipated to occur during ISR extraction at the proposed Moore Ranch site. Therefore, the potential environmental impacts to geology from subsidence during operations would be SMALL.

4.4.1.3 Aquifer Restoration Impacts

Section 4.3.3.3 of the GEIS describes aquifer restoration, which typically uses a combination of (1) groundwater transfer; (2) groundwater sweep; (3) reverse osmosis, permeate injection, and recirculation; (4) stabilization; and (5) water treatment and surface conveyance (NRC, 2009a). The groundwater sweep and recirculation process does not remove rock matrix or structure and; therefore, no significant matrix compression or ground subsidence would be expected. The water pressure in the aquifer is decreased during restoration. A negative water balance is maintained in the wellfield being restored to ensure that water flows into the wellfield from its edges, thus reducing the potential to spread contamination. However, the change in reservoir pressure is limited by the recirculation of treated groundwater; therefore, it is very unlikely that ISR operations would reactivate any local faults and extremely unlikely that any earthquakes would be generated. Therefore, in the Wyoming East Uranium Milling Region, where the proposed Moore Ranch Project is located, the potential impact on the geology from aquifer restoration would be SMALL.

Environmental Impacts and Mitigative Actions

For the proposed Moore Ranch Project, the potential for and response to spills would be comparable to that described during the operations phase in Section 4.4.1.2 of this SEIS since the same plant and wellfield infrastructure and the same spill and leak detection program would be used during aquifer restoration. Therefore, the potential impact to soils from spills could range from SMALL to LARGE, depending on the magnitude of the spill and the volume of affected soil. Based on the proposed design of the central plant which includes containment, the proposed monitoring capabilities and plans for immediate spill response, and the proposed spill recovery actions, the impact from a potential spill would be temporary, and the long-term impact to soils would be expected to be SMALL consistent with GEIS conclusions. Furthermore, the staff has not identified any new and significant information during its independent review that would change the expected environmental impact beyond those discussed in the GEIS.

With respect to the potential impact on the rock matrix, the processes described in the GEIS also apply to the planned aquifer restoration activities at the proposed Moore Ranch Project. The proposed aquifer restoration activities would not remove rock matrix; therefore, no significant matrix compression or ground subsidence would occur because the volume of fluid (bleed) withdrawn from the formation would typically be one percent or less; therefore, neither subsidence nor collapse of the overlying rock strata in the extraction zone would result from the proposed aquifer restoration activities. Therefore, the potential impact on geology from subsidence during the aquifer restoration phase would be SMALL.

4.4.1.4 Decommissioning Impacts

Section 4.3.3.4 of the GEIS discussed decommissioning of ISR facilities, which includes (1) dismantling process facilities and associated structures, (2) removing buried piping, and (3) plugging and abandoning wells using accepted practices. The main impacts on geology and soils at the proposed project site during decommissioning would be from land reclamation activities and cleanup of contaminated soils.

The GEIS also noted that before decommissioning and reclamation activities could begin, the licensee would be required to submit a decommissioning plan to NRC for review and approval. Any areas potentially impacted by operations would be surveyed to ensure areas with elevated soil concentrations are cleaned up, in accordance with NRC regulations at 10 CFR Part 40, Appendix A, Criterion 6 (6) (NRC, 2009a).

The GEIS concluded that the impacts on geology and soils from decommissioning would be SMALL. Disruption and/or displacement of existing soils would be temporary and occur for the duration of decommissioning activities (NRC, 2009a).

During the decommissioning phase of the proposed Moore Ranch Project, as described in Section 2.1.1.1.5 of this SEIS, the applicant proposes to restore all disturbed lands to their prior land use of livestock grazing and wildlife habitat. The magnitude of proposed soil disturbing decommissioning activities at the proposed Moore Ranch Project are within the bounds of the soil disturbing decommissioning activities evaluated in the GEIS based on the amount of land disturbance for the proposed action (Section 2.1.1.1 of this SEIS) relative to the GEIS analyses (Section 2.11.1 in NRC, 2009a,). The central plant and storage facilities would be decontaminated as required to meet regulatory standards, and either demolished and disposed of or turned over to the land owner if desired. Baseline soils, vegetation, and radiological data would be used to guide and evaluate final reclamation. As discussed in Section 2.1.1.1.5.3, stockpiled topsoil would be redistributed in the disturbed surface areas. These areas would be

Environmental Impacts and Mitigative Actions

recontoured to match existing topography, and seeding and revegetation activities would be conducted. Twelve months prior to the planned decommissioning of either a wellfield or portion of the project, a decommissioning plan would be provided to the NRC for review and approval (NRC, 2003b).

Short-term impacts to soil could occur as reclamation progressed; however, the outcome of this phase of the proposed Moore Ranch Project would be to return the area to land uses that existed prior to the start of proposed ISR activities. Based on the aforementioned analysis, the site-specific conditions are consistent with the conditions evaluated in the GEIS, therefore, the potential environmental impacts to geology and soils from decommissioning the proposed Moore Ranch Project would be SMALL.

4.4.2 No-Action (Alternative 2)

Under the No-Action alternative, neither the soils would be disturbed from earth-moving activities nor would the subsurface geology potentially be affected by the injection of fluids, because no license would be issued to authorize construction, operations, aquifer restoration, and decommissioning of the proposed Moore Ranch Project. No buildings would be constructed; no wells would be drilled; no wellfields would be developed, including laying pipeline to connect the wellfields to the central plant; thus, no soils would be disturbed from earth-moving activities that could change the existing topography.

Grazing and CBM operations in the proposed license area would continue, which could produce localized impacts to soil and the existing topography, but there would be no impact from the proposed action.

4.5 Water Resources Impacts

4.5.1 Surface Water and Wetlands Impacts

Potential environmental impacts on surface water and wetlands at the proposed Moore Ranch Project would be limited because of the lack of surface water bodies and because the channels at the site have intermittent flow, depending on the amount of precipitation and the volume of CBM discharges in the area. Surface water could potentially be impacted during all phases of the ISR facility lifecycle. Impacts could result from road construction and crossings, erosion runoff, spills or leaks of fuel and lubricants, stormwater discharges, potentially process-related fluids, and discharge of wellfield fluids as a result of pipeline or wellhead leaks.

The potential environmental impacts to surface water from construction, operation, aquifer restoration, and decommissioning are provided in the following sections.

4.5.1.1 Proposed Action (Alternative 1)

Section 4.3.4.1.1 of the GEIS identified potential impacts to surface water and wetlands that result from construction of road crossings, filling, erosion, runoff, and spills or leaks of fuels and lubricants for construction equipment. These potential impacts would be mitigated through proper planning, design, construction methods, and BMPs. U.S. Army Corps of Engineers (USACE) permits could be required when filling and crossing jurisdictional wetlands. Authorization from the WDEQ could also be required when filling or crossing wetlands. The GEIS concluded that temporary changes to spring and stream flow from grading and changes in

Environmental Impacts and Mitigative Actions

topography and natural drainage patterns could be mitigated or restored after the construction phase. The GEIS also noted that even though accidental spills of drilling fluids could flow into surface water bodies, these flows would be temporary and mitigated, in accordance with an NRC-approved contingency plan for spill response (NRC, 2003b) and a Wyoming Pollutant Discharge Elimination System (WYPDES) permit. The GEIS concluded that construction impacts to surface water would be SMALL in most cases, but could potentially be MODERATE if a USACE permit was required. As indicated in Section 3.5.1.4, the USACE determined on May 10, 2010, that there would be no jurisdictional areas at Wellfield 1, and an authorization would not be required for construction activities within Wellfield 1. Installation of wells and associated roads and pipelines within Wellfield 2 are authorized by Nationwide Permit (NP) 12 (U.S. Army Corps of Engineers, 2010).

4.5.1.1.1   Construction Impacts

As noted in Section 2.1.1.1.2 of this SEIS, the ground surface at the proposed Moore Ranch Project would be disturbed during construction to build the central plant, develop the wellfields (which would include both laying pipeline and drilling the wells), construct access roads, and install electric lines. Section 3.5.1 and Figure 2-2 of the SEIS show the location of these activities with respect to the intermittent channels and potential wetlands located on the proposed Moore Ranch Project. As shown in Figure 2-2, only activities at Wellfield 2 would likely have the potential to impact Upper Wash #2 to Simmons Draw (see Figure 3-10) because this intermittent channel bisects the proposed wellfield. The applicant indicated that construction work would occur during the summer and fall months when the intermittent channels were dry. Furthermore, as noted in Table 1-2 of this SEIS, Uranium One would obtain construction and industrial stormwater WYPDES permits in accordance with WDEQ regulations. The applicant would also implement BMPs to reduce potential erosion impacts as described next.

The central plant and support buildings would be constructed landward of all intermittent channels and above their peak flood elevation. The specific plant location was chosen because of the relatively flat terrain and the minimum amount of soil movement that would be necessary to create a level pad. Surface water runoff from precipitation (rain and snowmelt) would flow from the central plant area to natural drainages. Furthermore, the applicant has indicated the central plant and chemical and fuel tanks would be located within a bermed area to provide secondary containment (Uranium One, 2009b). Locally, surface water drainage would be directed away from facilities, roads, and topsoil stockpiles using shallow ditches and berms (Uranium One, 2009b).

One culvert would be installed during the development of the site access roads to maintain existing site surface drainage conditions (Uranium One, 2009b). This culvert would be located on the access road connecting the central plant to the main access road. The new temporary access roads to Wellfields 1 and 2 cross an intermittent channel at several locations in Wellfield 2, as noted previously. Within this impacted area, there would be no new road crossings. However, there would be one trunkline pipe crossing and 14 small (approximately 1-in. diameter) pipeline crossings to connect individual injection and production wells to a header house. The small lines would be combined into common trenches wherever possible (Uranium One, 2009b). Vehicles would only cross channels if either there was low flow or the channels were dry (Uranium One, 2009b). The existing culvert crossings would be used when the channels were flowing. The applicant would implement sedimentation and erosion-control

Environmental Impacts and Mitigative Actions

measures and disturbed soil would be reseeded to minimize surface water runoff into channels (Uranium One, 2009b).

Placement of wood poles to support power lines installed to transmit electricity to the central plant and other facilities would be installed landward of any intermittent channel. During the development of Wellfield 2, no wells would be installed in existing ponds; and, the placement of wells directly within the channel, washes, and delineated wetlands would be avoided, if possible (Uranium One, 2009b). The USACE has authorized activities in Wellfield 2 under NP 12. As previously noted, the applicant has proposed to avoid installing wells in the main channels of the ephemeral drainages, thereby avoiding potential impacts to wetlands. Furthermore, the applicant has stated that properly-sized culverts and drainage crossings to withstand a 25-year flood event and rip rap and rock on embankments and around culverts and drainage crossings would be used in drainages in accordance with WDEQ regulations.

A single trunkline would connect the pipeline network to the header houses. This trunkline would traverse uplands and would not bisect any tributary washes or channels. After the trunkline was installed, the soil would be backfilled to preconstruction contours and seeded with native grass seed to stabilize loose soil. If an accidental spill were to occur during the construction phase of the proposed Moore Ranch Project, it would be promptly mitigated in accordance with the applicant's site-specific ERP.

The proposed construction activities described before would have the potential to generate a limited amount of surface water runoff. The activities in Wellfield 2 have been authorized under USACE NP 12 indicating the potential impact on surface water could be MODERATE by the GEIS criteria. Because the occurrence of surface water at the proposed Moore Ranch Project is limited and surface water flow in the channels is intermittent, the USAC, determination (USACE, 2010), and because of the applicant's proposed mitigation measures, the NRC staff conclude that the potential impact to surface water and wetlands from the construction phase for Wellfield 1 of the proposed Moore Ranch Project would be SMALL. Because the USACE determined that construction activities in Wellfield 2 would be authorized by NP 12, the potential impacts from installation of wells and development of the related infrastructure to the wetlands would be MODERATE; however, the applicant's implementation of BMPs such as the central plant and shop buildings being self-contained and all exterior chemical and fuel tanks having a means of secondary containment and the applicant's commitment to reseed disturbed areas after wellfield construction would reduce the impact to SMALL.

4.5.1.1.2   Operation Impacts

Section 4.3.4.1.2 of the GEIS stated that federal and state agencies regulate the discharge of both stormwater runoff and the discharge of wastewater to surface waters through their permitting process (NRC, 2009a). The potential impact from these discharges would be mitigated through permit conditions. The expansion of facilities or the addition of pipeline during the operations phase would result in impacts comparable to that during the construction phase described in Section 4.5.1.1.1. If a spill occurred, the potential impact would depend on the size of the spill, the success of remediation, the proximity to and use of surface water, and the relative contribution of aquifer discharge to surface water. For these reasons, the GEIS concluded that impacts to surface waters during operations could range from SMALL to MODERATE.

Environmental Impacts and Mitigative Actions

No impact to surface water would be expected during the operations phase of the proposed Moore Ranch Project because there would be no permitted discharge of wastewater to surface waters, the infrastructure would be in place to manage stormwater discharge, and no large earth-moving activities that could generate surface water runoff would occur. The occurrence of surface water within the proposed license area is limited and there is intermittent surface water flow in the channels and washes as previously noted. Lixiviant injection and subsequent extraction of the uranium-rich groundwater would occur within a closed and pressurized system of pipes at or near the ground surface. Processing of the uranium into yellowcake would be performed within the enclosed central plant. Accidental spills would be collected and maintained in storage tanks for later disposal via deep well injection, and the applicant would be required to have in place a spill prevention and response plan (NRC, 2003b).

The applicant would construct the central plant on a concrete slab surrounded by a protective berm to contain and control accidental spills, and they would implement a stormwater management plan in accordance with WDEQ requirements to detain or treat runoff from the central plant. Runoff would be diverted away from the facility, where it would be absorbed into the soil. No wastewater discharge to surface water channels would occur.

Because of the limited occurrence of surface water at the proposed Moore Ranch Project, the design of the central plant and wetlands to minimize potential spills, the stormwater permit and stormwater management plan requirements to ensure that runoff from disturbed areas meets Wyoming Pollutant Discharge Elimination System (WYPDES) permit limits, the implementation of a site-specific ERP to address accidental spills, and the applicant's commitment to conduct operations in accordance with standard operating procedures for spill prevention and control, the potential impact to surface water and wetlands from the operations phase of the proposed Moore Ranch Project would be SMALL by the GEIS criteria. The anticipated impacts to surface water from operations at the proposed Moore Ranch ISR facility would be SMALL and would be further reduced by the applicant's proposed mitigation measures described above.

4.5.1.1.3 Aquifer Restoration Impacts

Section 4.3.4.1.3 of the GEIS identified aquifer restoration activities that could potentially impact surface water (NRC, 2009a). These activities included management of produced water, stormwater runoff and accidental spills, and management of brine reject from the reverse osmosis system. The GEIS concluded the potential impacts from these activities would be similar to that which could occur during the operations phase because of the occurrence of either the same infrastructure or similar activities (e.g., wellfield operation, transfer of fluids, water treatment, stormwater runoff). For these reasons, the GEIS determined aquifer restoration impacts to surface water and wetlands could range from SMALL to MODERATE.

The aquifer restoration phase at the proposed Moore Ranch Project would generate wastewater that would be disposed of via deep well injection. Automated sensors would monitor the injection process to detect potential leaks or pipe or well ruptures that could result in a surface discharge. In addition, the applicant would be required to have an NRC-approved spill response plan in place (NRC, 2003b). No wastewater would be discharged to surface water; therefore, there would be a SMALL impact to surface water during the aquifer restoration phase of the proposed Moore Ranch Project.

Environmental Impacts and Mitigative Actions

4.5.1.1.4   Decommissioning Impacts

Section 4.3.4.1.4 of the GEIS discussed impacts from the decommissioning phase of an ISR project and concluded the impacts from this phase would be similar to the impacts from construction (NRC, 2009a). Cleaning up, recontouring, and reclamation of disturbed lands during decommissioning would mitigate long-term impacts to surface water. The GEIS concluded that the potential impact to surface water from decommissioning could range from SMALL to MODERATE, depending on the site-specific conditions.

During the decommissioning phase of the proposed Moore Ranch Project, the central plant, storage facilities, and pipelines would be removed. The wells would be plugged and abandoned. The impact from the removal of the building and infrastructure would have impacts similar to those described in Section 4.5.1.1.1.

Temporary soil disturbances would occur during building and pipeline removal. This work would require temporary soil disturbance within the channel that bisects Wellfield 2 and would be conducted during the dry season to minimize potential sedimentation. Stockpiled topsoil would be returned to the disturbed areas, graded, and revegetated, in accordance with WDEQ-LDQ rules and regulations to mitigate soil erosion.

During the decommissioning phase, the applicant would replace topsoil in previously disturbed areas, and the land surface would be recontoured to restore it to a surface configuration to blend in with the natural terrain. The replaced topsoil and other disturbed wellfield areas would be revegetated with a native seed mix, in accordance with WDEQ regulations and guidelines.

The proposed decommissioning activities would have the potential to generate a limited amount of surface water runoff since the occurrence of surface water at the proposed Moore Ranch Project is very limited, surface water flow in the channels is intermittent, and given the USACE determination (USACE, 2010) the potential impact to surface water and wetlands from the decommissioning phase for Wellfield 1 of the proposed Moore Ranch Project was determined to be SMALL. Because the USACE authorized construction activities in Wellfield 2 under NP 12, the potential impacts from installation of wells and development of the related infrastructure to wetlands would be MODERATE. However, the applicant's implementation of the mitigative measures described previously during the ISR decommissioning phase would reduce the impact to SMALL. The NRC staff concludes that site-specific impacts for the proposed Moore Ranch Project would be SMALL. No new and significant information was reviewed by NRC staff to change the expected environmental impact beyond that discussed in the GEIS.

4.5.1.2   **No-Action (Alternative 2)**

Under the No-Action alternative, there would be no impact to either surface water or wetlands from the construction and operation of the proposed Moore Ranch Project. The current land use in the proposed license area is primarily livestock ranching and CBM activities, as discussed in Section 3.2.1.1 of this SEIS. There are no residences and no recreational activities that occur on the proposed license area. The combination of ranching and CBM activities could result in specific actions that could potentially affect surface water or wetlands.

The proposed Moore Ranch Project has an existing network of gravel and two-track roads for access to ranching and CBM facilities, which cross channels and washes at various locations. Under the No-Action alternative, these two-track roads would continue to be used in their

Environmental Impacts and Mitigative Actions

current conditions resulting in localized fugitive dust emissions that could potentially settle on surface water. Under the No-Action alternative there would be no impact to surface water quality from the proposed Moore Ranch Project because the facility would not be licensed and no central plant, wellfields, additional access roads would be constructed nor would pipeline be laid.

Livestock would continue to graze in channels and washes and would have unrestricted access to those channels and washes that would have limited access under the proposed action. Two companies have CBM claims within the proposed Moore Ranch Project (Uranium One, 2008). CBM production in the proposed license area would continue. Under the No-Action alternative, the 31 CBM wells located within the proposed license area would continue to operate and discharge surface water to channels and washes via outfalls permitted through the WYPDES program.

4.5.2 Groundwater Impacts

Potential environmental impacts to groundwater at the proposed Moore Ranch Project could occur during all phases of the ISR facility lifecycle but primarily during operations and aquifer restoration. A detailed discussion of the potential environmental impact to groundwater from the construction, operation, aquifer restoration, and decommissioning of the proposed Moore Ranch Project is described in the following sections.

ISR activities can impact aquifers that occur at various depths (separated by confining layers, also known as aquitards) above and below the uranium-bearing aquifer, as well as adjacent surrounding aquifers in the vicinity of the uranium-bearing aquifer. Surface or near-surface activities that can introduce contaminants into soils are more likely to impact shallow aquifers during ISR operations, and aquifer restoration would have the potential to impact the deeper uranium-bearing aquifer and to potentially impact overlying and underlying aquifers and adjacent surrounding aquifers.

ISR facility impacts to groundwater resources can occur from surface spills and leaks releases from shallow surface piping, consumptive water use, horizontal and vertical excursions of lixiviant from production aquifers, degradation of water quality from changes in production aquifer chemistry, and waste management practices involving deep well injection. The potential impacts to groundwater resources from construction, operations, aquifer restoration, and decommissioning is discussed in the following sections.

4.5.2.1 Proposed Action (Alternative 1)

4.5.2.1.1 Construction Impacts

Section 4.3.4.2.1 of the GEIS concluded that potential impacts to groundwater during construction would primarily be from consumptive use of groundwater, injection of drilling fluids and mud during well drilling, and the potential spills of fuels and lubricants from construction equipment. The GEIS further indicated that groundwater use during the construction phase would be limited, and that groundwater would be protected by implementing BMP, such as spill prevention and cleanup. The volume of drilling fluids and mud introduced into the environment during well installation would be limited. Thus, the construction impacts to groundwater would be SMALL, based on the limited nature of construction activities and the implementation of BMP to protect shallow groundwater (NRC, 2009a).

Environmental Impacts and Mitigative Actions

The consumptive water use during construction would be generally limited to routine activities such as dust control, drilling support, and cement mixing. The amount of consumptive water use in these activities would be small compared to consumptive water uses during ISR operation and restoration; therefore, its potential impacts on groundwater resources would be temporary and small.

The volume of drilling fluids and muds used during well installation would be limited; the Wyoming UIC Program would require the applicant to implement BMPs to prevent, identify, and correct impacts to soils and the surficial 72 Sand aquifer at the proposed Moore Ranch Project. Drilling fluids and muds would be placed into mud pits to control the spread of the fluids, to minimize the area of soil contamination, and to enhance evaporation. According to Figure CR3.4.3.2 in the applicant's technical report, the depth to the water table in the surficial 72 Sand at the site ranges from 9 to 61 m [30 to 200 ft] below ground surface (EMC, 2007b). Therefore, any leakage from either the pits or spills during drilling would result in a small amount of infiltration; however, no changes to the water quality of the 72 Sand surficial aquifer would be expected. The introduction of drilling fluids into the 72 Sand, 70 Sand, 68 Sand, and 60 Sand aquifers would occur during the drilling of production and monitoring wells but the effect would be small, since drilling muds are designed to seal boreholes to set casing.
As wells are installed, water would be pumped from aquifers for hydrologic testing, such as pumping tests. This water would be discharged to the surface in accordance with approved WYPDES permits obtained by applicant. The surface discharge permits would protect near-surface aquifers by limiting the discharge volume and prescribing concentration limits to discharged waters.

During construction and wellfield installation at the proposed Moore Ranch Project, the groundwater quality of near-surface aquifers would be protected both by permit and BMPs that would include the applicant's implementation of a spill prevention and cleanup program, extracting water from deeper, more prolific aquifers to minimize consumptive use impacts, and complying with WDEQ-approved discharge permits. The potential volume of stored fuels and lubricants in the proposed license area is expected to be small, and any leaks or spills would result in an immediate cleanup response to prevent soil contamination or infiltration to groundwater.

The types of construction activities for the proposed Moore Ranch Project would have an anticipated SMALL impact on groundwater resources, based on the GEIS criteria. Based on this analysis, consumptive groundwater use during the construction phase would be limited and would have a SMALL and temporary impact. The impact to groundwater resources during wellfield and facility construction would be SMALL, based on the limited nature of construction activities and the applicant's implementation of WYPDES permit requirements and the BMPs described above to protect soils and shallow groundwater.

4.5.2.1.2   Operations Impacts

Section 4.3.4.2.2 of the GEIS discussed potential impacts to shallow (near-surface) aquifers during ISR operations. During this phase of an ISR operation, shallow aquifers could potentially be affected by lixiviant leaks from pipelines, wells, or header houses and from waste management practices such as the use of evaporation ponds and disposal of treated wastewater by land application. Potential environmental impacts to groundwater resources in the production and surrounding aquifers also include consumptive water use and changes to water quality that could result from normal operations in the production aquifer and from

Environmental Impacts and Mitigative Actions

possible horizontal and vertical lixiviant excursions beyond the production zone. Disposal of processing wastes by deep well injection during ISR operations could also potentially impact groundwater resources (NRC, 2009a).

4.5.2.1.2.1 Operations Impacts to Shallow (Near-Surface) Aquifers

Section 4.3.4.2.2.1 of the GEIS discussed the potential impacts to shallow aquifers during ISR operations. A network of buried pipelines is used during ISR operations to transport lixiviant between the header house and the satellite or main processing facility and also to connect injection and extraction wells to manifolds inside the pumping header houses. Failure of pipeline fittings or valves, or failures of well mechanical integrity in shallow aquifers, could result in leaks and spills of pregnant and barren lixiviant, which could impact water quality in shallow aquifers. The potential environmental impact of such pipeline, valve, or well integrity failure depends on a number of factors, including the depth to shallow groundwater, the use of shallow groundwater, and the degree of hydraulic connection between shallow aquifers and regionally important aquifers. The GEIS concluded that potential environmental impacts could range from MODERATE to LARGE, if

(1) The groundwater in shallow aquifers is close to the ground surface;

(2) The shallow aquifers are important sources for local domestic or agricultural water supplies; and

(3) Shallow aquifers are hydraulically connected to other locally or regionally important aquifers.

The potential environmental impacts could be SMALL if either shallow aquifers have poor water quality or noneconomic production yields and if they are hydraulically separated from other locally and regionally important aquifers.

Hydrogeologic data from the proposed Moore Ranch Project indicates that the 72 Sand is the first aquifer encountered below the land surface. In some small areas, isolated occurrences of perched water are encountered in the 80 Sand, which overlies the 72 Sand across the proposed license area. The 72 Sand is not saturated in the southern portion of the proposed license area. In these areas, the 70 Sand is the surficial aquifer.

Because of the shallow depth to groundwater in the 72 Sand and 80 Sand where they occur, they could potentially be impacted by releases at or near the ground surface during operations. A surface release could potentially impact groundwater, depending on the depth to the water table, the permeability of the materials in the unsaturated zone, the potential adsorption of constituents in the unsaturated zone, and the volume of a potential release. The 72 and 80 Sand aquifers could also be impacted by potential well casing leaks during operations.

As indicated in the GEIS, the potential impact of a surface release on shallow groundwater would be greatly reduced by NRC-required leak detection programs. All wells would be tested for mechanical integrity every five years to detect casing leaks. Wells that failed mechanical integrity tests would be either corrected or removed from operation. A licensee would also follow an aggressive leak-detection and spill-cleanup program during operations. High- and low-flow alarms for individual wells would be the primary means for timely identification of a pipe rupture. Header houses would be equipped with a "wet building" alarm to detect the presence of liquids in building sumps. In addition, daily visual inspections of wellfield monitoring would occur. Spills exceeding 1,590 L [420 gal] would be reported to the WDEQ, accompanied by a report to NRC. Following repair of wellfield leaks, contamination surveys would be performed; and contaminated soils would either be immediately remediated if concentrations exceeded regulatory requirements or left in place and cleaned up during decommissioning. The applicant would design the concrete curb around the perimeter of the central plant and the underlying concrete pad at the proposed Moore Ranch Project to contain the contents of the largest tank within the central plant, in the event of a rupture. Plant fluid spills would be contained, drained to the sump system, and pumped to the waste disposal system. Thus, the applicant BMP would mitigate the impact of a potential release (EMC, 2007a,b) and its potential impact on shallow groundwater.

Because the 72 Sand aquifer overlies the 70 Sand production zone, it would be monitored by wells installed to detect vertical excursions. The applicant's proposed monitoring of the 72 Sand would provide an extra level of wellfield surveillance to detect impacts from either potential surface spills or casing leaks. The 72 Sand monitoring wells would be spaced at a density of one well to every 1.6 ha [4 ac] in the wellfields, and sampled for excursion parameters every 2 weeks to detect the presence of production fluids.

No water wells for either domestic or agricultural use are completed in the 72 Sand within the proposed Moore Ranch license area. Therefore, the shallow aquifer in the 72 Sand is not an important source for local domestic or agricultural water supplies. Furthermore, the 72 Sand is a perched aquifer over the majority of the proposed license area; therefore, it is not hydraulically connected to other locally or regionally important aquifers. At locations where the 72 Sand aquifer is not perched, it is underlain by a sufficiently thick shale layer to prevent a hydraulic connection to other significant aquifers.

Based on the previous analysis, the impact on shallow groundwater in the 72 Sand and on groundwater in the perched 80 Sand aquifer from operations at the proposed Moore Ranch Project would be SMALL. Impacts to the surrounding aquifer from the 72 Sand would also be expected to be SMALL.

4.5.2.1.2.2 Operations Impacts to Production and Surrounding Aquifers

The potential environmental impact to groundwater supplies in the production and other surrounding aquifers are related to consumptive water use and groundwater quality.

Water Consumptive Use

Section 4.3.4.2.2.2 of the GEIS discussed groundwater being withdrawn and reinjected into the production zone during ISR operations (NRC, 2009a). Most of the water withdrawn from the aquifer is returned to the aquifer. That portion of groundwater that is not returned to the aquifer is referred to as consumptive use. The consumptive use is primarily from production bleed and also includes other smaller losses. Production bleed is the net withdrawal maintained to ensure that groundwater flow is toward the production network to minimize the potential movement of lixiviant and its associated nonradiological constituents out of the wellfield.

The portion of an aquifer where production occurs must be designated as an exempt aquifer by the EPA, pursuant to the Federal UIC regulations, before any production begins. The exempt aquifer designation should be in compliance with 40 CFR 146.4, which states that an aquifer may be exempted if it is not, it cannot now, nor would it ever be a source of drinking water in the

Environmental Impacts and Mitigative Actions

location covered by the exemption, if the total dissolved solids content of the groundwater is more than 3,000 ppm and less than 10,000 ppm, and the aquifer is not reasonably expected to supply a public water system. Moreover, under the Federal UIC regulation, the exempted aquifer would no longer be protected under the Safe Drinking Water Act (SDWA) as an Underground Source of Drinking Water (USDW). At the proposed Moore Ranch Project, portions of the 70 Sand, where production operations would occur, and typically a buffer zone would potentially be exempted by EPA. Groundwater in the aquifer outside the designated exempt zone would be considered a possible source of drinking water, if of appropriate quality.

Consumptive water use during ISR operations could potentially impact local water users who also extract water from wells completed in the production aquifer outside the exempted zone. This potential impact would result from drawing down water levels in nearby wells, thus potentially reducing the well yields. Furthermore, if the production zone was hydraulically connected to other aquifers above and/or below, consumptive use could impact the water levels in both overlying and underlying aquifers and create a drawdown in water level, thus reducing the potential yield from nearby wells completed in these aquifers (NRC, 2009a).

As discussed in Section 3.5.2, the 70 Sand production zone at the proposed Moore Ranch Project is not completely saturated over the proposed license area. Therefore, it is an unconfined aquifer. The unconfined conditions in the production zone help to reduce the potential impact from the consumptive use anticipated during ISR operations. For a given net withdrawal, an unconfined aquifer exhibits substantially less drawdown in water level over a smaller area, relative to that exhibited in a confined aquifer. As shown in Figure 4-1, the water produced from a well in an unconfined aquifer (water level below overlying aquitard) comes from dewatering of the aquifer pore space in the production zone.

To assess the potential drawdown in the unconfined aquifer at proposed Moore Ranch Project, the applicant developed an unsaturated groundwater flow model for the 70 Sand that covered the entire proposed license area. The model was created within the Groundwater Vistas platform and used MODFLOW–SURFACT Version 3.0, an industry standard unsaturated groundwater flow code. The applicant calibrated the model to site-specific conditions and

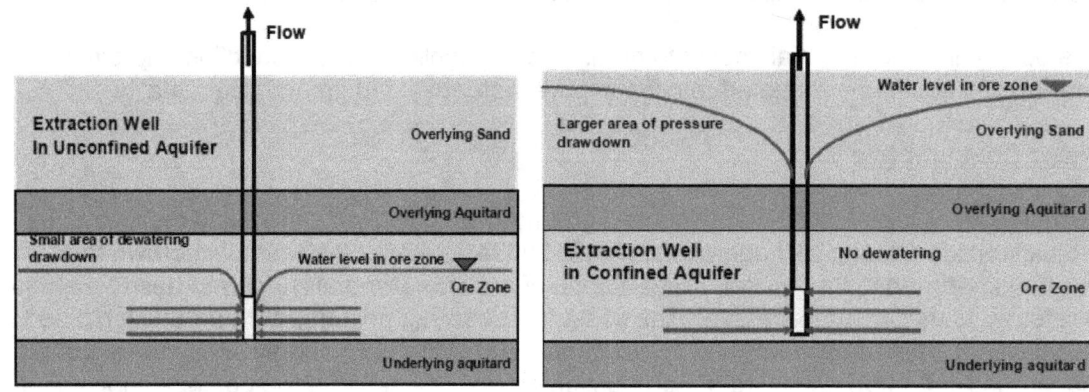

Figure 4-1. Difference In Size and Type of Drawdown in an Unconfined Aquifer and Confined Aquifer from an Extraction Well Operating at a Same Rate

verified it by use of site-specific Moore Ranch Project field pumping test data. The modeling, which the NRC staff reviewed and found acceptable, is presented in the applicant response to a request for additional information dated October 27, 2008, (Uranium One, 2008).

The model analyzed drawdowns during various phases of ISR production and aquifer restoration. For production operations, it assumed production rates of approximately 11,360 L/min [3,000 gal/min] and production bleeds ranging between 0.8 and 1.3 percent, which translates into a production consumptive use rate of 90.9 L/min to 147.6 L/min [24 gal/min to 39 gal/min].

The model drawdown simulations in the 70 Sand at the end of production operations for the previous consumptive use rates were provided in Figure CR4.4.2.1-1b (Uranium One, 2008). The end of production provides the best estimate of the maximum drawdown for total consumptive use during the operations phase. The results of the drawdown simulations show that the cone of depression created by the consumptive use at the end of operations results in a drawdown of about 0.30 to 1.2 m [1 to 4 ft] at the proposed license area boundary. The drawdown contour of 0.30 m [1 ft] extends outside the proposed license area to the north, northwest, west, and southwest for approximately 1.6 to 3.2 km [1 to 2 mi].

To estimate the potential impact of the simulated drawdown on private wells, private well users with wells completed in the 68–70 Sand, 70 Sand, and 70–72 Sand located within 3.2 km [2 mi] of the proposed Moore Ranch Project boundary were identified. Their locations are shown in Figure CR4.4.2.1-1a (Uranium One, 2008). The drawdown in each of these private wells was simulated and the results shown in Table CR4.4.2-1-2b (Uranium One, 2008). Only one private well, located just outside the northwest portion of the proposed license area and drilled to a total depth of 108 m [355 ft] below ground surface and completed in the 70 Sand, had a static water level of 46 m [150 ft] below ground surface and indicated a drawdown of more than one foot. The estimated drawdown in this well was 1.08 m [3.53 ft] at the end of the operations phase; this amount of drawdown would have a negligible impact on well yield.

Given the hydraulic isolation separating the 72 Sand and 70 Sand production zone, there appears to be little potential impact on water levels in the 72 Sand resulting from the consumptive use in the 70 Sand production zone. However, there appears to be hydraulic interconnection between the underlying 68 Sand and the 70 Sand production zone in the portion of Wellfield 2 where the 68 and 70 Sands coalesce (see Section 3.5.2). In this portion of Wellfield 2, the 68 Sand would be included as part of the production zone, although no production wells would be installed.

To determine the impact of production operations on water levels in the 68 Sand and surrounding users, the applicant assumed a worst-case-estimate scenario. In this scenario, the drawdown in the 68 Sand, which would have no operating wells completed in it, was assumed to be the same as that simulated for the 70 Sand at the end of production operations. This is a conservative drawdown estimate because data from pumping tests indicate that there would be less drawdown in the 68 Sand where the 70 and 68 Sands coalesce. Using this assumption, Table CR4.4.2.1-2b shows that three wells completed in the 68 Sand and located to the northeast and southeast outside of the proposed license area would have nominal drawdowns of 0.02, 0.07, and 0.003 m [0.08, 0.23, and 0.01 ft] (Uranium One, 2008). This amount of drawdown would not be expected to affect well yields.

Environmental Impacts and Mitigative Actions

Based on the consumptive use and groundwater modeling drawdown predictions in the 70 Sand and the 68 Sand during operations, private wells within 3.2 km [2 mi] radius surrounding the proposed license area would experience a small or nominal drawdown in their private wells, which would not impact well yields. Therefore, the potential environmental impact to groundwater supplies and users in the production and other surrounding aquifers would be SMALL.

Excursions and Groundwater Quality

As discussed in the GEIS (NRC, 2009a), groundwater quality in the production zone would be degraded as part of ISR operations. The production portion of the aquifer would be recommended for exemption as a USDW by WDEQ to EPA. After production operations are completed, the licensee would be required to return water-quality parameters to the standards in 10 CFR Part 40, Appendix A, Criterion 5B(5). As stated in 10 CFR Part 40, Appendix A, Criterion 5B(5), "at the point of compliance, the concentration of a hazardous constituent must not exceed—(a) the Commission approved background concentration of that constituent in the groundwater; (b) the respective value given in the Table 5C if the constituent is listed in the table and if the background level of the constituent is below the value listed; or (c) an alternate concentration limit (ACL) established by the Commission." Only after demonstrating that it cannot restore a particular hazardous constituent to the background concentration or maximum contaminant level (MCL), a licensee can request a license amendment from the NRC for an ACL for a particular hazardous constituent. Appendix C explains the process for granting an ACL. For proposed ACLs to be approved, they must be shown to protect public health at the site. For these reasons, potential impacts to the water quality of the uranium-bearing production zone aquifer as a result of ISR operations would be SMALL.

In Section 2.11.4 of the GEIS, the NRC staff documented historical information from operating ISR facilities at which excursions have occurred. Separately, the NRC staff analyzed the environmental impacts from both horizontal and vertical excursions at three NRC-licensed ISR facilities (NRC, 2009b). In that analysis, which considered 60 events at three facilities, the NRC staff found that the licensees were able to control and reverse the excursions through pumping and extraction at nearby wells for most events. Most excursions were short-lived, although a few continued for several years. In all cases, environmental impacts were SMALL and temporary (NRC, 2009b).

Poor well integrity involving a cracked well casing or leaking joints between casing sections could cause vertical excursions (NUREG–1910, Section 2.4.1.3). The applicant would be required to take preventive measures against vertical excursions prior to operations, including well integrity tests (Section 1 of NUREG-1910). The applicant is required to conduct mechanical integrity testing of each well to check for leaks or cracks in the casing, in compliance with 10 CFR 146.8. The conduct of mechanical integrity testing reduces the likelihood of poor well integrity and potential excursions. Therefore, the impacts from such excursions would be SMALL.

To prevent horizontal excursions, inward hydraulic gradients are maintained in the production aquifer during ISR operations (NRC, 2009a). These inward hydraulic gradients are created by the net groundwater withdrawals (production bleeds) maintained through continued pumping during ISR operations. Groundwater flows in response to these inward hydraulic gradients, thus ensuring that groundwater flow is toward the production zone. This inward groundwater flow

Environmental Impacts and Mitigative Actions

toward the extraction wells prevents horizontal excursions of lixiviant solutions away from the production zone.

NRC also requires a licensee to take preventive measures to reduce the likelihood and consequences of potential excursions. A ring of monitoring wells outside of and encircling the production zone is required for early detection of horizontal excursions. If excursions are suspected, corrective actions are required. The impacts from these excursions would therefore be expected to be SMALL.

The GEIS also discussed the potential for vertical excursions into aquifers overlying or underlying the production zone aquifer. The GEIS concluded that the potential for leaching solution to migrate into an overlying or underlying aquifer was SMALL, if the thickness of the aquitard separating the production zone from the overlying and underlying aquifer was of sufficient thickness and the aquitard has low permeability (NRC, 2009a). The vertical hydraulic gradient between the production zone and overlying or underlying aquifers is also used to determine the potential for vertical excursions. NRC also requires monitoring in the overlying and underlying aquifers. Corrective action is also required if any vertical excursions are detected (NRC, 2003b).

At the proposed Moore Ranch Project, the 70 Sand aquifer would be designated as an exempt aquifer before production operations began, in accordance with 40 CFR Part 146. The applicant reported that the Sand 70 and the Sand 68 coalesce in Wellfield 2 near the monitoring well MW-2 location. Therefore, NRC expects the 68 Sand would also be exempted. However, the WDEQ and the EPA would make the decision to exempt the Sand 68 aquifer within the proposed license area. The groundwater chemistry would be changed as lixiviant is injected to mobilize uranium for extraction. At the end of operations, aquifer restoration using Best Practicable Technology would be initiated to return the 70 Sand aquifer to baseline conditions, or the maximum contaminant levels provided in Table 5C of 10 CFR Part 40, Appendix A, or to ACLs. Restoration to these standards would ensure that groundwater quality within the exemption boundary after restoration would not pose a threat to surrounding groundwater. For these reasons, potential impacts to the water quality of the 70 Sand production zone aquifer and surrounding aquifers as a result of ISR operations would be SMALL.

The occurrence of an unconfined aquifer in the 70 Sand production zone at the proposed Moore Ranch Project requires special consideration when evaluating the appropriate inward hydraulic gradient, the reliability of monitoring around the wellfield periphery, and evaluating the capability to pull back a potential horizontal excursion. As discussed earlier, the applicant developed an unsaturated numerical groundwater model to simulate drawdown in the unconfined portion of the 70 Sand production zone. The model was calibrated using site-specific hydraulic data and presented in Appendix B–4 of the applicant's technical report (EMC, 2007b). The model simulations indicate that it would be possible to maintain the necessary inward gradient during ISR operations to prevent horizontal excursions (EMC, 2007b).

To detect horizontal excursions from the proposed wellfields at the proposed Moore Ranch Project, monitoring well rings with the wells completed in the 70 Sand production zone would be installed at each wellfield. The monitoring wells would be located approximately 152 m [500 ft] from the edge of each wellfield and spaced 152 m [500 ft] apart around the perimeter of the wellfield. The wells would be sampled biweekly for chloride, alkalinity, and conductivity, the excursion parameters that are indicative of the presence of production fluids. Any well samples containing more than two of these excursion indicators at prescribed levels (derived based on

Environmental Impacts and Mitigative Actions

baseline values) would be placed on excursion status and the applicant would notify WDEQ and NRC within 24 hours. All wells on excursion status would be monitored every seven days until the indicators returned to nonexcursion levels. The applicant would modify wellfield operations, as necessary, to correct the excursion. If a well remained on excursion for more than 60 days, the applicant would provide a plan to NRC to correct the excursion. The applicant will be able to confirm the behavior of the unsaturated aquifer during wellfield testing and before operations begin. By license condition, the testing results will be provided to the NRC for review and approval in a wellfield hydrologic test data package that NRC will evaluate to ensure the field data support the simulation results.

Given the applicant's maintenance of an inward hydraulic gradient to prevent excursions, the NRC requirement to implement a monitoring well ring to detect excursions, and a plan to correct them, the potential impact from horizontal excursions at the proposed Moore Ranch Project would be SMALL.

The 72 Sand aquifer overlies the 70 Sand production aquifer. The water table within the 72 Sand is perched on the underlying aquitard that separates the 72 Sand from the 70 Sand production aquifer. The water levels in the 72 Sand are generally much higher than that in the 70 Sand. The combination of the perched water table and the high water levels in the 72 Sand, relative to the 70 Sand, demonstrate the absence of a hydraulic interconnection between the 72 and 70 Sands. The unconfined conditions in the 70 Sand further support this conclusion. Pumping tests conducted to date have not demonstrated any hydraulic connection between the 70 and 72 Sands. When this lack of hydraulic connection is considered along with the requirement for the applicant to conduct mechanical integrity testing for all production wells (Section 2.1.1.2.3.3), the potential for vertical excursions from the production zone into the overlying 72 Sand would be SMALL.

A relatively thick and low permeability aquitard separates the 70 Sand production aquifer from the underlying 68 Sand throughout much of the proposed license area. Pumping tests conducted to date indicate that the 68 Sand is hydraulically isolated in Wellfield 1. Therefore, the potential for vertical excursions from the production zone into the underlying 68 Sand in Wellfield 1 is SMALL. In portions of Wellfield 2; however, the aquitard separating the 68 and 70 Sand is missing, as previously discussed in Section 3.5.2. The 68 and 70 Sands coalesce where the aquitard is missing, and the two aquifers appear to be hydraulically interconnected. The 68 Sand would be included as part of the production zone in the area where the 68 and 70 Sands coalesce. The underlying 60 Sand, which is separated by a continuous shale layer, would be treated as the underlying aquifer. The actual well spacing may be adjusted based on the results from the Wellfield 2 hydrologic test package discussed previously. The potential for vertical excursions from the production zone in Wellfield 2 into the underlying 60 Sand would be SMALL.

To detect potential vertical excursions at the proposed Moore Ranch Project, the aquifers that overly and underlie the 70 Sand, which include the 72, 68, and 60 Sands, would be monitored at a spacing of 1 well per 1.6 ha [4 ac]. The same sample constituents and process for horizontal excursions previously described would be followed to monitor for vertical excursions.

Given the isolation of the overlying and underlying aquifers from the 70 Sand production zone by low permeability shale layers, required mechanical integrity testing for all production wells, and the use of monitoring wells to detect excursions and correct them, the impact from a potential vertical excursion at the proposed Moore Ranch Project would be SMALL.

Environmental Impacts and Mitigative Actions

4.5.2.1.2.3 Operations Impacts to Deep Aquifers Below the Production Aquifers

Potential environmental impacts to confined deep aquifers below the production aquifers could occur from deep well injection of wastewater into deep aquifers. Under different environmental laws, such as the Clean Water Act, the Safe Drinking Water Act, and the Clean Air Act, EPA has statutory authority to regulate activities that may affect the environment. Underground injection requires a permit from EPA or from an authorized state UIC program. The WDEQ has been authorized to administer the UIC program in Wyoming and would be responsible for issuing permits for deep well disposal at the proposed Moore Ranch Project.

WDEQ would only permit Class I disposal wells if the groundwater quality in the injection zone would not be suitable for domestic or agricultural uses (e.g., high salinity), could not be designated as a USDW, and if the injection zone was confined above by sufficiently thick and continuous low-permeability layers.

The GEIS (Section 4.3.4.2.2.3) indicated that in the Wyoming East Uranium Milling Region, where the proposed Moore Ranch Project is located, the deep Paleozoic aquifers are hydraulically separated from the proposed aquifer sequence where ISR operations would occur. The stratigraphic sequence, from shallowest to the deepest, includes the Wasatch Formation, the Fort Union Formation, the Lance Formation, and the Fox Hills Formation. Thick, low-permeability confining layers separate the aquifer sequence, including the Pierre Shale, the Lewis Shale, and the Steele Shale (Whitehead, 1996). Hence, nonkarstic Paleozoic aquifers (e.g., Tensleep Sandstone) can be investigated for their suitability for deep well disposal of leaching solutions. The GEIS concluded that in the Wyoming East Uranium Milling Region, considering the relatively poor water quality in and the reduced water yields from nonkarstic Paleozoic aquifers and the occurrence of thick and regionally continuous aquitards confining them from above, the potential environmental impacts from deep injection of leaching solution into nonkarstic Paleozoic aquifers could be SMALL. Regionally, the Pierre Shale was reported to be fractured in some places (Whitehead, 1996). Considering potential heterogeneities in the hydrogeologic properties of the Pierre Shale, the potential impacts could range from SMALL to MODERATE in locations where the Pierre Shale might be fractured.

Up to four Class I wells could be drilled at the proposed Moore Ranch Project for deep disposal of liquid effluent, depending on the production rates and the capacity of each disposal well. The State of Wyoming is reviewing a permit application for up to four Class I disposal wells at the proposed Moore Ranch Project. The application includes injection into the Teapot-Teckla-Parkman Formation with an injection depth of 2,413 m to 2,929 m [7,916 to 9,610 ft]. Injection rates for this interval are expected to be about 30 gal/min. This aquifer may be a candidate for exempt aquifer status, if (1) it does not currently serve as a source of drinking water, (2) it cannot now, or will not in the future, serve as a source of drinking water because of contamination or economic or technical impractability; and (3) the TDS concentrations are greater than 3,000 ppm and less than 10,000 mg/L, and not reasonably expected to supply a public water system (40 CFR 146.4). The application also includes injection into the Lance Formation at depths of 1,615 m to 2,286 m [5,300 to 7,500 ft]; because the Lance Formation has a much greater injection capacity, only two Class I disposal wells would be required to support the proposed Moore Ranch Project operations. However, the water quality in the Lance Formation could be less than 10,000 mg/L TDS, and if the aquifer is not exempted, it could potentially be an underground source of drinking water as defined in 40 CFR 144.3, which would eliminate it from consideration as an injection zone for a Class I deep disposal well.

Environmental Impacts and Mitigative Actions

The WDEQ will evaluate the suitability of the formations proposed for deep well injection and would only grant such a permit to the applicant if it can be demonstrated that liquid effluent could be safely isolated in a deep aquifer. Consequently, it has been assumed that if WDEQ approved the permit application, the potential environmental impact to deep aquifers from deep well injection at the proposed Moore Ranch Project would be SMALL.

4.5.2.1.3 Aquifer Restoration Impacts

The potential environmental impacts to groundwater resources during aquifer restoration are related to groundwater consumptive use and waste management practices, including deep well injection of wastewater. In addition, aquifer restoration directly affects groundwater quality in the vicinity of the wellfield being restored. As discussed in the GEIS, the impacts of consumptive groundwater use during aquifer restoration are generally greater than during ISR operations because a larger volume of groundwater is generally withdrawn if groundwater sweeps are used during the aquifer restoration phase. Larger withdrawals could produce larger drawdowns in the production aquifer, resulting in a greater impact on the yields of nearby wells. However, the rate of groundwater consumptive use during aquifer restoration would be lowered during the reverse osmosis phase, because up to 70 percent of the pumped groundwater treated by reverse osmosis can be reinjected into the aquifer.

The impact from consumptive use during ISR production operations was previously discussed in Section 4.5.2.1.2.2 of this SEIS, which describes the applicant's use of an unsaturated numerical groundwater flow model to estimate drawdown for the production phase consumptive groundwater use of the 70 Sand production zone. The same model was used to predict the drawdowns in Wellfield 1 and Wellfield 2 at the end of aquifer restoration using assumed consumptive use rates for each phase.

The predicted drawdown in the 70 Sand from model simulation of the end of aquifer restoration in Wellfield 1 is shown in Figure CR4.4.2.1-1c of the applicant's technical report (EMC, 2007b). The drawdown simulation results indicate that the cone of depression created by consumptive use during aquifer restoration would drawdown the water level about 0.3 to 2.7 m [1 to 9 ft] at the proposed license area boundary. The drawdown contour of 0.3 m [1 ft] extends outside the proposed license area to the north, northwest, west, and southwest for approximately 1.6 to 6.4 km [1 to 4 mi].

The drawdown-model simulation in the 70 Sand at the end of aquifer restoration in Wellfield 2 is shown in Figure CR4.4.2.1-1d of the applicant's technical report (EMC, 2007b). The drawdown simulation results show that the cone of depression created by consumptive use during aquifer restoration would result in a drawdown of about 0.3 to 1.8 m [1 to 6 ft] at the proposed license area boundary. The drawdown contour of 0.3 m [1 ft] extends outside the proposed license area to the north, northwest, west, and southwest for approximately 1.6 to 6.4 km [1 to 4 mi.]

To estimate the impact of the simulated drawdown on private well users, the applicant identified all private wells completed in the 68 to 70 Sand, 70 Sand, and 70 to 72 Sand within 3.2-km [2-mi] radius of the proposed Moore Ranch Project. Only one private well, located just northwest of the proposed license area and completed in the 70 Sand, would have a drawdown of more than 0.3 m [1 ft] during the restoration phases of both Wellfields 1 and 2. The drawdown in this well at the end of Wellfield 1 aquifer restoration was predicted at 2.4 m [7.87 ft]. The drawdown at the end of Wellfield 2 restoration was predicted at 1.80 m [5.90 ft]. The well was completed to a depth of 108 m [355 ft] below ground surface (bgs) with a static water level of 46 m [150 ft] bgs, indicating an operating water level of 62 m [205 ft]. A decrease of 1.8 to 2.4 m [5.9 to 8.0 ft] would likely have a negligible impact on well yield.

Given the hydraulic isolation of the overlying 72 Sand from the production zone, there would be little potential to impact water levels in the 72 Sand from groundwater consumptive use in the production zone during aquifer restoration. However, as previously noted because there is an apparent hydraulic interconnection between the underlying 68 Sand and the 70 Sand production zone in Wellfield 2 where the 68 and 70 Sands coalesce (see Section 3.5.2), the 68 Sand would be included as part of the production zone at this location.

To determine the impact of restoration operations on water levels in the 68 Sand and to surrounding users, the applicant simulated a worst-case scenario in which the drawdown in the 68 Sand (where no operating wells would be completed), was assumed to be the same as that for the 70 Sand at the end of aquifer restoration. The results showed that three wells completed in the 68 Sand, located outside of the proposed license area to the northeast and southeast, would have nominal drawdowns of 0.21, 0.28, and 0.001 m [0.68, 0.91, and 0.04 ft] in Wellfield 1 at the end of aquifer restoration, and drawdowns of 0.3, 0.4, and 0.001 m [1.08, 1.20, and 0.04 ft] in Wellfield 2 at the end of aquifer restoration. These drawdowns would not likely impact well yields.

The potential impacts to groundwater from aquifer restoration could range from SMALL to MODERATE based on the GEIS criteria. Based on consumptive use and groundwater modeling predictions of drawdown in the 70 and 68 Sands in Wellfield 2 resulting from aquifer restoration, private wells within 3.2 km [2 mi] surrounding the proposed license area would experience a small or nominal drawdown in water level, which would not likely impact well yields. Therefore, the potential environmental impact on groundwater supplies and to other users of both the production and other surrounding aquifers would be SMALL.

Aquifer restoration should directly impact groundwater quality in the production zone. As discussed previously," the purpose of restoration is to return the groundwater quality in the production zone to groundwater protection standards in 10 CFR Part 40, Appendix A, Criterion 5B(5). These standards state the concentration of a hazardous constituent must not exceed (a) the Commission approved background concentration of that constituent in groundwater, (b) the respective value in the table in paragraph 5C if the constituent is listed in the table and if the background level of the constituent is below the value listed, or (c) an alternative concentration limit established by the Commission."

The restoration of the 70 Sand production zone, including the potentially-impacted portion of the 68 Sand in Wellfield 2 would restore the groundwater quality to standards that are protective of human health and the environment and that do not impact surrounding aquifers. Therefore, the impact of aquifer restoration on groundwater quality would be SMALL.

4.5.2.1.4 Decommissioning Impacts

Section 4.3.4.2.4 of the GEIS discussed potential impacts to groundwater during construction as being primarily from consumptive use of groundwater, potential spills of fuels and lubricants, and well abandonment. The consumptive use of groundwater during decommissioning would be much less than during either ISR operations or aquifer restoration. Fuel and lubricant spills during decommissioning activities could potentially impact shallow groundwater. Implementation of BMP as part of an NRC-approved decommissioning plan

Environmental Impacts and Mitigative Actions

(NRC, 2003b), would reduce the likelihood of such spills and the impact to groundwater resources in shallow aquifers from decommissioning would be SMALL.

Furthermore, prior to NRC termination of the ISR source material license, the licensee must demonstrate that there would be no long-term impacts to a USDW. NRC review and approval for the completion of wellfield restoration at the site would have determined that the restoration standards were protective of public health and safety.

As part of the restoration and reclamation activities, all monitor, injection, and recovery wells at the proposed Moore Ranch Project would be plugged and abandoned in accordance with the Wyoming UIC program requirements. If this process was properly implemented and the abandoned wells were properly isolated from the flow domain, the potential environmental impacts would be SMALL.

4.5.2.2 No-Action (Alternative 2)

Under the No-Action alternative, no construction, operations, aquifer restoration, or decommissioning activities would occur that could potentially impact shallow groundwater. No lixiviant would be injected into the subsurface; therefore, there would be no affect on the aquifer and no consumptive use of groundwater. No liquid effluents would be generated; therefore, there would be no Class I well constructed for disposal of liquid wastes. Therefore, under the No-Action alternative, there would be no impact to groundwater above the baseline described in Section 3.5.2.

4.6 Ecological Resources Impacts

Potential environmental impacts to ecological resources, including both flora and fauna at the proposed Moore Ranch Project, could occur during all phases of the ISR facility lifecycle. Impacts could include the removal of vegetation from the site (with the associated reduction in wildlife habitat and forage productivity and an increased risk of soil erosion and weed invasion); the modification of existing vegetative communities as a result of site activities; the loss of sensitive plants and habitats; and the potential spread of invasive species and noxious weed populations. Impacts to wildlife could include loss, alteration, or incremental fragmentation of habitat; displacement of and stresses on wildlife; and direct and/or indirect mortalities. Aquatic species could be affected by disturbance of stream channels, increases in suspended sediments, fuel spills, and habitat reduction.

The potential environmental impacts to ecological resources from construction, operation, aquifer restoration, and decommissioning are discussed in the following sections.

4.6.1 Proposed Action (Alternative 1)

Section 4.3.5.1 of the GEIS discussed potential impacts to terrestrial vegetation from construction through (1) the removal of vegetation from the milling site (and associated reduction in wildlife habitat and forage productivity and an increased risk of soil erosion and weed invasion), (2) the modification of existing vegetative communities, (3) the loss of sensitive plants and habitats as a result of clearing and grading, and (4) the potential spread of invasive species and noxious weed populations. Potential impacts to wildlife include (1) habitat loss or alteration and incremental habitat fragmentation, (2) displacement of wildlife from proposed

Environmental Impacts and Mitigative Actions

project construction, and (3) direct or indirect mortalities from proposed project construction and operation (NRC, 2009a).

As further indicated in the GEIS, the percentage of vegetation removed and land disturbed by construction activities would disturb a SMALL portion (from less than 1 percent up to 20 percent) of the total licensed area and surrounding plant communities. The clearing of herbaceous vegetation in an open grassland or shrub steppe community would have a short-term, SMALL impact, given the rapid colonization by annual and perennial species in the disturbed areas and restoration of the vegetative cover. The clearing of wooded areas could have a long-term impact given the pace of natural succession, and such impacts could range from SMALL to MODERATE, depending on the acreage of the surrounding wooded area. Noxious weeds would be expected to be controlled with appropriate spraying techniques, and therefore, impacts would be SMALL (NRC, 2009a).

The GEIS also noted that construction impacts to wildlife habitat would be minimized with the timely reseeding of disturbed areas following construction. In general, wildlife species would be expected to disperse from the proposed license area as construction activities approached, although smaller, less mobile species could perish during clearing and grading. Habitat fragmentation, temporary displacement, and direct or indirect mortalities would be possible; thus, the potential impact from construction could range from SMALL to MODERATE. The potential impact to sage-grouse and big-game species could be mitigated using measures such as those outlined in the U.S. Bureau of Land Management (BLM) and Wyoming Fish and Game Department guidelines. Impacts to raptor species from power distribution lines could be mitigated if the Avian Power Line Interaction Committee guidance was followed and disturbing areas near active nests and prior to the fledgling of young was avoided.

In-stream channel activities would temporarily disturb aquatic species; therefore, the impacts would be SMALL. Sediment loads would taper off quickly in time and distance; therefore, long-term impacts would be SMALL. The use of the State of Wyoming Game and Fish Department (WGFD) standard management practices would help limit impacts to aquatic life.

If threatened or endangered species were identified on the proposed project site during surveys, the impacts could range from SMALL to LARGE, depending on site conditions. Mitigation plans to avoid and reduce impacts to potentially affected species would be developed.

4.6.1.1   Construction Impacts

As noted previously, ecological resources could be affected by land disturbance during ISR facility construction. The construction phase of the proposed Moore Ranch Project could potentially impact ecological resources from clearing vegetation; constructing the central plant; developing the wellfields, including drilling wells; building header houses; constructing access roads; and clearing field laydown areas. The applicant projected construction at the proposed Moore Ranch Project would take 9 months to complete, and impacts are considered accordingly in the following section.

4.6.1.1.1   Impacts to Vegetation

The wellfields and central plant at the proposed Moore Ranch Project would be constructed within the upland grassland vegetation communities (see Figure 3-11). Direct impacts would include the short-term loss of vegetation (modification of structure, species composition, and

Environmental Impacts and Mitigative Actions

areal extent of cover types). An estimated 61 ha [150 ac] of upland grassland would be affected by construction disturbance under current development plans. Indirect impacts would include the short-term and long-term increased potential for nonnative species invasion, establishment, and expansion; potential soil erosion; shifts in species composition or changes in vegetative density; reduction of wildlife habitat; reduction in livestock forage; and changes in visual aesthetics.

The construction activities, increased soil disturbance, and increased traffic during construction could stimulate the introduction and spread of undesirable and invasive, nonnative species within the proposed license area. Two State-designated weeds, Canada thistle and field bindweed, were observed on the proposed Moore Ranch Project during the baseline surveys conducted by the applicant, along with other undesirable annual grass species such as cheat grass brome. The applicant would conduct weed control as needed to limit the spread of undesirable and invasive, nonnative species on disturbed areas (EMC, 2007a).

To mitigate the potential impact to vegetation, disturbed areas could be both temporarily and permanently revegetated, in accordance with WDEQ-LQD regulations and the WDEQ mine permit. The applicant would seed disturbed areas to establish a vegetative cover to minimize wind and water erosion and the invasion of undesirable plant species (EMC, 2007b).

The impact from vegetation removal and surface disturbance would affect approximately 61 ha (150 ac) of land, or about 2 percent of the total licensed area and less than 0.05 percent of the upland grassland area within the proposed license area; therefore, the impact would be SMALL. Some individual plants would be affected, but since construction activities would not affect 61 ha [150 ac] contiguous acres and the upland grasslands cover approximately 70 percent of the site, the impact would not affect a sizeable segment of the species population over a relatively large area.

4.6.1.1.2   Impacts to Wildlife

There are three primary impacts of ISR uranium recovery facility construction on terrestrial wildlife: (1) habitat loss or alteration and incremental habitat fragmentation; (2) displacement of wildlife from project construction; and (3) direct or indirect mortalities from project construction.

ISR facility construction and operation can have direct and indirect impacts on local wildlife populations. These impacts are both short-term (lasting until successful reclamation is achieved) and long-term (persisting beyond successful completion of reclamation). However, substantial long-term impacts would not be expected because of the relatively limited habitat disturbance associated with use of the ISR extraction method. The likelihood of injury or mortality to wildlife would be greatest during the construction phase because of increased traffic levels and physical disturbance during that ISR phase. The applicant would impose and enforce speed limits during all construction and maintenance operations to reduce impacts to wildlife throughout the year and particularly during the breeding season (EMC, 2007a).

The applicant estimated the area to be disturbed during the construction phase at 61 ha [150 ac] (noncontiguous). This area is comprised of two wellfields covering a combined area of 23 ha [57 ac], the central plant and associated storage facilities that cover approximately 2.4 ha [6 ac], approximately 3.2 km [2 mi] of new access road to the central plant and within the proposed wellfields, and the infrastructure supporting the wellfields (e.g., header houses that

Environmental Impacts and Mitigative Actions

consolidate pipelines from individual wells to a trunkline that connects to the central plant). Most of the habitat disturbance would consist of scattered, confined drill sites for wells in the wellfields that would not result in large expanses of habitat being dramatically transformed from its original character as in surface mining operations.

Indirect impacts could occur from displacement of wildlife from increased noise, traffic, or other disturbances associated with the development of the proposed Moore Ranch Project and from small reductions in existing or potential cover and forage due to habitat alteration, fragmentation, or loss. Indirect impacts typically persist longer than direct impacts. However, ISR uranium extraction does not involve large-scale habitat alteration.

Certain vegetative communities that exist in the proposed license area could be difficult to reestablish through artificial plantings, and natural seeding and recruitment could take many years. Consequently, wildlife species associated with specific habitats, such as blue grama grasslands, birdsfoot sagebrush, and big sagebrush, could be reduced in number or replaced by generalist species with broader habitat requirements until natural reseeding of certain vegetation occurs or reclamation matures to its target mix. However, as shown in Figure 3-11, most of the proposed applicant activities would occur in upland grass communities that cover 70 percent of the proposed license area. Therefore, the impact from the construction phase of the proposed Moore Ranch Project would have a SMALL impact on wildlife because the affected area would be small and noncontiguous, and the primary plant community to be affected is widespread throughout the proposed license area. In addition, the applicant would use temporary fencing around all open mud pits to protect wildlife from this hazard.

4.6.1.1.2.1 Impacts to Big Game

Pronghorn antelope and mule deer are the only two big-game species that regularly occur in the proposed license area. No crucial, big-game habitat or migration corridors occur on or within several kilometers of the proposed Moore Ranch Project (WGFD, 2009).

Direct impacts to pronghorn antelope and mule deer from project activities could include the disturbance of a portion of yearlong range, loss of forage, and vehicular collision accidents. An estimated 61 ha [150 ac] would be incrementally disturbed during the approximate 12-year life of the ISR facility. Because of these habitat disturbances, the yearlong range-carrying capacity for big game would be reduced over the life of the ISR facility and for several years thereafter, until the affected areas had revegetated and become productive enough to support big game. No significant increase in the potential for vehicle collision with big game would be expected because of the short distances traveled and the applicant enforcement of speed limits on the access roads. Direct impacts to pronghorn antelope and mule deer would be SMALL because only a few individual animals would be affected and the continued existence of the species would not be threatened.

Indirect impacts to pronghorn antelope and mule deer could include displacement into surrounding areas because of increased human activity and the increased potential for poaching. Human presence during construction could affect pronghorn antelope and mule deer use of adjacent areas. Some short-term disturbance (during the lifecycle of the ISR facility) of big-game habitat could occur because of the proposed project construction. However, the construction phase of the proposed action has been estimated to last nine months. Adequate habitat for pronghorn antelope and mule deer exists in the surrounding area; these species could return to the areas affected by construction when these activities were completed

Environmental Impacts and Mitigative Actions

(EMC, 2007b). The proposed staged reclamation of disturbed areas would provide grass and forage within a few years of habitat disturbance. The number of employees and the nature and intensity of the proposed activities would be comparable to those occurring from CBM production in the same vicinity. The movement of big game through the proposed license area would not be impacted by implementing the proposed action. The limited use of security fencing to impede ingress to and egress from the restricted area around the central plant and the use of fencing around wellfields to limit access to sheep would mitigate the potential impact from wildlife use of the area. Fencing that is preferred by the WGFD has been previously documented (WGFD, 2004).

Furthermore, the applicant's mitigative actions, such as enforcing speed limits, would further reduce big-game conflicts associated with the proposed Moore Ranch Project. Because pronghorn antelope and mule deer are highly mobile species, the potential impact to these species would be SMALL.

4.6.1.1.2.2 Impacts to Other Mammals

A variety of small- and medium-sized mammal species occur in the vicinity of the proposed license area, although not all have been observed on the proposed Moore Ranch Project itself. These mammals include coyotes, red foxes, swift foxes, bobcats, striped skunks, weasels, badgers, muskrats, raccoons, rodents (e.g., mice, rats, voles, gophers, ground squirrels, chipmunks) and rabbits.

Medium-sized mammals (e.g., coyotes, foxes) could be temporarily displaced to other habitats during construction activities. Direct losses of limited mobility small-mammal species (e.g., voles, ground squirrels, mice) could be higher than for other wildlife because of the likelihood they would retreat into burrows if disturbed, and thus potentially be impacted by topsoil scraping or staging activities. However, given the limited, noncontiguous area that could be disturbed, approximately 61 ha [150 ac], no major changes or reductions in small- or medium-sized mammalian populations would be expected. The species that occur in the area have shown an ability to adapt to human disturbance in varying degrees, as evidenced by their occurrence in areas of CBM development (EMC, 2007b). Small-mammal species in the area also have a high reproductive potential and tend to reoccupy and adapt to altered or reclaimed areas quickly (EMC, 2007b).

Since only a few individuals would be affected and most mammal species would likely travel to suitable habitat adjacent to the construction areas, the proposed Moore Ranch Project would have a SMALL impact on these mammals.

4.6.1.1.2.3 Impacts to Avian Species

Upland Game Birds. The only upland game bird prevalent in the vicinity of the proposed Moore Ranch Project is the mourning dove, which is a relatively common breeder in Campbell County, and is the most prevalent upland game bird in the area (EMC, 2007a). The proposed construction activities could affect approximately 61 noncontiguous ha [150 ac] of potential foraging and nesting habitat for mourning doves. While woody corridors are not abundant in the vicinity of the proposed project area, they also are not unique to the Moore Ranch Project proposed license area. Habitat that could support mourning doves occurs to the immediate south of the proposed license area where no ISR recovery has been proposed; therefore, the proposed Moore Ranch Project would not impact the occurrence of mourning doves.

Environmental Impacts and Mitigative Actions

As discussed in Section 3.6.1.2.2, sage-grouse neither occur nor is the appropriate habitat present within the proposed license area to support their occurrence. Therefore, there would be no expected impacts to sage-grouse from the proposed action.

Waterfowl and Shorebirds. Because surface water occurs only intermittently at the proposed Moore Ranch Project, little habitat exists to support large groups or populations of either waterfowl or shorebirds. Therefore, there would be no impact to these species.

Raptors. Three species of raptors occur within the proposed license area: the ferruginous hawk, the red-tailed hawk, and the great horned owl. The populations of these three species are common and are believed to be stable in the local vicinity. Nesting success by resident raptors could be reduced from disturbances caused by milling operations and traffic. Two nest sites occur within close proximity of the wellfields, but no trees with nests would be removed. Other nest sites occur within the southern half of the proposed license area. Use of the nest sites may continue as birds habituate to milling activities, and the potential impact to the raptor population could range from SMALL to MODERATE. If the applicant adhered to the WGFD and BLM seasonal noise, vehicular traffic, and human proximity guidelines (WGFD, 2010; BLM, 2008a and 2010), these mitigation measures could support the continued nesting success of area raptors, and the impact would be SMALL.

The applicant would conduct a raptor nest survey in late April or early May each year the proposed Moore Ranch Project operated to identify new raptor nests and to assess whether existing nests were being used (Uranium One, 2009b). The purpose of this program would be to protect against unforeseen conditions, such as the construction of a nest in an area that could be potentially affected by the operation of the proposed Moore Ranch Project (EMC, 2007b). If nests were discovered during these surveys, the applicant would take appropriate mitigation measures, such as moving the nest, to ensure the protection of the species.

4.6.1.1.2.4   Impacts to Reptiles and Amphibians

The only herpetological species recorded within the proposed Moore Ranch Project during the applicant's 2006 and 2007 baseline studies was the common bullsnake. Because the potential habitat for reptiles and amphibians is limited within the proposed license area, no impact to reptiles or amphibian populations would be expected.

4.6.1.1.3   Impacts to Aquatic Resources

Because of the limited occurrence of surface water, the potential habitat for aquatic species is also limited within the proposed license area. Aquatic species habitat occurs primarily as intermittent habitat in the small, scattered stock ponds or drainages in the area. Portions of Pine Tree Draw, Simmons Draw, Ninemile Creek, and their intermittent tributaries, occur within the proposed license area, but they are not reliable water sources as discussed in Section 3.5.1.2. No aquatic habitat exists on the proposed Moore Ranch Project to support fish or macroinvertebrates; therefore, there would be no impact to aquatic wildlife.

4.6.1.1.4   Impacts to Threatened and Endangered Species

No federal- or state-listed sensitive plant species, endangered or threatened plant species, or designated critical habitats occur within the proposed license area; therefore, there would be no expected impact to these species.

Environmental Impacts and Mitigative Actions

The bald eagle (formerly listed as threatened, currently delisted) and black-footed ferret (endangered) are the federally-listed, previously-listed, or candidate wildlife species that could occur in the proposed license area (FWS, 2008). However, the potential habitat for bald eagle nesting and roosting activities is limited within the proposed license area and a surrounding 1.6-km [1-mi] perimeter. The nearest documented winter roosting area is approximately 13.7 km [8.5 mi] to the north (BLM, 2008b). Project lands disturbed by uranium ISR activities would be unavailable for foraging bald eagles until these areas were reclaimed and prey species returned. On September 11, 2009, the U.S. Fish and Wildlife Service (FWS) published a rule concerning eagle take permits (74 FR 46836). The NRC contacted the FWS on March 15, 2010, to discuss whether the proposed Moore Ranch Project would require an eagle permit per this rulemaking. The FWS concluded that the NRC would neither need to further consult with FWS nor obtain an eagle take permit for the proposed Moore Ranch Project because no trees with nests would be disturbed (NRC, 2010).

The black-footed ferret occurs in active prairie dog colonies, none of which occur either on or within a 1.6 km [1 mi] radius of the proposed Moore Ranch Project boundary. Therefore, there would be no expected impacts to either the bald eagle or black-footed ferret from construction activities at the proposed Moore Ranch Project.

4.6.1.1.4.1   Impacts to Species of Concern

The proposed Moore Ranch Project has the potential to impact 14 avian species of concern (8 Level I species and 6 Level II species) known to occur or potentially be present as seasonal or year-round residents. Direct impacts such as injury or mortality could occur from vehicle or heavy equipment encounters during construction. Indirect impacts could result from habitat loss or fragmentation and increased noise and activity that could deter use of the area by some species. Surface disturbance would be limited to a total of approximately 61 noncontiguous ha [150 ac] out of 2,879 ha [7,100 ac] and would be greatest during construction. The applicant's enforcement of speed limits during all phases of the proposed Moore Ranch Project would reduce wildlife impacts throughout the year, particularly during the breeding season. Impacts to avian species of concern could potentially be MODERATE. Since construction would occur for only nine months, the potential impact could be reduced to SMALL if the applicant adhered to the mitigation measures outlined in the WGFD and BLM guidelines for seasonal noise vehicular traffic and human proximity (WGFD, 2010; BLM, 2008a and 2010). Furthermore, the proposed activities would not threaten the continued existence of these species in the proposed license area.

4.6.1.2   **Operations Impacts**

Section 4.3.5.2 of the GEIS discussed the alteration of wildlife habitats from operations (fencing, traffic, noise), and noted that individual takes could occur due to conflicts between species habitat and operations. Access to crucial wintering habitat and water could be limited by fencing. The WGFD specifies fencing construction techniques to minimize impediments to big-game movement.

The GEIS further noted the occurrence of temporary contamination or alteration of soils from operational leaks and spills. However, detection and response to leaks and spills (e.g., soil cleanup) and eventual survey and decommissioning of all potentially impacted soil would limit the magnitude of overall impacts to terrestrial ecology. Spill detection and response plans would reduce the potential impact to aquatic species from spills around wellheads and leaks from pipelines. Mitigation measures such as perimeter fencing, netting, leak detection and spill response plans, and periodic wildlife surveys would also limit the impact; therefore, the overall impact would be SMALL.

4.6.1.2.1 Impacts to Vegetation

During the operations phase of the proposed Moore Ranch Project, the wellfields and central plant would be frequently accessed using the defined road network. The installation and operation of the wellfields would involve the excavation of trenches for trunk lines and utilities. Surface disturbance would increase the susceptibility of the disturbed area to invasive and noxious weeds. However, surface disturbance would be minimized and vehicular access would be restricted to specific roads. Disturbed areas would be reseeded with WDEQ-approved seed mixtures to prevent the establishment of competitive weeds. The applicant would monitor invasive and noxious weeds; if they became an issue, other control alternatives, such as the application of an herbicide, would be considered (EMC, 2007b).

Impacts to vegetation resulting from spills around wellheads and leaks from pipelines during facility operations would be SMALL, and the applicant's use of BMPs to handle them (EMC, 2007b), such as leak detection systems and spill response plans to remove affected soils and capture released fluids would reduce the impact.

4.6.1.2.2 Impacts to Wildlife

Wildlife use of areas adjacent to ISR operations would likely initially decline because of human disturbances during milling operations and steadily increase to near-normal levels as animals became habituated to the activity. Some wildlife habituation to human activities may have occurred in the area because of ongoing CBM activities. Because wildlife may be in close proximity to the central plant, roads, and wellfields, some impacts to wildlife would be expected from direct conflict with vehicular traffic and the presence of onsite personnel. These impacts would be SMALL because only a few individual animals would be affected and the continued existence of any particular species in the proposed license area would not be affected. During facility operations, spills around wellheads and leaks from pipelines could expose wildlife to potentially toxic levels of chemicals. Leak detection systems and spill response plans to remove affected soils and to capture released fluids would minimize the impact. The applicant's use of BMP to handle spills or leaks would result in SMALL impact to wildlife. Further mitigation such as the applicant's use of fencing described in Section 4.6.1.1.2 of this SEIS could be used to maintain a SMALL impact on wildlife.

4.6.1.2.2.1 Impacts to Big Game

The potential impact to big game during the operations phase would either be similar to or less than that described for the construction phase because limited earth-moving activities would occur. Therefore, there could be SMALL impacts to big-game species during the operations phase.

4.6.1.2.2.2 Impacts to Other Mammals

The potential impact to other mammals during operations would be similar to or less than that described for the ISR construction phase. Because only a few individuals would be affected

Environmental Impacts and Mitigative Actions

and most mammal species would likely travel to suitable habitat adjacent to the operating areas, the proposed Moore Ranch Project would have a SMALL impact on these mammals.

4.6.1.2.2.3 Impacts to Avian Species

The potential impact to upland game birds, waterfowl and shorebirds, and to raptors would either be the same or less than that described for the construction phase because earth-moving activities would be more limited during the operations phase; therefore, the potential impact would be SMALL.

4.6.1.2.2.4 Impacts to Reptiles and Amphibians

The potential impact to reptiles and amphibians from the operations phase would be comparable to that described for the construction phase. Because the potential habitat for reptiles and amphibians is limited within the proposed license area, the potential impact would be limited and SMALL.

4.6.1.2.3 Impacts to Aquatic Resources

Because of the limited occurrence of surface water on the proposed Moore Ranch Project and because the operating plans do not require surface water discharge, the potential impact to aquatic resources would be SMALL.

4.6.1.2.4 Impacts to Threatened and Endangered Species

No impacts to federally-listed threatened and endangered species would occur during the operations phase because these species have not been identified within the proposed license area.

Continued mitigation would be implemented to ensure potential impacts to threatened and endangered species remain SMALL. Examples of mitigation that would benefit threatened and endangered species are applicant spill procedures, fencing around the central plant and wellfields, and activity timing restrictions (EMC, 2007b).

4.6.1.2.4.1 Impacts to Species of Concern

As described in Section 4.6.1.1.4.1, the operation of the proposed Moore Ranch Project has the potential to impact 14 avian species of concern (8 Level I species and 6 Level II species) known to either occur or potentially occur as seasonal or year-round residents. Impacts to species of concern during facility operation would either be similar to or less than, those impacts described during construction because the facilities and infrastructure would remain in place for the life of the milling operation. If the applicant continued to follow the BMP described for the ISR construction phase in Section 4.6.1.1, the potential impact to species of concern during the operations phase would be SMALL. Only a few individual species would be affected during operations and the continued existence of any particular species would not be threatened. Therefore, the potential impact to avian species of concern would be SMALL.

Environmental Impacts and Mitigative Actions

4.6.1.3 Aquifer Restoration Impacts

Section 4.3.5.3 of the GEIS discussed the potential impacts to ecological resources during the aquifer restoration phase. These impacts could include habitat disruption, but because existing (in-place) infrastructure would be used, little additional ground disturbance would be expected.

The GEIS also discussed contamination of soils and surface waters that could result from leaks and spills. However, detection and response techniques, and eventual survey and decommissioning of all potentially impacted soils and sediments, would limit the magnitude of overall impacts to terrestrial and aquatic ecology. Implementation of mitigation measures such as perimeter fencing, netting, and leak detection and spill response plans would maintain SMALL impacts.

There would be no expected impacts to threatened and endangered species beyond that which occurred during the construction phase because the existing infrastructure from the operations phase would continue to be used. Therefore, the overall impact to threatened and endangered species would be SMALL.

Because the existing infrastructure would be in place, the potential impact to ecological resources from aquifer restoration activities would be similar to that experienced during the operations phase; therefore, the potential impact to vegetation and wildlife would be SMALL. If the applicant adhered to the WGFD and BLM seasonal noise, vehicular traffic, and human proximity guidelines (WGFD, 2010; BLM, 2008a and 2010), these mitigation measures would further reduce the potential impact from noise, vehicular traffic, and human proximity.

4.6.1.4 Decommissioning Impacts

Section 4.3.5.4 of the GEIS discussed temporary land disturbance during decommissioning and reclamation from soil excavation, recovery and removal of buried piping, and the demolition and removal of structures. However, revegetation and recontouring would restore habitat previously altered during construction and operations. Wildlife would be temporarily displaced, but could return upon completion of decommissioning and reclamation and the reestablishment of vegetation and habitat. Decommissioning and reclamation activities could result in temporary increases in sediment load in local streams, but aquatic species would recover quickly as sediment load decreased. For these reasons, the GEIS concluded the overall potential impact during decommissioning would be SMALL.

As stated in the GEIS, with respect to threatened and endangered species, potential impacts resulting from individual takes would occur due to conflicts with decommissioning activities (e.g., equipment, traffic). Temporary land disturbance would occur as structures are demolished and removed and the ground surface recontoured. An inventory of threatened or endangered species developed during the site-specific environmental review of the detailed decommissioning plan would identify unique or special habitats, and consultation with the FWS under the Endangered Species Act would help to minimize impacts. Upon completion of decommissioning, revegetation, and recontouring, the habitat would be reestablished; therefore, the potential impact to threatened and endangered species could range from SMALL to LARGE, depending on site conditions.

Impacts to ecological resources during decommissioning of the proposed Moore Ranch Project would be similar to those experienced during the construction phase with respect to noise, traffic

Environmental Impacts and Mitigative Actions

flow, and earth-moving activities. However, the decommissioning phase would not disrupt as much natural habitat as would have occurred during the construction phase of the ISR process since activities would be conducted in the previously disturbed areas of the site. The applicant estimated a 12-month duration for the ISR decommissioning phase, which would be reduced with time as decommissioning and reclamation progressed.

Decommissioning would involve abandonment of the central plant, office and maintenance buildings, and wellfields and removal of surface equipment consisting of the injection and production feed lines and buried wellfield piping. Stockpiled topsoil would be used to regrade the land to preconstruction contours, as required, and seeded with native vegetation when the buildings are removed. No loss of vegetative communities beyond that disturbed during construction would occur. Piping removal would impact vegetation that has reestablished itself, although this, too, would be temporary when the disturbed soil is reseeded. The decommissioning process would create increased noise and traffic as buildings are taken down and hauled away. During this time, wildlife could either come in conflict with heavy equipment or could move elsewhere on the property due to higher-than-normal noise. As required, the applicant would submit a decommissioning plan for NRC review and approval. Temporarily displaced wildlife could return to the area when decommissioning and reclamation were completed.

Decommissioning impacts would be temporary and SMALL. The applicant's implementation of the previously discussed mitigation measures would further reduce the potential impact.

4.6.2 No-Action (Alternative 2)

Under the No-Action alternative, there would be no ISR facility construction associated with this project, and therefore no land disturbance or vegetation removal associated with construction, operation, aquifer restoration, or decommissioning. The area would continue to support vegetation communities and wildlife habitat typical of the region. Land would continue to be used for pastureland, and grazing leases would continue. Grazing of existing vegetation, particularly in the grassland communities, would continue. Existing wildlife within the proposed license area could be affected if continued cattle grazing destroyed wildlife habitat or if species are displaced by cattle populations due to lack of forage and cover. However, only a few individual species would be affected, and they could relocate to suitable adjacent habitats. There would be no impacts to ecological resources under this alternative compared to the proposed action.

4.7 Air Quality Impacts

Potential environmental impacts to air quality at the proposed Moore Ranch Project could occur during all phases of the ISR facility lifecycle. Nonradiological air emission impacts primarily involve fugitive road dust from vehicles used throughout the facility lifecycle and combustion engine emissions from diesel equipment associated with construction, operation, and decommissioning activities. Other air emissions may be associated with radon releases from well system relief valves, resin transfer, or elution. Potential radiological air impacts, including radon release impacts, are addressed in the Public and Occupational Health and Safety Impacts analyses in Section 4.13.

Factors used by the NRC staff in determining the significance of the potential impacts are described in Section 1.7.2 of the GEIS and include (1) whether the air quality for the site region

Environmental Impacts and Mitigative Actions

of influence is in compliance with the National Ambient Air Quality Standards (NAAQS) and (2) whether the facility can be classified as a major source under the New Source Review or operating (Title V of the Clean Air Act) permit programs. An additional concern would be the presence of Prevention of Significant Deterioration (PSD) Class I areas within the region that could be impacted by emissions from the proposed action. All three of these criteria would be met for the proposed Moore Ranch ISR Project, as discussed in the following paragraphs.

Air emissions from the proposed Moore Ranch Project would be expected to comply with the conditions of a WDEQ-approved construction air permit (application under review at the time of writing, as shown in Table 1-2) and a WDEQ minor source operating permit, if required. In addition, all of the nonradiological emissions estimates evaluated by the NRC staff (Section 2.1.1.1.6.1) support the conclusion that the proposed action would not be comparable to, nor considered, a major source of emissions and that such emissions (i.e., well below the major source thresholds) in an area with meteorology that is often favorable for dispersion (Section 3.7.1.2) would be unlikely to impact attainment of ambient air quality standards in the region surrounding the location. The NRC staff expects that emissions at levels well below the major source thresholds would not destabilize local air quality, although localized, short-term and intermittent visible air emissions would be possible in the surrounding area (i.e., when vehicles travel on unpaved roads).

As described in Section 3.7.2, the air quality of the region where the proposed Moore Ranch Project is located is classified as being in attainment for all of the NAAQS primary pollutants. The nearest PSD Class 1 area, Wind Cave National Park, located about 188 km [117 mi] east of the Moore Ranch site and Cloud Peak Wilderness Area, the closest Class II area located about 124 km [77 mi] northwest of the Moore Ranch site, are both classified as attainment areas. The attainment status of the air quality surrounding the proposed license area provides a measure of current air quality conditions and affects considerations for allowing new emission sources.

Based on construction air quality permits obtained for similar projects (WDEQ, 2009; 2010), the NRC staff expects that WDEQ would not consider the proposed facility to be a major source for emissions. In addition, the NRC staff estimated mobile nonroad emissions from construction equipment (Section 2.1.1.1.6.1) that are not addressed by WDEQ air permitting and found these emissions were also well below major source threshold levels. The low magnitude of emissions directly affects the potential for air quality impacts and, therefore, the level of detailed review NRC considered necessary to adequately evaluate potential impacts.

All phases of the proposed Moore Ranch Project would also result in greenhouse gas emissions, principally carbon dioxide (CO_2), however, the majority of these emissions would be from the use of diesel-powered equipment (including well drilling rigs) during the construction and decommissioning phases (Section 2.1.1.1.6.1 and Appendix D). Based on methods described in detail in Appendix D, the NRC staff calculated a maximum annual CO_2 emission from this diesel-powered equipment of 852 t/yr [940 T/yr] and cumulative CO_2 emissions (total facility lifecycle emissions) for the proposed facility lifecycle as 2400 t/yr [2600 T/yr]. For comparison, these calculated emissions from the proposed action are a small fraction of the net total of greenhouse gases produced annually in Wyoming at 20,000,000 t [22,000,000 T] (Center for Climate Strategies, 2007) and for the United States at 6,000,000,000 t [6,600,000,000 T] (EPA, 2009). Based on its assessment of the relatively small carbon footprint of the proposed facility as compared to the annual CO_2 emissions in both the State of Wyoming and the United States, the NRC staff concluded that the atmospheric impacts of greenhouse gases from the proposed facility lifecycle would not be noticeable, and additional mitigation would not be warranted.

Environmental Impacts and Mitigative Actions

In general, nonradiological emissions from pipeline system venting, resin transfer, and elution would be rapidly dispersed into the atmosphere and would be small, primarily due to the low volume of effluent produced. Such emissions were not considered in the following analysis. Detailed discussion of the potential environmental impacts to air quality from construction, operation, aquifer restoration, and decommissioning the proposed Moore Ranch Project are provided in the following sections.

4.7.1 Proposed Action (Alternative 1)

4.7.1.1 Construction Impacts

As discussed in the GEIS (Section 4.3.6.1), fugitive dust and combustion (vehicle and diesel equipment) emissions during land-disturbing activities associated with construction would be short term and reduced through BMPs (e.g., wetting of roads and cleared land areas to reduce dust emissions). The GEIS also estimated fugitive dust emissions during ISR construction would be well below the NAAQS for Particulate Matter$_{2.5}$ ($PM_{2.5}$) and for Particulate Matter$_{10}$ (PM_{10}). Additionally, the GEIS concluded particulate, sulfur dioxide, and nitrogen dioxide emissions from ISR facilities would be at a small percentage (1 to 9 percent) of the PSD Class II allowable increments. For NAAQS attainment areas, like the area around the proposed Moore Ranch Project, nonradiological air quality impacts would be SMALL.

The proposed Moore Ranch Project would meet the conditions pertaining to air quality specified in the GEIS as discussed in Section 4.7, and therefore, impacts would be SMALL. This conclusion is further supported by the limited footprint of the construction area relative to the proposed project area, the low volume of traffic generated by the proposed action (Section 4.3.1.1), and the short length {0.8 km [0.5 mi]} of the facility access road to connect to SR 387 (Section 3.3). The applicant proposes to apply water or other agents to control fugitive dust emissions (Uranium One, 2009b). Despite the use of controls, short-term and intermittent visible air emissions are possible to the local area surrounding the proposed project site when vehicles travel on unpaved roads. Therefore, short-term and intermittent MODERATE impacts from fugitive road dust are possible, however, the average air quality is expected to remain in compliance with ambient standards and overall impacts would be SMALL.

Emissions from diesel combustion engines in drilling rigs and construction equipment used during the construction phase were calculated by the NRC staff and are discussed in Section 2.1.1.1.6.1 and Appendix D. These calculations addressed emissions of Nitrogen Oxides (NO_x), Carbon Monoxide (CO), Sulfur Oxides (SO_x), Particulate Matter (PM_{10}), Formaldehyde, Volatile Organic Compounds (VOC), and Carbon Dioxide (CO_2). The results show Nitrogen Oxides and Carbon Monoxide are the highest emissions of the criteria pollutants evaluated. Estimated emissions of these pollutants are well below major source threshold levels. The calculated annual pollutant emissions for NO_x is 18.1 t/yr [20 T/yr] assuming two of the four proposed deep disposal wells were drilled in the same year as the first wellfield. If the NRC staff assumed all four deep disposal wells were drilled in the same year as the first wellfield, the annual NO_x emission result increases to 30 t/yr [33 T/yr]. This higher level of calculated emissions is still below the 91 t/yr [100 T/yr] major source threshold and is considered by the NRC staff to represent a single-year peak while all proposed deep wells would be completed.

The diesel combustion engine emissions calculated by the NRC staff for the proposed action are below those reported in the GEIS from a prior NRC Environmental Impact Statement (EIS)

(NRC, 1997) for a proposed ISR facility in Crownpoint, New Mexico; and therefore, the potential impacts to air quality from the proposed action would be less than those reported in the GEIS. The NRC staff considers the emissions and associated potential air impacts from constructing the Crownpoint facility to bound the emissions from constructing the proposed Moore Ranch Project based on the following considerations. First, the Crownpoint facility proposed a higher maximum annual production rate than the proposed Moore Ranch Project. The ore deposits at the Crownpoint facility are at a much greater depth and, therefore, would require longer drilling times per well during wellfield construction. For example, the Crownpoint ISR facility has ore occurring at an approximate depth of 561 m [1,840 ft] below ground surface, whereas the proposed Moore Ranch Project has ore occurring at depths that range from 76.2 to 91.5 m [250 to 300 ft] below the ground surface (EMC, 2007b).

Second, the meteorology used at the Crownpoint site to estimate average annual air concentrations of emitted pollutants is also more stable than at the proposed Moore Ranch Project, based on the NRC staff review of available joint frequency data for each site (NRC, 1997; EMC, 2007b), which indicated winds that fall within stability classes E and F occur about two times as frequently at the Crownpoint site than in the region surrounding the proposed Moore Ranch Project. The annual average air concentrations for the Crownpoint emissions are also based on a mixing height of 1 km [1.6 mi] (NRC, 1997), which is within the range of mixing heights reported for the State of Wyoming of 659 m [718 yd] (morning average) and 4,074 m [4,440 yd] (afternoon average) (EPA, 2010). Based on the information reviewed, the NRC staff expects the dispersion conditions at the Crownpoint site would be less favorable than at the proposed Moore Ranch Project and; therefore, based on the combination of dispersion conditions and higher emissions estimates for the Crownpoint facility, the NRC staff concluded the calculated annual average air concentration emission values reported in the GEIS are conservative and, therefore, applicable to the proposed Moore Ranch Project. As a result, the GEIS conclusions that particulate, sulfur dioxide (SO_2), and nitrogen dioxide (NO_2) emissions from ISR facilities would be expected to be well below the major source threshold for NAAQS attainment areas and account for a small percentage (1 to 9 percent) of the PSD Class II allowable increments would also be applicable to the proposed Moore Ranch Project.

The NRC staff considered the calculated magnitude of construction emissions, in an area that meets current air quality standards, is not sufficient to justify conducting additional detailed quantitative air quality modeling analysis of potential consequences. Considering (i) the aforementioned analyses in the GEIS, (ii) that other recently proposed ISR facilities have received WDEQ construction air quality permits that subject the facility to minor source operational permitting (WDEQ, 2009; 2010), and (iii) the conditions of the site area and region, the NRC staff concludes that such emissions (i.e., well below the major source thresholds) in an area with meteorology that is often favorable for dispersion would be unlikely to impact air quality locally, regionally, or in the nearest Class I or II areas. The nearest Class I area, Wind Cave National Park, is located approximately 188 km [117 miles] east of the Moore Ranch site. While the prevailing wind directions of west-southwest and west (Section 3.7.1.2) could transport pollutants in the direction of the Class I area, because of the long distance, the potential emissions would disperse. The Cloud Peak Wilderness Area is the closest Class II area to the proposed action, located about 124 km [77 mi] northwest of the proposed Moore Ranch Project. In addition to the low magnitude of emissions and distance, the prevailing wind direction would carry emissions to the northeast and east, away from the direction of this Class II area.

Environmental Impacts and Mitigative Actions

The NRC staff concludes that the site-specific conditions at the proposed Moore Ranch Project are comparable to those described in the GEIS for air quality and incorporates by reference the GEIS conclusions that the impacts to air quality during construction would be SMALL. The NRC staff has not identified any new and significant information during its independent review that would change the expected environmental impact beyond those discussed in the GEIS.

4.7.1.2 Operation Impacts

The GEIS (Section 4.3.6.2) noted that operating ISR facilities are not major point source emitters and are not expected to be classified as major sources under the operation (Title V) permitting program. Additionally, the GEIS concluded that although excess vapor pressure in pipelines could be vented throughout the system, such emissions would be rapidly dispersed in the atmosphere; therefore, potential impacts would be SMALL, due in part to the expected low volume of gaseous effluent produced. The GEIS also stated that other potential nonradiological emissions during operations include fugitive dust and combustion engine emissions from equipment, transport trucks, and other vehicles. For NAAQS attainment areas, the GEIS concluded that nonradiological air quality impacts would be SMALL.

Since the number of commuting vehicles and equipment operating at the proposed Moore Ranch Project would not exceed that evaluated in the GEIS, the potential impact would be SMALL. The applicant's proposed mitigative measures described in Section 4.7.1.1 would further reduce the potential impact.

The NRC staff concludes that the site-specific conditions at the proposed Moore Ranch Project are comparable to those described in the GEIS for air quality and incorporates by reference the GEIS conclusions that the impacts to air quality during operations would be SMALL. The NRC staff has not identified any new and significant information during its independent review that would change the expected environmental impact beyond that described in the GEIS.

4.7.1.3 Aquifer Restoration Impacts

As discussed in the GEIS (Section 4.3.6.3), because the same infrastructure is used during aquifer restoration as during operations, air quality impacts from aquifer restoration would be similar to, or less than, those during operations (NRC, 2009a). Additionally, fugitive dust and combustion engine emissions from vehicles and equipment during aquifer restoration would be similar to, or less than, the dust and combustion engine emissions during operations. For NAAQS attainment areas, nonradiological air quality impacts would be SMALL.

Vehicular traffic during the aquifer restoration phase would be limited to delivery of supplies and commuting staff, with a decreasing frequency of offsite yellowcake shipments as restoration proceeds. Therefore, fewer trips would occur than during the operation phase.

Air quality at the proposed Moore Ranch Project would not be substantially affected by the aquifer restoration activities because fewer vehicles would be required during this phase of the project; therefore, the potential impact would be SMALL. The applicant's proposed mitigative measures described in Section 4.7.1.1 would further reduce the potential impact.

The NRC staff concludes that the site-specific conditions at the proposed Moore Ranch Project are comparable to those described in the GEIS for air quality and incorporates by reference the GEIS conclusions that the impacts to air quality during aquifer restoration would be SMALL.

The NRC staff has not identified any new and significant information during its independent review that would change the expected environmental impact beyond that discussed in the GEIS.

4.7.1.4 Decommissioning Impacts

Section 4.3.6.4 of the GEIS noted that fugitive dust, vehicle emissions, and diesel emissions during land-disturbing activities associated with decommissioning would come from many of the same sources as used during construction. In the short term, emission levels would be expected to increase given the activity (demolishing of process and administrative buildings, excavating and removing contaminated soils, grading of disturbed areas). However, such emissions would decrease as decommissioning proceeds, and therefore, overall impacts would be similar to, or less than, those associated with construction, would be short term, and would be reduced through BMP (e.g., dust suppression). Based on the NRC staff calculated emission estimates discussed in Section 2.1.1.1.6.1 and Appendix D, the emissions from diesel-powered construction equipment during the decommissioning phase would be less than the diesel emissions during the construction phase. As discussed in Section 4.7.1.1, considering the minor source classification of emissions indicated by the WDEQ for other recently proposed ISR facilities (WDEQ, 2009, 2010), and the conditions of the site area and region, the NRC staff conclude that such emissions (i.e., well below the major source thresholds) in an area with meteorology that is often favorable for dispersion would be unlikely to impact air quality locally, regionally, or in the nearest Class I or II areas to the proposed action. Therefore, for NAAQS attainment areas, nonradiological air quality impacts would be SMALL.

The NRC staff concludes that the site-specific conditions at the proposed Moore Ranch Project are comparable to those described in the GEIS for air quality and incorporates by reference the GEIS conclusions that the impacts to air quality during decommissioning would be SMALL. The NRC staff has not identified any new and significant information during its independent review that would change the expected environmental impact beyond that disclosed in the GEIS.

4.7.2 No-Action (Alternative 2)

Under the No-Action alternative, in the next few years, there would be no change in air quality at the proposed Moore Ranch Project or at any surrounding receptors. While oil and gas extraction activities would continue and perhaps expand in the future (along with CBM operations), these activities have been shown to have a small impact – direct, indirect, or cumulative – on air quality, regardless of geographic scale (BLM, 2003). The generation of fugitive dust is currently limited by the fact that existing roads are shared and maintained by the natural resource extraction and ranching operations that occur in the area. Roads are also maintained in good repair by these entities and restricted from unpermitted uses.

This area currently meets the NAAQS for attainment status (Section 3.7.2), and because there are no significant air pollution sources at the proposed site, it is expected that this area would continue to meet the NAAQS. Current projections of air quality for the broader Powder River Basin area and the surrounding region over the next decade are discussed in Section 5.7 of this SEIS.

Environmental Impacts and Mitigative Actions

4.8 Noise Impacts

Potential environmental impacts from noise at the Moore Ranch site could occur during all phases of the ISR facility lifecycle. These impacts would be associated with the operation of equipment such as trucks, bulldozers, and compressors; from traffic due to commuting workers or material and waste shipments; and wellfield and central processing plant activities and equipment. These impacts could affect both humans and wildlife in the vicinity of the site.

As stated in the GEIS, the Occupational Safety and Health Administration (OSHA) has set permissible exposure limits for workplace noise levels (NRC, 2009a). The proposed Moore Ranch Project would be required to limit worker exposure in accordance with these regulations; therefore, occupational noise exposure is not discussed in this section but rather in Section 4.13. Instead, this section discusses the potential dispersion of noise impacts to off-site receptors described in Section 3.8 (NRC, 2009a).

The noise analysis evaluated both mobile and stationary noise sources to assess the potential to impact sound levels adjacent to the proposed Moore Ranch Project and to determine the site-specific impact. The GEIS concluded that the noise impact at an ISR facility could range from SMALL to MODERATE during all four phases of an ISR project, depending on the distance between the nearest resident and the activities occurring at the ISR facility (NRC, 2009a). Detailed discussion of the potential environmental impacts from noise due to construction, operation, aquifer restoration, and decommissioning are provided in the following sections.

4.8.1 Proposed Action (Alternative 1)

4.8.1.1 Construction Impacts

As discussed in Section 4.3.7.1 of the GEIS, potential noise impacts would be greatest during construction of an ISR facility, because of the heavy equipment involved and given the likelihood that these facilities would be built in rural, previously undeveloped areas where background noise levels are lower. The use of drill rigs, heavy trucks, bulldozers, and other equipment used to construct and operate the wellfields, drill wells, construct access roads, and build the production facilities would generate noise that would be audible above the undisturbed background levels. Noise levels would likely be higher during daylight hours when construction is more likely to occur and more noticeable in proximity to the operating equipment. Administrative and engineering controls would maintain noise levels in work areas below OSHA regulatory limits and mitigated by use of personal hearing protection. For individuals living in the vicinity of the site, ambient noise levels would return to background levels at a distance greater than 300 m [1,000 ft] from the construction activities. Wildlife would be expected to avoid areas where noise-generating activities were occurring; although for certain wildlife (e.g., sage-grouse) continuous elevated noise levels could reduce their breeding success. Overall, these types of noise impacts would be SMALL, given the distance to the nearest resident.

Additionally, as stated in the GEIS, traffic noise during construction (e.g., commuting workers; truck shipments to and from the facility; and construction equipment such as trucks, bulldozers, and compressors) would be localized, and limited to highways in the vicinity of the site, access roads within the site, and roads in the wellfields. Relative short-term increases in noise levels associated with passing traffic would be SMALL for the larger roads, but could be MODERATE for lightly traveled rural roads through smaller communities.

Environmental Impacts and Mitigative Actions

As noted in Section 3.8 of this SEIS, the construction phase of the proposed Moore Ranch ISR Project would involve the use of heavy equipment to create and improve road surfaces, furnish supplies, excavate footings, erect buildings, and install the wells and pipelines at the wellfields. Equipment such as bulldozers, graders, tractor trailers, excavators, cranes, and drill rigs would create noise that would be audible onsite above background noise levels estimated as 40 decibels. However, since the nearest residence is about 4.5 km (2.8 mi) from the center of the proposed Moore Ranch Project (Uranium One, 2009a), they would not notice a change in background noise. The applicant would enforce site speed limits to further mitigate traffic noise impacts (EMC, 2007a).

Truck transport of construction materials would be the primary noise source that could potentially affect the public. However, because of the limited traffic volume associated with the proposed project, as discussed in Section 4.3 of the SEIS, this impact would be minor. The incremental increase in project-related traffic on the relatively well-traveled public roadways in the area (e.g., I-25, SR 387, SR 50, and SR 59) would not be noticeable. Thus, project-related transportation noise impacts would be SMALL.

The NRC staff concludes that the site-specific conditions at the proposed Moore Ranch Project are comparable to that considered in the GEIS. The NRC staff concludes that site-specific impacts for the proposed Moore Ranch Project would be SMALL. Furthermore, the NRC staff has not identified any new and significant information during its independent review that would change the expected environmental impact beyond that discussed in the GEIS.

4.8.1.2 Operation Impacts

Section 4.3.7.2 of the GEIS discussed ISR operations activities that could generate noise. These activities would occur indoors within the central uranium processing facility; therefore, offsite noise from plant operations would be less than could be heard during the construction phase of an ISR project. Wellfield equipment (e.g., pumps, compressors) would be contained within structures (e.g., header houses, satellite facilities), also reducing the potential for noise to be heard by offsite individuals. Traffic noise from commuting workers, truck shipments to and from the facility, and facility equipment would likely be localized, limited to highways in the vicinity of the site, access roads within the proposed license area, and wellfield roads. Relative short-term increases in noise levels associated with this traffic would be SMALL for the larger roads, but could be MODERATE for lightly traveled rural roads through smaller communities.

As noted in the GEIS and described above, the staff assumed that a variety of mechanical equipment located at the central plant at the proposed Moore Ranch Project would generate noise. However, because the nearest residence is located about 4.5 km [2.8 mi] from the center of the proposed Moore Ranch Project, this person would not notice any change in sound from the operations phase. Taking into account the relatively small increase in traffic (see Section 4.3.1.2), the potential noise impacts during operations at the proposed Moore Ranch Project would be SMALL.

4.8.1.3 Aquifer Restoration Impacts

Section 4.7.3.3 of the GEIS stated that general noise levels during aquifer restoration would be expected to be similar, or less than, noise levels during operations. Additionally, workplace noise exposure during aquifer restoration would use the same administrative and engineering controls used during operations. Pumps and other wellfield equipment contained in buildings

Environmental Impacts and Mitigative Actions

would reduce sound levels to offsite receptors. Existing operational infrastructure would be used, and traffic levels would be expected to be less than during construction and operation phases of an ISR facility. Therefore, the potential impacts could range from SMALL to MODERATE, depending on the location of the nearest resident.

The types of activities described in the GEIS would occur at the proposed Moore Ranch Project. Vehicular traffic would be limited to delivery of supplies and staff travel to and from the site; therefore, fewer trips would occur than during the operations phase. Because the nearest residence is located about 4.5 km (2.8 mi) from the center of the proposed Moore Ranch Project (Uranium One, 2009a), this person would not notice a change in background noise. Taking into account the relatively small increase in traffic, (see Section 4.3.1.3), the potential noise impacts during aquifer restoration at the proposed Moore Ranch Project would be SMALL.

4.8.1.4 Decommissioning Impacts

Section 4.7.3.4 of the GEIS discussed the potential noise impact from decommissioning activities at an ISR facility. Noise levels generated during decommissioning and reclamation would be expected to be similar to or less than, noise levels during the ISR construction phase. Equipment used to dismantle buildings and milling equipment, remove potentially contaminated soils, or for surface grading during reclamation would generate above background noise levels. This noise would be temporary and once decommissioning and reclamation activities were complete, noise levels would return to baseline, with occasional vehicle traffic for any longer term monitoring activities. Like the construction phase of an ISR project, noise levels would be higher during daylight hours when decommissioning and reclamation would be more likely to occur, and would be more noticeable in proximity to the operating equipment. Workplace noise exposure would be managed using the same administrative and engineering controls implemented for the construction phase, and given the likely distance to nearby residents (i.e., greater than 300 m [1,000 ft]), the GEIS concluded that the noise from decommissioning activities would not be discernable to offsite residents or communities.

The noise during decommissioning of the proposed Moore Ranch Project would be similar to that experienced during construction activities and would be generated by earthmoving, excavation, and building demolition activities. Therefore, the noise impacts would either be similar to, or less noise than, during the construction activities at the site. Decommissioning activities would result in a large, but temporary noise impact onsite and just beyond the plant boundary. At the location of the nearest resident, located at a distance of about 4.5 km (2.8 mi) from the center of the proposed Moore Ranch Project (Uranium One, 2009a), there would be no change in background noise; therefore, anticipated noise impacts during decommissioning would be SMALL. As noted in Section 4.3.1.4 of this SEIS, the increase in truck traffic associated with decommissioning activities would be less than 2 percent. Therefore, transportation related noise impacts associated with the transfer of solid waste to the Midwest-Edgerton Landfill or transfer station and of byproduct material to a licensed facility would result in a SMALL impact above background noise levels.

4.8.2 No-Action (Alternative 2)

Under the No-Action alternative, there would be no change in the sound levels either within the proposed license area or to surrounding receptors. While natural resource exploration activities would continue and could potentially expand in the future, they would typically be of short duration and would involve few vehicles and no permanent, noise emitting infrastructure. The rural setting of the proposed project area and the continuation of ongoing natural resource exploration activities would result in sound levels remaining at or below 40 dBA.

4.9 Impacts to Historical and Cultural Resources

Potential environmental impacts to historic and cultural resources at the proposed Moore Ranch Project could occur during all phases of the ISR facility lifecycle. These impacts would predominantly result from the loss of or damage to historical, cultural, and archaeological resources and from temporary access restrictions to these resources.

Detailed discussion of the potential environmental impacts to historic and cultural resources from construction, operation, aquifer restoration, and decommissioning of the proposed Moore Ranch Project are provided in the following sections.

4.9.1 Proposed Action (Alternative 1)

Under the proposed action, NRC would issue a license for ISR uranium milling and processing at the proposed Moore Ranch Project. The area that could be directly disturbed by the proposed action would be within approximately 61 ha [150 ac] of the 2,879 ha [7,110 ac] proposed license area. For archaeological sites, the impacts from various actions are linked to the physical footprints associated with the proposed action. The potentially impacted areas are described in detail in Section 2.1.1.1.2 and discussed in the following paragraphs. At the proposed Moore Ranch Project, a central plant, two wellfields and access roads would be constructed and pipeline would be laid.

The construction of the central plant and storage facilities would disturb approximately 2.4 ha [6 ac]. Construction of the wellfields and a new access road would disturb approximately 23 ha [57 ac]. An existing two-track access road would connect SR 387 to service both Wellfields 1 and 2. However, a new secondary access road would be constructed to connect Brown Road to the central plant (Figure 2-1). This new road would extend east from the main two-track road (Brown Road) to service the central plant. This secondary access road would encompass about 0.7 ha [1.77 ac]. An ISR trunkline would connect Wellfield 2 and the central plant. Various wellfield-specific service roads also exist and they would connect the header houses, injection wells, and monitoring wells.

4.9.1.1 Construction Impacts

Section 4.3.8.1 of the GEIS discussed the potential impact to historic and cultural resources from excavation during the construction phase of an ISR facility (NRC, 2009a). Access to historical, cultural, and archaeological resources could also be temporarily restricted during the construction phase.

An applicant would be expected to conduct the appropriate historic and cultural resource surveys as part of prelicense application activities. The GEIS also noted that eligibility determination for listing in the *National Register of Historic Places* (NRHP) under criteria in 36 CFR 60.4(a)–(d) or as traditional cultural properties (TCPs), or both, would be conducted as part of the site-specific review.

TCPs are historic and cultural resources that are important for a group to maintain its cultural heritage and are most often associated with Native American religious or cultural practices.

Environmental Impacts and Mitigative Actions

Most TCPs can be identified only through consultation with federally-recognized Native American Tribes. To determine the presence of significant cultural resources and to mitigate potential impacts, consultation amongst the NRC, the applicant, the State Historic Preservation Offices (SHPO), other government agencies (e.g., BLM and State Environmental Departments), and Native American Tribes [Tribal government or designated Tribal Historic Preservation Office (THPO)] would be conducted as part of the site-specific review. In order for a property to be eligible as a TCP (National Register Bulletin 38, Guidelines for Evaluating and Documenting Traditional Cultural Properties), it must be eligible under one of the four eligibility criteria stipulated in 36 CFR 60.4. In addition, as discussed in the GEIS, an NRC licensee shall be required (under conditions in its license), to stop work upon discovery of previously undocumented historic or cultural resources and to consult with the appropriate federal, tribal, and state agencies with regard to the appropriate mitigation measures. The GEIS concluded that the potential impact to historic and cultural resources during the construction phase of an ISR project could range from SMALL to LARGE, depending on the site-specific conditions.

Brunette (2007) reported the results of various Class III surveys that have been conducted at the proposed Moore Ranch Project. The archaeological sites and isolated finds that could be affected during construction were determined to be ineligible for listing on the NRHP (Brunette, 2007). No sites recommended as eligible for listing on the NRHP would be affected by the proposed action. The Wyoming SHPO concurred that the sites located within the project area are ineligible for listing on the NRHP (Wyoming SHPO, 2009). It is recommended by the Wyoming SHPO that the NRC allow the proposed project to proceed in accordance with State and Federal laws, subject to the following stipulation: If any cultural materials are discovered during construction, work in the area shall halt immediately, the federal agency and SHPO staff be contacted, and the materials be evaluated by an archaeologist or historian meeting the Secretary of the Interior's Professional Qualification Standard (48 FR 22716, Sept. 1983). Additionally, if any future disturbance is planned at the locations of sites 48CA964, 48CA6694, 48CA6696 that evaluative testing be completed and submitted to our office with a determination of site eligibility and project effect. If eligible and adversely affected, a Memorandum of Agreement implementing appropriate mitigative measures would be required (Wyoming SHPO, 2009).

The applicant has agreed to condition the license, if issued, to include a stop-work provision should resources be encountered during construction. Since no sites potentially eligible for listing on the NRHP were identified that could be affected by construction phase activities at the proposed Moore Ranch Project, the potential impact to historical and cultural resources would be SMALL.

4.9.1.2 Operation Impacts

Section 4.3.8.2 of the GEIS concluded that potential impacts to historical, cultural, and archaeological resources from the operations phase of an ISR project would be less than during construction because the infrastructure would be in place and less land disturbance would occur. Upon the discovery of any previously undocumented historic or cultural resources, the licensee would stop work and notify the appropriate federal, tribal, and state agencies with regard to mitigation and because of the limited land disturbance during the operations phase, the GEIS concluded that impacts to historic and cultural resources from ISR operations would be SMALL.

Environmental Impacts and Mitigative Actions

There would be no impacts to historic properties from operations at the proposed Moore Ranch Project. Therefore, the potential impact to historical and cultural resources from operations at the proposed Moore Ranch Project would be SMALL. In accordance with the stipulations identified by the Wyoming SHPO, should ground-disturbing activities (maintenance activities) occur outside of previously surveyed areas, then archaeological surveys would be conducted prior to the activity.

4.9.1.3   Aquifer Restoration Impacts

Section 4.3.8.3 of the GEIS concluded that impacts to historical and cultural resources from aquifer restoration would be either similar to, or less than, potential impacts during the operations phase because aquifer restoration activities would generally be limited to the existing infrastructure and previously disturbed areas (e.g., access roads, central processing facility). Therefore, the GEIS concluded that the potential impact to historic and cultural resources from aquifer restoration activities would be SMALL.

As noted in Section 4.9.1.2 of this SEIS, the impact to historic and cultural resources during the aquifer restoration phase would be similar to that during operations. There would be no impacts on historic properties from the aquifer restoration phase of the proposed Moore Ranch Project; therefore, the impact would be SMALL.

4.9.1.4   Decommissioning Impacts

Section 4.3.8.4 of the GEIS discussed the potential impact to historic and cultural resources from decommissioning. Since decommissioning and reclamation activities would focus on previously disturbed areas, the historic and cultural resources would be known from the investigations conducted prior to construction. Therefore, the GEIS concluded that the potential impacts to historical, cultural, and archaeological resources from decommissioning and reclamation actions would be SMALL.

As noted in Section 4.9.1.2 of this SEIS, the impact to historical and cultural resources during the decommissioning phase would be similar to that during operations. There would be no impacts to historic properties from decommissioning actions; therefore, the impact would be SMALL.

4.9.2   No-Action (Alternative 2)

If the No-Action alternative is selected, there would be no impacts to subsurface or surface historic and cultural resources.

The impact to the resources resulting from the selection of the No-Action alternative considers only the consequences of the proposed action. It does not evaluate impacts to the resource categories that may be occurring from other, nonrelated actions. Other actions that are ongoing in the general area include oil and gas exploration and production and cattle ranching. The impact from cattle ranching on the cultural resources is ongoing. Cultural and ethnographic resources do not have to be inventoried or evaluated for this action to occur. This is not the case, however, for oil and gas exploration and production. State and Federal permits are required and cultural and ethnographic resources are routinely identified and evaluated as part of the permitting process.

Environmental Impacts and Mitigative Actions

One of the archaeological sites identified within the proposed project area (48CA970) has been impacted by oil and gas exploration and production. However, it is likely that most of the archaeological sites have been disturbed by routine cattle grazing. Some sites had been obviously disturbed by two-track roads and cattle fences, for example. If the proposed action is not selected, then impacts to the cultural and historical resources would continue as they have in the past.

4.10 Visual and Scenic Resources Impacts

Potential visual and scenic impacts from the proposed Moore Ranch Project could occur during all phases of the ISR facility lifecycle. These impacts primarily would be associated with the use of equipment such as drill rigs, dust and other emissions from such equipment; construction of the central plant and storage structures, site and wellfield access roads; land clearing and grading activities, and lighting for nighttime operations. Such impacts could be mitigated by rolling topography, color considerations for structures, and dust suppression techniques.

As described in Section 3.10, the BLM Visual Resource Management (VRM) classification of the proposed Moore Ranch Project was VRM Class IV, which allows an activity to contrast with basic elements of the characteristic landscape to a much greater extent (BLM, 2007).

4.10.1 Proposed Action (Alternative 1)

The proposed action would result in temporary, SMALL impacts to the visual and scenic resources of the area. The nature of the impacts would be in keeping with the BLM VRM classification as a Class IV area (see Section 3.10 of this SEIS).

4.10.1.1 Construction Impacts

As discussed in Section 4.3.9.1 of the GEIS, visual impacts during construction can result from equipment (drill rig masts and cranes), dust and diesel emissions from construction equipment, and hillside and roadside cuts. Depending on the location of a proposed ISL facility relative to viewpoints such as highways, process facility construction and drill rigs could be visible. For nighttime operation, the drill rigs would be lighted, and this would create a visual impact because the drill rigs would be most visible and provide the most contrast if they were located on elevated areas. Most impacts would be short term because the construction and drilling equipment would be removed when activities conclude at a specific location. Additionally, because these sites are expected to be in sparsely populated areas and there would be generally rolling topography of the region, most visual impacts during construction would not be expected to be visible from more than about 1 km [0.6 mi]. As previously discussed, Prevention of Significant Deterioration (PSD) Class I areas require more stringent air quality standards that can affect visual impacts; however, there are no PSD Class I areas in the Wyoming East Uranium Milling Region. Finally, proposed ISR facilities are expected to be located more than 16 km [10 mi] from the closest VRM Class II area, and the visual impacts associated with ISR construction would be consistent with the predominant VRM Class III and IV classifications. Therefore, visual impacts associated with ISR construction would be SMALL.

Since the land use surrounding the proposed Moore Ranch Project currently is a VRM Class IV area that has pipelines, wellfields, and utility lines that have previously disturbed the landscape, implementing the proposed action would not change the existing character of the landscape. Because more than 900 wells would be installed to support the ISR operations in Wellfields 1

Environmental Impacts and Mitigative Actions

and 2, multiple drill rigs would likely be operating during wellfield construction. Once a well was completed and conditioned for use, the drill rig would be moved to a new location to drill the next hole. Because temperatures drop below freezing during the winter, the applicant would cover the wellheads for completed wells to prevent freezing and protect the well. These covers would be small, low structures {1 m [3 ft] high and 0.6 m [2 ft] in diameter} and present only a slight contrast with the existing landscape. Unless the topography is extremely flat and void of vegetation, it is likely that these structures would not be visible from distances about 1 km [0.6 mi] or more.

Visual and scenic impacts associated with earthmoving activities during construction would be short term. Roads and structures would be more long lasting but would be removed and reclaimed after operations cease. As noted in Section 3.10, the proposed license area has been classified as BLM VRM Class IV, which allows an activity to have higher contrast with basic elements of the characteristic landscape. Wellfield development would occur first in Wellfield 2 and then in Wellfield 1. Restoration in Wellfield 2 would occur concurrently with operations in Wellfield 1.

The visible surface structures for the proposed Moore Ranch Project include wellhead covers, 13 header houses, electrical distribution lines {6 m [20 ft] wooden poles}, and the central plant {122 × 30 m [400 × 100 ft]}. The proposed project would use both existing and new roads to access each header house and the central plant. Temporary and short-term visual impacts from dust emissions during the construction period in each wellfield would result from header house construction, well drilling, and construction of access roads and electrical distribution lines. Following completion of wellfield installation, temporarily disturbed areas would be reclaimed. The applicant has indicated it would spray water to reduce dust emissions (Uranium One, 2009b), but short-term visibility from dust emissions could be MODERATE (see Section 4.7.1.1). In the longer term (>1 year), however, as major construction activities are completed, dust emissions would decrease. Taking into consideration the VRM Class IV classification for the area surrounding the proposed Moore Ranch Project, overall longer-term construction impacts to visual and scenic resources would be SMALL.

4.10.1.2 Operation Impacts

Section 4.3.9.2 of the GEIS stated that visual impacts during operations would be expected to be less than those associated with construction. Most of the wellfield surface infrastructure would have a low profile, and most piping and cables would be buried. The tallest structures would be expected to include the central uranium processing facility {10 m [30 ft]} and power lines {6 m [20 ft]}. Because these sites are in sparsely populated areas and the topography is typically generally rolling, most visual impacts during operations would not be visible from more than about 1 km [0.6 mi] away. Irregular layout of wellfield surface structures such as wellhead protection and header houses would further reduce visual contrast. The uranium districts in the four regions evaluated in the GEIS are all located more than 16 km [10 mi] from the closest VRM Class II region, and the visual impacts associated with ISR construction would be consistent with the predominant VRM Class III and IV classification. Therefore, the GEIS considered visual and scenic impacts from operations to be SMALL.

Because uranium deposits are typically irregular in shape, the network of pipes, wells, and powerlines (6 m [20 ft] tall) would not be regular in pattern or appearance (i.e., not a grid), reducing visual contrast and associated potential impacts. Each wellhead cover approximately 0.9 m [3 ft] high and 0.6 m [2 ft] in diameter typically consists of a weatherproof structure placed

4-57

Environmental Impacts and Mitigative Actions

over the well and each header house would be a small metal building. The central plant at the proposed Moore Ranch Project would be approximately 122 m [400 ft] by 30 m [100 ft] in size. In addition, maintenance, warehouse, and office structures would be constructed. A disturbance area around each header house would be necessary to provide an adequate area for turnaround of operations and maintenance vehicles. Electrical distribution lines would connect header houses to existing electrical distribution lines.

Extensive CBM development has occurred in the vicinity of the proposed Moore Ranch Project and future development is planned. CBM installations are similar in visual impact to those associated with ISR uranium extraction. CBM wells are installed in a network of approximately 3 wells/km^2 [8 wells/mi^2] connected by underground pipelines to collection and pumping structures that appear similar to ISR header houses. Overhead power lines are installed to each well. As a result of this activity, the BLM has identified the area in the vicinity of the proposed Moore Ranch project as VRM Class IV.

Even though the operations phase of the proposed Moore Ranch Project is estimated to take 12 years, the impacts to visual and scenic resources would be SMALL because of the BLM VRM Class IV classification, the existing natural resource extraction activities ongoing in the area, and the remoteness of the area.

4.10.1.3 Aquifer Restoration Impacts

Section 4.3.9.3 of the GEIS addressed visual and scenic impacts from aquifer restoration. The GEIS stated that aquifer restoration activities would be expected to take place some years after the facility had been in operation and that restoration activities would use in-place infrastructure. As a result, potential visual impacts would be similar to, or less than, those experienced during operations. Therefore, such impacts were expected to be SMALL (NRC, 2009a).

Visual resource impacts from aquifer restoration at the proposed Moore Ranch Project would be similar to those seen in the operations phase described in Section 4.10.1.2 of this SEIS because the buildings and equipment the applicant would use to support restoration would be the same as those used for the operations phase. No modifications to either scenery or topography would occur during restoration. Therefore, impacts to visual and scenic resources from aquifer restoration would be SMALL.

4.10.1.4 Decommissioning Impacts

Section 4.3.9.4 of the GEIS discussed the impact to visual and scenic resources from decommissioning. Because similar equipment use and decommissioning activities would be conducted as those occurring during the construction phase, the potential impact to visual and scenic resources would be similar to, or less than, those experienced during construction. Most potential visual impacts during decommissioning would be temporary and would diminish as equipment removal proceeded and disturbed surfaces became revegetated. NRC licensees are required to conduct final site decommissioning and reclamation under an NRC-approved decommissioning plan, with the goal of returning the landscape to preconstruction conditions. While some roadside cuts and hill slope modifications may persist beyond decommissioning and reclamation, the GEIS analysis determined that visual and scenic impacts from decommissioning would be SMALL (NRC, 2009a).

No modifications to scenery or topography would persist after restoration was complete. When project operations cease (the lifecycle of the proposed Moore Ranch Project is estimated at about 12 years), the applicant would return all lands disturbed by the facility to their preoperation land use for livestock grazing and wildlife habitat unless an alternative use (should the landowner wish to retain any structures) is justified and is approved by the State and the landowner. Reclamation would return the visual landscape to baseline contours and would reduce the visual impact by removing buildings and the associated infrastructure. After reclamation activities were completed, there would be no restrictions on surface use. The applicant would submit a decommissioning plan to the NRC, in accordance with 10 CFR Part 40, prior to final site decommissioning.

During decommissioning and reclamation, temporary impacts to the visual landscape would be similar to, or less than, those during the construction period. For example, equipment used to dismantle buildings and milling equipment, remove contaminated soil, or grade the surface as part of reclamation activities would generate temporary visual contrasts. Visual and scenic resources could also be affected by fugitive dust emissions from decommissioning activities. The applicant has indicated it would spray water to reduce dust emissions (Uranium One, 2009b). Overall impacts to the visual landscape would be temporary, and short-term impacts to visibility from dust emissions could be MODERATE (see Section 4.7.1.1). Once decommissioning and reclamation activities were complete, however, the visual landscape would be returned to baseline except for any required monitoring. Therefore, long-term decommissioning impacts to visual and scenic resources would be SMALL.

4.10.2 No-Action (Alternative 2)

Under the No-Action alternative, there would be no ISR facility construction and, therefore, no change to existing visual and scenic resources at the proposed project area or in the region. The existing pipelines, wellfields, and utility lines within the proposed project area from CBM and gas extraction activities would remain. No additional structures or uses associated with the proposed Moore Ranch Project would be introduced to affect the existing viewscapes and the existing scenic quality would be unchanged. The visual resource classification would remain as BLM Class IV, as defined in Section 3.10 of this SEIS.

Because there would be no ISR facility construction under the No-Action alternative, there would also be no facility operation, restoration, or decommissioning. The existing visual and scenic resources would remain.

4.11 Socioeconomic Impacts

Socioeconomic impacts are defined in terms of changes to the demographic and economic characteristics and social conditions of a region. For example, the number of jobs created by the proposed action could affect regional employment, income, and expenditures. Job creation is characterized by two types: (1) construction-related jobs, which are transient, short in duration, and less likely to have a long-term socioeconomic impact on the region; and (2) operation-related jobs in support of facility operations, which have the greater potential for permanent, long-term socioeconomic impacts in the region.

The socioeconomic region of influence (ROI) represents a geographic area where ISR facility employees and their families would reside, spend their income, and use their benefits, thereby affecting the economic conditions of the region. As previously discussed, the focus of the

Environmental Impacts and Mitigative Actions

analysis in this SEIS is on the impacts of constructing and operating the proposed ISR facility in Campbell County.

Socioeconomic impacts would occur from the construction, operation, aquifer restoration, and decommissioning of the proposed Moore Ranch Project. A discussion of the potential socioeconomic impacts from these actions is presented in the following sections.

4.11.1 Proposed Action (Alternative 1)

4.11.1.1 Construction Impacts

Construction of the proposed Moore Ranch Project is expected to last 9 months and employ 50 workers (EMC, 2007a). Section 4.3.10.1 of the GEIS describes the potential socioeconomic impact from construction of an ISR facility (NRC, 2009a). The GEIS estimated total peak construction employment at an ISR facility to be about 200 people. The GEIS also estimated an additional 140 indirect jobs could be created to support the construction of an ISR facility. The NRC staff concludes that the site-specific impacts of constructing the proposed Moore Ranch Project would be smaller than the impacts described in the GEIS.

It was assumed in the GEIS that most construction workers would stay in larger communities with access to more services. Although some construction workers would commute from outside the county to the construction site, skilled employees (e.g., engineers, accountants, managers) would come from outside the local work force. During construction, workers would temporarily relocate to the proposed project area and contribute to the local economy through the purchase of goods and services and the payment of taxes.

4.11.1.1.1 Demographics

Because of the short duration (9 months) and small size of the construction workforce (50 workers), the impact on demographic conditions in Campbell County from the proposed Moore Ranch Project would be SMALL.

4.11.1.1.2 Income

It is expected that construction workers would be paid at rates typical of the region. Therefore, impacts would be SMALL.

4.11.1.1.3 Housing

The number of construction workers would cause a short-term increase in the demand for temporary (rental) housing units in the county. However, the number of available housing units has been keeping pace with the increase in county population (see SEIS Sections 3.11.1, Demographics, and 3.11.3, Housing). Any changes in employment would have little to no noticeable effect on the availability of housing in Campbell County. Due to the short duration of the construction activity and the availability of housing in the region, there would be little or no employment-related housing impacts. Therefore, the impact would be SMALL.

Environmental Impacts and Mitigative Actions

4.11.1.1.4   Employment Structure

Construction of the proposed Moore Ranch Project would create employment opportunities for 50 construction workers, with the potential of up to 35 jobs being generated to support this activity in the local economy. Because of the short duration (9 months) and small size of the construction workforce (50 workers), employment impacts from the construction of the proposed Moore Ranch Project would be SMALL.

4.11.1.1.5   Local Finance

Construction of the proposed Moore Ranch Project would generate some tax revenue in the local economy through the purchase of goods and services as well as contributing to county and State tax revenues. Because of the short duration (9 months) and small size of the construction workforce (50 workers), construction of the proposed Moore Ranch Project would have a SMALL impact on local finances.

4.11.1.1.6   Education

Because of the short duration of the construction activity (9 months; the GEIS assumed 12 to 18 months), workers would not be expected to bring families and school-age children with them, and therefore, there would be no impact on educational services during construction of the proposed Moore Ranch Project.

4.11.1.1.7   Health and Social Services

The number of construction workers would cause a short-term increase in the demand for health and social services in the county. However, because of the short duration of the construction activity and the small size of the construction workforce (50 workers), there would be little or no impact on health and social services.

4.11.1.2   Operation Impacts

Operation of the proposed Moore Ranch Project is expected to last 12 years and employ from 40 to 60 workers (EMC, 2007a). Section 4.3.10.2 of the GEIS discussed employment levels during ISR facility operations and assumed 50 to 80 workers would support this phase of the ISR lifecycle (NRC, 2009a).

According to the GEIS, the effects on community services (e.g., education, healthcare, utilities, shopping, and recreation) during facility operations would be similar to the effects experienced during construction, except fewer people would be employed but for a longer duration (NRC, 2009a).

The operations phase of the proposed Moore Ranch Project is expected to last for approximately 12 years, although each wellfield would be operational for about 3.25 years each. The operations workforce would impact the local economy through creating jobs, purchasing local goods and services, as well as increasing county and State tax revenues. Severance tax on the uranium extracted would also be collected at the State level and would contribute to the State of Wyoming general fund.

Environmental Impacts and Mitigative Actions

The NRC staff concludes that the site-specific impacts of operating the proposed Moore Ranch Project would be comparable to the impacts described in the GEIS. The potential impact to each component of the socioeconomic system is discussed in the following paragraphs.

4.11.1.2.1 Demographics

Because of the small size of the operations workforce (40 to 60 workers) and the potential addition of 30 to 40 (indirect) workers in support of facility operations, demographic conditions in Campbell County are not likely to change. The combined effect of 70 to 100 new jobs in the region (assuming that all of the direct and indirect workers would relocate to the ROI) constitutes less than one percent of the current civilian labor force in Campbell County. Therefore, the impact on demographic conditions would be SMALL.

4.11.1.2.2 Income

The average annual salary for all full-time employees would be roughly $50,000. The total annual payroll is estimated at $2,900,000. This is slightly above the Wyoming average of $48,205 (USCB, 2008). Impacts to income during ISR facility operations are expected to be SMALL.

4.11.1.2.3 Housing

Demand for permanent housing is anticipated to increase in the communities surrounding the proposed Moore Ranch Project leading up to the startup of ISR facility operations. The surrounding towns of Wright, Edgerton, and Midwest, as well as larger cities such as Gillette and Casper, are within commuting distance. Vacancy rates are currently low in some of the nearby towns and cities (EMC, 2007a), and the added workforce could have an impact on the small housing inventory. Residents earning less than the median income and those on fixed incomes could be affected by the increased demand for housing. Because of the small size of the operations workforce (40 to 60 workers) and the potential addition of 30 to 40 (indirect) workers in support of facility operations, impacts to housing during ISR facility operations could range from SMALL to MODERATE.

4.11.1.2.4 Employment Structure

As previously discussed, ISR facility operations at the proposed Moore Ranch Project would generate 40 to 60 new jobs such as project managers, plant operators, lab technicians, and drill contractors. Some skilled positions are likely to be filled by people moving into the area rather than providing employment opportunities for people living in nearby communities. The proposed Moore Ranch Project could provide some jobs in the local economy. However, since it is likely that most skilled workers would be drawn from areas outside of the ROI, ISR facility operations at the Moore Ranch Project would not noticeably affect employment rates in Campbell County. Therefore, the impact on the employment structure would be SMALL.

4.11.1.2.5 Local Finance

Campbell County would receive tax revenue during ISR facility operations. Personal property tax would be applied to the value of all equipment used by the project. In addition, a State mineral severance tax would be applied to the milled uranium; however, this tax would not be

Environmental Impacts and Mitigative Actions

directly returned to Campbell County. A county ad valorem tax for production would also contribute to local government revenue.

Campbell County would indirectly benefit from the increased sales tax revenue from the increased number of workers relocating to the ROI and from increased demand for goods and services. The tax revenue-related impact from ISR facility operations on Campbell County is expected to be SMALL.

4.11.1.2.6 Education

An increase in the number of school-aged children because of 40 to 60 workers and their families relocating to Campbell County during ISR facility operations could have an impact on local public schools and education-related services. The average family size in Wyoming is 2.97; therefore, a conservative estimate for the number of school-aged children that could relocate to the ROI would be 40 to 60 children. Children of various ages spread across 24 schools and classrooms (kindergarten and grades 1–12). This small number of children is not likely to have a noticeable effect on student-to-teacher ratios in Campbell County School District #1. However, county planners indicated that the schools could accommodate a small increase in the number of students. Schools and education-related service impacts during ISR facility operations are expected to be SMALL.

4.11.1.2.7 Health and Social Services

A small increase in demand would be expected for health and social services during ISR facility operations from workers and their families relocating to the ROI. Operational impacts are not expected to differ significantly from those during the construction phase of the ISR facility. Impacts to health and social services during operations would be SMALL.

4.11.1.3 **Aquifer Restoration Impacts**

Section 4.3.10.3 of the GEIS indicated that the socioeconomic impact from aquifer restoration would be similar to the impacts experienced during ISR facility operations. This is because employment levels and demand for services would not change. The GEIS determined potential impacts to socioeconomics would be SMALL.

Socioeconomic impacts from the aquifer restoration phase at the Moore Ranch Project would be similar to those experienced during ISR facility operations. Because aquifer restoration would be short-term and would not require specialized skills, some ISR facility operations workers would likely remain and other workers may be drawn from the local area. Impacts on demographics, income, housing, employment, tax revenue, as well as, health, social and educational services would remain unchanged because workers would likely have already relocated their families to the area and temporary workers would not relocate their families. The overall socioeconomic impact of aquifer restoration would be SMALL.

The NRC staff concludes that the site-specific conditions at the proposed Moore Ranch Project are comparable to those described in the GEIS and incorporates by reference the GEIS conclusions that the socioeconomic impacts on the ROI from aquifer restoration would be SMALL.

Environmental Impacts and Mitigative Actions

4.11.1.4 Decommissioning Impacts

The applicant projects that during the 15 years following the completion of the operations and aquifer restoration, the number of direct employees at the proposed Moore Ranch Project would be about 27, with another 26 indirect and induced jobs (EMC, 2007a). The applicant anticipates that decommissioning the central plant would require about the same number of workers as the construction phase (50 workers). Both of these phases of the ISR lifecycle would require fewer employees than the workforce of 200 considered in the GEIS.

NRC has regulations and guidance for decommissioning. These regulations are found in 10 CFR Part 40.42. Additional guidance on how to decommission a nuclear facility is available in the *Consolidated NMSS Decommissioning Guidance*, NUREG–1757. Decommissioning of the proposed ISR facility would be subject to a separate safety and environmental review. The decommissioning process commences when the licensee informs NRC that it intends to decommission the facility or has ceased principal activities at the entire site or in any building or outdoor area. The licensee prepares a decommissioning plan and submits it to NRC for review. Upon approval of the decommissioning plan, NRC amends the existing license to allow decommissioning to proceed. At the completion of decommissioning, the licensee conducts a final status survey to demonstrate compliance with criteria established in the decommissioning plan. When NRC confirms that the criteria in the decommissioning plan for releasing the site or any part of the site, has been met, NRC either terminates or amends the license, depending on the intended use of the site.

Socioeconomic impacts from decommissioning the proposed Moore Ranch Project are expected to be SMALL, especially if a number of the ISR facility operations workers remain to assist in this activity.

The NRC staff concludes that the number of workers (20 to 30) anticipated during the decommissioning phase of the proposed Moore Ranch Project is less than the number (200) considered in the GEIS. Therefore, the anticipated socioeconomic impacts would be SMALL, within the range of impacts discussed in the GEIS.

4.11.2 No-Action (Alternative 2)

Under the No-Action alternative, the ISR facility would not be constructed and operated at the proposed Moore Ranch Project. Socioeconomic conditions in Campbell County would not change.

4.12 Environmental Justice Impacts

Under Executive Order 12898, Federal Actions to Address Environmental Justice in Minority Populations and Low-Income Populations, federal agencies are responsible for identifying and addressing, as appropriate, disproportionately high and adverse human health and environmental impacts on minority and low-income populations. In 2004, the Commission issued a Policy Statement on the Treatment of Environmental Justice Matters in NRC Regulatory and Licensing Actions (69 FR 52040), which states, "The Commission is committed to the general goals set forth in Executive Order 12898, and strives to meet those goals as part of its NEPA review process."

Environmental Impacts and Mitigative Actions

The Council of Environmental Quality (CEQ) provides the following information in *Environmental Justice: Guidance Under the National Environmental Policy Act* (1997):

> **Disproportionately High and Adverse Human Health Effects.** Adverse health effects are measured in risks and rates that could result in latent cancer fatalities, as well as other fatal or nonfatal adverse impacts on human health. Adverse health effects may include bodily impairment, infirmity, illness, or death. Disproportionately high and adverse human health effects occur when the risk or rate of exposure to an environmental hazard for a minority or low-income population is significant (as employed by NEPA [National Environmental Policy Act]) and appreciably exceeds the risk or exposure rate for the general population or for another appropriate comparison group (CEQ 1997).
>
> **Disproportionately High and Adverse Environmental Effects.** A disproportionately high environmental impact that is significant (as defined by NEPA) refers to an impact or risk of an impact on the natural or physical environment in a low-income or minority community that appreciably exceeds the environmental impact on the larger community. Such effects may include ecological, cultural, human health, economic, or social impacts. An adverse environmental impact is an impact that is determined to be both harmful and significant (as employed by NEPA). In assessing cultural and aesthetic environmental impacts, impacts that uniquely affect geographically dislocated or dispersed minority or low-income populations or American Indian tribes are considered (CEQ 1997).

The environmental justice analysis assesses the potential for disproportionately high and adverse human health or environmental effects on minority and low-income populations that could result from the construction and operation of the proposed ISR facility at Moore Ranch Project. In assessing the impacts, the following CEQ (1997) definitions of minority individuals and populations and low-income population were used:

> **Minority individuals.** Individuals who identify themselves as members of the following population groups: Hispanic or Latino, American Indian or Alaska Native, Asian, Black or African American, Native Hawaiian or Other Pacific Islander, or two or more races meaning individuals who identified themselves on a Census form as being a member of two or more races, for example, Hispanic and Asian.
> **Minority populations.** Minority populations are identified when (1) the minority population of an affected area exceeds 50 percent or (2) the minority population percentage of the affected area is meaningfully greater than the minority population percentage in the general population or other appropriate unit of geographic analysis.
> **Low-income population.** Low-income populations in an affected area are identified with the annual statistical poverty thresholds from the Census Bureau's Current Population Reports, Series PB60, on Income and Poverty.

4.12.1 Methodology

NRC addresses environmental justice matters for license reviews through (1) identification of minority and low-income populations that may be affected by the proposed construction and operation of the proposed Moore Ranch Project and (2) examining any potential human health

Environmental Impacts and Mitigative Actions

or environmental effects on these populations to determine if these effects may be disproportionately high and adverse.

The 2000 Census provides race and poverty characteristics for Census Tracts and Block Groups in Campbell County. The proposed Moore Ranch Project and a 3.2-km [2-mi] perimeter are contained within five Census Tracts and one additional Block Group that encompass portions of Campbell, Converse, Johnson, and Natrona Counties.

Campbell County was selected as the geographic area for comparison of demographic data for the affected Census Tract populations. This comparison was made to determine the concentration of minority or low-income populations in the affected Census Tracts relative to the State.

Census Block Group data are available from the 2000 Census. Table 4-5 shows the percent of people living in poverty and the minority population in the United States, Wyoming, Campbell, Johnson and Natrona Counties and the block groups closest to the proposed Moore Ranch Project.

The 2008 population of Campbell County has been estimated at 41,473 residents (see Table 3-11). Eleven percent of the Wyoming population is classified as being minority (Table 4-5). For the 2000 census, the minority population in the census tracts surrounding the proposed Moore Ranch Project ranged from 3.3 to 5.9 percent, approximately 5.3 to 7.9 percent below the state average of 11 percent. The U.S. population living below the poverty level was identified as 13 percent, and 11.4 percent of the population in Wyoming was determined to be living below poverty level. The percentage of people living below the poverty level within the census block groups surrounding the proposed Moore Ranch Project ranged from 12.4 to 15.9 percent.

The percentage of minority populations living in the affected block groups are similar to the percentage of minority populations recorded at the State and county level and well below the national level. No minority populations were identified as residing near the proposed Moore Ranch Project; most of the minority population block groups are located near Gillette, about 80 km [50 mi] to the northeast, and communities along the I-25 corridor to the south of Gillette.

The environmental justice impact analysis evaluates the potential for disproportionately high and adverse human health and environmental effects on minority and low-income populations that could result from the construction and operation of the proposed Moore Ranch Project. Adverse health effects are measured in terms of the risk and rate of fatal or nonfatal adverse impacts on human health. Disproportionately high and adverse human health effects occur when the risk or rate of exposure to an environmental hazard for a minority or low-income population is significant and exceeds the risk or exposure rate for the general population or for another appropriate comparison group. Disproportionately high environmental effects refer to impacts or risk of impact on the natural or physical environment in a minority or low-income community that are significant and appreciably exceeds the environmental impact on the larger community.

Environmental Impacts and Mitigative Actions

Table 4-5. Percent Living in Poverty and Percent Minority in 2000		
Geographic Unit	Percent Living in Poverty	Percent Minority
U.S	13.0	30.9
Wyoming	11.4	11.2
Campbell County	7.6	6.5
Campbell County Project Block Group 1-1	12.4	3.3
Johnson County	10.1	3.8
Johnson County Project Block Group 9551-1	12.5	4.0
Natrona County	11.4	9.1
Natrona County Block Group14.01-2	15.9	5.9
Source: US Census Bureau, 2009		

Such effects may include biological, cultural, economic, or social impacts (CEQ, 1997). Some of these potential effects have been identified in resource areas discussed in Chapter 4 of this SEIS. For example, increased demand for rental housing during construction of the ISR facility could disproportionately affect low-income populations. Minority and low-income populations are subsets of the general public residing around the proposed Moore Ranch site, and all would be exposed to the same health and environmental effects generated from construction, operations, decommissioning, and aquifer restoration activities.

4.12.2 Proposed Action (Alternative 1)

Potential impacts to minority and low-income populations due to the construction and operation of the proposed Moore Ranch Project would mostly consist of environmental and socioeconomic effects (e.g., noise, dust, traffic, employment, and housing impacts). Noise and dust impacts would be short-term and limited to onsite activities. Minority and low-income populations residing along site access roads could experience increased commuter vehicle traffic during shift changes. As construction and operations employment increases at the proposed Moore Ranch Project, employment opportunities for minority and low-income populations may also increase. Increased demand for rental housing during peak construction could disproportionately affect low-income populations. However, according to the latest census information, there were over 1,000 vacant housing units in Campbell County (see Section 3.11.3).

As part of addressing environmental justice associated with license reviews, NRC also analyzed the risk of radiological exposure through the consumption patterns of special pathway receptors, including subsistence consumption of fish, native vegetation, surface waters, sediments, and local produce; absorption of contaminants in sediments through the skin; and inhalation of plant materials. The special pathway receptors analysis is important to the environmental justice analysis because consumption patterns may reflect the traditional or cultural practices of minority and low-income populations in the area.

Subsistence Consumption of Fish and Wildlife

Executive Order 12898 (1994) directs federal agencies, whenever practical and appropriate, to collect and analyze information on the consumption patterns of populations who rely principally on fish or wildlife, or both, for subsistence and to communicate the risks of these consumption

Environmental Impacts and Mitigative Actions

patterns to the public. In this SEIS, NRC considered whether there were any means for minority or low-income populations to be disproportionately affected by examining impacts to American Indian, Hispanic, and other traditional lifestyle special pathway receptors. Special pathways that took into account the potential levels of contaminants in native vegetation, crops, soils and sediments, surface water, fish, and game animals on or near the proposed Moore Ranch Project were considered.

Potential impacts to minority and low-income populations would mostly consist of radiological effects; however, radiation doses from ISR facility operations are expected to be well below regulatory limits (see Section 4.13). Based on this information and the analysis of human health and environmental impacts presented in this SEIS, the proposed construction, operation, and decommissioning of the proposed ISR facility and aquifer restoration would not have disproportionately high and adverse human health and environmental effects on minority and low-income populations residing in the vicinity of the proposed Moore Ranch Project.

4.12.3 No-Action (Alternative 2)

Under the No-Action alternative, the ISR facility would not be constructed and operated at the proposed Moore Ranch Project. The relative conditions affecting minority and low-income populations in the vicinity of the Moore Ranch Project would remain unchanged. Therefore, there would be no disproportionately high and adverse impact to minority and low-income populations under the No-Action alternative.

4.13 Public and Occupational Health and Safety Impacts

Potential radiological and nonradiological impacts to public and occupational health and safety from ISR activities at the proposed Moore Ranch Project could occur during all phases of the ISR facility lifecycle. Such impacts could occur from normal operations or from accidents.

Detailed discussion of the potential environmental impacts to public and occupational health and safety from construction, operation, aquifer restoration, and decommissioning the proposed Moore Ranch Project are provided in the following sections.

4.13.1 Proposed Action (Alternative 1)

4.13.1.1 Construction Impacts

Section 4.3.11.1 of the GEIS discussed construction activities at an ISR facility, which would include installation of wellfields (and associated piping) and construction of surface processing structures, access roads, and supporting utilities. Fugitive dust generated from construction and vehicle traffic is expected but would likely be of short duration. The construction phase at the proposed Moore Ranch Project has been estimated to last 9 months. Based on radiological environmental monitoring data for the proposed Moore Ranch Project (see discussion in Section 6.2 of the SEIS), no significantly elevated levels of radioactive materials in soils above natural background levels have been identified (EMC, 2007b). Therefore, inhalation of fugitive dust with these background levels does not pose a radiological dose significantly different than that from natural background exposure (NRC, 2009a).

Construction equipment would likely be diesel powered and would result in diesel exhaust that includes small particles. The impacts and potential human exposures from these emissions would be SMALL because the releases are usually of short duration and are readily dispersed into the atmosphere (NRC, 2009a). Appendix D of the final SEIS describes the emissions inventory evaluated for the proposed Moore Ranch Project.

The NRC staff has not identified any new and significant information during its independent review of the proposed Moore Ranch Project environmental report, the site visit, or evaluation of other available information. Therefore, the NRC staff has determined that there would be no significant impacts to public and occupational health and safety from construction beyond those discussed in the GEIS. The construction phase of the proposed Moore Ranch Project would have a SMALL impact on workers and the general public.

4.13.1.2 Operation Impacts

Section 4.3.11.2 of the GEIS discussed potential occupational radiological impacts from normal operations that could result from (1) exposure to radon gas from the wellfields, (2) IX resin transfer operations, and (3) venting during processing activities. Workers could also be exposed to airborne uranium particulates from dryer operations and maintenance activities. Potential public exposures to radiation could occur from radon releases from the wellfields and uranium particulate releases (i.e., from facilities without vacuum dryer technology). Both worker and public radiological exposures are addressed in NRC regulations at 10 CFR Part 20, which require licensees to implement an NRC-approved radiation protection program. Measured and calculated doses for workers and the public are commonly only a fraction of regulated limits. For these reasons, the GEIS determined that potential radiological impacts to workers and the public would be SMALL (NRC, 2009a).

Nonradiological worker safety would be addressed through occupational health and safety regulations and practices.

Radiological accident risks could involve processing equipment failures leading to yellowcake slurry spills, or radon gas or uranium particulate releases. The GEIS stated that the consequences of these accidents to workers and the public would generally be low, except for a dryer explosion, which could result in a worker dose above NRC limits. The likelihood of such an accident would be expected to be low, due to design considerations and operational monitoring, and therefore the GEIS concluded the risk would also be low.

The potential impact from nonradiological accidents include high consequence chemical release events (e.g., of ammonia) that could expose workers and nearby populations. However, the GEIS stated that the likelihood of such a release would be low, based on historical operating experience at NRC-licensed facilities, primarily because operators follow chemical safety and handling protocols. Therefore, the GEIS concluded that radiological and nonradiological impacts from accidents during operations could range from SMALL to MODERATE.

4.13.1.2.1 Radiological Impacts to Public and Occupational Health and Safety From Normal Operations

As discussed in the GEIS, some amount of radioactive materials would be released to the environment during ISR operations. The potential impact for these releases can be evaluated by the MILDOS-AREA computer code (MILDOS), which was developed by Argonne National Laboratory for calculating radiation doses to individuals and populations from releases that occur at uranium recovery facilities. MILDOS uses a multipathway analysis for determining

Environmental Impacts and Mitigative Actions

external dose; inhalation dose; and dose from ingestion of soil, plants, meat, milk, aquatic foods, and water. The primary radionuclide of interest at an ISR facility is radon-222; other key radionuclides that may also be released, which are also in the uranium decay scheme, include uranium, thorium-230, radium-226, and lead-210. MILDOS uses a sector-average Gaussian plume dispersion model to estimate downwind concentrations. This model typically assumes minimal dilution and provides conservative estimates of downwind air concentrations and doses to human receptors.

The GEIS presented historical data for ISR operations, providing a range of estimated offsite doses associated with six current or former ISR facilities. For these operations, doses to potential offsite exposure (human receptor) locations range between 0.004 mSv [0.4 mrem] per year for the Crow Butte facility located in Nebraska and 0.32 mSv [32 mrem] per year for the Irigaray facility located in Johnson County, both well below the 10 CFR Part 20 annual radiation dose limit of 1 mSv/yr [100 mrem/yr] (NRC, 2009a).

The GEIS also provides a summary of doses to occupationally exposed workers at ISR facilities. As stated, doses would be similar regardless of the facility's location and are well within the 10 CFR Part 20 annual occupational dose limit of 0.05 Sv [5 rem] per year. The largest annual average dose to a worker at a uranium recovery facility over a 10-year period [1994–2006] was 7 mSv [700 mrem]. More recently, the maximum total dose equivalents reported for 2005 and 2006 were 6.75 mSv [675 mrem] and 7.13 mSv [713 mrem].

The license application for the proposed Moore Ranch Project addresses several normal operations activities that have the potential to expose workers and members of the public to sources of radiation. The primary source of exposure would be from the release of radon-222 during operations, which include extraction of the uranium onto IX columns from the pregnant lixiviant from the wellfield extraction, the elution of the uranium from the IX columns and subsequent precipitation of uranium, followed by the drying and packaging of the yellowcake for shipment to an offsite facility for further processing.

As described in the GEIS, and discussed in Section 2.1.1.1.6.1 of this SEIS, the drying and packaging of the precipitated uranium would be conducted under vacuum, thereby limiting release of airborne radioactive materials (uranium and short-lived particulate progeny) to zero or near zero. The applicant has proposed to dispose of radioactive and potentially toxic liquid effluent from the operations phase via deep well injection. Therefore, there would be no anticipated routine liquid releases or pathways of exposure from routine operations. Leaks and spills are evaluated as abnormal conditions in Section 4.13.1.2.2. No routine releases of radioactive liquids during operations have been reported.

For normal operations, radon-222 would be the only significant radionuclide anticipated to be released; the primary sources would be from wellfield venting and releases from within the central plant for process operations (predominantly via vent stacks on the IX columns and various tanks). As discussed in Section 2.1.1.6.1 the applicant has proposed using pressurized downflow IX columns that are expected to significantly reduce the radon emissions from the processing circuit.

The potential source term (i.e., radiological releases to the atmosphere) for normal operations were calculated by the applicant using the NRC approved methodology of Regulatory Guide 3.59 for releases from the production fluids and NUREG–1569 for the processing of resins from satellite facilities. The application of this methodology for the Moore Ranch Project

Environmental Impacts and Mitigative Actions

and the resultant source term is discussed in Section 4.12 of the applicant's environmental report (Uranium One, 2009a,b). Table 4-6 summarizes these releases.

Based on this source term, radiation doses at the site boundary in each of the 16 meteorological sectors (e.g., N, NNE, NE, ENE, and E) and at the locations of nearby residences were calculated using the MILDOS-AREA code (Argonne National Laboratory, 1989). The MILDOS-AREA code was also used to assess radiation dose in the GEIS. The principal exposure pathways modeled include inhalation, ingestion, and direct exposure. The highest dose at the site boundary was 0.008 mSv [0.8 mrem] per year TEDE at the northwest property boundary, which is 0.8 percent of the 1 mSv [100 mrem] per year dose limit for a member of the public specified in 10 CFR 20.1301. The maximum exposed individual for a resident located 4.5 km [2.8 mi] to the east of the facility is calculated to be 0.007 mSv [0.7 mrem] per year, also a small fraction of the 1 mSv per year regulatory limit. These doses are consistent with the doses identified for other ISR facilities considered in the GEIS, where the range was from a high

Table 4-6. Estimated Radon-222 Releases (Ci yr^{-1})* from the Proposed Moore Ranch Project				
Location	Production	Restoration	Drilling	Resin Transfer
Wellfield 1	85.3	20	0	0
Wellfield 2	85.3	20	0	0
Main Plant Stack	230	53	0	5.3
New wellfield	0	0	0.43	0

*Based on the phased approach to wellfield development, production, and restoration the annual emissions for any specific year would vary based on the degree of overlap in planned activities

Source: EMC, 2007a

of 0.317 mSv [31.7 mrem] per year for the Crow Butte facility to 0.004 mSv [0.4 mrem] per year for the Irigaray facility.

The applicant also calculated the collective dose using MILDOS-AREA for the population residing within 80 km [50 mi] of the facility. This dose, which is a measure of the total radiological impact from routine operations for the potentially affected communities, was estimated at 0.0009 person-Sv [0.09 person-rem] per year.

The applicant also evaluated the deposition of the radon-222 particulate decay products (polonium-210, lead-210, bismuth-214, and lead-210) and the potential exposure to flora and fauna. The calculated soil concentrations were less than 0.01 pCi/g at the surface, which is a small fraction of that normally present in the soil from the natural background levels of uranium and decay products. Therefore, any impact from increased soil radioactivity levels from airborne releases of radon during normal operations would be SMALL.

Based on typical occupational injury and illness rates for the Wyoming mining industry, the applicant estimated that operations at the proposed Moore Ranch Project could potentially result in 1.9 nonfatal occupational injuries and illnesses per year of operation (EMC, 2007a).

Environmental Impacts and Mitigative Actions

In summary, potential radiation doses to occupationally exposed workers and members of the public from operation of the proposed Moore Ranch Project would be SMALL. Calculated radiation doses from the modeling of releases of radioactive materials to the environment are small fractions of the limits of 10 CFR Part 20 that have been established for the protection of the public health and safety. The NRC staff have not identified any new and significant information during its independent review of the Moore Ranch environmental report, the site visit, or evaluation of other available information. Therefore, the NRC staff have determined that there would be no significant radiological impacts from normal operations to the public or occupational exposed workers beyond those discussed in the GEIS.

4.13.1.2.2   Radiological Impacts to Public and Occupational Health and Safety from Accidents

The GEIS provides an identification, discussion, and consequence assessment for accident conditions that could occur with an ISR operation (NRC, 2009a). As discussed, a radiological hazard assessment (Mackin, et al., 2001) considered three types of accidents, representing the sources containing the higher levels of radioactivity for all aspects of operation:

- Thickener failure and spill
- Pregnant lixiviant and loaded resin spills (radon release)
- Yellowcake dryer accident release

Table 4-7. Generic Accident Dose Analysis for ISR Operations		
Accident Scenario	**Maximum Dose to Workers**	**Maximum Dose to Public**
Thickener spill	50 mSv [5,000 mrem]	0.25 mSv [25 mrem]
Pregnant lixiviant, resin spill	13 mSv [1,300 mrem]	<0.13 mSv [<13 mrem]
Yellowcake dryer release	0.088 Sv [8.8 rem] Generic <0.01 Sv [1 rem] Moore Ranch Project	<1 mSv [<100 mrem]
Data adapted from GEIS (NRC, 2009a)		

The following discussion presents an overview for each of these accident scenarios, as evaluated in the GEIS, along with a specific application to the proposed Moore Ranch Project. Table 4-7 summarizes the potential dose to workers and the public from the accident scenarios described in the following paragraphs.

Thickener Failure and Spill. Thickeners are used to concentrate the yellowcake slurry before it is transferred to the dryer or packaged for off-site shipment. Radionuclides could be inadvertently released to the atmosphere through a thickener failure or spill. The accident scenario evaluated in the GEIS assumed a tank or pipe leak that releases 20 percent of the thickener outside of the processing building. The analyses included a variety of wind speeds,

Environmental Impacts and Mitigative Actions

stability classes, release durations, and receptor distances. A minimum receptor distance of 152 m [500 ft] was selected because it was found to be the shortest distance between a processing facility and an urban development for current operating ISR facilities. Off-site, unrestricted doses from such a spill could result in a dose of 25 mrem, or 25 percent of the annual public dose limit of 100 mrem y^{-1} with negligible external doses based on sufficient distance between the facility and receptor (NRC, 2009a). Because the nearest resident to the proposed Moore Ranch Project is located 4.5 km [2.8 mi] east of the proposed license area, the potential dose would be even less.

As discussed in the GEIS, doses to unprotected workers inside the facility have the potential to exceed the annual dose limit of 5 rem y^{-1} if timely corrective measures were not taken for protecting workers and remediating the spill. Typical protection measures, such as monitoring, respiratory protection, and material control, which would be a part of the applicant Radiation Protection Program, would reduce worker exposures and the resulting doses to a small fraction of those evaluated.

Pregnant Lixiviant and Loaded Resin Spills. Process equipment (IX columns, drying and packing facilities) would be located on curbed concrete pads, as discussed in Section 2.1.1.2, to prevent any liquids from spills or leaks from exiting the building and contaminating the outside environment. Therefore, except for wellfield leaks, as further evaluated, the potential for an accidental liquid release with exposure from a liquid pathway was not considered realistic. The primary radiation source for liquid releases within the facility would be the resulting airborne radon-222 as released from the liquid or resin tank spill.

The radon accident release scenario assumes a pipe or valve of the IX system, containing pregnant lixiviant, develops a leak and releases (almost instantaneously) all present radon-222 at a high activity level (8×10^5 pCi L^{-1}). For a 30-minute exposure, the dose to a worker located inside the central plant performing light activities without respiratory protection was calculated to be 13 mSv [1,300 mrem], which is below the 10 CFR Part 20 occupational annual dose limit. The analysis did not evaluate public dose; however, because atmospheric transport offsite would reduce the airborne levels by several orders of magnitude, any dose to a member of thepublic would be less than the 1 mSv [100 mrem] public dose limit of 10 CFR Part 20. Radiation Protection Program controls and monitoring measures would be expected to minimize the magnitude of any such release and further reduce the consequences of this type accident.

Yellowcake Dryer Accident Release. Dryers used to produce yellowcake powder from yellowcake slurry are another source for accidental release of radionuclides. A multiple-hearth dryer is capable of releasing yellowcake powder inside the processing building as a result of an explosion and was evaluated in the GEIS as a bounding condition for this type of accidental scenario. The analysis assumes about 4,300 kg [9,500 lb] of uranium yellowcake is released within the building area housing the dryer and of this, 1 kg [2.2 lb] is subsequently released as an airborne effluent to the outside atmosphere as a 100 percent respirable powder. Due to the nature of the material, most of the yellowcake would rapidly fall out of airborne suspension. For the occupationally exposed worker using respiratory protection, which is the normal mode during dryer access and drum-filling operations, the dose was calculated to be 0.088 Sv [8.8 rem], which exceeds the annual occupational dose limit of 0.05 Sv [5 rem)]. The amount assumed to remain airborne and to be transported outside the building for atmospheric dispersion to an offsite location would be 1 kg [2.2 lbs] of yellowcake. The rapid fallout within the building and the atmospheric dispersion to an offsite location would significantly reduce the exposure to members of the public, where the calculated dose was less than 100 mrem.

Environmental Impacts and Mitigative Actions

The applicant proposes to use two rotary vacuum dryers with the use of heat-transfer fluid that circulates through the dryer shell. This configuration separates the heater combustion source from the dryer itself, thereby mostly eliminating the possibility of an explosion, which is the initiating event for the assumed catastrophic failure and significant release of dryer radioactive content. The removal of the driving force for the resuspension of the yellowcake greatly reduces consequences. Additionally, the applicant would have emergency response procedures in place to provide proper directions for mitigating worker exposures; and emergency training drills, dosimetry, respiratory protection, and contamination control and decontamination are required as part the applicant Radiation Protection program. Both of these would further reduce the consequences of this type accident.

Accident Analysis Conclusions. With the addition of site-specific consideration for the yellowcake dryer accident, the GEIS evaluations appropriately encompass the type of accidents and consequences for the proposed Moore Ranch Project. The NRC Staff has not identified any new and significant information during its independent review of the Moore Ranch environmental report, the site visit, or evaluation of other available information. Therefore, there would be no significant radiological impacts from potential accidents to the public or occupationally exposed workers beyond those considered in the GEIS. The impacts to workers would be SMALL, if radiation safety and incident response procedures in the applicant's NRC approved radiation protection plan were followed; the impacts to the general public would also be SMALL.

4.13.1.2.3   Nonradiological Impacts to Public and Occupational Health and Safety From Normal Operations

The GEIS identified the various chemicals, hazardous and nonhazardous, along with typical quantities that are generally used at ISR facilities. The use of hazardous chemicals at ISR facilities are controlled under several regulations that are designed to provide adequate protection to workers and the public. The primary regulations applicable to the use and storage include:

- 40 CFR Part 68, Chemical Accident Prevention Provisions. This regulation includes a list of regulated toxic substances and threshold quantities for accidental release prevention.

- 29 CFR 1910.119, OSHA Standards (which includes Process Safety Management [PSM]). This regulation provides a list of highly hazardous chemicals, including toxic and reactive materials that have the potential for a catastrophic event at or above the threshold quantity.

- 40 CFR Part 355, Emergency Planning and Notification. This regulation contains a list of extremely hazardous substances and their threshold planning quantities (TPQs) for the development and implementation of ERPs. A list of reportable quantity (RQ) values is also provided for reporting releases.

- 40 CFR 302.4, Designation, Reportable Quantities, and Notification– Designation of Hazardous Substances. This regulation provides a list of Comprehensive Environmental Response, Compensation, and Liability Act (CERCLA) hazardous substances compiled from the Clean Water

Act, Clean Air Act, Resource Conservation and Recovery Act, and the Toxic Substances and Control Act.

The following lists the hazardous chemicals and their associated protective provisions expected to be used at the proposed Moore Ranch Project (EMC, 2007a; 2007b):

- Sulfuric acid (H_2SO_4)–Due to the quantities that would be used, ERPs would be required per 40 CFR Part 355. The storage tank would be located away from other process tanks to preclude accidental mixing with other chemicals.

- Oxygen (O_2)–Oxygen would be stored near, but a safe distance from, the central plant or within wellfield areas. The oxygen storage facility would be designed to meet industry standards contained in National Fire Protection Association 50–Standards for Bulk Oxygen Systems at Consumer Sites. (National Fire Protection Association, 2001) Procedures would be developed for spills or fires in the oxygen system.

- Liquid hydrogen peroxide–50 percent (H_2O_2)–Because the concentration would be <52 percent, no additional regulatory protective measures would be required.

- Carbon dioxide (CO_2)–Carbon dioxide would be stored adjacent to the central plant. Floor-level ventilation and low-point carbon dioxide monitors would be installed to preclude a buildup of carbon dioxide in occupied areas.

- Sodium carbonate (Na_2CO_3) and Sodium Chloride (NaCl)–Systems utilizing these chemicals would be designed to industry standards.

- Sodium Sulfide (Na_2S)–Sodium Sulfide would be stored outside of process areas and separate from hydrogen peroxide and sulfuric acid

- Hydrochloric acid (HCl)–Due to the quantities that would be used, reporting quantities would be required per 40 CFR Part 302.4. The hydrochloric acid storage tank would be located away from other process tanks to preclude accidental mixing with other chemicals.

The typical on-site quantities for some of these chemicals exceed the regulated, minimum reporting quantities and trigger an increased level of regulatory oversight regarding possession (type and quantities), storage, use, and disposal practices. Compliance with applicable regulations reduces the likelihood of a release. Off-site impacts would be SMALL and do not typically pose a significant risk to the public, while workers involved in a response and cleanup could experience MODERATE impacts if the proper emergency and cleanup procedures and worker training was not available or was inadequate.

In general, the handling and storage of chemicals at the facility would follow standard industrial safety standards and practices. As identified in Section 4.12.1.2.1 of the applicant's environmental report, industrial safety aspects associated with the use of hazardous chemicals at Moore Ranch Project are regulated by OSHA (EMC, 2007a). Section 3.2.3 of the applicant technical report provides an overview of storage practices (EMC, 2007b). Chemical storage facilities would include hazardous and nonhazardous material storage areas. Bulk hazardous materials would be stored outside and segregated from areas where licensed materials are processed and stored to minimize potential impact on radiation safety. Bulk storage of hazardous chemicals would be separated to avoid mixing of incompatible materials; and outside storage areas would be located at a sufficient distance from facilities to minimize hazards to people during an accidental release. Other nonhazardous bulk process chemicals (e.g., sodium carbonate) that do not have the potential to impact radiological safety could be stored within the central plant facilities.

The applicant has proposed an overall chemical safety program that includes:

- Risk Management Planning, as required in 40 CFR Part 68
- Process Safety Management of Highly Hazardous Chemicals standard contained in 29 CFR 1910.119
- Threshold Planning Quantities as contained in 40 CFR Part 355
- Reportable quantities for spills from CERCLA in 40 CFR 302.4

In the State of Wyoming, industrial safety at ISR operations is regulated by the Wyoming State Mine Inspector.

The types and quantities of chemicals (hazardous and nonhazardous) for proposed use at Moore Ranch Project do not differ from those evaluated in the GEIS. Information provided for the proposed Moore Ranch Project does not contain any new or significant information that is either contrary to or varies from the information in the GEIS conclusions regarding potential impacts to the public or occupational health and safety. Therefore, the nonradiological impacts during normal operations at Moore Ranch Project would be SMALL.

4.13.1.2.4 Nonradiological Impacts to Public and Occupational Health and Safety From Accidents

The risks from accidents associated with the use of the typical hazardous and nonhazardous chemicals for an *in-situ* uranium recovery operation are not different than those for other typical industrial applications. In general, these risks are deemed acceptable as long as design and facility safety policies and practices meet industry and regulatory standards. Past history at current and former ISR facilities has shown these facilities can be designed and operated with appropriate measures to ensure proper safety for workers and the public.

Appendix E of the GEIS, Hazardous Chemicals, provides an accident analysis for the more hazardous chemicals (NRC, 2009a). That analysis indicates chemicals commonly used at ISR facilities can pose a serious safety hazard if not properly handled. The GEIS does not evaluate potential hazards to workers or the public due to specific types of high-consequence, low-probability accidents (e.g., a fire or large magnitude sudden release of chemicals from a major tank or piping system rupture). The application of common safety practices for handling and use of chemicals would be expected to lower the likelihood of these severe release events and therefore lower the risk to acceptable levels.

Spills of reportable quantities from chemical bulk storage areas would be reported to the WDEQ in accordance with WDEQ-WQD Rules and Regulations, Chapter 17, Part E and 40 CFR 302 (CERCLA).

Environmental Impacts and Mitigative Actions

The types and quantities of chemicals (hazardous and nonhazardous) for proposed use at Moore Ranch Project do not differ from that evaluated in the GEIS. Information provided for the proposed Moore Ranch Project does not contain any new or significant information that is either contrary to or varies from the information and conclusions in the GEIS regarding potential nonradiological impacts on public and occupational health and safety from chemical accidents. Offsite impacts would be SMALL and do not typically pose a significant risk to the public, while workers involved in a response and cleanup could experience MODERATE impacts. Based on this finding and the GEIS conclusions, the impacts from potential accidents for both occupationally exposed workers and members of the public would be SMALL.

4.13.1.3 Aquifer Restoration Impacts

Section 4.3.11.3 of the GEIS discussed potential radiological and nonradiological impacts from aquifer restoration. Activities occurring during aquifer restoration would overlap similar activities occurring during operations (e.g., operation of wellfields, wastewater treatment and disposal). Therefore, the potential impact on public and occupational health and safety would be bound by the operational impacts. The GEIS also stated that the reduction of some operational activities (e.g., yellowcake production and drying, remote IX) as aquifer restoration proceeded would be expected to limit the relative magnitude of potential worker and public health and safety hazards. The GEIS concluded that the overall impacts from aquifer restoration would be SMALL.

Aquifer restoration activities for the proposed Moore Ranch Project involve activities similar to those during operations (e.g., operation of wellfields, wastewater treatment and disposal); therefore, the potential impact on public and occupational health and safety would be expected to be similar to the operational impacts. The reduction or elimination of some operational activities (e.g., yellowcake production and drying, remote IX) would further limit the relative magnitude of potential worker and public health and safety hazards. The radiation doses associated with restoration are included in the operations assessment in Section 4.13.1.2.1. Similarly, nonradiological hazards during aquifer restoration are assessed in Section 4.13.1.2.3. Accident consequences would be expected to be smaller than those evaluated in Sections 4.13.1.2.2 and 4.13.1.2.4. Therefore, aquifer restoration would be expected to have a very localized SMALL occupational impact to workers (primarily from radon gas) and to the general public for the duration of the aquifer restoration phase, which is estimated to last for 3.5 years at Wellfield 1 and 5.25 years at Wellfield 2.

4.13.1.4 Decommissioning Impacts

Section 4.3.11.4 of the GEIS discussed potential radiological and nonradiological impacts to public and worker health and safety during the decommissioning phase of an ISR facility. Worker and public health and safety would be addressed in an NRC-required and approved decommissioning plan. This plan would be prepared in compliance with 10 CFR Part 40.42 and discusses implementation of the safety program to ensure worker safety and protection of the public during decommissioning and compliance with applicable safety regulations. An ISR licensee would conduct decommissioning activities in accordance with the approved plan, and compliance would be enforced through NRC inspections.

The GEIS also assumed that as decommissioning proceeded, the potential environmental impact would be expected to decrease because the hazard would be removed, soils and structures would be decontaminated, and disturbed lands would be reclaimed.

Environmental Impacts and Mitigative Actions

As discussed in the GEIS, the environmental impact from decommissioning an ISR facility would be SMALL. The degree of potential impact would decrease as the hazards were either reduced or removed, soils and facility structures were decontaminated, and lands were restored to preoperational conditions. Typically, the initial decommissioning steps would include removal of hazardous chemicals. As such, the majority of safety issues to be addressed during the decommissioning phase would involve radiological hazards at the facility.

To ensure the safety of the workers and the public during decommissioning, NRC requires licensed facilities to submit a decommissioning plan for review. The plan would include details of the radiation safety program that would be implemented during decommissioning to ensure that the workers and public would be adequately protected and that their doses are compliant with 10 CFR Part 20, Subpart C and Subpart D limits. An approved plan would also provide ALARA provisions to further ensure best safety practices are being use to minimize radiation exposures. Finally, adequate protection of workers and the public during decommissioning is further ensured through NRC plan approval, license conditions, and inspection and enforcement.

The decommissioning of the proposed Moore Ranch Project and any subsequent NRC approval for release of the site for unrestricted access would have to be in conformance with the NRC radiation protection standards for decommissioning of uranium recovery facilities. Therefore, any potential radiation dose to members of the public would also be in conformance with standards established for protecting public health and safety.

Applicant-provided information does not contain any new or significant information that is contrary to or varying significantly from the GEIS information and conclusions regarding the potential impact to public and occupational health and safety. Therefore, the potential impact from and following decommissioning would be short term and SMALL.

4.13.2 No-Action (Alternative 2)

Under the No-Action alternative, there would be no occupational exposure. There would be no additional radiological exposures to the general public from project related effluent releases, and there would be no impact on long-term environmental radiological conditions. Radiation exposure and risk to the general public would continue to be determined by exposure from natural background, medical-related exposures, consumer products, and exposures from existing residual contamination. Under the No-Action alternative, the existing residual radioactivity would remain in these areas and would not be remediated.

4.14 Waste Management Impacts

Potential environmental impacts from waste management at the proposed Moore Ranch Project could occur during all phases of the ISR facility lifecycle. ISR facilities generate radiological and nonradiological liquid and solid wastes that must be handled and disposed of properly. The types of waste streams to be disposed of at the proposed Moore Ranch Project are discussed in Section 2.1.1.1.6 of this SEIS. See the text box in Section 2.1.1.1.6 for a list of liquid and solid waste types. The primary radiological wastes to be disposed of at the proposed Moore Ranch Project are process-related liquid wastes and process-contaminated structures and soils all of which are classified as byproduct material. Before operations could begin, NRC requires an ISR facility to have an agreement in place with a licensed disposal facility to accept byproduct material. The applicant has committed to disposing of byproduct material at a licensed disposal

Environmental Impacts and Mitigative Actions

site. This disposal agreement would be in place prior to the start of operations as required by a license condition.

Detailed discussion of the potential environmental impacts from waste management actions during the construction, operations, aquifer restoration, and decommissioning phases of the proposed Moore Ranch Project are discussed as follows. Discharges of storm water runoff and wastewater discharges to surface waters are discussed in Section 4.5.1.

4.14.1 Proposed Action (Alternative 1)

Under the proposed action, the applicant would dispose of liquid effluent via a Class I injection well discussed in Section 4.14.1.1. Alternative wastewater disposal options, including evaporation ponds, surface water discharge, land application, and disposal via Class V injections wells, are discussed in Section 4.14.1.2.

4.14.1.1 Disposal Via Class I Injection Well

4.14.1.1.1 Construction Impacts

Section 4.3.12.1 of the GEIS concluded that waste management impacts from the construction phase of an ISR facility would be SMALL. Because construction activities would be on a relatively small scale and the wellfields would be developed incrementally, a low volume of construction waste would be generated. The primary wastes to be disposed of during this phase of the ISR lifecycle would be expected to be nonhazardous solid wastes, such as building materials and piping. As discussed in Section 3.13.2, the applicant has proposed to dispose of nonhazardous solid wastes at the City of Casper landfill in Casper, Wyoming, approximately 97 km [60 mi] from the proposed Moore Ranch Project site. The applicant would transport the wastes either directly to the Casper facility or to a permitted transfer station located at the Midwest/Edgerton landfill, located approximately 39 km [24 mi] from the proposed Moore Ranch Project. Any small amounts of hazardous wastes generated during construction (e.g., batteries, solvents, waste equipment oil) would be stored, in accordance with WDEQ regulations, and ultimately transported to the Casper landfill special waste and diversion facility for disposal.

The relatively small scale of construction activities and incremental development of the wellfields at ISR facilities generate low volumes of construction waste. No byproduct material would be generated during the construction phase at Moore Ranch Project. Therefore, impact on waste management during the construction phase would be SMALL.

The NRC staff concludes that the site-specific conditions at the proposed Moore Ranch Project are comparable to the generic conditions described in the GEIS for waste management. Therefore, this SEIS incorporates by reference the GEIS conclusions that the impacts to waste management during construction are expected to be SMALL. Furthermore, the NRC staff has not identified any new and significant information during its independent review that would change the environmental impacts estimated in the GEIS.

4.14.1.1.2 Operation Impacts

Section 2.7 of the GEIS indicated that wastes generated during the operations phase would primarily be liquid waste streams consisting of process bleed (1 to 3 percent of the process flow rate). Wastes would also be generated from flushing of eluant to limit impurities, resin transfer

Environmental Impacts and Mitigative Actions

wash, filter washing, uranium precipitation process wastes, and plant washdown water. The method used to handle and process these wastes (disposal by deep well injection) reduces the solid waste volume that must be disposed of at an approved facility. State permitting actions, NRC license conditions, and NRC inspections ensure proper practices would be used to comply with safety requirements to protect workers and the public; therefore, the waste management impact would be SMALL (NRC, 2009a).

At the proposed Moore Ranch Project, liquid wastes from operations (Section 2.1.1.1.6.2) are classified as byproduct material and would be disposed of via deep well injection, which is regulated by WDEQ. The WDEQ permit and approval process would specify the concentrations of hazardous constituents to maintain acceptable safe levels for discharge through deep well injection and to ensure the feasibility of deep well disposal in the selected geologic formations. Proper installation and operating procedures would be used to ensure adequate protection of the public and environmental health and safety. Class I disposal wells are designed to protect all potentially useable underground sources of drinking water, and injection of liquid waste would isolate liquid effluent from the accessible environment. By definition, the WDEQ could not issue a permit for Class I injection if a complete exposure pathway existed that could result in public consumption.

Nonhazardous solid wastes generated during operations could include facility trash, septic solids, and other uncontaminated solid wastes (e.g., piping, valves, instrumentation, and equipment). As appropriate, solid wastes would be reused, recycled, or disposed of at the Casper landfill, as discussed in Section 3.13.2. The potential impact would be SMALL because a small volume of material would be disposed of in comparison to the size of the Casper landfill. This is consistent with the discussion of solid waste generation and disposal impacts in the GEIS.

Solid byproduct material that could be generated during operations (i.e., material that does not meet NRC criteria for unrestricted release) would likely include maintenance and housekeeping rags and trash, packing materials, replacement components, filters, protective clothing, and solids removed from process pumps and vessels. The applicant has estimated approximately 77 m^3 [100 yd^3] of this solid byproduct material would be generated per year and stored onsite within a restricted area until sufficient volume was generated for disposal. The applicant has identified the Pathfinders Mines Corporation–Shirley Basin site, located approximately 213 km [132 mi] from the proposed Moore Ranch Project, for disposal of byproduct material. The applicant is negotiating an NRC-required preoperational disposal agreement with the Pathfinder Mines facility. The applicant has identified the Energy Solutions disposal site in Clive, Utah, as an alternate location, and is also negotiating a disposal agreement with this site, as a contingency measure (Uranium One, 2010). Based on the disposal options currently available, and the disposal agreement that NRC requires prior to operations, the staff concludes that the potential waste management impacts associated with the generation of byproduct material would be SMALL.

As discussed in Section 2.1.1.1.6.3, the applicant indicated that it would likely be classified as a Conditionally Exempt Small Quantity Generator (CESQG) of hazardous waste. A CESQG: (1) must determine if their waste is hazardous; (2) must not generate more than 100 kilograms of hazardous waste per month or, except with regard to spills, more than 1 kilogram of acutely hazardous waste; (3) may not accumulate more than 1,000 kilograms of hazardous waste onsite at any time; and (4) must treat or dispose of their hazardous waste either in a nonsite or offsite U.S. treatment storage or disposal facility that meets specific requirements of

Environmental Impacts and Mitigative Actions

40 CFR 261.5. The applicant would transport its hazardous wastes to the Casper landfill special waste and diversion facility (Uranium One, 2010).

Based on the type and quantity of expected waste generation, and the availability of disposal options, the operations phase of the proposed Moore Ranch Project would have a SMALL impact on waste management.

Furthermore, the NRC staff has not identified any new and significant information during its independent review that would change the environmental impacts estimated in the GEIS.

4.14.1.1.3   Aquifer Restoration Impacts

Section 4.3.12.3 of the GEIS discussed waste management activities that would occur during the aquifer restoration phase of an ISR project and noted that the same treatment and disposal options would be implemented as used during operations. Therefore, the waste management impacts would be similar to that during the operations phase of an ISR project. Some increase in wastewater volumes could be experienced, but the increase in volume would be offset by the decrease in production capacity. The impact to waste management from aquifer restoration would be SMALL.

At the proposed Moore Ranch Project, produced water from aquifer restoration (Sections 2.1.1.1.4.1.2; and 2.1.1.1.6.2) would be treated through the combination of ion exchange and reverse osmosis processes and injected back into the production aquifer. The proposed water treatment and reinjection into the aquifer would help limit the amount of water that is permanently withdrawn from the production aquifer. The concentrated waste solutions resulting from this treatment would be classified as byproduct material and would be disposed of in the deep disposal wells. The potential impacts associated with the use of the deep disposal wells during aquifer restoration would be the same as previously discussed for the operations phase in Section 4.14.1.1.2. Other waste management activities during aquifer restoration would also be similar to what is done for the operations phase, and therefore impacts would be SMALL. Furthermore, the NRC staff has not identified any new and significant information during its independent review that would change the environmental impacts estimated in the GEIS.

4.14.1.1.4   Decommissioning Impacts

Section 2.6 of the GEIS indicated wastes generated from decommissioning an ISR facility would be predominantly byproduct material and nonhazardous solid waste (NRC, 2009a, Section 2.6). Section 4.3.12.4 of the GEIS stated that decommissioning byproduct material (including contaminated facility demolition materials, process and wellfield equipment, and contaminated soils) would be disposed of at a licensed facility (NRC, 2009a). As discussed previously, NRC requires a preoperational agreement with a licensed disposal facility to accept radioactive wastes to ensure that sufficient disposal capacity is available for byproduct material generated by decommissioning activities. Safe handling, storage, and disposal of decommissioning wastes would be addressed in a decommissioning plan required for NRC review prior to the initiation of decommissioning. The decommissioning plan would describe how a 10 CFR Part 20-compliant radiation safety program would be implemented to ensure the safety of workers and the public. The GEIS concluded that volumes of radioactive, chemical, and solid wastes generated during decommissioning would be SMALL, and the waste management impacts would also be expected to be SMALL (NRC, 2009a).

Environmental Impacts and Mitigative Actions

For decommissioning the proposed Moore Ranch ISR facility, the applicant proposes to recycle or reuse a large portion of the process equipment and materials. Materials would be surveyed for residual radioactivity. Materials that are not radiologically contaminated or that meet NRC release limits would be removed for reuse, recycling, or disposal. Contaminated materials would be decontaminated, transferred to another licensed facility for use, or disposed of as byproduct material. The NRC staff expects the applicant proposed use of wellfield monitoring instrumentation and wellfield visual inspection to support timely identification and remediation of potential leaks and spills, thereby reducing the potential for generating large volumes of contaminated soil that would need to be excavated and disposed of as byproduct material at a licensed facility. Any hazardous wastes generated during decommissioning would be stored, in accordance with WDEQ regulations, and transported to the Casper landfill special waste and diversion facility for proper disposal (Uranium One, 2010).

As discussed in Section 2.1.1.1.6.3, the staff's cumulative estimate for byproduct material from decommissioning the plant facilities and all wellfields (over a planned 2 year period) is 11,010 m^3 [14,390 yd^3] plus 512 t [565 T] of concrete. This estimate is above the decommissioning byproduct material volume evaluated in the GEIS. As discussed in the draft SEIS Section 2.1.1.1.6.3, the applicant does not presently have an agreement in place with a licensed site to accept its solid byproduct material for disposal. The applicant preferred destination for disposal of byproduct material is at the Pathfinder-Shirley Basin site in Mills, Wyoming. If that facility does not have sufficient capacity at the time the request for an agreement is made, then the applicant could engage other low-level radioactive waste disposal facilities that are licensed to accept byproduct material. Another existing NRC-licensed facility to accept byproduct material for disposal is the Rio Algom Ambrosia Lake uranium mill tailings impoundments near Grants, New Mexico. Additionally, three sites are licensed by NRC Agreement States to accept byproduct material for disposal (i.e., the Energy Solutions site in Clive, Utah; the White Mesa uranium mill site in Blanding, Utah; and the Waste Controls Specialists site in Andrews, Texas). Based on the disposal options currently available, and the disposal agreement that NRC requires prior to operations, NRC concludes that the potential waste management impacts associated with the generation of byproduct material would be SMALL.

The staff's cumulative estimate of nonhazardous solid waste that would be generated from decommissioning is 482 m^3 [630 yd^3] plus 6,102 t [6,730 T] of concrete. This material would be generated in a single year when the plant facilities are decommissioned. Assuming 1.96 T/yd^3 for the concrete waste, the annual solid waste volume would be approximately 4,064 yd^3. This estimated volume of solid waste is higher than what was analyzed in the GEIS and, therefore, the NRC staff considered additional site-specific information to evaluate potential impacts. While local disposal capacity is limited based on the planned closing of the Midwest-Edgerton landfill in year 2010, that facility has been permitted to continue to accept waste and transfer the material to a much larger regional balefill facility (i.e., a landfill that bales the waste) near Casper, Wyoming, in Natrona County. Considering year 2005 estimates of waste disposed of at the Casper balefill from the Wyoming Office of State Lands and Investments (2007) of 93,804 t [103,466 T], the aforementioned cumulative decommissioning nonhazardous solid waste volume resulting from the proposed action would constitute approximately 4 percent of the annual volume of waste disposed at the Casper balefill. Based on this comparison, the NRC staff concludes the region has sufficient capacity to dispose of the nonhazardous solid waste generated from the proposed action. Therefore, the waste management impacts for disposal of decommissioning nonhazardous solid waste would be SMALL.

Environmental Impacts and Mitigative Actions

4.14.1.2   Alternative Wastewater Disposal Options

In most of the alternative wastewater disposal options considered in the following sections, the footprint of the disposal system would increase, as compared to disposal via a UIC Class I injection well (Section 4.14.1.1). Increasing the size of the proposed facility would lead to more land disturbance and a heavier use of construction equipment, with an anticipated increase in potential impacts to resource areas such as ecological and wetland systems, cultural and historical resources, and nonradiological air quality. The applicant would have to amend their license application to select one of these alternative wastewater disposal options. The NRC would perform an additional environmental and safety review prior to deciding whether to grant or deny the licensing application with the new wastewater disposal option. The applicant would survey the areas to be affected prior to construction, and the applicant and NRC would consult with agencies such as the Wyoming SHPO, the WGFD, and the FWS. Mitigation measures, such as avoidance of sensitive areas or documentation of cultural resources, would be established as part of these consultations, as necessary. With these mitigation measures in place, it is anticipated that potential adverse impacts would be SMALL.

4.14.1.2.1   Evaporation Ponds

The types of waste streams and the infrastructure necessary for using evaporation ponds as a wastewater disposal option are described in Section 2.1.1.2. The types and amounts of wastewater that would be disposed in an evaporation pond would be the same as described in the previous section for disposal by deep injection into a Class I UIC well. Before the applicant could begin disposing wastewater into an evaporation pond system, the NRC staff would review the design and construction of the ponds and monitoring system against the criteria in 10 CFR Part 40, Appendix A (NRC, 2003; 2008), taking into consideration EPA criteria in 40 CFR Part 61, Subpart W. The applicant would be required to demonstrate that the evaporation ponds could be designed, operated, and decommissioned to prevent migration of wastewater to subsurface soil, surface water, or groundwater. Applicants would also be required to demonstrate that monitoring requirements would be established to detect any migration of contaminants to the groundwater. The NRC staff would establish any license conditions needed to ensure that the applicant meets the necessary requirements.

Individual evaporation ponds could have a surface area of up to 2.5 ha [6.25 ac], and the total pond system could be as much as 40 ha [100 ac]. During the period of operations for the proposed Moore Ranch ISR facility, this area would be fenced to exclude wildlife and livestock. This would provide a footprint that is less than about 1.5 percent of the total permitted area {2,873 ha [7,100 ac]} for the proposed Moore Ranch ISR facility, but it would be much larger than the footprint {2.4 ha [6 ac]} for a central processing plant without evaporation ponds (see Section 4.2.1). The additional land disturbance required to install an evaporation pond system for wastewater disposal would be similar in scale to the current proposed action {61 ha [150 ac]} for the proposed Moore Ranch ISR facility (Section 4.2.1). It is also anticipated that the applicant would need to have at least one other wastewater disposal option or additional storage capacity during the winter months in Wyoming when evaporation rates would be low. Although a wastewater disposal option that uses an evaporation pond system would roughly double the facility footprint relative to UIC Class I injection wells, the total amount of disturbed and fenced land would be small compared to the permitted area and comparable to the generic conditions evaluated in the GEIS with respect to land use. For these reasons, the overall impacts to land use associated with an evaporation pond system would be SMALL.

Environmental Impacts and Mitigative Actions

Construction of an evaporation pond system would require earthmoving equipment such as bulldozers, backhoes, and trucks, and to prepare the site and construct the impoundment. The equipment would produce diesel emissions and fugitive dust emissions during construction that could have a temporary and adverse effect on nonradiological air quality. Depending on how the applicant elected to phase in the pond system, these effects could extend into the operational phase of the facility as well. BMP such as wetting unpaved roads would minimize fugitive dust, and the anticipated impacts to nonradiological air quality would be SMALL. The applicant may also need to obtain a National Emission Standards for Hazardous Air Pollutants review to evaluate whether the anticipated radiological releases to air from the evaporation ponds would meet the criteria in 40 CFR Part 61, Subpart W. The applicant would also be required to have an NRC-approved air monitoring system for the wastewater disposal system. Keeping the pond wet to reduce dust and radon emissions would effectively reduce potential air emissions, and the anticipated impacts to radiological air quality would be SMALL.

As described in NRC (2008), the evaporation ponds would be designed and constructed with clay or geotextile liners to reduce the potential for infiltration into the subsurface. An NRC-approved monitoring system would be installed to detect leaks from the ponds, and the applicant would also implement an NRC-approved inspection plan for the ponds (NRC, 2008). Based on these measures, it is anticipated that potential impacts to surface water and groundwater resources would be SMALL. As described in Section 4.6, the proposed Moore Ranch Project with one or more UIC Class I injection wells could potentially have MODERATE impacts on avian species of concern. A wastewater disposal option that uses an evaporation pond system would roughly double the facility footprint. The evaporation ponds, however, would be constructed at the same time and with the same mitigation measures described in Section 4.6 for the construction of the rest of the facility. Additional measures such as netting could be used to prevent birds from landing on the ponds. For these reasons, the potential impact from an evaporation pond disposal system would be the same as identified in Section 4.6 and could be reduced to SMALL.

At the end of the operational phase of the facility, all of the pond liners and berms, as well as accumulated precipitates and sludges, would be classified as solid byproduct material. For example, the GEIS indicates that about 52 m^3 [68 yd^3] of byproduct material would be generated during evaporation pond decommissioning. These solids would need to be transported to a licensed facility for disposal as part of the decommissioning program. This would increase the total amount of byproduct decommissioning wastes, increasing the number of truck trips needed to transport the materials to a disposal facility. Given the potential limitations on available byproduct waste disposal capacity in the local area, it is anticipated that the impacts from an evaporation pond wastewater disposal system to waste management would be SMALL to MODERATE during the decommissioning phase of the facility. It is important to note that at the conclusion of operations, the licensee would be required to provide a decommissioning plan for NRC review that demonstrates it has a disposal path for any decommissioning wastes, including those related to the wastewater disposal system. The NRC staff would conduct detailed technical and environmental reviews of the proposed decommissioning program for the facility at that time.

4.14.1.2.2   Land Application

For the land application of process wastewater, the applicant would be required to meet the regulatory provisions in 10 CFR Part 20, Subparts D, K, and Appendix B (NRC, 2003). The applicant would also be required to analyze the chemical toxicity of radioactive and

Environmental Impacts and Mitigative Actions

nonradioactive constituents, including an assessment of projected concentrations of radioactive contaminants in the soil and projected impacts on groundwater and surface-water quality and on land uses, especially crops and vegetation. The applicant would also be required to obtain NRC approval of a monitoring program that would include periodic soil surveys to verify that contaminant levels in the soil would not exceed those projected, and it should also include a remediation plan that can be implemented if projected levels are exceeded. The applicant would also need to treat the wastewater to quality requirements for surface discharge under a WYPDES permit from WDEQ. Finally, the applicant would also need to demonstrate that the soils in the land application area would meet the criteria in 10 CFR Part 40, Appendix A, at the time of decommissioning. Practices would be subject to NRC license conditions and enforced through the NRC inspection program to ensure protection of public health and safety and the environment.

Land application typically requires large areas to ensure that soil concentrations do not exceed regulatory levels. Typical land application areas are on the order of about 40 ha [100 ac]. During the period of operations for the proposed Moore Ranch ISR facility, this area would be fenced to exclude wildlife and livestock. Like a wastewater disposal system using evaporation ponds, land application would provide a footprint that is less than about 1.5 percent of the total permitted area {2,873 ha [7,100 ac]} for the proposed Moore Ranch ISR facility, but it would be much larger than the footprint {2.4 ha [6 ac]} described in the proposed action (Section 4.2.1) for a central processing plant without land application (Uranium One, 2009). The additional land disturbance required to install a land application system for wastewater disposal would be similar in scale to the current proposed action {61 ha [150 ac]} for the proposed Moore Ranch ISR facility (Section 4.2.1). It is also anticipated that the applicant would need to have at least one other wastewater disposal option or additional storage capacity during the winter months in Wyoming when evaporation rates would be low and the ground would be covered by snow. Like the evaporation pond system discussed in the previous section, a wastewater disposal option that uses land application would roughly double the facility footprint relative to the proposed action using UIC Class I injection wells. The amount of disturbed and fenced land, however, would be small compared to the permitted area and comparable to the generic conditions evaluated in the GEIS with respect to land use. For these reasons, the overall impacts to land use associated with w2astewater disposal by land application would be SMALL.

Establishing the land application area would not require extensive use of earthmoving equipment other than to install pipelines, small berms, access roads, and fencing, and the potential impacts to land use would be anticipated to be SMALL. The wastewater, however, would likely require additional treatment to meet WYPDES standards, including facilities such as an IX circuit, reverse osmosis, one or more radium settling basins {0.1 to 1.6 ha 0.25 to [4 ac]}, purge storage reservoirs {4 ha [10 ac] or more}. Constructing these treatment facilities, basins, and storage reservoirs would require earth moving equipment such as bulldozers, scrapers, backhoes, and trucks to prepare the site and construct the impoundments. The equipment would produce diesel emissions and fugitive dust emissions during construction that could have a temporary and adverse effect on nonradiological air quality. BMP such as wetting unpaved roads would minimize fugitive dust, and the anticipated impacts to nonradiological air quality would be SMALL. The applicant may also need to consider potential radiological releases to air from the land application area(s). Given the low radionuclide content anticipated for the wastewater, and low calculated radon fluxes for similar application areas (NRC, 1997, 2003b), the anticipated impacts to radiological air quality would be SMALL. As described in Section 4.6, the proposed Moore Ranch Project with one or more UIC Class I injection wells could potentially have MODERATE impacts on avian species of concern. A wastewater disposal option that

Environmental Impacts and Mitigative Actions

uses a land application system would roughly double the facility footprint. The land application system, however, would be constructed at the same time with the same mitigation measures described in Section 4.6 for the construction of the rest of the facility. For these reasons, the potential impact associated with wastewater disposal by land application would be the same as identified in Section 4.6 and could be reduced to SMALL.

As described previously, the applicant would be required to demonstrate that the soil in a land application area would meet 10 CFR Part 20 requirements. In addition, during operations the applicant would be required to routinely monitor the soil to ensure that predicted concentrations were not exceeded. For these reasons, it is not anticipated that decommissioning the land application area would produce any additional solid byproduct material for disposal, and the potential impacts on waste management would be SMALL during the decommissioning phase of the facility. For decommissioning the wastewater treatment facility, all pond liners and berms associated with radium settling basin(s), as well as accumulated precipitates and sludges generated at an estimated rate annual of about 22.4 m^3/yr [29.3 yd^3/yr] (see Section 2.1.1.2.2), would be classified as solid byproduct material. These solids, as well as any other solid byproduct material generated by the wastewater treatment process (e.g., spent resins, contaminated building debris) would need to be transported to a licensed facility for disposal as part of the decommissioning program. This would increase the total amount of byproduct decommissioning wastes, increasing the number of truck trips needed to transport the materials to a disposal facility. Given the potential limitations on available byproduct waste disposal capacity, the potential impacts to waste management from decommissioning the radium-settling basin(s) and other storage facilities associated with treating wastewater for disposal by land application would be SMALL to MODERATE.

It is important to note that at the conclusion of proposed operations, the licensee would be required to provide a plan for decommissioning any wastewater treatment facilities for NRC review (NRC, 2003b). The decommissioning plan would include final radiological surveys to identify whether there were any areas of soil contamination that would require disposal as byproduct material. The NRC staff would conduct detailed technical and environmental reviews of the proposed decommissioning program for the facility at that time.

4.14.1.2.3 Surface Water Discharge

For the surface discharge wastewater, the applicant would be required to meet the regulatory provisions in 10 CFR Part 20, Subparts D, K, and Appendix B. The applicant would also be required to obtain a zero-release WYPDES permit from WDEQ. The applicant would be required to distinguish between "process wastewater" generated during uranium recovery operations, and "mine wastewater" generated during aquifer restoration (NRC, 2003b). In accordance with EPA regulations, the applicant would not be allowed to discharge "process" wastewater to navigable waters of the United States (NRC, 2003a). The applicant would either need to develop storage capabilities depending on whether it intended to maintain separate wastewater streams, or commingle (mix) "process" and "mine" wastewater prior to treatment to 10 CFR Part 20 standards. In addition, the applicant would need to address any radioactivity at the discharge point or from storage facilities (tanks, impoundments), radium-settling basins, and related liners and sludges as part of the decommissioning of the facility (NRC, 2003b; Cohen and Associates, 2008b).

Establishing the discharge point for the treated effluent would be likely to require short-term use of earthmoving equipment to install pipelines, small berms, access roads, and fencing to exclude livestock and wildlife. The amount of land to be fenced for the discharge point alone would be limited (see Section 2.1.1.2), and the potential impacts to land use would be anticipated to be SMALL. As is the case with land application, however, the wastewater would likely require additional treatment to meet WYPDES zero-release permit requirements, including facilities such as an IX circuit, reverse osmosis, one or more radium settling basins {0.1 to 1.6 ha [0.25 to 4 ac]}, purge storage reservoirs {4 ha [10 ac] or more}. These treatment facilities would also be fenced to exclude wildlife and livestock and limit access to the public. The amount of land needed for the wastewater treatment facilities would be similar to that for land application, but if the applicant chose to segregate "process" and "mine" wastewaters to meet the WYPDES permit requirements, the involved land area would be greater, to provide separate storage facilities. As with evaporation ponds and land application, the increased footprint of the additional wastewater treatment facilities needed to meet WYPDES requirements would be small relative to the entire permitted area {2,873 ha [7,100 ac]}, but large relative to the 2.4 ha [6 ac], for a central processing plant as described in the proposed action (Section 4.2.1) (Uranium One, 2009a,b). The current proposed action identifies about 61 ha [150 ac] of disturbed land for the proposed Moore Ranch ISR facility. Overall, the increase in the amount of disturbed land to accommodate the addition of a wastewater treatment facility would be about 10 to 20 percent, and would have a SMALL impact on land use.

Constructing the wastewater treatment facilities (e.g., radium-settling basins) would require earthmoving equipment such as bulldozers, backhoes, trucks, and to prepare the site and construct the impoundment(s). The equipment would produce diesel emissions and fugitive dust emissions during construction that could have a temporary and adverse effect on nonradiological air quality. BMP such as wetting unpaved roads would minimize fugitive dust. Taking into consideration the likely short-term duration of the construction period, the anticipated impacts to nonradiological air quality would be SMALL. The applicant may also need to consider emissions of radionuclides such as radon from the surface discharge points. Given that the WYPDES permit would require the applicant to monitor and maintain low radionuclide concentrations for the treated wastewater, the anticipated impacts to radiological air quality would be SMALL.

The proposed Moore Ranch ISR facility and any surface water discharge points would be entirely within the Antelope Creek drainage basin (Section 3.5.1.4). This drainage basin is classified by WDEQ as a Class 3B surface water, with no known fish populations or use as a drinking water supply. However, surface water discharge would create more reliable water flow, and could lead to the development of aquatic habitat, and surface discharge could lead to an increase in erosion and suspended sediments in existing stream channels. Sediment loads would be expected to taper off quickly both in time and distance; therefore, long-term impacts would be SMALL. The applicant would use WFGD standard management practices to limit impacts to aquatic life.

As noted previously, the applicant would not be allowed to discharge treated wastewater into navigable waters of the United States. The disconnected and isolated nature of existing ponds and wetlands systems in the vicinity of the proposed Moore Ranch ISR facility are not considered jurisdictional waters under Section 404 of the Clean Water Act because there is no connection to navigable waters. However, surface discharge of treated wastewater could create a more continuous system, and the applicant would need to obtain the necessary jurisdictional determination and permits from the USACE.

Environmental Impacts and Mitigative Actions

The applicant would be required to demonstrate that any soil affected by the surface discharge of treated wastewater would meet 10 CFR Part 20 requirements. In addition, during operations, the applicant would be required to routinely monitor the soils and discharged water to ensure that predicted concentrations were not exceeded. For these reasons, it is not anticipated that decommissioning the surface discharge point would produce any additional solid byproduct material for disposal, and the potential impacts on waste management would be SMALL during the decommissioning phase of the facility. As with the land application wastewater disposal option, however, decommissioning wastewater treatment facilities may produce solid byproduct materials such as spent resins, sludges and liners from radium-settling basin(s), or contaminated building debris. These solids would need to be transported to a licensed facility for disposal as part of the decommissioning program. This would increase the total amount of byproduct decommissioning wastes, increasing the number of truck trips needed to transport the materials to a disposal facility. Given the potential limitations on available byproduct waste disposal capacity, it is anticipated that the potential impacts to waste management from decommissioning the radium settling basin(s) and other storage facilities associated with treating wastewater for surface water discharge would be SMALL to MODERATE.

It is important to note that at the conclusion of operations, the licensee would be required to provide a plan for decommissioning any wastewater treatment facilities for NRC review (NRC, 2003b). The decommissioning plan would include final radiological surveys to identify whether there were any areas of soil contamination that would require disposal as byproduct material. The NRC staff would conduct detailed technical and environmental reviews of the proposed decommissioning program for the facility at that time.

4.14.1.2.4   Class V Injection Well

The potential impacts associated with wastewater disposal through a UIC Class V deep injection well would be similar to those associated with the proposed action (Disposal via a UIC Class I deep injection well). Under the terms of a UIC Class V permit issued by WDEQ, however, the wastewater would require additional treatment to meet class of use or federal drinking water standards (whichever is more stringent) prior to injection.

The potential impacts associated with constructing, operating, and decommissioning the necessary wastewater treatment facilities would be similar to those described in the previous sections for land application (Section 4.14.1.2.2) and surface water discharge (Section 4.14.1.2.3) disposal options. For example, although the footprint of the Class V well itself would be small {0.1 ha [0.25 ac]}, the wastewater would likely require additional treatment to meet the necessary discharge requirements (Class of Use or Federal drinking water standards). This treatment would require facilities such as an IX circuit, reverse osmosis, one or more radium settling basins {0.1 to 1.6 ha [0.25 to 4 ac]}, and purge storage reservoirs {4 ha [10 ac] or more}. These treatment facilities would be fenced to exclude wildlife and livestock and limit access of the public. The amount of land needed for the wastewater treatment facilities would be similar to that for land application or surface discharge. The increased footprint of the additional wastewater treatment facilities would be small relative to the entire permitted area {2,873 ha [7,100 ac]}, but large relative to the 2.4 ha [6 ac], for a central processing plant as described in the proposed action (Section 4.2.1) (Uranium One, 2009a,b). The current proposed action identifies about 61 ha [150 ac] of disturbed land for the proposed Moore Ranch ISR facility. Overall, the increase in the amount of disturbed land to accommodate addition of a wastewater treatment facility would be about 10 to 20 percent and would have a SMALL impact on land use.

Constructing the wastewater treatment facilities (e.g., radium-settling basins) would require earth moving equipment such as bulldozers, backhoes, trucks, and to prepare the site and construct the impoundment(s). The equipment would produce diesel emissions and fugitive dust emissions during construction that could have a temporary and adverse effect on nonradiological air quality. BMP such as wetting unpaved roads would minimize fugitive dust. Taking into consideration the likely short-term duration of the construction period, the anticipated impacts to nonradiological air quality would be SMALL. The applicant may also need to consider emissions of radionuclides such as radon during the wastewater treatment process. These emissions would be included as part of the NRC-approved monitoring plan for the facility, and the anticipated impacts to radiological air quality would be SMALL.

As with the land application and surface discharge wastewater disposal options, the solid wastes generated by decommissioning wastewater treatment facilities associated with a UIC Class V injection well, such as piping, spent resins, sludges, and liners from radium settling basin(s), or contaminated building debris would need to be disposed as byproduct material. These solids would need to be transported to a licensed facility for disposal as part of the decommissioning program. This would increase the total amount of byproduct decommissioning wastes, increasing the number of truck trips needed to transport the materials to a disposal facility. Given the potential limitations on available byproduct waste disposal capacity, it is anticipated that the potential impacts to waste management from decommissioning the radium-settling basin(s) and other storage facilities associated with treating wastewater for surface water discharge would be SMALL to MODERATE.

It is important to note that at the conclusion of operations, the licensee would be required to provide a plan for decommissioning any wastewater treatment facilities for NRC review (NRC, 2003). The decommissioning plan would include final radiological surveys to identify whether there were any areas of soil contamination that would require disposal as byproduct material. The NRC staff would conduct detailed technical and environmental reviews of the proposed decommissioning program for the facility at that time.

4.14.2 No-Action (Alternative 2)

Under the No-Action alternative no radioactive or nonradioactive liquid or solid waste would be generated because the proposed Moore Ranch Project would not be licensed. No earthmoving activities that could result in the generation of nonhazardous solid waste would occur, no buildings would be constructed, no wellfields would be developed, and no wastewater would be injected into the subsurface. No arrangements would be made for the management of any wastes.

4.15 References

10 CFR Part 20. *Code of Federal Regulations*, Title 10, *Energy*, Part 20, "Standards for Protection Against Radiation."

Environmental Impacts and Mitigative Actions

10 CFR Part 40. Appendix A. *Code of Federal Regulations*, Title 10, Energy, Part 40, Appendix A, "Criteria Relating to the Operations of Uranium Mills and to the Disposition of Tailings or Wastes Produced by the Extraction or Concentration of Source Material from Ores Processed Primarily from their Source Material Content."

40 CFR Part 144. *Code of Federal Regulations*, Title 144, *Protection of Environment*, Part 144, "Underground Injection Control Program."

40 CFR Part 146. *Code of Federal Regulations*, Title 40, *Protection of Environment*. Part 146, "Underground Injection Control Program: Criteria and Standards."

69 FR 52040, 2004. "Policy Statement on the Treatment of Environmental Justice matters in NRC Regulatory and Licensing Actions." *Federal Register*. Vol. 69, No. 163. August 24.

Argonne National Laboratory, 1989. "MILDOS-AREA (Computer Code)–Calculation of Radiation Dose from Uranium Recovery Operations for Large-Area Sources." Argonne, Illinois: Argonne National Laboratory.

BLM (U.S. Bureau of Land Management, 2010. Instruction Memorandum No. 2010-071. "Subject: Gunnison and Greater Sage-grouse Management Considerations for Energy Development (Supplement to National Sage-Grouse Habitat Conservation Strategy)." Washington, DC: U.S. Department of Interior Bureau of Land Management. March 5.

BLM, 2008a. "Proposed Resource Management Plan and Final Environmental Impact Statement for Public Lands Administered by the Bureau of Land Management, Rawlins Field Office." Rawlins, Wyoming. U.S. Department of Interior, BLM, Rawlins Field Office. January 2008.

BLM, 2008b. "Bureau of Land Management Buffalo Field Office Environmental Assessment for Yates Petroleum Corporation All Day POD Plan of Development WY-070-08-026." Cheyenne, Wyoming: BLM. August 28.

BLM. 2007. "Visual Resource Management." Manual 8400. Washington, DC: BLM. <http://www.blm.gov/nstc/VRM/8410.html>.

BLM, 2004. "National Sage-Grouse Habitat Conservation Strategy." November. <http://www.blm.gov/pgdata/etc/medialib/blm/wo/Planning_and_Renewable_Resources/fish__wildlife_and.Par.9151.File.dat/Sage-Grouse_Strategy.pdf>.

BLM, 2003. "Final Environmental Impact Statement and Proposed Amendment for the Powder River Basin Oil and Gas Project (WY–070–02–065)." January.

Brunette, J.A., 2007. "Class III Cultural Resources Inventory for the Energy Metals Corporation, Moore Ranch *In-Situ* Uranium Project." (nonpublic due to sensitive information on cultural resources).

Center for Climate Strategies, 2007. "Wyoming Greenhouse Gas Inventory and Reference Case Projections
1990–2020." Cheyenne, Wyoming: Center for Climate Strategies.

Environmental Impacts and Mitigative Actions

CEQ (Council of Environmental Quality, 1997. "Environmental Justice Guidance under the National Environmental Policy Act." Executive Office of the President. December.

Cohen and Associates, 2008. "Final Report Review of Existing and Proposed Tailings Impoundment Technologies." Vienna, Virginia: S. Cohen and Associates. <http://www.epa.gov/radiation/docs/neshaps/subpart-w/tailings-impoundment-tech.pdf> <25 May 2010>.

EMC (Energy Metals Corporation U.S.), 2007a. "Application for USNRC Source Material License, Moore Ranch Uranium Project, Campbell County, Wyoming, Environmental Report." Casper, Wyoming: Uranium 1 Americas Corporation. ADAMS Accession Nos. ML072851222, ML072851229, ML072851239, ML07285249, ML07285253, and ML07285255. October 12.

EMC, 2007b. "Application for USNRC Source Material License, Moore Ranch Uranium Project, Campbell County, Wyoming, Technical Report." Casper, Wyoming: Uranium 1 Americas Corporation. ADAMS Accession Nos. ML072851222, ML072851258, ML072851259, ML072851260, ML072851268, ML072851350, ML072900446. October 12.

EPA (U.S. Environmental Protection Agency), 2010. "SCRAM Mixing Height Data: Wyoming." Washington DC: EPA. <http://www.epa.gov/scram001/mixingheightdata.htm> (07 June 2010).

EPA, 2009. "Global Greenhouse Gas Data." <http://www.epa.gov/climatechange/emissions/globalghg.html.> (15 December 2009). ADAMS No. ML100221499.

FWS (U.S. Fish and Wildlife Service, 2008. "Response to Request for Additional Information Regarding Endangered or Threatened Species and Critical Habitat for the Proposed License Application for Energy Metals Corporation's Moore Ranch Uranium Recovery Project." ADAMS Accession No. ML081420589. May 7.

Griffin, M., 2009. <mike.griffin@uranium1.com>"Phases of the Moore Ranch Project" 28 September 2009 [email communication]. ADAMS Accession No. ML092720144. (28 September 2009).

Mackin, P.C., D. Daruwalla, J. Winterle, M. Smith, and D.A. Pickett, 2001. NUREG/CR–6733, "A Baseline Risk-Informed Performance-Based Approach for *In-Situ* Leach Uranium Extraction Licensees." Washington, DC: NRC. September.

National Fire Protection Association, 2001. "NFPA 50: Standard for Bulk Oxygen Systems at Consumer Sites."

NRC (U.S. Nuclear Regulatory Commission), 2009a. NUREG–1910, "Generic Environmental Impact Statement for *In-Situ* Leach Uranium Milling Facilities." Washington, DC: NRC. May.

NRC, 2009b. "Staff Assessment of Groundwater Impacts from Previously Licensed *In-Situ* Uranium Recovery Facilities." Memorandum from C. Miller to Chairman Jaczko, et al. ADAMS Accession No. ML091770402. July 10.

NRC, 2009c. Memo to A. Kock, Branch Chief, from I. Yu, B. Shroff, and A. Bjornsen, Project Managers, Office of Federal and State Materials and Environmental Management Programs. "Subject: Informal Meetings with Local, State, and Federal Agencies in Wyoming Regarding the

Environmental Reviews Being Conducted on the Moore Ranch, Nichols Ranch, and Lost Creek *In-Situ* Leach Applications for Source Material Licenses (Docket Nos. 040-09073, 040-09067, 040-09068." ADAMS Accession No. ML090500544. March 2.

NRC, 2008. Regulatory Guide 3.11, "Design, Construction, and Inspection of Embankment Retention Systems at Uranium Recovery Facilities." Washington, DC: NRC. November.

NRC, 2003a. NUREG–1748, "Environmental Review Guidance for Licensing Actions Associated With NMSS Programs." Washington, DC: August.

NRC, 2003b. NUREG–1569, "Standard Review Plan for *In-Situ* Leach Uranium Extraction License Applications." Final Report. Washington, DC: June.

NRC, 1997. NUREG–1508, "Final Environmental Impact Statement To Construct and Operate the Crownpoint Uranium Solution Mining Project, Crownpoint, New Mexico." Washington, DC: NRC. February.

Uranium One (Uranium One Americas), 2010. "Responses to NRC Waste Management Questions." ADAMS Accession No. ML101330379. May 12.

Uranium One, 2009a. "Response to Request for Additional Information for the Moore Ranch *In-Situ* Uranium Recovery Project License Application (TAC JU011)." ADAMS Accession No. ML091900402. June 19.

Uranium One, 2009b. "Response to Request for Additional Information for the Moore Ranch *In-Situ* Uranium Recovery Project License Application (TAC JU011)." ADAMS Accession No. ML092450317. August 31.

Uranium One, 2008. "Response to Request for Additional Information for the Moore Ranch *In-Situ* Uranium Recovery Project License Application (TAC JU011), Second Set of Responses." ADAMS Accession No. ML090400079. October 28.

USACE (U.S. Army Corps of Engineers), 2010. "Subject: Response to a Preconstruction Notification (PCN)." Letter to J. Winter from M.A. Bilodeau, Program Manager, Department of the Army Corps of Engineers. May 10.

USCB (U.S. Census Bureau), 2009. "State and County QuickFacts for Campbell, Johnson, and Natrona Counties Wyoming." <http://quickfacts.census.gov> (10 September 2009).
USCB, 2008.

USCB (U.S. Census Bureau), 2008. 2008 Population Estimates for Wright, Edgerton, Midwest, Gillette, and Casper in Wyoming. <http://factfinder.census.gov> (16 October 2009). ADAMS No. ML092940144.

WDEQ (Wyoming Department of Environmental Quality) 2010: "Re: Permit No. CT-7896." Letter (January 4) from D.A. Finley to J. Cash, Lost Creek ISR, LLC. Cheyenne, Wyoming: WDEQ-Quality, Air Quality Division. 2010.

WDEQ, 2009. "Re: Permit No. CT-8644." Letter (October 2) from D.A. Finley to M. P. Thomas, Uranerz Energy Corporation. Cheyenne, Wyoming: WDEQ-Air Quality Division.

Environmental Impacts and Mitigative Actions

WDEQ, 2000. "Guideline No. 4–*In-Situ* Mining, Attachment III Topsoil and Subsoil Management and the Associated Erosion Control at Uranium *In-Situ* Leaching Operations." March.

WGFD (Wyoming Game and Fish Department), 2010. "Recommendations for Development of Oil and Gas Resources Within Important Wildlife Habitats." Version 6.0. Cheyenne, Wyoming. April.

WGFD, 2009. "Big Game Crucial Range Maps, Cheyenne, Wyoming." April. <http://gf.state.wy.us/habitat/index.asp>

WGFD, 2004. "Fencing Guidelines for Wildlife." Revised Version, Habitat Extension Bulletin No. 53—Habitat Extension Services. November.

Whitehead, R.L., 1996. "Ground Water Atlas of the United States: Segment 8, Montana, North Dakota, South Dakota, Wyoming." U.S. Geological Survey Hydrologic Investigations Atlas 730-I. <http://pubs.er.usgs.gov/usguspubs/ha/ha730I>.

Wyoming Department of Employment, Research and Planning, 2009. Wyoming Unemployment Rate Increases to 5.0 percent in May. <http://wydoe.state.wy.us/LMI/news.htm>.

Wyoming Office of State Lands and Investments, 2007. "Wyoming Biomass Inventory: Animal Waste, Crop Residue, Wood Residue, and Municipal Solid Waste." Cheyenne, Wyoming: Office of State Lands and Investments, Wyoming State Forestry Division. March.

Wyoming SHPO (Wyoming State Historic Preservation Office), 2010. "Subject: Status of Site 48CA962." E-mail to J. Davis from R.L. Currit, Senior Archaeologist, Wyoming State Historic Preservation Office. June 15. ADAMS Accession No. ML101660667.

Wyoming SHPO, 2009. "Subject: Uranium One, Inc. More Ranch *In-Situ* Uranium Recovery Project Cultural Resources Inventory (SHPO File# 0608RLC007)." Letter to A. Kock from R.L. Currit, Senior Archaeologist, Wyoming State Historic Preservation Office. November 3. ADAMS Accession No. ML093170805.

5 CUMULATIVE IMPACTS

5.1 Introduction

The Council on Environmental Quality (CEQ) *National Environmental Policy* Act (NEPA) regulations, as amended (40 CFR 1500 to 1508) define cumulative effects as "...the impact on the environment that results from the incremental impact of the action when added to other past, present, and reasonably foreseeable future actions regardless of what agency (federal or nonfederal) or person undertakes such other actions." Cumulative effects or impacts[1] can result from individually minor but collectively significant actions taking place over a period of time. The proposed Moore Ranch Project could contribute to cumulative effects when its environmental impacts overlap with those of other past, present, or reasonably foreseeable future actions. For this supplemental environmental impact statement (SEIS), other past, present, and future actions in the area include, but are not limited to, coal mining, oil and gas production, coal bed methane (CBM) operations, other mining (i.e., sand, gravel, bentonite, clinker), *in-situ* recovery (ISR) operations, conventional uranium mining, and wind farms.

The cumulative impact analysis of the proposed action was based on publicly available information on existing and proposed projects, information from the GEIS (NRC, 2009a), general knowledge of the conditions in Wyoming and in the nearby communities, and reasonably foreseeable actions that could occur. The primary activity in the area is a resurgence, within the last few years, of mineral mining and oil and gas development, although this interest has not necessarily translated into active projects. Within 8 km [5 mi] of the proposed Moore Ranch Project and within the proposed license area, coal bed methane operations are occurring. No long-term changes from the proposed action within the projected license area are anticipated because the applicant plans to return the proposed license area to its preextraction use following restoration and reclamation activities. There are several ISR and conventional uranium projects within the vicinity of the proposed Moore Ranch Project that are either in the prelicensing stage or decommissioning. Oil and gas operations are ongoing throughout the area. At distances beyond 8 km [5 mi], NRC assumed that the resurgence in extractive industries would continue, along with government and industry support, to develop infrastructure.

The GEIS (NRC, 2009a) provides an example methodology for conducting a cumulative impacts assessment. Section 5.1.1 describes other past, present, and reasonably foreseeable future actions considered in the cumulative impact analysis. The methodology to conduct the cumulative impacts analysis for this SEIS is provided in Section 5.1.2.

5.1.1 Other Past, Present, and Reasonably Foreseeable Future Actions

The proposed Moore Ranch Project, which covers an area of approximately 28 km^2 [11 mi^2], is located in the middle of the Powder River Basin, which covers an area of approximately 26,000 km^2 [10,000 mi^2] and spans large portions of northeastern Wyoming and southeastern Montana. Therefore, the proposed activities at the Moore Ranch Project would affect less than 0.1 percent of the area within the Powder River Basin. The Powder River Basin contains the

---

[1] For the purposes of this analysis, "cumulative impacts" is synonymous with "cumulative effects"

Cumulative Impacts

largest deposits of coal in the United States, as well as significant reserves of other natural resources including uranium, oil, and gas. As such, there has been, and continues to be, substantial mining activities throughout the Powder River Basin. CBM extraction continues to be the most prolific mining activity in the region, which is a form of natural gas extraction from coal beds. Several environmental impact statements (EISs) issued by the U.S. Bureau of Land Management (BLM) and studies by environmental groups in the Powder River Basin that date back to the 1970s have looked at the various effects that coal-related mining activities have had on the affected environment.

The various past, present, and reasonably foreseeable future actions in the vicinity of the proposed Moore Ranch Project are discussed as follows.

5.1.1.1 Uranium Recovery Sites

Along with the proposed Moore Ranch Project, there are other ISR and conventional uranium (underground and pit) operations in various stages of the licensing process within the Powder River Basin. Uranium-related exploration in the area include the Smith Ranch/Highland Uranium Project, an ISR project operated by Power Resources, Inc., and the Irigaray/Christensen Ranch Project, operated by Cogema Mining, Inc., are located approximately 59 km [37 mi] south-southeast and 32 km [20 mi] north-northwest of the proposed Moore Ranch Project (Table 5-1). The applicant has indicated that rather than constructing the central processing plant evaluated as part of the proposed action, it is also considering operating the Moore Ranch project as a satellite facility, along with the Ludeman and Allemand-Ross projects (Uranium One, 2010). In this reasonably foreseeable future action, the pregnant lixiviant from the wellfields at these ISR facilities would be pumped to remote satellite ion exchange facilities for initial stripping of uranium. On a daily basis during operations, Uranium One would load the uranium-saturated resins that result from the ion exchange process into 15,140 L [4,000 gal] sole-use tanker trucks. For the Moore Ranch facility, the trucks would carry the uranium-bearing resins west over State Highway 387, north over State Highway 192, and east over the Streeter Ranch Road to the Irigaray/Christensen Ranch Central Plant (Table 5-1) (Uranium One, 2010). There, the resins would be added to the elution circuit for uranium stripping and yellowcake production in a process similar to that described in Section 2.1.1.1.3.5 (NRC, 1998). Barren resin would be transferred into the tanker truck and transported by Uranium One back to the satellite facility (Uranium One, 2010). The tanker trucks used to transport ion exchange resins to and from the satellite facility would be labeled in accordance with U.S. Department of Transportation requirements at 49 CFR 171–189 and NRC regulations at 10 CFR Part 71 (Uranium One, 2010). The amount of uranium that could be produced in this fashion would be subject to the NRC license condition (NRC License No. SUA–1341) limiting annual yellowcake production at the Irigaray/Christensen Ranch Central Plant to 1,134,000 kg [2.5 million lb] (NRC, 1998). Any increase in this limit would require an amendment to the Irigaray/Christensen Ranch NRC license, with an independent evaluation of the environmental impacts (NRC, 2003).

The NRC staff is aware that several companies are actively investigating the potential for ISR extraction, as well as other types of mining and milling, in areas near the proposed Moore Ranch Project. These projects are in various stages of development, will be monitored by the NRC staff and other local government agencies, and will be discussed within the context of cumulative impacts in this SEIS based on available information.

The current uranium-recovery sites in the Powder River Basin pertaining to potential new uranium-recovery sites are listed in Table 5-1.

Table 5-1. Uranium Mining and Milling Projects—Distance and Direction From the Proposed Moore Ranch Project

Site Name	Company/Owner	Type[a]	County, State	Status[b]	Distance km (mi)	Direction
Ruby Ranch	Power Resources, Inc.	ISR	Campbell, WY	Potential site	40(25)	NNW
Reno Creek	NCA Nuclear Corp.	ISR	Campbell, WY	Potential site	22(13.5)	NW
Ludeman	Uranium One	ISR	Converse, WY	Potential site	50(31)	SSE
Allemand-Ross	Uranium One	ISR	Johnson, WY	Potential site	26(16)	SW
Ruby Ranch	Conoco	ISR[1]	Campbell, WY	Not licensed - application withdrawn	40(25)	NNW
Reno Creek 1	Rocky Mountain Energy Co.	ISR[1]	Campbell, WY	License terminated	22(13.5)	NW
Collins Draw	Cleveland Cliffs Iron Co.	ISR[1]	Campbell, WY	License terminated	13(8)	NW
Peterson Ranch	Arizona Public Service Co. Malapai Resources	ISR[1]	Converse, WY	Not pursued	80(49.5)	SSE
South Powder River Basin	Kerr-McGee	ISR[1]	Converse, WY	License terminated with approval of Smith Ranch license	59(36.5)	SSE
Willow Creek	J&P Corp. Western Nuclear	ISR[1]	Johnson, WY	License terminated with approval of Irigaray license	-	-
North Platte	Uranium Resources	ISR[1]	Platte, WY	License terminated	75(46.5)	SSE
Reynolds Ranch	Power Resources, Inc.	ISR[2]	Converse, WY	Licensed, but not operational	50.5(31.5)	SSE
Reno Creek 2	International Uranium Corp.	ISR[3]	Campbell, WY	Not licensed - application withdrawn	22(13.5)	NW
Moore Ranch	Uranium One	ISR[3]	Campbell, WY	Potential site - license application under review by NRC	Not Applicable	Not Applicable

Cumulative Impacts

Table 5-1. Uranium Mining and Milling Projects–Distance and Direction From the Proposed Moore Ranch Project (continued)

Site Name	Company/Owner	Type[a]	County, State	Status[b]	Distance km (mi)	Direction
Highland	Exxon Minerals	ISR[3]	Converse, WY	Licensed, but not pursued	63(39)	SSE
Highland 2	Everest Minerals	ISR[3]	Converse, WY	Licensed - later combined with Smith Ranch facility license	59(36.5)	SSE
Smith Ranch-Highland	Power Resources, Inc.	ISR[3]	Converse, WY	Operating	59(36.5)	SSE
Leuenberger	Teton Exploration Drilling	ISR[1,3]	Converse, WY	License terminated	72.5(45)	SSE
Nichols Ranch & Hank	Uranerz Energy Corp.	ISR[2,3]	Campbell & Johnson, WY	Potential site - license application under review by NRC	17.5(11)	NW
North Butte & Ruth	Power Resources, Inc.	ISR[2,3]	Campbell, WY	Licensed - but not operational	37(23)	WNW
Irigaray/ Christensen Ranch	Cogema Malapai Resources	ISR[2,3]	Johnson, WY	Licensed for operations	32(20)	NNW
Shirley Basin South	DOE	Conv.	Carbon, WY	UMTRCA Title II disposal site	137(85)	SSW
Bear Creek	Bear Creek Uranium Co.	Conv.	Converse, WY	Decommissioning	38(23.5)	SSE
Spook	Department of Energy	Conv.	Converse, WY	UMTRCA Title I disposal site	43.5(27)	SSE
Shirley Basin	Pathfinder Mines Corp.	Conv.	Natrona, WY	Decommissioning	140(87)	SSW

Source: NRC, 2009b
a. Type:
1 = Research and Development/Pilot
2 = Satellite
3 = Commercial scale
Conv. = Conventional uranium mill
b. Status: Uranium Mill Tailings Radiation Control Act (UMTRCA) Title I and Title II sites are uranium mill processing or tailings sites that have been decommissioned. The U.S. Department of Energy is the long-term custodian of these sites.

Cumulative Impacts

5.1.1.2 Coal Mining

The Powder River Federal Coal Region was decertified as a federal coal production region by the Powder River Regional Coal Team in 1990, which allowed leasing to occur in the region on an application basis. Because of decertification, United States coal production increased 11 percent, from 1,029.1 million tons in 1990 to 1,145.6 million tons in 2007 (BLM, 2009a). From 1990 to 2008, the BLM Wyoming State Office held 25 competitive lease sales and issued 19 new federal coal leases containing more than 5.7 billion tons of coal using the "lease by application" process (BLM, 2005a,b,c). Powder River basin coal mines make up over 96 percent of the coal produced in Wyoming each year (BLM, 2005a,b,c). In 2003, the cumulative disturbed land area attributable to coal mines within the Powder River Basin totaled nearly 28,000 ha [70,000 ac]. Reasonably foreseeable future development projects contributing to the estimate of the cumulative acreage disturbed range from 47,400 to 50,600 ha [117,000 to 125,000 ac] in the year 2015. Other development related to coal includes railroads, coal-fired power plants, major (230 kV) transmission lines, and coal technology projects. The total land area of other coal-related disturbance in the Powder River Basin in 2003 was nearly 2,000 ha [5,000 ac].

Table 5-2 lists surface coal mines within the Powder River Basin in Wyoming. The Wyoming East Uranium Milling Region where the proposed Moore Ranch Project is located, includes 16 surface mines. Surface mining of coal can impact land use, geology and soils, water resources, ecology, air quality, noise, historical and cultural resources, visual and scenic resources, socioeconomics, and waste management.

5.1.1.3 Oil and Gas Production

There are approximately 472 oil and gas production units evenly dispersed throughout the Powder River Basin in various stages of production. The Wyoming Oil and Gas Conservation Commission reported that in 2003, oil and gas wells in the Powder River Basin produced approximately 13 million barrels of oil and 1.1 billion m^3 [40 billion ft^3] of conventional gas (BLM, 2005a,b,c).

Most of Wyoming current oil production is from old oil fields with declining production and the level of exploration drilling to discover new fields has been low (WSGS, 2002, as cited in BLM, 2008a). From 1992 to 2002, oil production from conventional oil and gas wells in Campbell and Converse Counties within the Powder River Basin decreased approximately 60.4 percent. Oil- and gas-related development includes major transportation pipelines and refineries. In 2003, the cumulative disturbed land area in the Powder River Basin from oil and gas, CBM, and related development was nearly 76,100 ha [188,000 ac]. The corresponding projection for the year 2015 is 123,000 ha [305,000 ac] (BLM, 2005a,b,c). The depth to producing gas and oil-bearing horizons generally ranges from 1,220 to 4,120 m [4,000 to 13,500 ft], but some wells are as shallow as 76 m [250 ft] (BLM, 2005a,b,c).

5.1.1.4 Coal Bed Methane Development Projects

Natural gas production has been increasing in Wyoming. In the Powder River Basin, this is from the development of shallow CBM resources (BLM, 2005a,b,c). Annual CBM production in the Powder River Basin increased rapidly between 1999 and 2003, with nearly 15,000

Cumulative Impacts

5-2. Coal Mining and Milling Projects - Distance and Direction From the Moore Ranch Project

Site Name	Company/Owner	Type	County, State	Production in 2008 (Tons)	Distance (mi)	Direction
Buckskin	Buckskin Mining Company	Surface	Campbell, WY	26,076,356	99.5(62)	NNE
Dry Fork	Western Fuels of Wyoming, Inc.	Surface	Campbell, WY	5,261,242	94(58.5)	NNE
Eagle Butte	Foundation Coal West	Surface	Campbell, WY	20,443,413	95(59)	NNE
KFx Plant	Evergreen Energy	Surface	Campbell, WY	0 (was in production 2006, 2007)	95(59)	NNE
Wyodak	Wyodak Resources Development Corp.	Surface	Campbell, WY	6,017,311	90(56)	NNE
Caballo	Powder River Coal Company	Surface	Campbell, WY	31,205,381	75(46.5)	NNE
Belle Ayr	Foundation Coal West	Surface	Campbell, WY	28,707,982	69(43)	NE
Cordero/Rojo Complex	Rio Tinto Energy America	Surface	Campbell, WY	40,033,283	67(41.5)	NE
Coal Creek	Thunder Basin Coal Company, LLC	Surface	Campbell, WY	11,453,547 (not in production 2001-2005)	63(39)	NE
Jacobs Ranch	Rio Tinto Energy America	Surface	Campbell, WY	42,145,705	51.5(32)	ENE
Black Thunder	Thunder Basin Coal Company, LLC	Surface	Campbell, WY	88,587,310	47.5(29.5)	ENE
North Antelope/ Rochelle Complex	Powder River Coal Company	Surface	Campbell & Converse, WY	97,578,499	46.5(29)	E
North Rochelle	Triton Coal Company	Surface	Campbell, WY	no data	48(30)	E
Antelope	Rio Tinto Energy America	Surface	Campbell & Converse, WY	35,795,491	42(26)	ESE
Seminoe #2	Arch Coal, Inc.	Surface	Carbon, WY	Final reclamation in 2006	191.5(119)	SSW
Medicine Bow	Arch Coal, Inc.	Surface	Carbon, WY	28,212, but 0 in 2005; relatively small operation	196(122)	SSW

Source: Wyoming Mining Association, 2008

producing CBM wells in the Powder River Basin in 2003 and a total production volume of
364 billion cubic feet (BLM, 2005a,b,c). In 2007, CBM production within Campbell County was
4.7-million m³ [167,000 million ft³] (BLM, 2009c). The BLM Buffalo Field Office, which
administers the area where the proposed Moore Ranch Project is located, has processed
approximately 3,000 applications for permits to drill since 2003; more than 98 percent of these
applications are for CBM recovery (BLM, 2009c).

The recovery of CBM involves the installation of facilities that include access roads; pipelines for
gathering gas and produced water; electrical utilities; facilities for measuring and compressing
recovered gas; facilities for treating, discharging, disposing of, containing, or injecting produced
water; and pipelines to transport gas high-pressure transmission pipelines (EMC, 2007a,b). The
wells are collocated on a well pad installed on a 32 ha [80 ac] spacing pattern (eight pads per
square mile). The overall life of each well is approximately 10 years: 7 years of production
followed by final reclamation, which takes about 2 to 3 years (EMC, 2007a,b).

There are 534 CBM wells within 3.2 km [2 mi] of the boundary of the proposed license area.
The largest CBM volumes are produced from depths greater than 60 m [200 ft] below ground
surface (bgs). The target formation occurs at depths from 305 to 366 m [1,000 to 1,200 ft] bgs.

5.1.1.5 Other Mining

Sand, gravel, bentonite, and clinker (or scoria) have been and continue to be mined in the
Powder River Basin. Bentonite is weathered volcanic ash that is used in a variety of products,
including drilling mud and making cat litter, because of its absorbent properties. There are three
major bentonite-producing districts in and around the Powder River Basin. Aggregate, which is
sand, gravel, and stone, is used for construction purposes. The largest aggregate operation is
located in the Powder River Basin in northern Converse County, and it has an associated total
disturbance area of approximately 27 ha [67 ac], of which 1.62 ha [4 ac] have been reclaimed.
Scoria, or clinker, is used as aggregate where alluvial terrace gravel or in-palace
granite/igneous rock is not available. Scoria generally is mined in the Converse and Campbell
Counties portion of the Powder River Basin (BLM, 2005a,b,c).

5.1.1.6 EISs as Indicators of Past, Present, and Reasonably Foreseeable Actions

Another indicator of present and reasonably foreseeable future actions is the number of draft
and final EISs prepared by federal agencies within a recent time period. Using information in
NUREG-1910, *Generic Environmental Impact Statement for In-Situ Leach Uranium Milling
Facilities* (GEIS) (NRC, 2009a), Section 5.1.1 and publicly available information, several
site-specific EISs evaluating actions in the Powder River Basin in addition to draft and final
programmatic EISs for large-scale actions related to several states including Wyoming were
considered (see NRC, 2009a, Tables 5.2-3 and 5.2-4). These projects could contribute to both
local and regional cumulative impacts on air quality, land usage, terrestrial plants and animals,
and groundwater and surface water resources.

5.1.2 Methodology

To determine the potential cumulative impact, the following methodology was developed, based
on the CEQ guidance (CEQ, 1997):

Cumulative Impacts

1. Identify for each resource area, the potential environmental impacts that would be of concern from a cumulative impacts perspective. These impacts are discussed and analyzed in Chapter 4.

2. Identify the geographic scope for the analysis for each resource area. This scope is expected to vary from resource area to resource area, depending on the geographic extent to which the potential impacts could be at issue.

3. Identify the time frame over which cumulative impacts would be assessed. The timeframe selected begins in 2007 when Uranium One submitted a license application to NRC for a new source material license for the Moore Ranch ISR project. The cumulative impact analysis timeframe will terminate in 2020 which represents the license termination at the end of the decommissioning period.

4. Identify existing and anticipated future projects and activities in and surrounding the project site. These projects and activities are identified in Table 5-1 and 5-2 of this chapter.

Assess the cumulative impacts for each resource area from the proposed action and reasonable alternatives, and other past, present, and reasonably foreseeable future actions identified in Table 5-1 and 5-2 of this SEIS. This analysis would consider the environmental impacts of concern identified in Step 1 and the resource area-specific geographic scope identified in Step 2.

The following terminology was used to define the level of cumulative impact:

SMALL: The environmental effects are not detectable or are so minor that they will neither destabilize nor noticeably alter any important attribute of the resource considered.

MODERATE: The environmental effects are sufficient to alter noticeably but not destabilize important attributes of the resource considered.

LARGE: The environmental effects are clearly noticeable and are sufficient to destabilize important attributes of the resource considered.

In conducting this assessment, the NRC staff recognized that for many aspects of the activities proposed by the applicant, there would be a SMALL impact on the affected resources. However, an impact that may be SMALL by itself could result in a MODERATE or LARGE cumulative impact when considered in combination with the impacts of other actions on the affected resource. Likewise, if a resource is regionally declining or imperiled, even a SMALL individual impact could be important if it contributes to or accelerates the overall resource decline. The NRC staff determined the appropriate level of analysis merited for each resource area potentially affected by the proposed action and alternatives. The level of detailed analysis was determined by considering the impact level to that resource, as described in Chapter 4, as well as the likelihood that the quality, quantity, or stability of the given resource could be affected.

Table 5-3 summarizes the cumulative impact from the proposed Moore Ranch ISR project on environmental resources, based on analyses conducted by the NRC staff and considering the other past, present, and reasonably foreseeable activities identified in Section 5.1.1.

Table 5-3. Cumulative Impacts on Environmental Resources		
Resource Category	**Cumulative Impacts**	**Comment**
Land Use	MODERATE	The proposed project is projected to have a SMALL incremental effect on land use when added to MODERATE cumulative land use impacts.
Transportation	MODERATE	The proposed project is likely to have a SMALL incremental effect on the MODERATE cumulative impacts to transportation.
Geology and Soils	MODERATE	The proposed project would have a SMALL incremental effect on the MODERATE cumulative impacts to geology and soils.
Water Resources		
Surface/Wetlands	MODERATE	The proposed project may have a SMALL incremental impact on surface water and wetland resources when added the MODERATE cumulative impacts.
Groundwater	SMALL to MODERATE	The proposed project may have a SMALL incremental impact on groundwater resources when added to the SMALL to MODERATE cumulative impacts.
Ecology		
Terrestrial Ecology	MODERATE	The proposed project may have a SMALL incremental impact on terrestrial ecological resources, when added to the MODERATE cumulative impacts.

Cumulative Impacts

Table 5-3. Cumulative Impacts on Environmental Resources (continued)		
Resource Category	**Cumulative Impacts**	**Comment**
Aquatic Ecology	SMALL	The proposed project would have a SMALL impact on aquatic resources when added to the SMALL cumulative impacts.
Threatened & Endangered Species	SMALL	The proposed project would have a SMALL impact on threatened and endangered species when added to the SMALL cumulative impacts.
Meteorology		
Air Quality	SMALL	The proposed project would have a SMALL impact on air quality when added to the SMALL cumulative impact.
Noise	SMALL	The proposed project is projected to have a SMALL incremental impact on noise when added to the SMALL cumulative impact.
Historic & Cultural	MODERATE	The proposed project is projected to have a SMALL incremental impact on historical and cultural resources when added to the MODERATE cumulative impact.
Visual	MODERATE	The proposed project is projected to have a SMALL incremental impact on visual and scenic resources when added to the MODERATE cumulative impact to the viewshed.
Socioeconomic	SMALL	The proposed project is projected to have a SMALL incremental impact on socioeconomic resources when added to the MODERATE cumulative impacts expected from other past, present, and reasonably foreseeable future activities.

Table 5-3. Cumulative Impacts on Environmental Resources (continued)		
Resource Category	**Cumulative Impacts**	**Comment**
Environmental Justice	SMALL	The proposed project is projected to have a SMALL incremental impact on environmental justice when added to the SMALL cumulative impacts expected from other past, present, and reasonably foreseeable future actions.
Public and Occupational Health and Safety	SMALL	The proposed project is projected to have a SMALL incremental impact on public and occupational health and safety when added to the SMALL cumulative impacts expected from other past, present, and reasonably foreseeable future actions.
Waste Management	SMALL	The proposed project is projected to have a SMALL incremental impact on waste management when added to the SMALL cumulative impacts expected from other past, present, and reasonably foreseeable future actions.

5.2   Land Use

The cumulative impact to land use was assessed within the planning area administered by the BLM Buffalo Field Office, which includes parts of Campbell, Johnson, and Sheridan Counties, because the proposed Moore Ranch Project is located in this planning area and a recent BLM Resources Management Plan provided information to assess cumulative land use impact. The timeframe for the cumulative effects analysis begins in 2007 and terminates in 2020. However, recent data from a BLM draft report forecasts the projected magnitude of oil and gas development from 2009 to 2029 in the BLM Field Office area (BLM, 2009d). Therefore, certain findings from this report were considered in the analysis of cumulative land use impacts.

Land use within the Powder River Basin is diversified and cooperative, with CBM and oil and gas extraction activities sharing land with livestock grazing. Although federal grasslands and forests cover approximately 21 percent of the Basin area, most rangeland is privately owned (68 percent) and is primarily used for grazing cattle and sheep. Figure 5-1 shows the extent of BLM pasture allotments in the region.

Cumulative Impacts

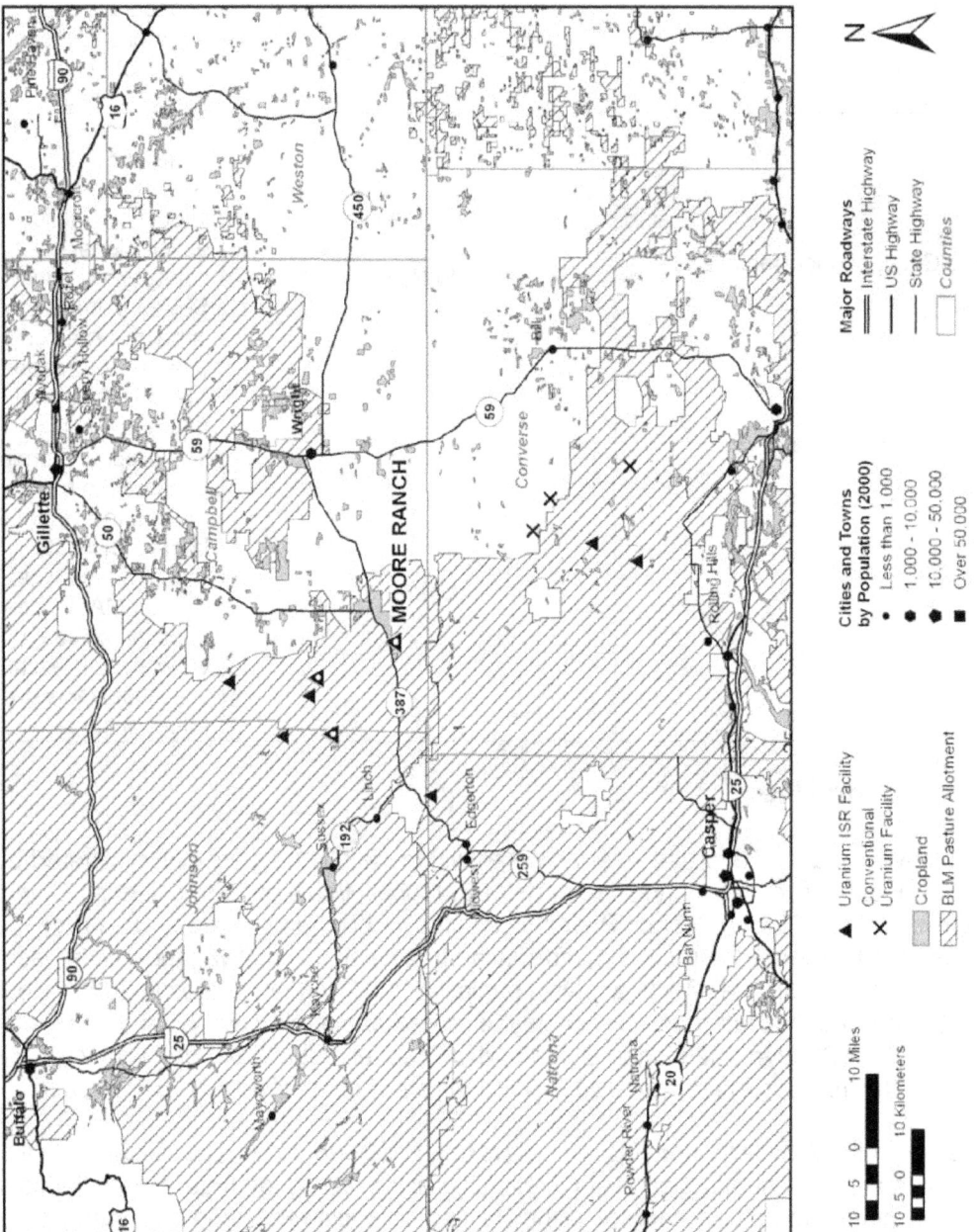

Figure 5-1. Conventional and ISR Uranium Facilities, BLM Pasture Allotments, and Croplands near the Proposed Moore Ranch ISR Facility. Source: BLM, 2010a; Neal et al., 1990; EIA, 2007; Wyoming Oil and Gas Conservation Commission, 2010.

Ranching in the area has occurred since the Civil War when Texas cattlemen moved their herds of Longhorn cattle north looking for open range. Mining in the form of coal, mineral, and oil and gas production is another important land use. The first commercial oil field discovery was made in 1948. Oil discoveries in 1956 touched off the first oil boom. Other major oil and gas discoveries were made in the 1960s and 1970s. Both conventional and CBM oil and gas production are expected to continue in upcoming years. As of 2009, there were a total of 6,421 conventional oil and gas wells in the BLM Buffalo Planning Area (Planning Area), of which 3,090 were active. An additional 1,359 wells are projected to be drilled from 2009 to 2028. Through 2008, 28,776 CBM wells had been drilled, while an additional 13,800 are forecast to be developed between the years 2009 and 2028. It is estimated that a total of 25,958 ha [64,144 ac] in the Planning Area would be disturbed by well pads and access roads related to all types of oil and gas drilling through 2028. This encompasses approximately 1.3 percent of all land in the Planning Area. Of this total, 5,519 ha [13,639 ac] would remain unreclaimed at the end of 2028, which is approximately 0.2 percent of the total Planning Area lands (BLM, 2009d).

Coal mining activity in the Powder River Basin began during 1883, and underground coal mines began operation during 1894. The Powder River Basin emerged as a major coal production area during the 1970s and early 1980s. The largest area, the Gillette coal field is approximately 24 km [15 mi] wide and extends from approximately 35 km [22 mi] north of Gillette, Wyoming, to approximately 40 km [25 mi] south of Wright, Wyoming. The second area is approximately 32 km [20 mi] wide and extends from Sheridan, Wyoming, north to the Wyoming-Montana state line. In 2007, the Powder River Basin was the single most productive coal basin in the United States, producing nearly 40 percent of the nation's coal. It accounts for approximately 97 percent of Wyoming's production and boasts the 10 largest coal mines in the United States by 2007 production. Coal production in the Wyoming portion of the basin is expected to grow at an annual rate of 2 to 3 percent per year. Additional coal leases and associated lands land may be required to keep up with demand (BLM, 2009e). Figure 5-2 shows the extent of coalfields in the region.

The total uranium mine production in the United States in the year 2007 was 2,059,762 kg [4,541,000 lbs], almost half of which occurred in the southernmost Powder River Basin. Uranium deposits in Wyoming are concentrated in southeastern Johnson and southwestern Campbell Counties, and exploration and production dates back to about 1918. Uranium was first mined in Wyoming in 1920, and the first uranium occurrence in the Planning Area was discovered in 1951. Continued uranium exploration resulted in discovery of additional sedimentary uranium deposits in the major basins of central and southern Wyoming. Most uranium production in the Planning Area occurred from 1955 to 1959. Uranium production declined in the mid-1960s but picked up again in the late 1960s and 1970s with the discovery of major uranium deposits in the Powder River Basin, including Christensen Ranch, Smith Ranch, Morton Ranch, and the Highland Mine. Conventional mine production peaked in 1980 and then through the early 1990s when many ISR operations were developed. During the 1980s ISR replaced conventional mine production as the preferred means for extracting uranium ores in the United States. Currently, only ISR milling operations are producing uranium in Wyoming. There is one active ISR uranium milling operation in the Wyoming portion of the Powder River Basin: the Power Resources' Highland/Smith Ranch facility located in Converse County. In addition to this ISR facility, there are two permitted uranium Plans of Operations within the BLM Buffalo Planning Area: the Ruth and Christensen Ranch ISR Projects. The Nichols Ranch ISR Project is currently in the permitting process (BLM, 2009f).

Cumulative Impacts

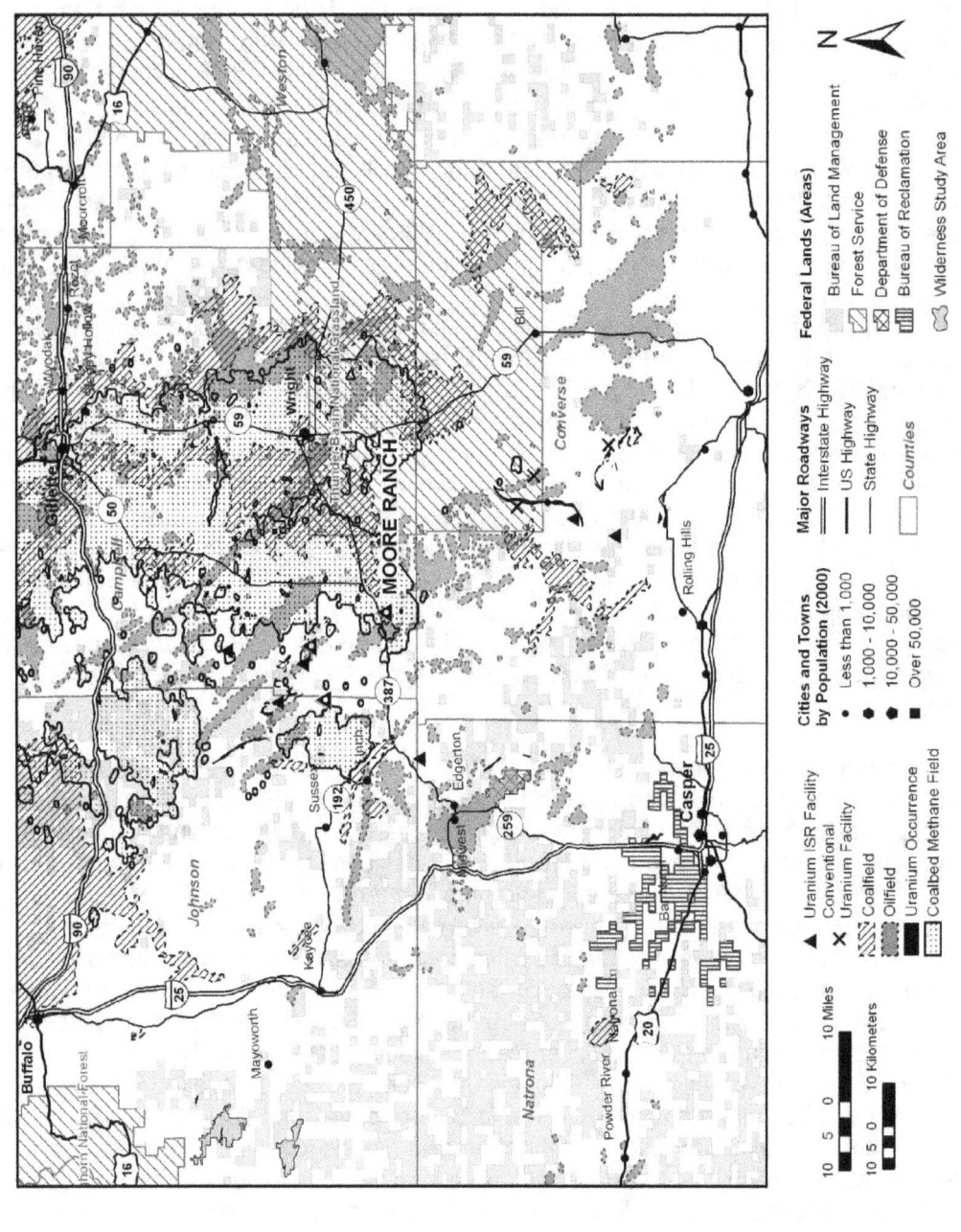

5-2. Oilfields, Coalfields, CBM Project Areas, Uranium Occurrences, and Uranium Facilities near the Proposed Moore Ranch ISR Facility. Source: BLM, 2010a; Neal et al., 1990; EIA, 2007; Wyoming Oil and Gas Conservation Commission, 2010.

Land use impacts result from interruption to, reduction or impedance of, livestock grazing areas, open wildlife areas, land access, and limitations placed on natural resource extraction activities. The potential land use impact from the proposed Moore Ranch Project would be SMALL through ISR phases, as discussed in Section 4.2 of this SEIS. Although the proposed license area encompasses 2,879 ha [7,100 ac], approximately 61 ha [150 ac] of the land area would be used for the operation of the facility, and an even smaller area, approximately 23 ha [57 ac] would be disturbed by earthmoving activities and would be fenced to exclude grazing and other activities as identified before. Because no central plant would be constructed, the amount of land disturbance would be reduced if the applicant elected to operate the Moore Ranch project as a satellite facility as described in Section 5.1.1.1, and the potential impacts to land use would also be less. The impact would be limited because the applicant has indicated the land would be returned to its preextraction condition, except for one access road. Therefore, the proposed Moore Ranch Project was projected to have a SMALL incremental effect on land use when added to the MODERATE cumulative land use impacts from the other past, present, and reasonably foreseeable future action identified in Section 5.1.1 that could affect tens of thousands of acres in the BLM Buffalo Field Office.

5.3 Transportation

Potential cumulative impacts on transportation were assessed that could result from past, present, and reasonably foreseeable development activities. Campbell, Johnson, and Natrona Counties served as the geographic boundary area (referred to herein as the cumulative effects study area). The cumulative effects timeframe is from 2007–2020.

Regional direct effects to roads and highways include increased vehicular traffic and the risk of traffic accidents in the cumulative effects study area from daily travel by workers and their families. Indirect effects include increased wear and tear on existing roads, air emissions, fugitive dust from roads, noise, increased potential access to remote areas, and an increased risk of vehicle collisions with livestock and wildlife. Direct effects on railroads, pipelines, and transmission lines would primarily include increased demand for capacity to move coal, oil and gas, and electricity from production locations in the cumulative effects study area to markets outside the area. The magnitude of any cumulative transportation impacts is largely tied to corresponding increases in population.

Local impacts in the immediate vicinity of the proposed Moore Ranch Project include transport of chemical supplies, yellowcake product, byproduct material, nonhazardous solid waste, the commuting facility workforce, and the potential for accidents. During the ISR phases of the proposed Moore Ranch Project, truck traffic was estimated to increase by 0.2 to 9.1 percent and car traffic was estimated to increase from 0.3 to 7.2 percent along SR 387, the main entrance to the proposed facility. Therefore, the transportation impact would be SMALL, as discussed in Section 4.3. Wellfield roads constructed as part of the proposed action would be removed upon decommissioning; the gravel road leading to the central plant from the main access road would remain in place for future use. If the applicant elected to operate the Moore Ranch, Ludeman, and Allemand-Ross projects as satellite facilities, the maximum annual number of round trip resin shipments to the Irigaray/Christensen Ranch would be 1,460 (4/day) (Uranium One, 2010). Assuming one daily round trip, resin transfer from a Moore Ranch satellite facility would represent about 3.6 times the 100 yellowcake shipments analyzed in Section 4.3, but it is still small compared to the average daily traffic on State Highway 387 in the vicinity of the proposed Moore Ranch ISR facility (see Table 4-2). Also, as described in the GEIS (NRC, 2009, the uranium loaded exchange resins would be less concentrated than yellowcake, and the number

Cumulative Impacts

of truck trips associated with constructing a smaller satellite facility would be reduced (see Table 4-1). For these reasons, impacts would still be SMALL.

As noted in Section 5.1.1, there are other ongoing or planned activities occurring within the Powder River Basin and within the vicinity of the proposed Moore Ranch Project to consider in the analysis of cumulative impacts. These activities, which include CBM development, oil and gas extraction activities, and large surface-mining operations that may have railways to support the transport of coal, all have associated transportation impacts. There is approximately 43-km [27-mi] of existing two-track, ungraveled road about the width of the wheels on a truck that traverse the proposed Moore Ranch Project, which is currently being used by active CBM operators in the area. Furthermore, there are six ISR sites either operating or planned within an 80-km [50-mi] radius of the proposed Moore Ranch Project, each with transportation requirements comparable to that for Moore Ranch Project. In addition, oil and gas exploration and production and coal mines continue to be developed on both public and private lands throughout the Powder River Basin.

The existing or planned ISR facilities would require the construction of new road surfaces or the improvement of existing roads within the vicinity of the proposed Moore Ranch Project. Therefore, the number of roads and road networks would be expected to grow concurrently with the natural resource exploration and extraction activities with a concomitant increase in traffic and the potential for accidents. The demand for railroads, pipelines, and transmission lines would increase to meet the increased demand for capacity to move coal, oil and gas, and electricity from production locations in the area to markets outside the area.

Any potential impacts to the transportation system in support of the proposed Moore Ranch Project would be reclaimed and, therefore, overall project-related transportation impacts would be relatively minor and SMALL.

The proposed Moore Ranch ISR Project is likely to have a SMALL incremental effect on transportation when added to the MODERATE cumulative transportation impact from the other past, present, and future ISR projects, CBM projects, oil and gas operations, surface coal mining activities, and other development with transportation requirements identified in Section 5.1.1.

5.4 Geology and Soils

Cumulative impacts to soils and geology were assessed within the BLM Buffalo Planning Area over the period from 2007 to 2020. The principal impacts on geology and soils from the proposed Moore Ranch Project would result from earthmoving activities associated with constructing surface facilities, access roads, wellfields, and laying pipeline. Earthmoving activities that would impact soils include clearing the ground surface of topsoil to build the central plant (or a smaller satellite facility) and develop the wellfields, which would include preparing a drilling pad to install wells, constructing headerhouses, building access roads, and laying pipeline. As discussed in Chapter 4, all phases of the proposed Moore Ranch Project would have a SMALL impact on geology and soils.

The other ISR projects either ongoing or planned within the vicinity of the proposed Moore Ranch Project, as described in Section 5.1.1, would impact geology and soils at an intensity comparable to what would be seen at the proposed Moore Ranch Project. The past, ongoing, and reasonably foreseeable actions to explore for and extract minerals within the region contribute to the cumulative impact on geology and soils (BLM, 2008b). Increased vehicle traffic, clearing vegetated areas, salvaging and redistributing soil, discharging CBM- and ISR-produced groundwater, and constructing and maintaining project-specific components (e.g., roads, well pads, industrial sites, and associated ancillary facilities) all contribute to the cumulative impact on soils (BLM, 2008b).

The main soil resource concerns within the area administered by the BLM Buffalo Field Office, where the proposed Moore Ranch Project is located, is wind erosion and water erosion that occur where the ground cover has deteriorated (BLM, 2009c). Long-term and short-term impacts to soils include accelerated wind or water erosion; declining soil quality factors; a decline in microbial populations, fertility, and organic matter; compaction; and the permanent removal of soil (BLM, 2005c). Soil composition can be affect by alkalinity changes from discharge of CBM-produced water.

Based on the BLM Powder River Basin Coal Review, approximately 90,140 ha [222,568 acres] (5 percent) of land area in the Powder River Basin has been disturbed by development activities as of 2007. Much of this disturbance is from coal mining and oil and gas production (BLM, 2009g). The proposed Moore Ranch Project would disturb about 61 ha [150 ac], which represents less than 0.00003 percent of the Basin surface area. Since the soil disturbance for an ISR project is relatively minor compared to conventional surface coal mining and oil and gas production, the potential incremental impact from activities at the proposed Moore Ranch Project would have a SMALL incremental effect when added to the MODERATE cumulative impact to geology and soils from the other past, present, and reasonably foreseeable actions described in Section 5.1.1 and illustrated in Figure 5-2 that include ISR projects, CBM projects, oil fields, and conventional mining/milling occurring within the BLM Buffalo Planning area that could affect tens of thousands of acres.

5.5 Water Resources

The cumulative impact on surface and groundwater resources was considered with regard to the area within a radius of 80 km [50 mi] of the proposed Moore Ranch Project (Figure 5.3).

5.5.1 Surface Waters and Wetlands

The proposed Moore Ranch Project is located in the BLM Upper Belle Fourche River watershed. No surface water would be discharged as part of the operations of the ISR facility (or a satellite facility) and the potential impact to onsite washes would be from potentially increased surface water runoff, primarily during the construction and decommissioning phases of the proposed Moore Ranch Project. The U.S. Army Corps of Engineers (USACE) determined that authorization was not required for any construction activities within Wellfield 1; however, activities within Wellfield 2 would be authorized by Nationwide Permit (NP) 12. Therefore, according to the GEIS criteria, the proposed Moore Ranch Project could have a MODERATE impact on wetlands (NRC, 2009). The potential impact would be mitigated through the industrial and construction WYPDES permits the applicant is required to obtain from WDEQ as part of license application before operations commence. Furthermore, the applicant has proposed to avoid installing wells in the main channels of ephemeral drainages, which would avoid potential impacts to wetlands. In addition, the applicant has stated that properly sized culverts would be installed for crossing drainages to withstand a 25-year flood event, and embankments and culverts and drainage crossings would be protected using best management practices (BMP), such as rip rap and rock, in accordance with State of Wyoming regulations. Therefore, the potential impact on surface waters and wetlands would be SMALL.

Cumulative Impacts

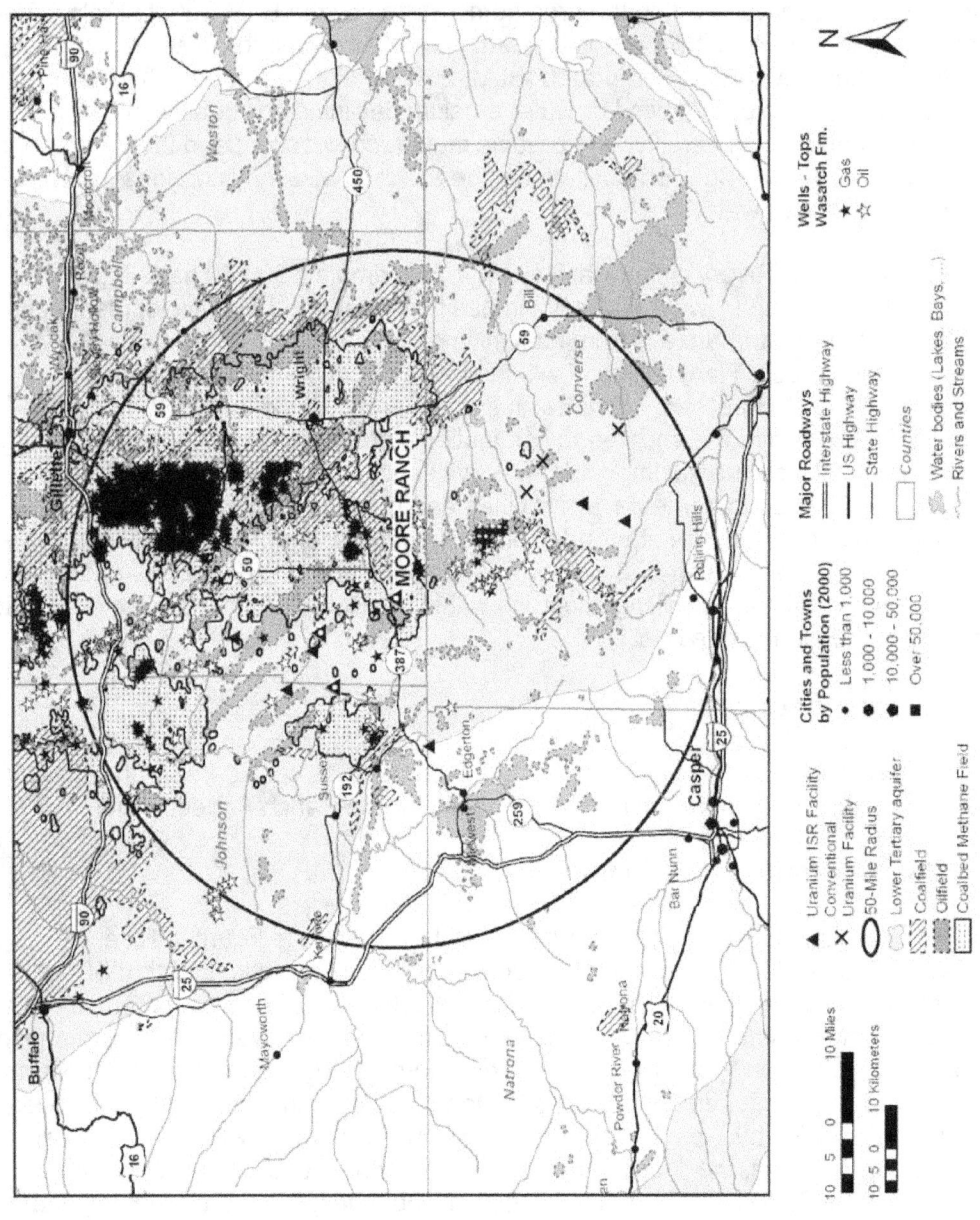

Figure 5-3. Energy Developments within 50 Mile Radius of Proposed Moore Ranch ISR Project./ Source: USGS, 2003; Neal et al., 1990; EIA, 2007; Wyoming Oil and Gas Conservation Commission, 2010

Cumulative Impacts

However, other activities occurring within the proposed license area, as well as within 80 km [50 mi] of the proposed Moore Ranch ISR Project also have the potential to impact surface water. The applicant indicated that CBM production has occurred and continues to occur within the proposed license area from the Roland coal within the Fort Union Formation, which occurs at a depth of approximately 396 m [1,300 ft] bgs. CBM-produced water from these operations is discharged through 22 permitted locations, seven of which are located upstream of the proposed Moore Ranch Project, to release water directly to the drainage or small impoundments specifically designed to facilitate infiltration to the groundwater. From 2000 to 2008, approximately 93 million gallons of CBM-produced water was discharged to the surface drainages and impoundments located within the proposed license area (EMC, 2007a,b).

The BLM estimated that 9 to 52 percent of CBM-produced water would contribute to surface water flows, and perennial flows would be likely to develop in former ephemeral channels (BLM, 2003). Moreover, nine wetlands, created in response to upstream CBM discharges, have been identified on the proposed Moore Ranch Project. Hence, CBM-produced water would increase the availability of surface waters for irrigation and other purposes for downstream users. BLM noted that noticeable changes in water quality would occur in the main channel drainages during periods of low flow and that sodicity and salinity are key water-quality parameters because of their impact on water used for irrigation. BLM projected that the concentrations of suspended sediment in surface water would likely rise above baseline levels from increased flow and surface water runoff from disturbed areas (EMC, 2007a,b). The WDEQ adopted the Most Restrictive Proposed Limit (MRPL) for sodicity and salinity into its WYPDES permitting process to mitigate potential water quality impacts for downstream users.

The CBM facilities operators located within the proposed license area have indicated that one of the CBM facilities would likely be at the end of its operational life around the time the applicant would begin construction, and a second operator has plans to install new facilities within the proposed license area. However, these locations would not coincide with the Moore Ranch Project wellfield locations (EMC, 2007a,b). BLM (2009b) reported 22,543 active permits for CBM operations within Campbell County.

Surface water quality within 80-km [50-mi] of the proposed Moore Ranch Project could also be impacted by conventional oil and gas development, minerals extraction, road maintenance, rangeland grazing, and agriculture (BLM, 2009c). The proposed action and past, present, and reasonably foreseeable future actions within an 80-km [50-mi] radius of the proposed Moore Ranch Project could have a MODERATE cumulative impact on surface water quality. However, prudent resource development and use, and the proper application of mitigation measures identified in site-specific management or development plans would help mitigate the impact (BLM, 2009c).

The Smith Ranch/Highland (operated by Power Resources, Inc. south of the proposed Moore Ranch Project) and the Irigaray/Christensen Ranch (operated by Cogema Mining Inc. north of Moore Ranch Project) ISR projects are the only licensed facilities within 30 to 60 km [18.5 to 37 mi] of the proposed Moore Ranch Project. The licensees are required to obtain USACE permits if surface waters and wetlands within their ISR permit boundaries are jurisdictional. Moreover, the licensees are required to obtain industrial and construction permits from the WDEQ. These permits require best management practices for spill prevention and control; therefore, potential cumulative impacts of produced water at the ISR facilities on surface waters and wetlands within 30 km [18.5 mi] of the Moore Ranch site would be SMALL.

Cumulative Impacts

Bear Creek Union Pacific (between Highland Ranch and Moore Ranch Project) and Highland Exxon Mobil (south of Highland Ranch) are the closest conventional mills located within 80-km [50-mi] of the proposed Moore Ranch Project. Both mills are currently in final reclamation and near completion. Because these conventional mills are in final reclamation stage and any potential spills to surface waters would be controlled and mitigated under the State-approved WYPDES permits, their expected potential cumulative impacts on surface water features at and near the Moore Ranch site would be SMALL.

Oil wells occur within an 80-km [50-mi] radius of the proposed Moore Ranch Project (Figure 5-3). Oil wells are largely clustered around Johnson City, west of Gillette City, between the proposed Nichols Ranch and Moore Ranch ISR projects, and south of the proposed Moore Ranch Project. Under the Clean Water Act, licensees at these oil well sites are also required to obtain construction and industrial WYPDES permits, in addition to USACE permits if wetlands and surface water within their permit boundaries are determined to be jurisdictional. These permits involve require best management practices for spill prevention and control, therefore, the potential cumulative impacts of produced water on surface waters and wetlands due to oil production units would be SMALL.

Based on the above analysis, the proposed Moore Ranch Project would have a SMALL incremental impact on surface water and a MODERATE impact on wetlands by GEIS criteria that if a permit was required, then the potential impact on wetlands could be MODERATE. However, the applicant's proposed mitigation measures described above would reduce the impact to SMALL. The incremental effect on surface water and wetlands from the proposed Moore Ranch Project would be SMALL when added to the MODERATE cumulative impact from the other past, present, and reasonably foreseeable future actions described above.

5.5.2 Groundwater

Potential environmental impacts to groundwater resources from the proposed Moore Ranch Project would occur primarily during the operations and aquifer restoration phases of the ISR facility lifecycle. The analysis of impacts to groundwater resources from operation of the proposed Moore Ranch Project in Section 4.5.2 showed that the potential drawdown in wells outside the license area from facility operation would be nominal and would not affect well yields. Moreover, because the applicant is required to install monitoring wells around and within the proposed facility, as part of the license application, for early detection, control, and reversal of potential horizontal and vertical excursions, potential groundwater quality impacts on nonexempted aquifers would be small. Therefore, the potential impact on groundwater resources from operating the proposed Moore Ranch Project would be SMALL. Because wellfield operations would not be affected, the impacts to groundwater would be the same if Moore Ranch were to be operated as a satellite facility.

However, within the proposed license area and within 80-km [50-mi] of the proposed Moore Ranch Project, there are other ongoing or planned activities that would contribute to a cumulative impact on groundwater resources. These include the operation of other ISR facilities (although production may be from a different ore-producing zone) and CBM production. The BLM estimated that CBM development in the Powder River Basin through the year 2018 would remove about 3 million acre-feet, less than 0.3 percent of the total recoverable groundwater (nearly 1.4 billion acre-feet) in the Wasatch and Fort Union Formations within the Powder River Basin. An estimated 15 to 33 percent of the removed groundwater would infiltrate the surface and recharge the shallow aquifers above the coals (BLM, 2003). The redistribution of pressure within the coals after water production ended would allow the hydraulic pressure

Cumulative Impacts

head to recover to within approximately 15 m [50 ft], or less, of preproject levels within 25 years after project completion. The complete recovery of water levels would take tens to hundreds of years, depending on the specific location. Wells completed in developed coals that are located within the areal extent of a 30 m [100 ft] drawdown contour induced by a CBM well could experience water level drops and possibly encounter methane (BLM, 2003). BLM (2003) noted that the areal extent and magnitude of drawdown effects on coal zone aquifers and overlying or underlying sand units in the Wasatch Formation would be limited by the discontinuous nature of different coal zones within the Fort Union Formation and sandstone layers within the Wasatch Formation.

Within the vicinity of the proposed Moore Ranch Project, the CBM-producing unit is the Roland Coal within the Fort Union Formation. The Fort Union Formation is separated from the 70 Sand ore production zone in the Wasatch Formation by more than 213 m [700 ft] of interbedded clays, siltstone, and discontinuous sands. BLM noted that coal methane fields would impact groundwater in the ore-bearing Wasatch Formation. As the coal zone is depressurized by pumping out groundwater, groundwater levels in the lower Wasatch Formation would be lowered because of the increased pumpage-induced vertical hydraulic gradient between the Wasatch and Fort Union Formations. However, BLM estimated that drawdown in the Wasatch Formation incurred from pumpage from the Fort Union Formation during coal methane field development and would be 10 percent of the drawdown in the Fort Union Formation. Because of the (i) expected larger vertical downward gradients from the relatively larger decline in groundwater levels in the underlying Fort Union Formation than in the overlying Wasatch Formation during coal bed methane productions, and (ii) the occurrence of a 213 m [700 ft] confining layer between the ore production zone in the Wasatch Formation and the coal layers in the Fort Union Formation, the potential for the water quality in the Wasatch Formation to be affected by the CBM production would be SMALL.

CBM-produced water historically has been and continues to be discharged at the surface in the proposed license area, potentially affecting the water quality of the 72 Sand surficial aquifer, which overlies the 70 Sand ore production zone. BLM predicts that the volume of groundwater in shallow layers of the Wasatch Formation would increase because of infiltration of CBM produced water. The applicant analyzed the rate of infiltration of CBM-produced water down to the 72 Sand surficial aquifer and determined that it would be possible for the 72 Sand surficial aquifer to receive infiltration from CBM-produced water within 1 to 10 years of its surface discharge. However, the quantity and quality of the produced water that would be surface discharged would be controlled by the issuance of a State of Wyoming WYPDES permit. Hence, potential cumulative impacts on the quality and quantity of groundwater in the Wasatch Formation because of CBM related activities would range from SMALL to MODERATE.

The Smith Ranch/Highland (operated by Power Resources, Inc. south of Moore Ranch Project) and the Irigaray/Christensen Ranch (operated by Cogema Mining Inc., north of Moore Ranch Project) projects are the only licensed ISR projects within 30 to 60 km [18.5 to 37 mi] of the proposed Moore Ranch Project. The Smith Ranch/ Highlands ISR operations occur in the deeper Fort Union Formation. Because of the distance between the Smith Ranch/Highlands operation and its location in a deeper formation, its potential to contribute to the cumulative impact on groundwater levels at and near the Moore Ranch site would be SMALL. The Irigary/Christensen Ranch ISR operations occur in the same Wasatch Formation as the proposed Moore Ranch Project. Although both ISR operations are in the same formation, the physical distance between them is sufficient to eliminate any significant drawdown; therefore the potential to contribute to the cumulative impact on groundwater levels at and near the Moore Ranch site would be SMALL. Moreover, because the licensees at both of these sites are

Cumulative Impacts

required to implement excursion detection, control, mitigation, and remediation plans under NRC regulations, their potential cumulative impacts on groundwater quality and yield at and near the Moore Ranch Project would be SMALL.

Oil wells located within 80-km [50-mi] of the proposed Moore Ranch Project, are completed at depths greater than ~3,050 m [10,000 ft] below the ground surface; the targeted aquifers for ISR uranium production is in the Wasatch Formation, which usually occurs at a depth that ranges from 42 to 135 m [137 to 440 ft] below the ground surface, thousands of feet above the oil-producing horizons. Therefore, the potential for oil production in the Powder River Basin to contribute to the cumulative impacts on the quality and quantity of groundwater at and near the Moore Ranch site would be SMALL.

Bear Creek Union Pacific (located between Highland Ranch and Moore Ranch Project) and Highland Exxon Mobil (south of Highland Ranch) are the closest conventional mills within 80-km [50-mi] of the proposed Moore Ranch Project. The Bear Creek operation conventionally mined uranium in the Wasatch Formation. The Highland Exxon Mobil Site mined uranium from the Ft. Union Formation. Both mills are Title II sites currently in final reclamation and near completion. When Title II sites are reclaimed, they are placed under the care of either DOE or the State in perpetuity and require NRC-approved, long-term groundwater surveillance monitoring plans to ensure they have no impact on the groundwater. Therefore, their expected potential cumulative water quality and yield impacts on groundwaters at and near the Moore Ranch site would be SMALL.

Deep disposal of process water is a disposal method used by CBM, ISR, and oil production facilities in the Powder River Basin. For deep well disposal, the applicant is required to obtain WDEQ UIC permits for the targeted deep aquifer. The permit would be granted if the deep disposal practice is protective of public health and safety and does not impose a risk to underground sources of water. Therefore, the proposed Moore Ranch ISR project deep disposal wells may have a SMALL incremental impact on groundwater resources when added to the SMALL to MODERATE cumulative impacts anticipated from the other past, present, and reasonably foreseeable future actions discussed in Section 5.1.1 and above.

5.6 Ecological Resources

The cumulative impact on ecological resources from the proposed Moore Ranch ISR project was considered. The area considered for the analysis of cumulative impacts is the Powder River Basin because grasslands and sagebrush shrubland habitats are important features of the Basin landscape and occur on the proposed project site. The timeframe for the analysis runs from 2007 to 2020 although older data are considered to demonstrate historical trends.

5.6.1 Terrestrial Ecology

Activities occurring in the area around the proposed Moore Ranch Project include grazing and herd management, hunting, and mineral exploration. Potential cumulative impacts to ecological resources, both flora and fauna, include reduction in wildlife habitat and forage productivity; modification of existing vegetative communities; and potential spread of invasive species and noxious weed populations. Cumulative impacts to wildlife could involve loss, alteration, and incremental habitat fragmentation; displacement of and stresses on wildlife; and direct and indirect mortalities. Land disturbance resulting from other development activities in the vicinity of the proposed license area have similar ecological impacts as described in Section 4.6 and would be SMALL, individually, if mitigative measures were employed. However, assuming that adjacent habitats for each disturbed land parcel would be at, or near, carrying capacity, and that there would be an unavoidable reduction or alteration of habitats, development activities in this portion of the Powder River Basin could cumulatively contribute to reduction in plant and wildlife populations and alteration of population structure. For some species that require specific habitat conditions, future use would be strongly influenced by the quality and composition of the remaining habitats. The BLM Powder River Basin Oil & Gas EIS concluded that continued natural resource development across the Basin has the potential to alter the distribution of various types of native vegetation, resulting in cumulative impacts to biodiversity (BLM, 2003).

Loss and degradation of native sagebrush shrubland habitats has affected much of this ecosystem type as well as sagebrush-obligate species including the Greater sage-grouse. Most of the sagebrush lands in the region have been changed by land use such as livestock grazing, agriculture, or resource extraction. These uses can influence habitats either directly or indirectly, or alter the disturbance regime by changing the frequency of fire (Naugle, et al., 2009). The long-term viability of the sage-grouse range wide continues to be at risk because of population declines related to habitat loss and degradation. Because of its spatial extent, oil and gas resource development is regarded as playing a major role in the decline of the species in the eastern portion of species range (Becker, et al., 2009). As of this writing, the United States Fish and Wildlife Service (USFWS) has designated the greater sage-grouse as a "candidate species" under the Endangered Species Act which will consider the bird on an annual basis for listing as a threatened or endangered species.

However, the impact to sagebrush shrubland communities at the proposed Moore Ranch ISR license area would be SMALL because only 61 ha [150 ac] or about 2 percent out of 2,879 ha [7,110 ac] would be disturbed. A smaller land disturbance would result if the Moore Ranch project were to be operated as a satellite facility as described in Section 5.1.1.1. Additionally, Table 3-4 shows that only 10 percent of the proposed project area consists of sagebrush habitat, while Figure 3-11 illustrates that plant facilities are located some distance from identified sagebrush areas. Most of the habitat disturbance would consist of scattered, confined drill sites for wells that would not result in large expanses of habitat being dramatically transformed from its original character as in other surface mining operations; no substantial long-term impact would be expected. The NRC staff acknowledges that sage-grouse is rare in the vicinity of the proposed Moore Ranch Project because of the limited habitat available to support its existence and no large expanses of contiguous sagebrush occur within several kilometers of the Moore Ranch Project. Additionally, few sage-grouse have been documented in the area and no grouse leks have ever been discovered either on or near the proposed Moore Ranch Project. The nearest known sage-grouse lek is located approximately 4.0-km [2.5-mi] to the northwest of the Moore Ranch Project area.

Regarding other species of concern, the proposed Moore Ranch ISR project has the potential to impact 14 avian species known to occur, or potentially be present, as seasonal or year-round residents. Impacts may occur to species during all phases of the proposed project. Impacts would be SMALL due to the limited footprint of the project across the entire proposed project site area. Additionally, potential SMALL impacts that occur at the proposed Moore Ranch ISR site (e.g., habitat loss, fragmentation, noise disturbance) would also be likely to occur at other mining and oil and gas facilities throughout the geographic boundary area and potentially impact other localized populations. The BLM Powder River Basin Oil & Gas EIS concluded that there could be cumulative impacts to certain species of raptors and migratory birds resulting from shifts in the habitat composition or distribution (BLM, 2003).

Cumulative Impacts

Therefore, the proposed action could have a SMALL incremental impact on terrestrial ecological resources, especially those that are presently in decline, when added to the MODERATE cumulative impacts anticipated in the Powder River Basin from other past, present, and reasonably foreseeable future actions discussed in Section 5.1.1 and above that could result in the disturbance of tens of thousands of acres.

5.6.2 Aquatic Resources

Since no aquatic habitat exists on the proposed Moore Ranch ISR project to support fish or macroinvertebrates, the proposed project would have a SMALL impact on aquatic resources when added to the SMALL cumulative impacts anticipated from the other past, present, and reasonably foreseeable future actions.

5.6.3 Threatened and Endangered Species

No Federal- or State-listed sensitive plant species, endangered or threatened plant species, or designated critical habitats occur within the proposed license area. Any potential impacts to these species during the various project phases would be limited. Therefore, the proposed project would have a SMALL impact on threatened and endangered species when added to the SMALL cumulative impacts anticipated from other past, present, and reasonably foreseeable future actions.

5.7 Air Quality

Potential impacts to air quality from the proposed Moore Ranch Project are anticipated to be SMALL and are discussed in detail in Section 4.7 of this SEIS. Nonradiological air emission impacts primarily involve fugitive road dust from vehicles used throughout the facility lifecycle and combustion engine emissions from diesel equipment that is used predominantly during the construction and decommissioning phases. The NRC staff concluded that the air quality for the region in the vicinity of the site is in compliance with the National Ambient Air Quality Standards (NAAQS), and based on emissions estimates described in Section 2.1.1.1.6.1, the facility would not be classified as a major source under the New Source Review or operating (Title V of the Clean Air Act) permit programs. The NRC staff analysis noted the presence of Prevention of Significant Deterioration (PSD) Class I and II areas within the region that could potentially be impacted by emissions from the proposed action; however, based on the magnitude of emissions, the prevailing wind direction, and distance from the proposed facility, the NRC staff concluded that impacts would be unlikely.

Downwind concentrations of emitted air pollutants are affected by a number of variables, including the ambient meteorological conditions and the magnitude of the emission rate. Based on the low magnitude of estimated emissions from the proposed action (Section 2.1.1.1.6.1), good air quality in the region (Section 3.7.2) and meteorology that is often favorable for dispersion (Section 3.7.1), the NRC staff concluded detailed quantitative air analyses were not necessary to evaluate potential air impacts. As a result, the extent of the geographic area that could be impacted by proposed emissions was not quantified; however, other regional air modeling studies addressing larger scale emission sources applicable to oil and gas activities, coal bed methane production, and conventional coal mining suggest the region of influence for air emissions could range from about 60 km [37 mi] (Stoeckenius, et al., 2006) to beyond 242 km [150 mi] (AECOM, 2009). Because of the low magnitude of proposed emissions, the distance where measurable air impacts could be detected from the proposed action would be closer to the proposed license area.

Cumulative Impacts

Past, present, and reasonably foreseeable future activities in the vicinity of the proposed Moore Ranch Project site that emit air pollutants include other uranium mining/milling activities, CBM, coal mining, and oil and gas operations. The past and present contributions of projects in the region that emit air pollutants are represented in the ambient air quality monitoring results described in Section 3.7.2. These monitoring results indicate the air quality is in attainment for all NAAQS. Emissions from projected development of future oil and gas exploration and production, including CBM and coal mining, have been evaluated for impacts to air in previous EISs and supporting documents for proposed developments in the Powder River Basin (BLM, 2003; 2009; ENSR International, 2006; AECOM, 2009) where the proposed Moore Ranch Project is located. While the concurrent activities emit a variety of pollutants, the principal emissions from the oil and gas industry that would overlap significantly with emissions from the proposed action are nitrogen oxides, volatile organic compounds, and fugitive road dust. The principal emissions from coal mining include fugitive dust (particulates including coal dust) and exhausts from diesel-powered equipment. Therefore, the focus of the NRC staff cumulative impact analysis on air quality is from nitrogen oxides and fugitive dust emissions from the proposed Moore Ranch Project, other proposed ISR facilities, and future oil and gas, and coal operations in the Powder River Basin of Wyoming.

Within an 80-km [50-mi] radius of the proposed Moore Ranch Project, there are at least six operating or planned ISR facilities (Table 5-1) that would generate emissions comparable to emissions projected for the proposed project. Because ISR facilities commonly use a phased approach to well drilling and wellfield construction, and all six facilities would not begin construction concurrently (as each must go through the average 2-year licensing process and obtain the necessary Federal, State, and local permits), the degree of overlap in construction activities would be most likely to occur for wellfield drilling activities because each facility would construct more than one wellfield over a period of years. To estimate the potential annual contribution of the six facilities to local air emissions, the NRC staff considered the emissions results in Appendix D and assumed two facilities were simultaneously starting construction of their first wellfield and drilling two deep disposal wells and the remaining four facilities each had an active wellfield and deep well under development. For that scenario, the total annual contribution of ISR facility nitrogen oxide emissions in the region that would add to the emissions from other past, present, and reasonably foreseeable future actions would be approximately 76 t/yr [84 T/yr]. The estimate of fugitive road dust emissions would scale directly with each new facility because emissions would occur during all phases of the facility lifecycle. Because the proposed Moore Ranch Project has a relatively short main access road, the NRC staff assumed an average value of 63.5 tons/yr [70 tons/yr] for each facility, resulting in a total of 381 tons/yr [420 tons/yr] of fugitive road dust. Since these facilities and their emissions would be spatially dispersed throughout the region, they would not represent a single point source. Because it considers the air quality impacts associated with the more extensive construction and decommissioning requirements and larger workforces associated with a central plant, the potential impacts of operating the Moore, Ludeman, and Allemand-Ross projects as satellite facilities are bounded by the analysis.

The potential air impacts from future CBM activities, coal mining, and oil and gas exploration in the Powder River Basin have been previously evaluated (BLM, 2003; 2009; ENSR International, 2006; AECOM, 2009). A recent BLM cumulative air analysis of the Powder River Basin was conducted to support review of coal development in the Powder River Basin (AECOM, 2009). That analysis involved executing a state-of-the-art EPA guideline dispersion model, CALPUFF, version 5.8 (Scire, et al., 1999) to calculate local-scale, short-range dispersion as well as region-scale, long-range dispersion of emissions assuming worst case meteorological conditions. Emissions were developed for base year 2004 (NO_2, SO_2, $PM_{2.5}$, and PM_{10}) and

Cumulative Impacts

were projected for year 2020. Emission sources included coal-related (mines, power plants, railroads, conversion facilities); permitted sources in Wyoming and Montana; CBM production sources; and miscellaneous roads, urban areas, conventional oil and gas, noncoal power plants). The estimated impacts from that study for the baseline year (2004) indicated calculated air concentrations were below NAAQS, except for short-term PM_{10} and $PM_{2.5}$ in the near field. Year 2020 projected impacts showed compliance with standards, except for short-term and annual $PM_{2.5}$ and PM_{10} in localized areas. Far-field visibility impacts were identified for downwind Class I areas (Northern Cheyenne Indian Reservation, Badlands National Park, Wind Cave National Park) as a consequence of power plant and CBM emissions, and visibility impacts to several Class II areas were projected from power plant and coal mine emissions. These modeling results suggest that local and regional air quality in the Powder River Basin and nearby areas is presently good but is degrading with time, primarily from particulate emissions, as various emissions sources are projected to increase to 2020. While NO_x projections for near-field receptors in 2020 were below the ambient standard, the calculated concentrations were at 80 percent of the limit compared to the base year calculation of 30 percent of the limit. Therefore, the margin for compliance is being reduced with the future projected development and the associated increase in emissions.

The licensing of proposed ISR facilities would contribute incremental increases to area emissions including, in particular, NO_x and fugitive dust and, therefore, incrementally impact air quality. Because the volume of ISR nonradiological emissions are low compared to existing and future proposed developments in the region, the NRC staff concluded the relative proportion of future air quality impacts from ISR operations would be SMALL. While detailed emissions data for specific projects, practices, or industries in the local area were not identified by the NRC staff to compare with the proposed action estimates, the NRC staff did identify general information to provide context and support for the NRC staff conclusion that proposed annual ISR nonradiological air emissions levels are relatively low as discussed in detail in the following paragraph.

At the state level, emissions inventory estimates (Russell and Pollack, 2005) for 2002 suggest the total amount of NO_x emitted from oil and gas drilling that year was approximately 4,500 t [4,964 T] from the construction of 2,948 wells. From these numbers, the NRC staff approximate an average of 1.54 t [1.7 T] of NO_x per well. For comparison, the calculated drill rig emissions for the proposed Moore Ranch Project wellfield development activities is 7.89×10^{-3} t [8.7×10^{-3} T] of NO_x per well (derived from emission calculation results reported in Appendix D), orders of magnitude less than wells being drilled in support of oil and gas exploration and production. The state average value is more comparable to the emissions calculated for a single proposed deep disposal well (calculated to emit approximately 5.89 t [6.5 T] per well as shown in Appendix D). The higher emissions from the deep well drilling is temporary, with two to five deep disposal wells expected per proposal and approximately 528 hours per well (Appendix D) or 66 8-hour drilling days each. Other regional sources of NO_x considered in the aforementioned air-modeling analysis included power plants and trains (for example, shipping coal that is mined locally). Year 2004 NO_x emissions from coal-fired power plants in Southwest Wyoming are reported as 31,116 and 12,004 t/yr [34,321 and 13,240 T/yr]. Maximum NO_x emissions from a proposed rail line from Miles City to Decker, Wyoming, was reported as 10.1 t per km per year [6.9 T per mile per year] (ENSR International, 2006) and therefore 184 t/yr [203 T/yr] along the 47.3-km [29.4-mile] route. Oil and gas drilling varies considerably in well depth and associated emissions. Examples of NO_x emissions from oil and gas drilling in the Jonah-Pinedale area of southwestern Wyoming (Stoeckenius, 2010) indicate large clusters of drilling rigs emit between approximately 0.91 and 5.9 t [1 and 6.5 T] of NO_x per day. While this area complies with the NO_x ambient air quality standard (NO_x is a precursor to ozone formation) it has experienced episodic exceedances of the ambient ozone standard based on a combination of specific factors (including strong temperature inversions, low winds, snow cover, bright sunlight, and emissions) that is resulting in more regulatory and research attention (Stoeckenius, 2010). The aforementioned NO_x emitted from drilling and construction equipment for six proposed ISR facilities that would be within an 80-km [50-mi] radius was calculated as 76 t/yr [84 T/yr] (approximately 0.2 and 0.6 percent of the aforementioned coal plant emissions, 40 percent of the 47.3-km [29.4-mi] rail spur estimate, and 3.5 to 23 percent of the Jonah-Pinedale oil and gas drilling cluster example). The contribution to annual NO_x emissions calculated for the proposed action 18 t/yr [20 T/yr] is approximately 24 percent of the six-facility estimate used in this analysis.

Fugitive dust emissions from the proposed action and other existing or proposed ISR facilities would contribute to the cumulative particulate matter emissions from power plant, CBM activities, and coal mining, in particular. As the projected emissions from these activities for 2020 indicate near-field exceedances of $PM_{2.5}$ and PM_{10} (AECOM, 2009) and potential far-field visibility impacts would be increasing, the NRC staff expect particulate emissions would continue to be an air-quality concern in future years. Because ISR facilities are not major sources of particulate emissions, the principal emission would be fugitive road dust. Based on the limited daily traffic expected from the proposed action (Section 4.3) and dust control measures the applicant proposes to implement, and could be required by the WDEQ construction air permit that presently is under review for the proposed action (previously a permit condition for similar facilities; WDEQ, 2009; WDEQ, 2010), and the magnitude of estimated emissions, the NRC staff conclude that the impacts to air would be localized, intermittent, and temporarily MODERATE (e.g., visible plumes of dust from traffic on unpaved roads) but would be predominantly SMALL.

While the proposed ISR emissions are relatively low, the actual cumulative effect of multiple new ISR facilities that could be licensed in the future would depend on the ambient air quality at the time of licensing, the continued development of other emission-generating activities in the area and region, and the timing and magnitude of emission-generating activities at each proposed ISR facility. As these ISR facilities would be licensed by NRC and permitted sequentially on a first-come, first-serve basis, the addition of emissions from each new facility would be incremental. This incremental development of uranium milling facilities in the region allows the NRC to evaluate each proposal and state air quality staff to evaluate potential impacts within the context of existing air quality

5.7.1 Global Climate Change and Greenhouse Gas Emissions

As discussed in the U.S. Global Change Research Program (GCRP) report (GCRP, 2009), it is the "... production and use of energy that is the primary cause of global warming, and in turn, climate change will eventually affect our production and use of energy. This assessment is focused on greenhouse gas emissions.

Greenhouse gas emissions associated with the construction, operation, and decommissioning of an ISR facility are addressed in Chapters 3 and 4 of this SEIS (see Table 5-4). Evaluating the cumulative impacts of greenhouse gas emissions (GHGs) is challenging.

Evaluation of cumulative impacts of GHGs requires the use of a global climate model. GCRP (2009) synthesized in a technical support document the results of numerous climate modeling studies as discussed in Section 3.7.1.5. Based on this study, the EPA determined that potential changes in climate caused by GHG emissions endanger public health and welfare. NRC

Cumulative Impacts

Table 5-4. Comparison of Carbon Dioxide Emissions by Source	
Source	CO_2 Emission
Global Emissions (t/yr)	28,000,000,000*
United States (t/yr)	6,000,000,000*
Single ISR (t/yr) Facility Lifecycle (12 years)	2,400
Current/Proposed ISR Facilities Lifecycles (12 years)	14,400
Average Annual U.S. Passenger Vehicle	5†

*EPA, 2009
†Federal Highway Administration, 2006

recognizes that the global cumulative impacts of greenhouse emissions as presented in the report, are the appropriate basis to evaluate cumulative impacts. Based on the impacts identified in the GCRP report, the NRC recognizes that the national and worldwide cumulative impacts of greenhouse gas emissions are noticeable but not destabilizing.

5.7.1.1 Greenhouse Gas (GHG) Emissions in the Region

The Center for Climate Strategies (CCS) prepared a report for the Wyoming Department of Environmental Quality (WDEQ) that provides an inventory and forecast of the Wyoming's GHG emissions (CCS, 2007). These emissions data were based on projections from electricity generation, fuel use, and other GHG-emitting activities. Emissions are reported as CO_2 equivalents (CO_2e), a conversion for the various gases emitted, (i.e., methane or nitrous oxides), into equivalent greenhouse effect to compare to CO_2 (BLM 2008). Gross carbon dioxide-equivalent (CO_2e) emissions in 2005 for Wyoming were 56 million metric tons (MMt).

This volume accounted for less than one percent (0.8%) of the total United States gross GHG emissions. This total is reduced to 20 MMt CO_2e as a result of annual sequestration (removal) due to forestry and other land uses (CCS, 2007).

Wyoming has a higher per capita emission rate than the national average (>4 times). This is due to the state's fossil fuel production industry and industries that consume high amounts of fossil fuels, as well as a large agricultural industry, large distances between cities, and a small population (CCS, 2007). The CCS report expects that the Wyoming GHG emissions would continue to grow with the increased demand for electricity, followed by emissions associated with transportation. These GHG projections are reflected in Table 5-5.

As of 2009, there are 13 active coal mines in the Wyoming portion of the Powder River Basin, and these mines produced approximately 496 million short tons (BLM, 2010b). According to the

Table 5-5. Wyoming Historical and Reference Case GHG Emissions					
Million Metric Tons CO2e	1990	2000	2005	2010	2020
Energy Sector	38.0	43.6	47.5	51.6	59.6
Electricity Production Based	39.8	43.3	44.2	47.8	54.2
Coal	39.8	43.2	44.1	47.7	53.9
Natural Gas	0.00	0.1	0.1	0.1	0.2
Petroleum	0.0	0.0	0.0	0.0	0.0
Geothermal, Biomass and Waste (CO2, CH4, and N2o)	0.0	0.0	0.0	0.0	0.0

*Million Metric Tons CO_2e
Source: CCS, 2007

Wyoming Oil and Gas Conservation Commission, there are over 33,000 active gas and oil wells in the State, 45 operational gas processing plants, 5 oil refineries, and over 9,000 miles of gas pipelines (CCS, 2007). Because there is no regulatory requirement to track carbon dioxide or methane emissions, there is a high degree of uncertainty associated with the Wyoming GHG emissions from this industry. However, the CCS (2007) estimated that approximately 13.5 MMt of CO_2e was emitted by fossil fuel industries. Of this amount, 80 percent was due to the natural gas industry. This amount is expected to grow an additional 8–10 percent in the next decade (CCS, 2007). No data currently exists for the nonfossil fuel industries, including uranium.

5.7.1.2 GHG Emissions from the Proposed Moore Ranch Project

In response to current concerns related to GHG emissions, NRC has focused on evaluating CO_2 emissions for the lifecycle of the proposed facility and compares this with other forms of extraction. The primary source of CO_2 emissions from ISR facilities are combustion engine emissions from construction equipment (including drill rigs). Construction equipment is used most during initial wellfield and facility construction and later during the decommissioning phase to remove buildings and equipment and reclaim land surfaces.

Annual and cumulative CO_2 emissions over the life of the facility from the proposed Moore Ranch Project for construction and decommissioning activities were estimated by the NRC staff and documented in Appendix D of this SEIS. Combustion engine exhaust calculations performed for the proposed Moore Ranch Project were based on a combination of proposal-specific and representative information appropriate to support a conservative emissions screening analysis (Appendix D). Only nonroad combustion emissions were considered. Diesel emissions, including drilling rigs, were estimated using emission factors provided by the EPA using different engine classes, based upon power output and operating time (Appendix D). The applicant (Uranium One) proposes to drill two wellfields of approximately 470 and 500 wells, respectively, during the first few years of the project. This includes injection and production wells (for the ISR process) as well as monitoring wells. In addition, the applicant proposes to have two to four UIC Class I wells (for deep well injection of byproduct material [wastewater]). Results show that drilling rigs and other construction equipment used during the construction phase have the highest annual CO_2 emissions for the proposed action. This amounts to 852 t [940 T] of CO_2 per year. The cumulative calculated CO_2 emissions from drilling and construction of all wellfields followed by decommissioning all wellfields and associated facilities was estimated as 2,400 t [2,600 T].

The majority of estimated annual CO_2 emissions are from drilling and nearly 81 percent of the calculated drilling CO_2 emissions are from deep disposal well-drilling activities. Well drilling activities would occur over a period of at least two years, perhaps longer, because there are two wellfields from which the applicant proposes to recover uranium, and the first wellfield could be operational for a few years. If the Moore Ranch project were to be operated as a satellite facility, the increased number of trips would lead to an increase in CO_2 emissions from transportation. The construction and decommissioning footprint would be less, however, and wellfield operations and aquifer restoration would not be affected by a satellite facility. For these reasons, the estimated annual CO_2 emissions would be about the same. The estimate did not consider sequestration (removal) due to forestry or other agricultural activities (EPA, 1996).

Cumulative Impacts

5.7.1.3 Moore Ranch ISR Facility GHG Emissions Impact

As described in Section 5.7.1.1, the total amount of GHGs produced in Wyoming in 2005 was 56 gross MMt without considering sequestration (CCS, 2007). If 36 gross MMt for sequestration of GHGs is considered, as estimated in the Wyoming Greenhouse Gas Inventory and Reference Case Projections 1990–2020 (CCS, 2007), the net total GHGs annually produced in Wyoming is 20 MMt. The proposed Moore Ranch Project would conservatively produce a maximum annual total of 852 t (940 T) of GHGs (as carbon dioxide), which equates to approximately 0.004 percent of the net total GHGs produced in Wyoming in 2005. This compares to approximately 2.2 percent from conventional mining operations as discussed in Section 5.7.1.6. If either GHG emissions has increased or sequestration has decreased since 2005, the effect from the proposed Moore Ranch Project would be even less. Therefore, the potential impact of GHGs from the proposed Moore Ranch Project would be SMALL.

5.7.1.4 Effect of Climate Change on the Moore Ranch ISR Facility

While there is general agreement in the scientific community that some change in climate is occurring, considerable uncertainty remains regarding the magnitude and direction of some of the changes, especially predicting trends in a specific geographic location. To predict the effect of climate change on the proposed Moore Ranch Project, temperature and precipitation data from two NWS stations located in and two stations located near the Powder River Basin were reviewed (NCDC, 2009). The data, daily records for both temperature and precipitation, covered a period of 50 years. Aside from the year-to-year fluctuations, there was no observable increase or decrease in either temperature or precipitation during the periods of record for the four NWS stations (NCDC, 2009). In another study, the National Climatic Data Center (NCDC) evaluated 105 years of climatological data from throughout the State of Wyoming, which revealed a slight upward trend temperature (0.16 °F per decade) (NCDC, 2010a). In the report, *Global Climate Change Impacts in the United States* (GCRP, 2009), the U.S. Global Change Research Team indicates that the temperatures in the past 15 years have risen even faster (1–2 °F for the Powder River Basin), most of which is attributed to warmer winters. This trend is expected to continue into the next decade, and by the end of this century, average annual temperatures in the Powder River Basin could rise as much as 4–8 °F (GCRP, 2009).

In the 50-year period from 1958 to 2008, there was no obvious observable change in annual precipitation. However, the NCDC, in a similar evaluation of 105 years of climatological data for the entire State of Wyoming, revealed a slight downward trend in precipitation (0.13 inches per decade) (NCDC, 2010b). Nevertheless, the U.S. Global Change Research Team is predicting that the Northern Great Plains Region (which includes the Powder River Basin) would receive increased precipitation in future decades. Most of the precipitation is expected to fall in the colder months (winter and spring), and the summer and fall are to become drier. In addition, with the colder months expected to warm over the next several decades, more precipitation would fall in liquid form, resulting in less snow pack in the higher elevations (GCRP, 2009).

The overall effect of projected climate change on the proposed Moore Ranch ISR facility is SMALL. The small predicted increases in temperatures and precipitation over the next decade would have no effect on any of the ISR phases. Because the major functioning of the facility is below ground, the effects of the surficial and atmospheric environments would not be expected to impact the target (ore body) aquifer. There could be an increase in recharge to the aquifer in future years, resulting from the projected increase in precipitation (and consequent infiltration into the groundwater) which could affect the proposed project by increasing the volume of groundwater in the ore body and improve the effectiveness of the aquifer restoration process.

Similarly, while potential changes to the site environment and resources such as ecology are plausible, the staff considers the small magnitude of the predicted climate changes over the period when the proposed action would occur would not be sufficient to alter the environmental conditions at the site in a manner that would significantly change the environmental impacts from what has been evaluated in this SEIS.

5.7.1.5 GHG Mitigation Measures

Best management practices (BMPs) and mitigation measures could be used to minimize the emission of GHGs at the proposed Moore Ranch Project. These include, but are not limited to:

- Use of [fossil-fuel] vehicles that meet latest emission standards

- Ensure that [diesel-powered] construction equipment and drill rigs are properly tuned and maintained

- Use low-sulfur diesel fuel

- Use newer, cleaner-running equipment

- Avoid leaving equipment idling or running unnecessarily

- Minimize the number of trips to well pads

5.7.1.6 Other Mining Activities in the Powder River Basin

Extensive research into the relative volumes of GHGs emitted by ISR facilities and other natural resource extraction methods has been performed. In support of the analysis for this final SEIS, the NRC staff surveyed the recent EISs issued for projects located in the Powder River Basin. Based on this survey, the NRC staff found that estimates and projections of the carbon footprint of the natural resource extraction activities vary widely.

West Antelope II Coal Lease Application FEIS

The FEIS for the West Antelope II Coal Lease Application also addressed greenhouse gas emissions as specifically related to the proposed action (Antelope Mine), the mine adjacent to the West Antelope II lease by application (LBA) tract. An inventory of expected greenhouse gas emissions in 2007 was conducted at Antelope Mine. Additionally, the applicant also projected emissions for a typical year of operations at Antelope Mine if the West Antelope II lands are leased and mined. Emissions are measured as CO_2 equivalents (CO_2e), a conversion to put any of the various gases emitted, (i.e., methane or nitrous oxides), into the equivalent greenhouse effect as compared to CO_2 (BLM, 2008a).

Emissions would be generated from the following: carbon fuels used in mining operations, electricity used onsite, blasting, methane released from mined coal, spontaneous combustion, onsite rail transport, and coal transported to purchasers (see Table 5-6).

Projected emission rates increase if the West Antelope II tract is added to mining operations. The increase in CO_2 emissions would result from the additional diesel fuel that would be used in consideration of the added haul distances and overburden hauling, as well as increased electricity and explosives related to increasing strip ratios (BLM, 2008a).

Cumulative Impacts

Table 5-6. Annual Greenhouse Gas Emissions at the West Antelope II Mine		
Source	2007*	Average year with West Antelope II LBA*
Fuel	110,877	195,173
Electricity	77,574	111,854
Mining Process	36,772	40,884
On-site Rail	1959	2251
Total At Mine	227,182	347,911
Other Rail†	656,444	754,338

*CO_2e in metric tons
†Assumes 10-percent increase, based on demand in eastern United States
Source: BLM, 2008a

The CCS estimated that activities in Wyoming accounted for 55.6 million metric tons of gross CO_2e emissions in 2005 (Center for Climate Strategies, 2007). Using that estimate, the 2007 Antelope Mine emissions total represents 0.41 percent of state-wide emissions. With the addition of the West Antelope II LBA tract, the projected total Antelope Mine emissions would represent 0.63 percent of state-wide emissions (BLM, 2008a).

Wright Area Coal Lease Application DEIS

The Wright Area Coal Lease Applications (BLM, 2009b) DEIS analyzes the environmental impacts of leasing six tracts of federal coal reserves adjacent to the Black Thunder, Jacobs Ranch, and North Antelope Rochelle mines. All are operating surface coal mines in the southern Powder River Basin (PRB), near the town of Wright, Wyoming. While BLM does not authorize mining through the issuance of a Federal coal lease, WDEQ, with oversight from the Office of Surface Mining (OSM), has regulatory authority in issuing permits to mine coal in Wyoming. However, BLM considered the impacts of mining coal because it is a logical consequence of issuing a maintenance lease to an existing coal mine. BLM analyzed GHG emissions specifically related to mining activities for the Black Thunder, Jacobs Ranch, and North Antelope Rochelle mines; adjacent to the North, South, and West Hilight Fields, West Jacobs Ranch, North Porcupine; and South Porcupine LBA tracts. The use of the coal after it is mined is not determined at the time of leasing. However, almost all coal that is currently being mined in the Wyoming PRB is being used to generate electricity by coal-fired power plants (BLM, 2009b).

CO_2e emissions are projected to increase at the Black Thunder, Jacobs Ranch and North Antelope Rochelle mines if these additional LBA tracts are added to the mining operations (see Table 5-7). The increase in CO_2e emissions are expected to result from the additional fuels (especially diesel) that would be used in consideration of the increased coal and overburden haul distances, as well as increased use of electricity and explosives related to increasing overburden thicknesses. Estimates assume that the combined annual production rate from these three mines is 270 million tons (BLM, 2009b).

The CCS estimated that activities in Wyoming will account for approximately 60.3 million metric tons of gross CO_2e emissions in 2010 and 69.4 million metric tons in 2020 (CCS, 2007). Using the CCS projections, the 2007 emissions from the three conventional mines identified in Table 5-7 would contribute 2.22 percent of the 2010 Wyoming state-wide emissions. The addition of six LBA tracts (the North Highlight Field, South Highlight Field, West Highlight Field, West

Cumulative Impacts

Table 5-7. Estimated Annual Equivalent CO_2 Emissions* at the Black Thunder, Jacobs Ranch, and North Antelope Rochelle Mines

Source	2007	With LBA Tracts
Fuel	577,463	1,429,582
Electricity	465,908	777,141
Mining Process	201,871	296,166
Total of Three Sources	1,245,241	2,502,889

*CO_2e in metric tons Source: BLM, 2009b

Jacobs Ranch, North Porcupine, and South Porcupine) together with the conventional mines identified in Table 5-7, would increase the projected 2020 state-wide emissions to 3.61 percent (BLM, 2009b).

5.8 Noise

Cumulative impacts from noise were assessed within an 8-km [5-mi] radius of the proposed Moore Ranch Project. This area served as the cumulative assessment geographic boundary and was chosen because noise dissipates quickly from the source. The cumulative effects time frame runs from 2007 to 2020.

The GEIS noted that noise would not be discernible to an offsite person at distances of greater than 300 m [1,000 ft] (NRC, 2009a). Section 4.8 of this SEIS evaluated potential noise impacts to the nearest resident who lives 4.5-km [2.8-mi] east of the center of the proposed license area. Because this person lives beyond 300 m [1,000 ft] of the proposed license area, there would be no noise impact above background levels.

Past, present, and reasonably foreseeable future noise-generating activities in the vicinity of the proposed Moore Ranch Project would primarily be from traffic noise, oil and gas operations, CBM operations, and uranium mining/milling operations. The FEIS for the PRB Oil and Gas Project noted that sound levels from CBM operations would be expected to be unnoticeable at distances of 490 m [1,600 ft] and beyond, and the FEIS concluded there would be no cumulative noise impact on the surrounding area (BLM, 2003). CBM operations are active within the proposed license area.

Although noise-related impacts are generally constrained to within a 610 m [2,000 ft] radius of activities associated with oil and gas development (e.g., drilling, operation of compressor stations) the level of energy-related development both on and around the proposed Moore Ranch Project has been increasing and is anticipated to continue to increase (BLM, 2003).

The licensing of the proposed Moore Ranch Project would have a SMALL impact on noise generated in the area, as discussed in Section 4.8. Although other noise-generating activities (i.e., CBM operations) occur within an 8-km [5-mi] radius of the Moore Ranch ISR project, they do not overlap the proposed Moore Ranch Project. The noise generated from either the proposed Moore Ranch Project or from CBM operations would be at background levels at distances ranging from 300 m [1,000 ft] to 610 m [2,000 ft], as discussed above. Because the nearest resident is located 4.5-km [2.8 mi] east of the center of the proposed license area, there would be a SMALL noise-related impact. As discussed previously in Section 5.3, truck traffic would increase if the Moore Ranch project were operated as a satellite, but it would still

Cumulative Impacts

represent only a small fraction of the average daily traffic on SR 387. Therefore, the proposed project (or satellite facility) would have a SMALL incremental impact on noise when added to the SMALL cumulative impacts expected from other past, present, and reasonably foreseeable future actions. Furthermore, noise levels would be mitigated by the use of administrative and engineering controls to maintain noise levels in work areas below OSHA regulatory limits.

5.9 Historical and Cultural Resources

The assessment of the cumulative impact on historical and cultural resources was defined with regard to activities occurring within the area of potential effects (APE), which includes the project site, access roads, and a 3.2-km [2 mi] buffer area outside the proposed license area. This area serves as the geographic study area for this resource.

The potential impact on cultural resources from implementing the proposed Moore Ranch Project (or satellite facility) was estimated to be SMALL because the ISR lifecycle was not expected to directly impact specific archaeological sites determined to be eligible for listing on the National Register of Historic Places (NRHP), as discussed in Section 4.9 of this SEIS.

BLM identified various actions that may have cumulative impacts within the greater Powder River Basin (2009b). These actions included coal extraction, oil and gas operations, utility transmission and distribution actions, other mining/milling actions including uranium, wind power activities, reservoir development, various nonenergy-related developments including transportation, and county-level economic development actions.

As previously noted, six other ISR facilities located within 80-km [50-mi] of the proposed Moore Ranch Project are either operating or planned. CBM activities are occurring, and are planned to occur, within the proposed license area, and oil and gas development within the planning area of the BLM Buffalo Field Office is expected to continue.

However, any potential impacts to historic and cultural resources would likely be minimized for projects occurring on Federal or State lands or which are funded in part by the government since these projects would be subject to the *National Historic Preservation Act* (NHPA), Section 106 consultation process, and applicable statutes.

The proposed Moore Ranch ISR project is projected to have a SMALL incremental impact on historical and cultural resources when added to the MODERATE cumulative impact to these resources expected from the other past, present, and reasonably foreseeable future actions described in Section 5.1.1. The proposed Moore Ranch Project is located in an archaeologically rich area; therefore, the activities described above could result in a cumulative loss of historic and cultural resources. However, any past, present, or future actions that occur on Federal lands or require a Federal permit would require a Section 106 Consultation, which would ensure that historical and cultural resources are adequately considered.

5.10 Visual and Scenic Resources

The cumulative impact on visual and scenic resources was considered within the area administered by the BLM Buffalo Field Office.

The development of the proposed Moore Ranch Project would have a SMALL impact on visual and scenic resources because the facility would be located in a BLM visual resource management (VRM) Class IV area, which allows an activity to contrast with the basic elements of the characteristic landscape to a much larger degree.

However, within the area administered by the BLM Buffalo Field Office and within the Powder River Basin region, energy development is expected to grow over the next 15 to 20 years and would involve constructing railroads, coal-fired power plants, major (230 kV) transmission lines, coal technology projects, oil and gas transportation pipelines and refineries, and CBM processing plants. Within the BLM Buffalo Field Office planning area, there are other ongoing or planned uranium recover projects, oil and gas developments, and CBM projects, all that have an impact on visual and scenic resources.

The proposed Moore Ranch ISR project is projected to have a SMALL incremental impact on visual and scenic resources when added to the MODERATE cumulative impact to the viewshed expected from other past, present, and reasonably foreseeable future actions. The lower profile and smaller footprint associated with operating the Moore Ranch project as a satellite facility would reduce visual impacts as compared to constructing and operating a Central Plant.

5.11 Socioeconomics

The cumulative socioeconomic impact was considered within the Powder River Basin because mineral extraction dominates activities within this area. The following socioeconomic indicators were evaluated as part of this analysis:

- Population
- Employment
- Housing
- School enrollment
- Public services
- Fiscal revenue

Socioeconomic impacts from the proposed Moore Ranch Project (or satellite facility) would range from SMALL to MODERATE, depending on the phase of the ISR lifecycle and the particular socioeconomic characteristic (e.g., finance, demographics) as discussed in Section 4.11 of this SEIS. The GEIS socioeconomic analysis is based on 2000 U.S. Census Bureau data. Data presented in this SEIS for the proposed Moore Ranch Project region of influence (ROI) is based on a combination of 2000 U.S. Census Bureau data, U.S. Census Bureau 2005–2007 American Community Survey 3-Year Estimates, and U.S. Census Bureau 2009 State and County QuickFacts. Though specific numbers may differ, the analysis of socioeconomics presented in Section 4.3.10 of the GEIS remains valid for the proposed Moore Ranch Project. Additionally, REMI Policy Insight (REMI), a professionally recognized regional economic model, was employed to develop the cumulative employment and population projections presented next, as part of the BLM *PRB Coal Review Task #C Report: Cumulative Social and Economic Effects* (BLM, 2005a,b,c). The model used two future scenarios that assume a lower and upper coal-production scenario in the PRB. The two scenarios represent a range of economic activity derived by combining the range of future coal production with other identified foreseeable activities, including oil and gas production and other mining operations. The timeframe for the analysis spans from 2007 to 2020, although older data was incorporated to demonstrate historical trends.

Cumulative Impacts

Population change over time is generally an excellent indicator of cumulative social and economic change in a given area. Wyoming population has grown from 332,416 in 1970 and is projected to increase modestly from 2010 to 2020 (from 539,740 to 578,730), as shown in Figure 5-4 (Wyoming Department of Administration and Information, 2008). Growth in Campbell County grew from 12,957 in 1970 to 43,440 in 2010 and is projected to be 52,130 in 2020; Johnson County grew from 5,587 in 1970 to 8,640 in 2010 and will be 9,990 in 2020. Natrona County population increased from 51,264 in 1970 to 74,050 in 2010 and is projected to grow to 79,650 in 2020. Based on the relatively low number of workers expected at the proposed facility, the proposed Moore Ranch Project would have a SMALL incremental impact on population.

These population projections do not take into account the current recession, climate change legislation (including cap and trade components), and future technological changes (e.g., clean coal innovations). Projected increases in employment across the entire six-county (Campbell, Johnson Converse, Crook, Sheridan, Wetson) Powder River Basin BLM coal review study area from coal mining operations and oil and gas development range from 12,120 to 28,625 jobs under the lower and upper production scenarios between 2003 and 2020. Most of this incremental gain is expected to take place in Campbell County, which would capture 60 percent of the new jobs under the lower and 65 percent under the upper production scenario (BLM, 2009c). Figure 5-5 shows employment and population trends for Campbell County from 2000 to 2020. Based on the relatively low number of workers expected at the proposed Moore Ranch Project would have a SMALL incremental impact on employment.

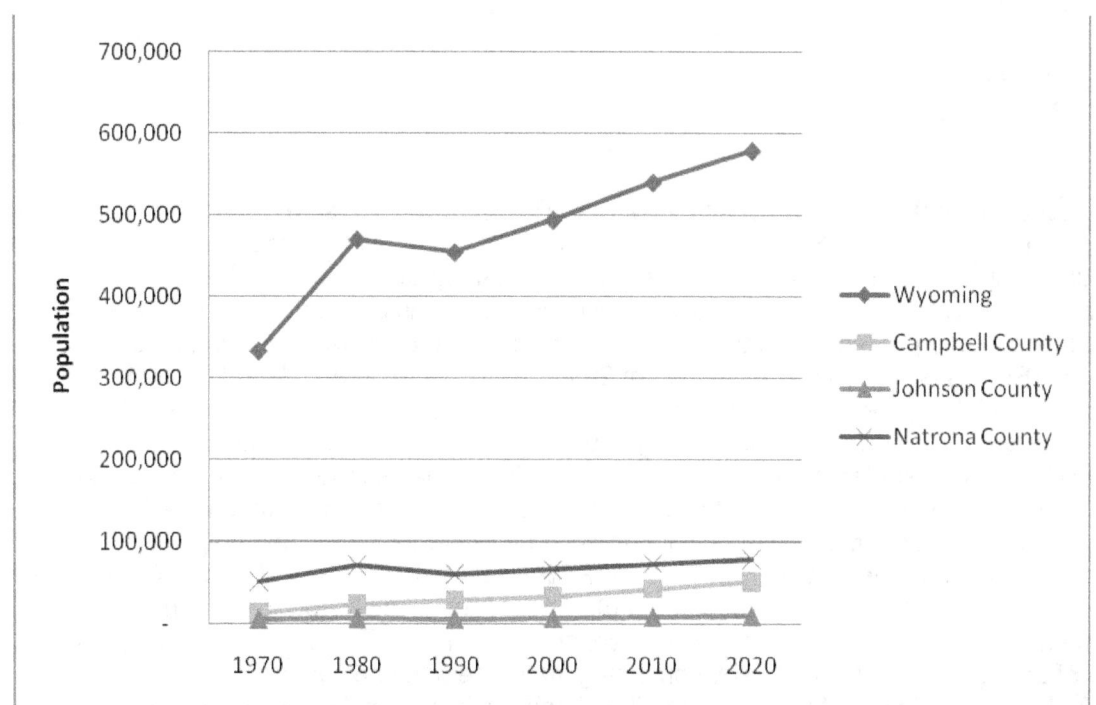

Figure 5-4. Wyoming, Campbell County, Johnson County, and Natrona County Population, 1970–2020. Source: Wyoming Department of Administration and Information, Economic Analysis Division, 2008

Cumulative Impacts

While Campbell County and the entire Powder River Basin have been characterized as being able to adapt to, respond to, and accommodate growth, periods of rapid growth can stress communities and their social structures, housing resources, and public infrastructure and service systems (BLM, 2005a,b,c).

Both the lower and upper projections indicate a strong demand for housing resources through the year 2020. Campbell County is expected to need a minimum 58 percent increase in total housing requirements between 2003 and 2020, according to the lower production scenario. This demand is anticipated to exert substantial pressure on housing markets, prices, and the real estate development and construction industries, all at a time when demand for labor and other resources would be high overall. Any incremental impacts to housing from the proposed Moore Ranch Project would be SMALL, based on the limited number of facility employees.

Short-term school capacity shortages are also anticipated to result from the increase in population. Under the lower production scenario, Campbell County is projected to experience a substantial increase in school-age children through 2020. An additional 1,587 students is forecasted for 2020, representing a 22-percent increase. Actual enrollment, however, is outpacing both the lower and upper production scenario projections. During the 2009 to 2010 school year, Campbell County school enrollment was 8,300 students. The lower production scenario estimated that by 2015, the county would be enrolling only 8,225 students. This accelerated growth is likely due to the ramping up of oil and gas production in the Powder River Basin in recent years. Any incremental impacts to schools from the proposed Moore Ranch Project would be SMALL, based on the limited number of permanent employees and respective families required for the different phases of the ISR project.

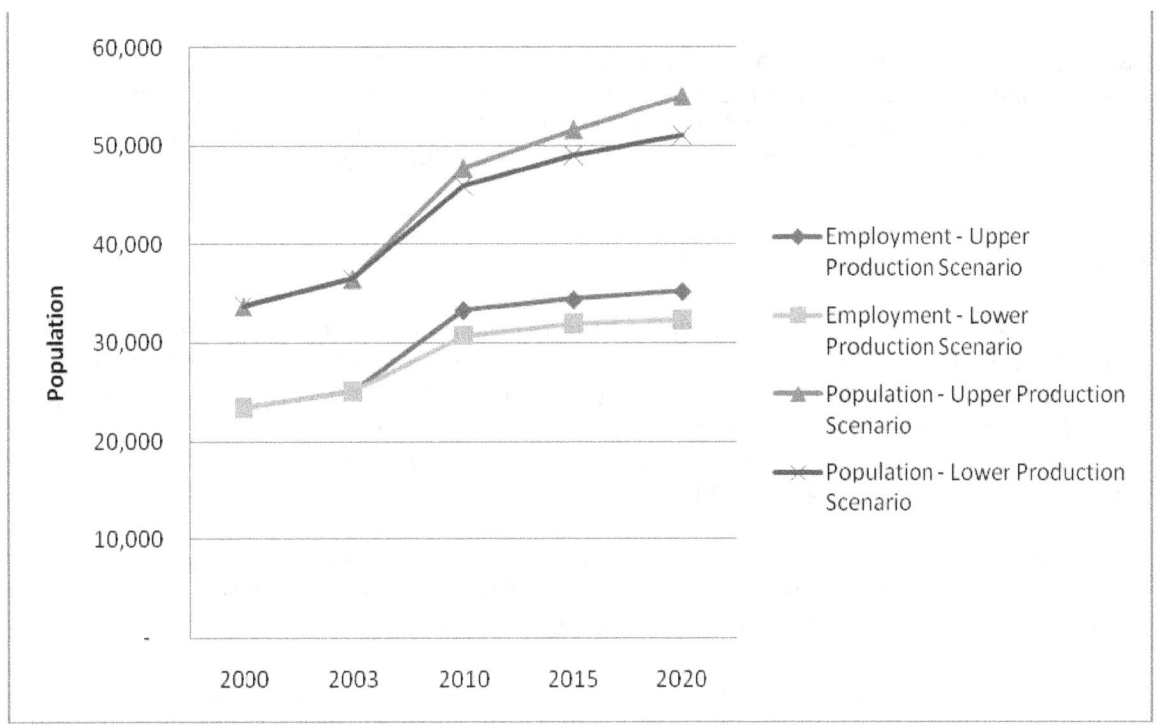

Figure 5-5. Projected Campbell County Population and Employment to 2020. Source: Wyoming Department of Administration and Information. 2008

Cumulative Impacts

There could be incremental impacts to local government facilities and public services. Population increases in affected counties and communities generally result in across-the-board increases in demand on services. Additionally, various reasonably foreseeable development activities may result in increased demand for specific services (e.g., road maintenance, law enforcement, and emergency response). Incremental impacts to government facilities and public services from the proposed Moore Ranch Project would be SMALL, based on the relatively low number of employees required for all phases of the project.

There could be additional cumulative impacts to socioeconomic resources if extractive industries and power production were to increase above average historic levels of growth. A cumulative negative impact could result if the housing supply and real estate market could not meet the labor demands in the extractive industries. There could be a long-term impact on local schools; healthcare facilities; fire and police services; and infrastructure, including waste management facilities, if large industrial projects created a demand for labor in the PRB.

If the population size were to remain stable or grow at a rate that the area could manage (approximately 2 percent per year based on past experience), the local economy could be positively affected by multiple mining/milling operations occurring simultaneously, which would generate local and State economic revenue. However, the maximum number of jobs associated with the proposed Moore Ranch Project would be approximately 50 employees during construction while 40 to 60 workers would be needed for operation and restoration phases. This represents a relatively small pool of workers compared to the sizable increases projected for employment under both BLM Coal Review scenarios, in addition to growth associated with CBM development. As such, any incremental impact on local finances would be SMALL.

Based on assessments of population, employment, housing, school enrollment, public services and local finances, the proposed project is projected to have a SMALL incremental impact on socioeconomic resources when added to the MODERATE cumulative impacts expected from other past, present, and reasonably foreseeable future activities described in Section 5.1.1 of this SEIS.

5.12 Environmental Justice

No concentrations of people living below the poverty level and no concentrated minority populations are located near the proposed project area; therefore, no disproportionately high and adverse environmental impacts would result to minority populations or those living below the poverty level from the proposed Moore Ranch ISR Project (or a satellite facility). Environmental justice impacts related to the proposed Moore Ranch ISR project are discussed in more detail in Section 4.12 of this SEIS.

The percentage of people living below the poverty level within the census block groups surrounding the proposed Moore Ranch Project is well within the 20 percent level of significance compared to the state and county proportion of those living in poverty.

The GEIS identified no minority populations in the Wyoming East Uranium Milling Region. SEIS Section 4.12 determined that the percentage of minority populations living in the affected block groups in the vicinity of the proposed Moore Ranch Project site is similar to the percentage of minority populations recorded at the State and county level and well below the national level. Therefore, no minority populations were identified as residing near the proposed Moore Ranch Project.

Cumulative Impacts

The relative homogeneity of Wyoming, despite 40 years of energy/natural resource development, indicates that environmental justice issues do not pose a problem. Because the economic base of the study area is largely ranching and resource extraction, low-income areas are widely dispersed within the study area. People with incomes below the poverty status may reside within the study area, but not disproportionately. At the present time, there is no significant concentration of people living below the poverty level and no significant concentration of minority populations located near the proposed project. Furthermore, human health and environmental impacts throughout the lifecycle of the proposed project are predicted to be SMALL for the general population. Therefore, the proposed project is projected to have a SMALL incremental impact on environmental justice when added to the SMALL cumulative impacts expected from other past, present, and reasonably foreseeable future actions.

5.13 Public and Occupational Health and Safety

The cumulative impact on public and occupational health and safety was considered within an 80-km [50-mi] radius of the proposed Moore Ranch Project. The public and occupational health and safety impacts from the proposed Moore Ranch Project would be expected to be SMALL, under normal operations because the potential radiological exposure would be consistent with background. Under accident conditions, the potential exposure could be MODERATE, as discussed in Section 4.13 of this SEIS. The annual dose to the population located within 80-km [50-mi] of the proposed Moore Ranch ISR Project would be far below applicable NRC regulations. If an accident were to occur, the potential impact could be MODERATE if the appropriate mitigation measures and other procedures that ensure worker safety were not followed.

The proposed Moore Ranch ISR Project site is located within the GEIS Wyoming East Uranium Milling Region, which contains 21 previous, current, or potential uranium-handling sites (NRC, 2009a). The GEIS (NRC, 2009a) identified eight draft or final EISs submitted from January 2005 to February 2008 for projects that could contribute to a cumulative impact on public and occupational health and safety within the Wyoming East Uranium Milling Region. In addition, the GEIS identified ten programmatic EISs affecting the entire State of Wyoming. No additional projects initiated since February 2008 were identified that would contribute to the cumulative impact on radiological public health and safety. As noted previously, the proposed Moore Ranch Project would have a SMALL impact on public health and safety consistent with background radioactivity under normal operations. Because the identified projects within an 80-km [50-mi] radius of the proposed Moore Ranch Project would not significantly contribute to the cumulative public and occupational health and safety effects from the identified projects, the cumulative impact would be SMALL.

The maximum expected exposure to any member of the public from the proposed Moore Ranch Project, as with other operating ISR facilities in the United States, would be expected to be less than 10 mrem per year at the site boundary (NRC, 2009a). This exposure, combined with exposures from other facilities, is expected to remain far below the regulatory public limit of 100 mrem/year and have a negligible contribution to the 620 mrem average yearly dose received by a member of the public from all sources. Therefore, the proposed project is projected to have a SMALL incremental impact on public and occupational health and safety when added to the SMALL cumulative impacts expected from other past, present, and reasonably foreseeable future actions described in Section 5.1.1 of this SEIS.

Cumulative Impacts

5.14 Waste Management

Waste management impacts from the proposed Moore Ranch Project would be SMALL, as discussed in Section 4.14 of this SEIS.

As discussed in Section 2.7 of the GEIS, all stages of the proposed Moore Ranch Project (construction, operation, aquifer restoration, and decommissioning) would generate effluents and waste streams, all of which must be handled and disposed of properly. These would include liquid wastes and solid wastes. Any wastewater generated during or after the uranium extraction phase of site operations are classified as byproduct material (NRC, 2000). Based on information provided by the applicant in EMC (2007) (Section 2.1.1.1.6), during the operations phase, the proposed Moore Ranch Project would be anticipated to annually produce approximately 170 to 550 Lpm [45 to 145 gpm] of liquid byproduct material, 76 m^3 [100 yd^3] of solid byproduct material, and 1,530 m^3 [2,000 yd^3] of nonhazardous solid waste. Because of demolition and disposal of buildings, equipment, and contaminated soil, the amount of solid waste [both byproduct material and nonhazardous waste] generated during decommissioning would be greater. The applicant indicates that solid wastes would be reused, recycled, or compacted, as appropriate. As discussed in Section 2.1.1.1.6, solid byproduct material that could be decontaminated would also be disposed of as nonhazardous solid waste, but all waste would ultimately need to be disposed of, with totals (Uranium One, 2008) on the order of

- 11,010 m^3 [14,390 yd^3] of solid byproduct material, the largest part of which is chipped trunkline, plant equipment, and contaminated soil from wellfields

- 512 t [565 T] of solid byproduct material in the form of contaminated demolition concrete from demolition of plant building foundations and structures

- 482 m^3 [630 yd^3] of nonhazardous solid waste (plant equipment)

- 6,102 t [6,730 T] of nonhazardous solid waste in the form of uncontaminated concrete from plant building foundations and structures

These decommissioning waste estimates are preliminary. When the applicant initiates decommissioning, they will be required to provide updated waste volume estimates. The NRC staff would conduct an environmental review at that point to evaluate potential waste management cumulative impacts from the proposed decommissioning activities. Given an operating period of about 12 years (EMC, 2007a,b), the total amount of solid byproduct material to be generated through the end of decommissioning would be about 14,395 m^3 [15,878 yd^3] and the total amount of nonhazardous solid waste for offsite disposal would be about 25,443 m^3 [28,064 yd^3]. Because of the smaller footprint, the amount of both nonhazardous solid waste and byproduct material resulting from decommissioning a Moore Ranch satellite facility would be reduced as compared to decommissioning a central processing plant. All other waste streams would remain largely the same because wellfield operations and aquifer restoration would not change for a satellite facility.

Past, present, and reasonably foreseeable future activities in the vicinity of the proposed Moore Ranch ISR project site that could generate nonhazardous solid, hazardous, or radioactive wastes include uranium mining/milling activities, CBM activities, and oil and gas exploration. Each of these facilities would generate solid and hazardous wastes and would be responsible for complying with applicable regulations and site-specific license agreements that manage generated wastes. Within an 80-km [50-mi] radius of the proposed Moore Ranch ISR

Project, there are at least six operating or planned ISR facilities that would generate waste volumes consistent with that projected for the proposed project. The cumulative effects on past, present, and reasonably foreseeable actions that would contribute to the total amount of 11e.(2) byproduct material generated by ISR facilities could therefore be as much as about 100,767 m^3 [111,146 yd^3] (i.e., 14,395 m^3 [15,878 yd^3] x 7 facilities). Similarly, the cumulative volume of nonhazardous solid waste that could be generated is approximately 178,103 m^3 [196,448 yd^3].

The applicant indicates that it will be seeking permits from the WDEQ for two or more deep disposal wells for liquid byproduct material (EMC, 2007a,b). Additional deep disposal well use in the region is anticipated as additional facilities are licensed. The state permitting process for these evaluates the suitability of proposals to ensure groundwater resources are protected and potential environmental impacts would be limited to acceptable levels. Based on the assumption that the state would not permit deep injection wells that would have a significant potential to impact groundwater resources, the staff conclude the cumulative impacts of using deep disposal wells for the proposed action along with the potential impacts from reasonably foreseeable future actions would be SMALL.

The applicant also indicates that it is pursuing several disposal options for the disposal of solid byproduct material, including both in-state and out-of-state disposal, but has not yet finalized a contract established with an NRC-or State-licensed disposal facility (EMC, 2007a,b). Available local capacity for disposal of byproduct material is at the Pathfinder-Shirley Basin site in Mills, Wyoming. As reasonably foreseeable additional ISR sites are licensed this local capacity may become limited. Future ISR applicants could engage other low-level radioactive waste disposal facilities that are licensed to accept byproduct material. Another existing facility that is licensed by NRC to accept byproduct material for disposal is the Rio Algom Ambrosia Lake uranium mill tailings impoundments near Grants, New Mexico. Additionally, three sites are licensed by NRC Agreement States to accept byproduct material for disposal (i.e., the Energy Solutions site in Clive, Utah; the White Mesa uranium mill site in Blanding, Utah; and the Waste Controls Specialists site in Andrews, Texas). Based on the disposal options currently available, and the disposal agreement that NRC requires prior to operations, the staff concludes that the potential cumulative waste management impacts associated with the generation of byproduct material would be SMALL.

Licensees must also comply with applicable State and Federal regulations with respect to disposing of solid and hazardous wastes. Based on the small projected quantities of hazardous wastes generated by the proposed action and other similar reasonably foreseeable ISR facilities that could be licensed in the region in the future, the staff conclude the potential cumulative impacts from the additional generation hazardous wastes would be SMALL.

The applicant indicates that nonhazardous solid waste would be disposed of at the Casper landfill in Casper, Wyoming. Considering year 2005 estimates of waste disposed of at the Casper balefill (landfill) from the Wyoming Office of State Lands and Investments (OSLI, 2007) of 93,804 t [103,466 T], the aforementioned cumulative decommissioning nonhazardous solid waste volume resulting from the proposed action would constitute approximately 16 percent of the volume of waste that would be disposed at the Casper balefill over the next 12 years if the regional waste generation continued at the 2005 level. Based on this comparison, the staff concludes the region has sufficient capacity to dispose of the nonhazardous solid waste generated from the proposed action. Therefore, the waste management impacts for disposal of decommissioning nonhazardous solid waste would be SMALL.

Assuming that the applicant obtains the necessary permits and contractual agreements for disposing of its byproduct material, the proposed Moore Ranch ISR project is projected to have a SMALL incremental impact on waste management when added to the SMALL cumulative impacts expected from other past, present, and reasonably foreseeable future action described in Section 5.1.1.

5.15 References

40 CFR Parts 1500 and 1508. *Code of Federal Regulations*, Title 40, *Protection of Environment*, Part 1500, "Purpose Policy and Mandate" and Part 1508, "Terminology and Index."

AECOM, 2009. "Update of Task 3A Report for the Powder River Basin Coal Review Cumulative Air Quality Effects for 2020." Prepared for U.S. Bureau of Land Management High Plains District Office, Wyoming State Office, and Miles City Field Office. Fort Collins, Colorado: AECOM, Inc. December.

Anderson, N.B., R.A. Beach, A.W. Gray, C.A. Roberts, D.A. Ferderer, M.J. Gobla, 1990. "Occurrences for Wyoming." U.S. Department of the Interior, U.S. Bureau of Mines. Denver, Colorado. <URL:http://www.wygisc.uwyo.edu/clearinghouse/mineral.html> (03 May 2010)

Becker, J.M., C.A. Duberstein, J.D. Tagestad, and J.L. Downs, 2009. "Sage-Grouse and Wind Energy: Biology, Habits, and Potential Effects of Development." Prepared for the U.S. Department of Energy Office of Energy Efficiency and Renewable Energy Wind & Hydropower Technologies Program under Contract DE–AC05–76RL01830. <http://www.pnl.gov/main/publications/external/technical_reports/pnnl-18567.pdf> (28 April 2010).

BLM (U.S. Bureau of Land Management, 2010a. Wyoming GIS Available Data <http://www.blm.gov/wy/st/en/resources/public_room/gis/datagis.html> (03 May 3 2010).

BLM, 2010b. "Powder River Basin Coal." <http://www.blm.gov/wy/st/en/programs/energy/Coal_Resources/PRB_Coal.html> (June 2010).

BLM, 2009a. "Final Environmental Impact Statement for the South Gillette Area Coal Lease Applications WYW172585, WYW173360, WYW172657, WYW161248." <http://www.blm.gov/wy/st/en/info/NEPA/HighPlains/Wright-Coal.html> (17 September 2009).

BLM, 2009b. "DRAFT Environmental Impact Statement for the Wright Area Coal Lease Applications." Vol 1 and 2. BLM/W/PL-09/026+1320. Casper, Wyoming: U.S. Department of Interior, BLM, High Plains District Office. June, 2009.

BLM, 2009c. "Summary of the Analysis of the Management Situation Buffalo Resource Management Plan Revision." BLM, Buffalo Field Office. <http://www.blm.gov/wy/st/en/programs/Planning/rmps/buffalo/docs.html> (September 2009).

BLM, 2009d. "Draft Reasonable Foreseeable Development Scenario for Oil and Gas Buffalo Field Office Planning Area, Wyoming." <http://www.blm.gov/pgdata/etc/medialib/blm/wy/programs/planning/rmps/buffalo/docs/rfd.Par.69667.File.dat/01_text.pdf> (23 April 2010)

BLM, 2009e. "Update of the Task 2 Report for the Powder River Basin Coal Review Past and Present and Reasonably Foreseeable Development Activities." <http://www.blm.gov/wy/st/en/programs/energy/Coal_Resources/PRB_Coal/prbdocs/coalreview/task_2_update__12.html> (24 April 2010).

BLM, 2009f. "Final Mineral Occurrence and Development Potential Report Buffalo Resource Management Plan Revision." BLM Buffalo Field Office, Wyoming. .<http://www.blm.gov/pgdata/etc/medialib/blm/wy/programs/planning/rmps/buffalo/docs.Par.90169.File.dat/RevisedFinalMineralReport_Part1.pdf> (23 April 2010).

BLM, 2009g. "Update Of The Task 3d Report For The Powder River Basin Coal Review Cumulative Environmental Effects." BLM High Plains District Office and Wyoming State Office. <http://www.blm.gov/wy/st/en/programs/energy/Coal_Resources/PRB_Coal/prbdocs/coalreview/task_3d_update__12.html> (24 April 2010).

BLM, 2008a. "Final Environmental Impact Statement for the West Antelope II Coal Lease Application WYW163340." <http://www.blm.gov/wy/st/en/info/NEPA/cfodocs/West_Antelope_II.html> (September 2009).

BLM, 2008b. "Fortification Creek Area Draft Resource Management Plan Amendment/Environmental Assessment." . <http://www.blm.gov/wy/st/en/info/NEPA/bfodocs/fortification_creek.html> (September 2009).

BLM, 2005a. "Task 1D Report for the Powder River Basin Coal Review Current Environmental Conditions." <http://www.blm.gov/wy/st/en/programs/energy/Coal_Resources/PRB_Coal/prbdocs/coalreview/Task1D.html> (September 2009).

BLM, 2005b. "Task 2 Report for the Powder River Basin Coal Review Past and Present and Reasonably Foreseeable Development Activities." <http://www.blm.gov/wy/st/en/programs/energy/Coal_Resources/PRB_Coal/prbdocs/coalreview/Task2.html> (September 2009).

BLM, 2005c. "Task 3D Report for the Powder River Basin Coal Review Cumulative Environmental Effects." <http://www.blm.gov/wy/st/en/programs/energy/Coal_Resources/PRB_Coal/prbdocs/coalreview/task3d.html> (September 2009).

BLM, 2003. "Final Environmental Impact Statement and Proposed Plan Amendment for the Powder River Basin Oil and Gas Project." <http://www.blm.gov/wy/st/en/info/NEPA/bfodocs/prb_eis.html> (September 2009).

CCS (Center for Climate Strategies), 2007. "Wyoming Greenhouse Gas Inventory and Reference Case Projections 1990–2020. Cheyenne, Wyoming: Center for Climate Strategies. Spring 2007.

CEQ (Council on Environmental Quality), 1997. "Considering Cumulative Effects Under the National Environmental Policy Act."

EIA (US. Energy Information Administration), 2007. <http://www.eia.doe.gov/pub/oil_gas/natural_gas/analysis_publications/maps/maps.htm#geodata> (03 May 2010)

Cumulative Impacts

EMC (Energy Metals Corporation U.S.), 2007a. "Application for USNRC Source Material License, Moore Ranch Uranium Project, Campbell County, Wyoming, Environmental Report." Casper, Wyoming: Uranium One Americas Corporation. ADAMS Accession Nos. ML072851222, ML072851229, ML072851239, ML07285249, ML07285253, ML07285255.

EMC, 2007b. "Application for USNRC Source Material License, Moore Ranch Uranium Project, Campbell County, Wyoming, Technical Report." Casper, Wyoming: Uranium 1 Americas Corporation. ADAMS Accession Nos. ML072851222, ML072851258, ML072851259, ML072851260, ML072851268, ML072851350, ML072900446. October 2.

ENSR International, 2006. "Task 3A Report for the Powder River Basin Coal Review Cumulative Air Quality Effects." Fort Collins, Colorado: ENSR International. February.

EPA, 1996. "Compilation of Air Emission Factors, Volume 1: Stationary and Point Sources." AP-42, Fifth Edition: Chapters 3.3 and 3.4. Washington, DC. EPA. October.

EPA, 2009a. "Climate Change—Climate Emission, Non-CO_2 Mitigation." Washington, DC: EPA. September. <http://www.epa.gov/climatechange/economics/mitigation.html>.

EPA, 2009b. "Endangerment and Cause or Contribute Findings for Greenhouse Gasses Under Section 202(A) of the Clean Air Act (40 CFR Chapter 1)." Final Rule 74 FR 66496. Washington, DC: EPA. December.

EPA, 2010a. "Climate Change, High GWP Gases." Washington, DC: EPA. March. <http://www.epa.gov/highgwp>.

EPA, 2010b. "Climate Change—Greenhouse Gas Emissions." Washington, DC: EPA. April. <http://www.epa,gov/climatechange/emissions/index.html>.
GCRP (U.S. Global Change Research Program), 2009. "Global Climate Change Impacts in the United States." Cambridge University Press. 2009

Federal Highway Administration, 2006. *Highway Statistics 2005*. Table VM–1. Washington, DC: Federal Highway Administrati9on.

NCDC (National Climatic Data Center), 2009. Record of Climatological Observations: Need to Insert NWS stations for Powder River Basin." Ashville, North Carolina: NCDC. August.
NCDC, 2010a. "Climate at a Glance—Annual Wyoming Temperature." Ashville, North Carolina: NCDC. May.

National Environmental Policy Act of 1969 (NEPA). 42 USC 4321 et seq.
Naugle, D.E, K.E. Doherty, B. Walker, M.J. Holloran, and H.E. Copeland, 2009. "Chapter 21. Energy Development and Greater Sage-Grouse." *Ecology and Conservation of Greater Sage-Grouse: A Landscape Species and Its Habitats.*" Studies in Avian Biology, Prepublication release. <http://sagemap.wr.usgs.gov/Docs/SAB/Chapter21.pdf> (28 April 2010).

NCDC, 2010b. "Climate at a Glance—Annual Wyoming Precipitation." Ashville, North Carolina: NCDC. May.

NRC (U.S. Nuclear Regulatory Commission), 2010. NUREG–1555, "Supplemental Staff Guidance to NUREG-1555, "Environmental Standard Review (ESRP) for Consideration of the Effects of Greenhouse Gases and of Climate Change." Washington, DC: NRC. March.

NRC (U.S. Nuclear Regulatory Commission), 2009a. NUREG–1910,"Generic Environmental Impact Statement for In-Situ Leach Uranium Milling Facilities." Washington, DC: NRC. May.

NRC, 2009b. "Expected New Uranium Recovery Facility Applications/Restarts/Expansions: Updated 3/11/2009" <http://www.nrc.gov/info-finder/materials/uranium/2008-ur-projects-list-public.pdf> (07 April 2009).

NRC, 2003. NUREG–1569, "Standard Review Plan for In-Situ Leach Uranium Extraction License Applications." Final Report. Washington, DC: June.

NRC. 1998. "Environmental Assessment for Renewal of Source Material License No. SUA–1341, Cogema Mining, Inc. Irigaray and Christensen Ranch Projects, Campbell and Johnson Counties, Wyoming." Docket No. 40-8502. Washington, DC: NRC. June 1998.

OSLI (Office of State Lands and Investments), 2007. "Wyoming Biomass Inventory: Animal Waste, Crop Residue, Wood Residue, and Municipal Solid Waste." Cheyenne, Wyoming: OSLI, Wyoming State Forestry Division. March.

Russell, J. and A. Pollack, 2005. "Final Report: Oil and Gas Emission Inventories for the Western States." Novato, California: Environ International Corporation. December 27.
Scire, J. S., D.S. Strimitis, and R.J. Yamartino, 1999. "A User's Guide for the CALPUFF Dispersion Model (Version 5.0)." Concord, Massachusetts: Earth Tech, Inc. October.

Stoeckenious. T., 2010. "Final Report: A Conceptual Model of Winter Ozone Episodes in Southwest Wyoming." 06–20441E. Novato, California: Environ International Corporation. January 29.

Stoeckenious. T., S. Lau, E. Tai, R. Morris, M. Jimenez, and J. Russell, 2006. "Final Report: Southwest Wyoming NO2 PSD Increment Consumption Modeling: Results for Sublette County." Novato, California: Environ International Corporation. May 22.

Uranium One, 2008. Response to Request for Additional Information for the Moore Ranch In Situ Uranium Recovery Project License Application (TAC JU011)," ADAMS Accession No. ML082060527. July 11, 2008.

UCSB (U.S. Census Bureau), 2009. "State and County QuickFacts for Campbell, Johnson, and Natrona Counties, Wyoming." <http://quickfacts.census.gov> (September 10, 2009).

USGS (U.S. Geological Survey), 2003. "Principal Aquifers of the 48 Conterminous United States, Hawaii, Puerto Rico, and the U.S. Virgin Islands." <http://nationalatlas.gov/atlasftp.html> (03 May 2010).

Uranium One, 2010. Email from J.F. Winter, Manager Environmental and Regulatory Affairs, Uranium One to R. Linton, Safety Review Project Manager, NRC. Subject: Response to Cumulative Impacts on Satellites Final 4.21.10 (1).doc. April 21, 2010. ADAMS Accession No. ML101160084.

Wyoming Department of Administration and Information, Economic Analysis Division, 2008. <http://eadiv.state.wy.us/pop/pop.html> (26 April 2010)

Cumulative Impacts

WDEQ (Wyoming Department of Environmental Quality), 2009. Re: Permit No. CT–8644." Letter (October 2) from D.A. Finley to M. P. Thomas, Uranerz Energy Corporation. Cheyenne, Wyoming: WDEQ, Air Quality Division. 2009.

WDEQ, 2010. "Re: Permit No. CT-7896." Letter (January 4) from D.A. Finley to J. Cash, Lost Creek ISR, LLC. Cheyenne, Wyoming: WDEQ, Air Quality Division.

USGS (U.S. Geological Survey), 2003. "Principal Aquifers of the 48 Conterminous United States, Hawaii, Puerto Rico, and the U.S. Virgin Islands." <http://nationalatlas.gov/atlasftp.html> (03 May 2010).

Wyoming Oil and Gas Conservation Commission, 2010. <http://wogcc.state.wy.us> (04 May 2010).

Wyoming Mining Association, 2008. "Wyoming Coal Data." <http://www.wma-minelife.com/coal/coalfrm/coaldat.htm> (28 October 2009).

6 ENVIRONMENTAL MEASUREMENTS AND MONITORING PROGRAMS

6.1 Introduction

As discussed in the Generic Environmental Impact Statement for *In-Situ* Leach Uranium Milling Facilities (GEIS, NUREG–1910, NRC, 2009) (Section 8.0), monitoring programs, in general, are developed for *in-situ* uranium recovery (ISR) facilities to verify compliance with standards for the protection of worker health and safety in operational areas and for protection of the public and environment beyond the facility boundary (NRC, 2009). Monitoring programs provide data on operational and environmental conditions so that prompt corrective actions can be implemented if an adverse condition is detected. In this regard, these programs help to limit the potential environmental impacts at ISR facilities and the surrounding areas.

Required monitoring programs can be modified to address unique site-specific characteristics by the addition of license conditions based on the U.S. Nuclear Regulatory Commission (NRC) site-specific safety and environmental reviews.

The discussion of the proposed monitoring programs for the proposed Moore Ranch Project is organized in the following manner:

- Radiological monitoring (Section 6.2)

- Physiochemical monitoring (Section 6.3)

- Ecological monitoring (Section 6.4)

6.2 Radiological Monitoring

This section describes the Uranium One Americas' (Uranium One) proposed radiological monitoring program as described in its license application (EMC, 2007a,b) and subsequent responses to the NRC requests for additional information (Uranium One, 2009). The purpose of this monitoring program is to characterize and evaluate the radiological environment, to provide data on measurable levels of radiation and radioactivity, and to provide data on the principal pathways of radiological exposure to the public and workers (NRC, 2003).

In accordance with NRC regulations in 10 CFR Part 40, Appendix A, Criterion 7, a preoperational monitoring program is required to establish the facility baseline conditions. Following this baseline program, operators of ISR facilities are required to conduct an operational monitoring program to measure or evaluate compliance with standards and to evaluate the environmental impact of an ISR facility under operational conditions. Although not a requirement, NRC Regulatory Guide 4.14 (NRC, 1980) provides guidance for implementing monitoring programs that are acceptable to the NRC staff for uranium mills, including ISR facilities.

The results of the Uranium One's baseline monitoring program are discussed in Section 3.12.1 of this SEIS. The following sections provide a brief description of the applicant's proposed operational monitoring program.

Environmental Measures and Monitoring

6.2.1 Airborne Radiation Monitoring

Uranium One proposes to implement an airborne radiation monitoring program to detect radon and air particulate releases from the central plant processes. Figure 6-1 shows the air sampling locations proposed by the applicant, which, based on the recommendations in Regulatory Guide 4.14, include a minimum of three air monitoring stations at or near the site boundaries, one station at or close to the nearest structure within 10 km [6.2 mi] of the site, and one station at a control or background location. These operational monitoring locations would be the same as those used to perform the baseline analysis described in Section 3.12.1. However, the NRC staff did not concur with the applicant's proposed air particulate sampling locations because the proposed sampling program would not monitor downwind locations from the predominant wind directions. Therefore, the NRC will impose license conditions to ensure air particulate sampling locations and the nearest resident location are sampled in accordance with Regulatory Guide 4.14.

Air particulate monitoring would be performed using low volume samplers. Filters would be collected weekly to help prevent dust loading and would be analyzed for average Ra-226, natural uranium, Th-230, and Pb-210 concentrations and detection levels. Results of the operational air particulate monitoring program would be reported in the semi-annual effluent reports to the NRC, as required by 10 CFR 40.65. Radon sampling would be conducted at the air particulate sampling locations using Track-Etch radon cups which would be exchanged semiannually and analyzed to determine radon concentrations (in pCi/L). In addition to the environmental monitoring, Uranium One would estimate the release of radon from process operations using MILDOS-AREA modeling and would report results in the semi-annual effluent reports required by 10 CFR 40.65.

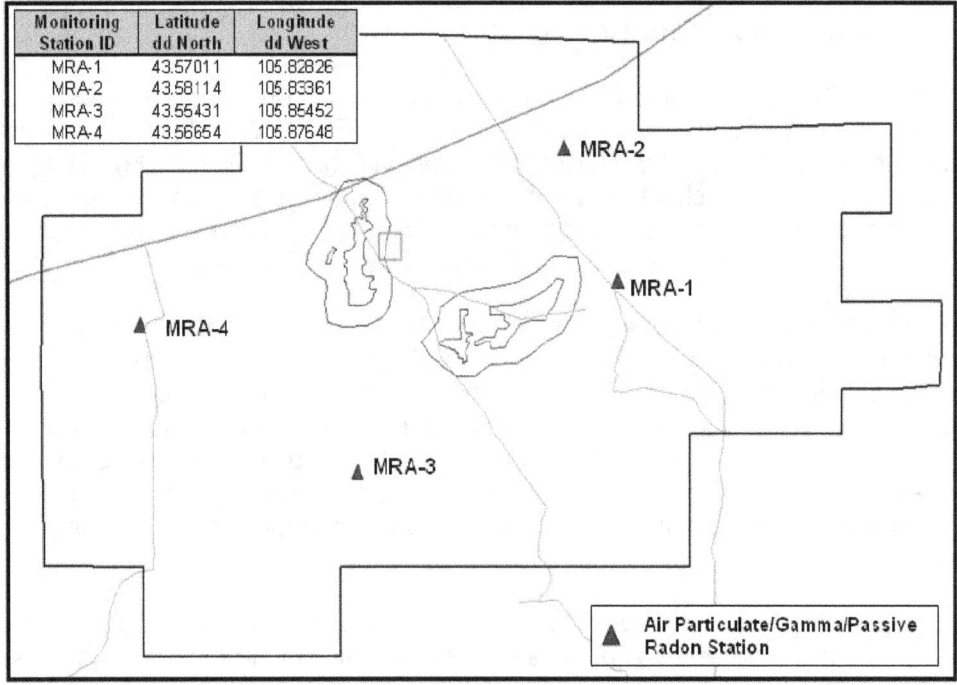

Figure 6-1. Proposed Moore Ranch Project Operational Environmental Monitoring Locations

6.2.2 Soils and Sediment Monitoring

During operations, Uranium One would conduct soil sampling on an annual basis. Samples would be collected to a depth of 5 cm for consistency with the baseline soil sampling surveys described in Section 3.12.1. Following Regulatory Guide 4.14, discrete grab samples of surface soils would be collected at the four air particulate sampling locations shown in Figure 6-1 and would be analyzed for natural uranium, Ra-226, and Pb-210. Prior to decommissioning, following the conclusion of operations, subsurface soil samples would be taken to compare the results with subsurface soil samples collected as part of the pre-operational monitoring program.

6.2.3 Vegetation, Food, and Fish Monitoring

As described in Section 3.12.1, Uranium One conducted pre-operational vegetation sampling in 2007 at various locations on the Moore Ranch site. The applicant evaluated the ingestion pathway to individuals from vegetation using the MILDOS-Area model and concluded that the ingestion pathway is not significant. Following Regulatory Guide 4.14, the applicant does not intend to conduct vegetation, food, or fish sampling because the predicted dose to an individual from these pathways would be less than five percent of the applicable radiation protection standard.

6.2.4 Surface Water Monitoring

The proposed license area contains only intermittent streams and natural runoff occurs during heavy rainfall and snowmelt events. Current CBM operations contribute surface water discharge, which maintains some ponding at select locations across the proposed license area for portions of the year. Surface water samples would be collected on a quarterly basis at the same locations sampled for the pre-operational baseline if surface water is present as shown in Figure 3.4.2-4 of the applicant's Technical Report (EMC, 2007b).

Surface water samples would be collected in appropriate containers and field measurements of pH and conductivity would be documented. A preservative (acid) would be added to the surface water sample immediately after collection and filtration, if required. The sampling volume, preservative, and holding times for the proposed analytes are summarized in Uranium One (2008). The samples would be analyzed for Pb-210, Ra-226, Th-230, natural uranium, and Po-210. Surface water monitoring results would be submitted to the NRC in the semi-annual environmental and effluent monitoring reports.

6.2.5 Groundwater Monitoring

Groundwater monitoring of wells located within 1 km [0.6 mi] of the boundary of an operating wellfield would be performed to detect any migration of contaminated groundwater. These wells would be sampled quarterly, and analyzed for natural uranium and Ra-226 with the landowner's consent, to identify potential impacts from the ISR operation. The sampling would be conducted in accordance with standard operating procedures. Furthermore, the water levels in private wells would be measured by the applicant every three months during operations (EMC, 2007a)

6.3 Physiochemical Monitoring

The ISR process significantly alters the water quality in the production zone aquifer. Therefore, before uranium extraction may occur in a production aquifer, the EPA must declare the production aquifer to be exempt. Appendix C of this final SEIS discusses the criteria

Environmental Measures and Monitoring

U.S. Environmental Protection Agency (EPA) uses for an aquifer exemption. During operations, physiochemical groundwater monitoring is conducted to help prevent and limit potential impacts to groundwater quality in any of the nonexempt aquifers surrounding the exempt production zone aquifer. Physiochemical monitoring provides data on operational and environmental conditions so that prompt corrective actions can be taken if an adverse condition is detected. The physiochemical monitoring program at the proposed Moore Ranch Project includes groundwater monitoring and wellfield and pipeline flow and pressure monitoring.

6.3.1 Wellfield Groundwater Monitoring

Section 8.3 of the GEIS discussed the potential for ISR production processes to affect groundwater in and near the operating wellfield. Hence, groundwater conditions are extensively monitored both before and during operations, and after restoration. The proposed pre-operational and baseline groundwater monitoring that would occur at the Moore Ranch Project is discussed below in Section 6.3.1.1. The groundwater quality monitoring that would occur during operation and restoration is discussed in Section 6.3.1.2.

6.3.1.1 Preoperational Groundwater Sampling

Section 8.3.1.1 of the GEIS discussed how a baseline groundwater quality program would be established prior to uranium production (NRC, 2009). The purpose of this program is to characterize water quality in monitoring wells used to detect lixiviant excursions from the ore production zone, to remediate excursions, and to establish Restoration Target Values (RTVs) for aquifer restoration after the operations phase is complete.

Groundwater monitoring wells were installed at the proposed Moore Ranch Project to evaluate pre-operational water quality as part of the site characterization discussed in Section 3.5.2. Four well groups, each with a well in the 70 Sand production zone aquifer, the overlying 72 Sand aquifer, and the underlying 68 Sand aquifer were installed across the proposed license area. Three wells were also completed in the 60 Sand aquifer. Four additional wells completed in the 70 Sand aquifer, installed in the 1980s by Conoco, Inc., and existing stock water wells completed in either the production zone aquifer or the underlying and/or overlying aquifers were also sampled for the Wyoming Department of Environmental Quality-Land Quality Division (WDEQ-LQD) Guideline 8 groundwater quality parameters to establish the WDEQ class of use as described in Section 3.5.2 of this SEIS.

This sampling program in combination with groundwater sampling data from the 1980 Conoco, Inc. project provided a preliminary baseline of groundwater quality. The purpose of the preoperational analysis is to evaluate the overall groundwater quality in the proposed license area under normal pre-operational conditions. It is not used to establish the baseline water quality which forms the basis for establishing restoration criteria for the individual wellfields.

To establish baseline water quality before operations at the proposed Moore Ranch Project, baseline monitoring wells would be installed in the 70 Sand production zone at a density of one well per 1.2 ha [3 ac] of the two planned wellfields. Each monitoring well would be analyzed for all WDEQ LQD Guideline 8, Appendix 1, parts III and IV parameters shown in Table 5.7-1 of the applicant's technical report (EMC, 2007b). The third and fourth sampling events would be analyzed for a reduced list of parameters defined by the previous sample results. If certain constituents were not detected during the first and second sampling events, then they would not be analyzed for again during the third and fourth sample events. Data for each water quality parameter would be averaged. If the collected wellfield data indicated that waters of different

Environmental Measures and Monitoring

underground water classes coexist (WDEQ-LQD Rules and Regulations, Chapter VIII), then the data would not be averaged but rather treated as sub-zones. Subzone specific data would also be averaged. A sub-zone boundary would be delineated half-way between the sampled well sets as appropriate.

Once the baseline water quality for each wellfield is established, it would be used to determine the appropriate restoration target values to assess the effectiveness of groundwater restoration on a wellfield-specific basis. The restoration target values are a combination of the average and range of baseline values for specific constituents in wells completed in the 70 Sand ore production zone. WDEQ would review and approve the baseline water quality assessment and restoration target values for each wellfield; NRC would also review and approve the restoration target values for specific constituents.

Monitoring wells would be installed in a ring around each wellfield in the 70 Sand production zone and in the overlying 72 Sand and underlying 68 and 60 Sand aquifers prior to the start of operations. Monitoring wells would be placed at a spacing of one well per 4 acres in the underlying 60 Sand in the areas where the 70 and 68 sand coalesce. The wells would be sampled to determine baseline water quality data to establish upper control limits (UCL) for operational excursion monitoring. The wells would be sampled four times, at least 2 weeks apart. The first sample would be analyzed for the full set of constituents required by the WDEQ; subsequent samples would only be analyzed for the UCL parameters (see Section 6.3.1.2).

The applicant's technical report (EMC, 2007b) provides detailed procedures for sampling and analysis, including methods for measuring water levels, well purging and sampling protocols, sample preservation and documentation, analytical methods, and quality assurance/quality control requirements (Uranium One, 2008).

6.3.1.2 Groundwater Quality Monitoring

Section 8.3.1.2 of the GEIS discussed the placement of monitoring wells around the perimeter of the wellfields, in the aquifers overlying and underlying the ore-bearing production aquifers to provide early detection of potential horizontal and vertical lixiviant excursions during production operations. Monitoring well placement is based on a number of factors including the nature and extent of the confining layer and the occurrence of drill holes, hydraulic gradient and aquifer transmissivity, and well abandonment procedures used in the region. The ability of a monitoring well to detect groundwater excursions is influenced by several factors, such as the thickness of the aquifer, the distance between the monitoring wells and the wellfield, the distance between adjacent monitoring wells, the frequency of groundwater sampling, and the magnitude of changes in lixiviant migration indicator parameters. Therefore, the spacing, distribution, and number of monitoring wells are site-specific and established by license conditions.

The groundwater monitoring program at the proposed Moore Ranch Project would be designed to detect excursions of lixiviant outside the wellfield under production and into the overlying and/or underlying water bearing strata. The groundwater monitoring is divided into four phases: pre-operational, baseline, production and restoration monitoring. Section 5.7.8 of the applicant's Technical Report documents the groundwater monitoring program that would be implemented at the proposed Moore Ranch Project (EMC, 2007b). Monitoring wells completed in the 70 Sand production zone would be installed around the perimeter of each wellfield. Approximately 24 groundwater wells would monitor the perimeter of Wellfield 1 and approximately 27 groundwater wells would be used to monitor the perimeter of Wellfield 2. Within the pattern area wells completed in the overlying 72 Sand aquifer and in the underlying

Environmental Measures and Monitoring

68 Sand aquifer would be spaced at one well per every 6.4 ha [4 ac] of pattern area resulting in approximately 6 monitor wells completed in the overlying and underlying aquifers in Wellfield 1 and about 9 monitor wells completed in overlying and underlying aquifers in Wellfield 2 since it covers a larger area. In the Wellfield 2 area where the 68 and 70 Sands coalesce, the sands would be treated as one aquifer and the underlying aquifer would be the 60 Sand. Additional monitoring wells would be placed in the 68 Sand in this area to detect potential impacts. The final number of such wells would be determined during final wellfield planning and submitted to the WDEQ-LQD and NRC for review and approval.

The distance between perimeter monitoring wells surrounding the production wells would be no more than 152 m [500 ft] and the distance between the perimeter monitoring wells and the production pattern would also be approximately 152 m [500 ft] based on the output from the applicant's groundwater flow model and the estimated hydraulic properties within the production area. The applicant used model simulations to demonstrate that if an excursion occurred, the perimeter monitoring ring would be able to detect an excursion in a timely manner. Appendix B-4 of the applicant's technical report discusses the groundwater model and analyzes the model results (EMC, 2007b).

At the proposed Moore Ranch Project, the constituents selected as lixiviant migration indicators for which UCLs would be established are chloride, conductivity, and total alkalinity. Chloride was selected because it has a low background concentration in native groundwater, it would be introduced into the lixiviant from the ion exchange process, and it is very mobile in groundwater. Conductivity was selected because it is an indicator of overall groundwater quality. Finally, total alkalinity was selected because bicarbonate is the major constituent added to the lixiviant during production; therefore, elevated concentrations of total alkalinity could be indicative of an excursion.

Per NRC guidance in NUREG-1569 and WDEQ requirements, the applicant must provide a field demonstration of the hydraulic interconnection between the monitoring wells and production pattern using pump tests before operations can be initiated. Because of the unconfined nature of the groundwater aquifer in the 70 Sand at the proposed Moore Ranch Project, more intensive pump tests were necessary to demonstrate hydraulic interconnections between the production zone and the perimeter monitor wells. The applicant therefore used the numerical groundwater model presented in Appendix B-4 of the technical report to develop a pump test strategy that could demonstrate the hydraulic connections between the monitoring wells and production pattern in the 70 Sand unconfined aquifer (EMC, 2007b). This pump test strategy would be implemented after the required monitoring and production wells were completed but prior to operations.

A Wellfield Hydrologic Data Package would be prepared by the applicant following the installation of the production pattern and monitoring well network in a wellfield. This package would provide the monitoring well locations, the pump test results, baseline water quality for all wells, and RTVs for each wellfield production zone. The applicant's Safety and Environmental Review Panel, responsible for monitoring any proposed change in the facility or process, would review the data package to ensure that the hydrologic testing results and planned ISR activities would be consistent with technical requirements and did not conflict with NRC regulatory requirements. The Wellfield Hydrologic Data Package would be submitted to the WDEQ for review and approval to ensure the acceptability of the baseline data and the RTVs. WDEQ would also review the monitoring well locations and the wellfield-specific monitoring program to ensure they would provide timely detection and correction of potential horizontal or vertical excursions. Based on the outcome of the safety review of the technical report, NRC will request

Environmental Measures and Monitoring

by license condition to review and approve the Wellfield Hydrologic Data Package for an individual wellfield if NRC determines that a safety review cannot be adequately completed without the local wellfield-specific information provided in the package.

After operations were completed, the wellfields would be restored. During restoration, lixiviant injection would be suspended; thereby reducing the potential for an excursion. The applicant therefore has proposed a reduced groundwater monitoring program during aquifer restoration. During the aquifer restoration phase, wells located in the perimeter monitoring ring and completed in the overlying and underlying aquifers would be sampled every 60 days for chloride, alkalinity and conductivity excursion parameters. An excursion would be defined in the same manner as during operations and subject to the same correction requirements.

6.3.2 Wellfield and Pipeline Flow and Pressure Monitoring

Section 8.3.2 of the GEIS discussed operator monitoring of injection and production well flow rates to manage the entire wellfield water balance. The pressure of each production well and the production trunk line in each wellfield header house would also be monitored. Unexpected pressure loss could indicate equipment failure, a leak, or well integrity problems.

The proposed Moore Ranch Project would have an extensive program of wellfield and pipeline flow and pressure monitoring as described in Section 3.1.3 of the applicant's technical report (EMC, 2007b). Injection well and production well flow rates and pressures would be monitored at each header house to balance injection and production in each pattern and throughout the wellfield. The production and injection flow rate in each well would be continuously individually monitored by electronic flow meters in each wellfield header house. The pressure of each production and injection well trunk line would also be monitored at the header house with electronic pressure gauges. Both flow meter and pressure gauges would tie into the header house control panel that would ultimately tie into the central plant control room. High and low pressure and/or flow alarms would alert wellfield and plant operators if specified ranges were exceeded. Automatic shutoff valves would stop the flow in the event of significant changes in volume or pressure. This monitoring would alert the operators to detect malfunctions that could lead to either wellfield infrastructure or pipeline failures, thus minimizing the potential to impact groundwater.

6.3.3 Surface Water Monitoring

Uranium One does not plan to conduct physiochemical monitoring of surface water since there would be no surface water discharges associated with the ISR process at the proposed Moore Ranch Project. To ensure the protection of surface water, each injection and production well would be monitored to detect a change in flow and/or pressure that might indicate a leak or rupture in the system. If a leak were to occur, the system would be shut down and remediation conducted as appropriate.

6.3.4 Meteorological Monitoring

Uranium One does not plan to conduct meteorological monitoring at the site. To describe the affected environment and assess air quality impacts resulting from the proposed project, the applicant used meteorological data from the Antelope Coal Mine meteorological station located approximately 40 km (25 mi) southeast of the proposed Moore Ranch Project. The Antelope coal mine site has similar topographic features to the proposed Moore Ranch Project and is characterized by mildly rolling hills covered with grass and sparse shrubs (Uranium One, 2008).

Environmental Measures and Monitoring

6.4 Ecological Monitoring

6.4.1 Vegetation Monitoring

As discussed in Section 6.2.3, the applicant concluded from its pre-operational vegetation sampling program and through modeling that the ingestion pathway for radiological dose is not significant. Therefore, Uranium One does not intend to conduct vegetation, food, or fish sampling because the predicted dose to an individual from these pathways would be less than five percent of the applicable radiation protection standard.

6.4.2 Wildlife Monitoring

Large game animals such as deer or pronghorn have extensive ranges and are not confined to the site. Therefore, the potential for bioaccumulation of radionuclides in these animals would be limited since they would likely derive only a small fraction of total sustenance from the proposed Moore Ranch Project. No fish species occur within the proposed license area since surface water is intermittent in nature and does not have a sufficient volume to support aquatic species.

The applicant has proposed wildlife studies that would include an annual raptor survey conducted in late April or early May to identify new nests and to assess whether known nests are being utilized (Uranium One, 2009). The survey would cover all areas of planned activity for the life of the ISR project (i.e., wellfields and the central plant) and within a one-mile area around the activity primarily to protect against unforeseen conditions such as the construction of a new nest in an area that could be affected by ISR operations (Uranium One, 2009).

6.5 References

10 CFR Part 40. *Code of Federal Regulations*, Title 10, *Energy*, Part 40, "Domestic Licensing of Source Material." Appendix A, "Criteria Relating to the Operation of Uranium Mills and the Disposition of Tailings or Wastes Produced by the Extraction or Concentration of Source Material From Ores Processed Primarily for Their Source Material Content."

EMC (Energy Metal Corporation US), 2007a "Application for USNRC Source Material License, Moore Ranch Uranium Project, Campbell County, Wyoming, Environmental Report." Casper, Wyoming: Uranium 1 Americas Corporation. ADAMS Accession Nos. ML072851222, ML072851229, ML072851239, ML07285249, ML07285253, ML07285255. October 2.

EMC, 2007b. "Application for USNRC Source Material License, Moore Ranch Uranium Project, Campbell County, Wyoming, Technical Report." Casper, Wyoming: Uranium 1 Americas Corporation. ADAMS Accession Nos. ML072851222, ML072851258, ML072851259, ML072851260, ML072851268, ML072851350, ML072900446. October 2, 2007.

NRC (U.S. Nuclear Regulatory Commission), 2009. NUREG–1910, Generic Environmental Impact Statement for *In-Situ* Leach Uranium Milling Facilities. Washington, DC: NRC. May.

NRC, 2003. NUREG–1569, Standard Review Plan for *In-Situ* Leach Uranium Extraction License Applications. Final Report. Washington, DC: NRC. June.

NRC, 1980. Regulatory Guide 4.14, "Radiological Effluent and Environmental Monitoring at Uranium Mills." Rev. 1. Washington, DC. April 1980.

Uranium One (Uranium One Americas), 2009. "Responses to Request for Additional Information for the Moore Ranch In Situ Uranium Recovery Project License Application (TAC JU011)." ADAMS Accession No. ML092450317. August 31, 2009.

Uranium One, 2008. "Additional Information Requested for the Moore Ranch In Situ Uranium Recovery Project License Application (TAC JU011), First Set of Responses." ADAMS Accession No. ML082060527. July 11, 2008.

7 COST-BENEFIT ANALYSIS

This chapter summarizes benefits and costs associated with the proposed action and the No-Action alternative. Chapter 4 of this supplemental environmental impact statement (SEIS) discusses the potential socioeconomic impacts of the construction, operation, aquifer restoration, and decommissioning of the proposed Moore Ranch Project by Uranium One Americas.

The implementation of the proposed action would primarily generate regional and local benefits and costs. The regional benefits of constructing and operating the proposed Moore Ranch Project would be increased employment, economic activity, and tax revenues in the region around the proposed license area. Some of these regional benefits, such as tax revenues, would be expected to accrue specifically to Campbell County, Wyoming, where the proposed facility would be located, and the town of Wright, located approximately 40 km [25 mi] from the proposed project. Other benefits could extend to the neighboring Wyoming counties of Johnson and Natrona. Costs associated with the proposed Moore Ranch Project are, for the most part, limited to the immediate area surrounding the site. Examples of these environmental impacts would include changes to current land use and increased road traffic.

7.1 No-Action Alternative

Under the No-Action alternative, the NRC would not grant a license for the proposed Moore Ranch Project. The No-Action alternative would result in the applicant not constructing, operating, restoring the aquifer, or decommissioning the proposed Moore Ranch Project. No facilities, road, or wellfields would be built; no pipeline would be laid as described in Section 2.1.1.1.2. No uranium would be recovered from the subsurface orebody; therefore, injection, production, and monitoring wells would not be installed in wellfields and the facility would not operate. Neither lixiviant would be introduced in the subsurface nor would buildings be constructed to process extracted uranium or for chemical storage. Because no uranium would be recovered, neither aquifer restoration nor decommissioning activities would occur. No liquid or solid effluents would be generated. As a result, the proposed site would not be disturbed by the proposed project activities, and ecological, natural, and socioeconomic resources would remain unaffected. All potential environmental impacts from the proposed action would be avoided. Similarly, all project-specific socioeconomic impacts (e.g., related to housing) would be avoided.

7.2 Benefits from Proposed Action in Campbell County

Under the proposed action, the applicant would construct, operate, restore the aquifer, and decommission the proposed Moore Ranch Project in Campbell County, Wyoming. The central plant, access roads, and initial development of the first of two wellfields for the proposed Moore Ranch Project would occur over a 9-month period, the second wellfield would be developed within the following 2-year period. Operation of the central plant for uranium recovery and processing would occur over 12 years, with approximately 3.25 years of uranium recovery in each of the two wellfields. Aquifer restoration activities and stability monitoring following restoration would occur over a 4.5- to 6.5-year period. The applicant expects to conduct final wellfield and site decommissioning within 1 year.

The principal socioeconomic impact or benefit from the proposed Moore Ranch Project would be an increase in the jobs in Campbell County, Wyoming. The applicant expects to employ 50 workers during construction and 40–60 workers during operations, aquifer restoration, and decommissioning. As discussed in Section 4.11, the construction and decommissioning workers would most likely not relocate in the area because of the short period of time (9 months to 1 year) over which these activities would occur.

However, during the 12-year period of operations and aquifer restoration, workers would be more likely to relocate to be nearer the facility. If the majority of operational requirements was filled by a workforce located outside the region, given a multiplier of about 0.7[1], there could be an influx of 28 to 42 jobs (i.e., 40 jobs × 0.7 = 28 jobs and 60 jobs × 0.7 = 42 jobs).

The closest town to the proposed Moore Ranch Project is the town of Wright with an estimated population of 1,462 (U.S. Census Bureau, 2008). However, employees supporting operations could prefer to reside in larger communities (NRC, 2009) and therefore, could choose to reside in the towns of Gillette and Casper. The influx of these jobs and a reduction of unemployment would have a MODERATE benefit to the businesses in the smaller towns such as Wright and a SMALL to MODERATE impact in the businesses of the larger towns within commuting distance from the proposed project site.

In addition to creating jobs, the operation of the proposed Moore Ranch Project and its employment opportunities would contribute to local, regional, and state revenues through the purchase of goods and services and through the taxes levied on such goods and services. Furthermore, severance taxes of 4 percent of taxable market value associated with uranium milling/mining in Campbell County are levied by the State of Wyoming, Mineral Tax Division of the Department of Revenue (Wyoming Department of Revenue, 2009). The applicant's current resource estimate for the proposed Moore Ranch Project is 5.8 million lbs of uranium. If the applicant is able to fully recover this resource and sell it at a nominal market price of $45 per pound of uranium, the severance tax would yield approximately $10,440,000 in net economic benefits over the life of the project. This figure excludes potential reserve resources and the potential benefit from taxes on royalties or lease payments to local landowners from the operation of the proposed Moore Ranch Project

7.2.1 Benefits From Potential Production

The employment generated by the proposed Moore Ranch Project and the taxes paid by the applicant depend on the production of yellowcake. The volume of yellowcake produced would depend on the market price for yellowcake (as uranium) and the cost of production. Since 2007, the spot-market price for U_3O_8 has fluctuated significantly, from a high of over $130 in 2007 to as low as $40 in 2009. As of July 12, 2010, the price was $41.50 per pound.

---

[1] The Economic Multiplier is used to summarize the total impact that can be expected from a change in a given economic activity. It is the ratio of total change to initial change. The multiplier of 0.7 was used as a typical employment multiplier for the milling/mining industry (Economic Policy Institute, 2003).

Cost-Benefit Analysis

The project's potential benefits to the local community depend on the applicant's operating costs being lower than the future price of uranium. If the price of uranium drops below the operating costs, then the operation of the facility would become uneconomic and the operations could be suspended and/or discontinued.

7.2.2 Costs to the Local Communities

Table 7-1 identifies both the towns and their population within 40 km (25 mi) of the proposed Moore Ranch Project and towns within commuting distance.

As noted in Section 7.2, the proposed Moore Ranch Project would employ 40 to 60 workers during the operations period; if the majority of these workers came from outside the region, there could be an influx of 28 to 42 jobs (given an economic multiplier of 0.7). Assuming that operations workers would tend to relocate to be closer to the site, the creation of these new jobs could result in an influx of 69 to 104 people, based on an assumption of 2.48 persons per household for the State of Wyoming (U.S. Census Bureau, 2000).

Chapter 4 of this SEIS states that because of the small relative size of the workforce at the proposed Moore Ranch Project, the potential impact on socioeconomics would be SMALL except for the impact on housing which could range from SMALL to MODERATE. As stated previously, operations employees could prefer to reside in larger communities (NRC, 2009) and, therefore, could choose to reside in larger towns. The influx of new jobs along with the reduction in unemployment would result in a SMALL to MODERATE increase in housing demand and in the construction of new homes within the region of influence. The population growth would have a SMALL impact on education infrastructure and health and social services.

The local communities would require a minimal increase in emergency response and medical treatment capabilities because of the small risk of an industrial accident from the proposed action.

Table 7-1. Towns Near the Proposed Moore Ranch Project		
Town	Population *	Distance From Project Site (km) [mi]
Towns Within 40 km [25 mi] From the Project Site		
Edgerton	176	38 [24]
Wright	1,462	40 [25]
Midwest	435	40 [25]
Towns greater than 40 km (25 mi) from the project site		
Kaycee	290	64 [40]
Gillette	26,871	80 [50]
Buffalo	4,832	100 [62]
Casper	54,047	100 [62]
*U.S. Census Bureau, 2008		

7.3 Evaluation Findings of the Proposed Moore Ranch Project

Implementation of the proposed action would have a SMALL to MODERATE overall economic impact on the region of influence. The implementation of the proposed action would generate primarily regional and local benefits and costs. The regional benefits from the operation of the proposed Moore Ranch Project would be increased employment, economic activity, and tax revenues to the region around the site. Some of these regional benefits, such as tax revenues, would accrue to Campbell County specifically. Other benefits could extend to neighboring counties in the State of Wyoming. Costs associated with the proposed Moore Ranch Project would be limited primarily to the area surrounding the site and the communities within commuting distance. Table 7-2 summarizes the costs and benefits.

Table 7-2. Summary of Costs and Benefits of the Proposed Moore Ranch Project	
Cost-Benefit Category	**Proposed Action**
BENEFITS	
Capacity Produced	5.8 million pounds of U_3O_8
Other Monetary	$10.44 million (estimated)
Non-Monetary (50% of jobs would be from Campbell County)	50 jobs-- during construction 40-60 jobs—during operations, aquifer restoration, and decommissioning 28-42 jobs—local jobs from economic multiplier during operations and aquifer restoration
COSTS	
Education Infrastructure	SMALL
Health and Social Services	SMALL
Housing Demand	SMALL to MODERATE
Emergency Response	SMALL

7.4 References

Economic Policy Institute, 2003. "Updated Employment Multipliers for the U.S. Economy," Washington, DC. <http://www.epi.org/page/-/old/workingpapers/epi_wp_268.pdf> (13 October 2009).

NRC (U.S. Nuclear Regulatory Commission), 2009. NUREG–1910, "Generic Environmental Impact Statement for *In-Situ* Leach Uranium Milling Facilities," Washington, DC. May.

U.S. Census Bureau, 2008. U.S. Census Bureau, 2008 Population Estimates, <http://factfinder.census.gov>

U.S. Census Bureau, 2000. State and County Quick Facts, <http://quickfacts.census.gov/qfd/states/56000.html>

Wyoming Department of Revenue. 2009. "State of Wyoming Department of Revenue 2009 Annual Report." http://revenue.state.wy.us/PortalVBVS/uploads/ Department%20of%20Revenue%20%2010.29> .2009.pdf> (11 November 2009).

8 SUMMARY OF ENVIRONMENTAL CONSEQUENCES

This chapter summarizes the potential environmental impacts and consequences of the proposed action. In doing so, the potential impacts and consequences of the proposed action are discussed in terms of (1) the unavoidable adverse environmental impacts, (2) the relationship between local short-term uses of the environment and the maintenance of long-term productivity, and (3) the irreversible and irretrievable commitment of resources. The information is presented for the proposed action for the 13 resource areas and discussed by stage of the proposed Moore Ranch Project's lifecycle (i.e., construction, operation, aquifer restoration, and decommissioning). These impacts are described in the table below.

NRC's NUREG-1748 (NRC, 2003) defines the following terms:

- **Unavoidable adverse environmental impacts:** impacts that cannot be avoided and for which no practical means of mitigation are available
- **Irreversible:** commitments of environmental resources that cannot be restored
    - **Irretrievable:** applies to material resources and will involve commitments of materials that, when used, cannot be recycled or restored for other uses by practical means
    - Short-term: represents the period from pre-construction to the end of the decommissioning activities, and therefore generally affect the present quality of life for the public.
    - Long-term: represents the period of time following the termination of the site license, with the potential to affect the quality of life for future generations.

As discussed in Chapter 4, the significance of potential environmental impacts is categorized as follows:

SMALL: The environmental effects are not detectable or are so minor that they will neither destabilize nor noticeably alter any important attribute of the resource.

MODERATE: The environmental effects are sufficient to alter noticeably, but not to destabilize, important attributes of the resource

LARGE: The environmental effects are clearly noticeable and are sufficient to destabilize important attributes of the resource

Table 8-1 summarizes the environmental consequences for the proposed action. Under the No-Action alternative there would be no unavoidable adverse impacts because no licensing action would occur. Likewise, there would be no irreversible and irretrievable commitment of resources since no materials would either be committed or consumed under the No-Action alternative. Similarly, there would be no short- or long-term impacts under the No-Action alternative. The proposed action and the resulting environmental impacts are discussed the impacts associated with the No-Action alternative are provided for comparison in Section 5.2.

Summary of Environmental Consequences

8.1 Proposed Action (Alternative 1)

Under the Proposed Action, the NRC would issue a license for the construction, operation, aquifer restoration, and decommissioning of the Moore Ranch Project. Construction is expected to last about 9 months. During this phase, buildings, access roads, wellfields, pipelines, and injection wells to be used for liquid effluent disposal would be constructed. These actions would disturb approximately 61 ha [150 ac] of the 2,879 ha [7,100 ac] proposed license area. Operations are expected to last about 12 years, however the wellfields would only be operational 3.25 years, injection wells would be used to inject lixiviant (recovery) solutions into the ore body to recover uranium. Production wells would be used to recover the dissolved uranium which then would be processed through the central plant. Finally monitoring wells would be installed to monitor the performance of the wellfield and to mitigate potential excursions from the production zone. Initially, approximately 2 to 3 million pounds of uranium would be produced per year. Aquifer restoration would be initiated to ensure that water quality and groundwater use from surrounding aquifers was not impacted by the proposed action. The process is expected to last about 3.5 years in Wellfield 1 and 5.25 years in Wellfield 2, and would involve transferring contaminated groundwater from one wellfield to the next, "sweeping" groundwater (i.e., replacing contaminated groundwater with cleaner baseline water through pumping action), and treating the groundwater to minimize the groundwater volume consumed during the restoration phase. During the decommissioning phase expected to last about 1 year, the disturbed lands would be returned to their preextraction use. The wells would be plugged and abandoned and the land surface would be reclaimed.

8.2 No-Action (Alternative 2)

Under the No-Action alternative, the NRC would not issue a license. No buildings, roads, wellfields and the supporting infrastructure would be built, no uranium would be recovered from the subsurface orebody; therefore, the aquifer would be unaffected by activities at the proposed Moore Ranch Project and there would be no need for restoring the aquifer or for decommissioning. The decision to not license the proposed Moore Ranch Project would leave a large resource unavailable for energy production supplies to fuel power generation facilities.

Under the No-Action alternative, there would be no impact to land use because the facility would not be constructed; there would neither be earthmoving activities to disturb the land nor restrictions put on the land for grazing or ranching. The existing land use would continue and the property would be available for other uses. There would be no impact on the local transportation system. The current volume and existing traffic patterns would continue as described in the affected environment. Since the land surface would not be disturbed under the No-Action alternative, there would be no impact to soils. Natural phenomena such as wind and water erosion (during storms and severe weather events) would remain the most significant variable associated with geology and soils at the site. The subsurface geology at the site would be unaffected by the injection of fluids.

Surface water and associated wetlands at the site would continue to occur intermittently in response to snowmelts, large precipitation events or from the discharge of surface water from upstream coal-bed methane operations. Under the No-Action alternative, groundwater would be unaffected by the proposed ISR operation. The groundwater quality in the aquifer and the water levels in wells surrounding the proposed license area would remain unaffected. Because there would be neither earthmoving nor grazing restriction activities under the No-Action alternative, the existing vegetation and wildlife communities would be undisturbed. There would be no impact to air quality since there would be no activities to generate either fugitive dust or gaseous emissions nor would there be any noise-generating activities.

Summary of Environmental Consequences

No historic or cultural resources would be disturbed under the No-Action alternative nor would there be any proposed activities that could affect the viewscape. The viewscape would consist of existing activities in the area, such as coal-bed methane extraction and oil and gas development. There would be no additional radiological exposure to the general public other than that from background radiation levels. No additional waste streams or materials such as sanitary waste or byproduct material would be generated.

Under the No-Action alternative, there would be no impact on the socioeconomics of the area. No new jobs would be created, no additional revenue would accrue to the tax base, there would be no impact on the availability of housing or public services. There would be no disproportionately high and adverse impact on minority and low-income populations.

8.3   Reference

NRC (U.S. Nuclear Regulatory Commission), 2003. NUREG–1748, "Environmental Review Guidance for Licensing Actions Associated with NMSS Programs." Washington, DC: August.

Summary of Environmental Consequences

Table 8-1. Summary of Environmental Consequences

The Proposed Action (Alternative 1)

Impact Category (as applicable)	Unavoidable Adverse Environmental Impacts	Irreversible and Irretrievable Commitment of Resources	Short-term Impacts and Uses of the Environment	Long-Term Impacts and the Maintenance and Enhancement of Productivity
Land Use 4.2.1	There would be a SMALL impact to land during the construction and decommissioning phases of the proposed Moore Ranch Project. During construction, approximately 60 ha [150 ac] of land would be fenced and disturbed by earthmoving activities to construct the central plant, wellfields and associated infrastructure, and to build the access roads. This area is less than 2 percent of the proposed license area. During decommissioning, the land would also be impacted by earthmoving activities to reclaim and reseed the area.	No impact. There would be no irreversible and irretrievable commitment of land resources from implementing the proposed action. The duration of the proposed action would last approximately 12 years after which time the land would be reclaimed and made available for other uses.	There would be a short-term impact to land use from implementing the proposed action. Approximately 60 ha [150 ac] of the proposed license area would be unavailable for other uses such as rangeland or grazing; coal bed methane (CBM) or oil and gas exploration could coexist with the applicant's proposed action. The impact would be SMALL.	There would be no long-term impact on land resources from implementing the proposed action. The land would be available for other uses at the end of the license period.
Transportation 4.3.1	There would be a SMALL impact on transportation. Increased truck and	No impact.	There would be a SMALL impact. Small increases in the numbers of traffic	There would be no transportation impacts attributable to the proposed

8-4

Summary of Environmental Consequences

Table 8-1. Summary of Environmental Consequences

The Proposed Action (Alternative 1)

Impact Category (as applicable)	Unavoidable Adverse Environmental Impacts	Irreversible and Irretrievable Commitment of Resources	Short-term Impacts and Uses of the Environment	Long-Term Impacts and the Maintenance and Enhancement of Productivity
	vehicle traffic along State Highway (SR) 387 would result in small changes in the current use of this local road.		accidents resulting in injuries or fatalities, and small increases in vehicle emissions that should not degrade local air quality. The generic environmental impact statement (GEIS) concluded the risk from transporting yellowcake, ion exchange resin, and byproduct material, and hazardous chemicals was small (NRC, 2009).	Moore Ranch Project following license termination.
Geology and Soils 4.4.1	There would be a SMALL impact on geology and soils. The construction and decommissioning phases would disturb surface soils during construction of the central plant, development of the wellfields, laying of pipelines, and construction of new access roads. These impacts would be temporary and at the end of the decommissioning	Topsoil salvaged during the construction phase of the project would be replaced during the reclamation and reseeding processes.	There would be a SMALL impact to geology and soils. No significant matrix compression or ground subsidence would be expected since the net withdrawal of fluid from the 70 Sand production zone would be about one percent or less. Earthmoving activities would disturb about 60 ha [150 ac] of soil.	There would be no long-term impacts to geology and soils following license termination.

8-5

Summary of Environmental Consequences

Table 8-1. Summary of Environmental Consequences

The Proposed Action (Alternative 1)

Impact Category *(as applicable)*	Unavoidable Adverse Environmental Impacts	Irreversible and Irretrievable Commitment of Resources	Short-term Impacts and Uses of the Environment	Long-Term Impacts and the Maintenance and Enhancement of Productivity
	phase topsoil would be replaced.			
Water Resources (Surface Water) 4.5	There would be a SMALL impact to surface water and wetlands from the construction of the central plant and the two wellfields from increased sediment yield in the disturbed areas. A small (<2 ac) wetland area in Wellfield #2 could be disturbed by constructing wells near the drainage area. However, the applicant would avoid installing wells in the main channels of ephemeral drainages; therefore, there would be a SMALL impact. The applicant would also use best management practices such as the use of rip rap and rock on embankments, culverts and drainage crossings.	There would be no irreversible and irretrievable commitment of either surface water or wetlands from implementing the proposed action. No drainage or body of water would be significantly altered during operations. The impact to wetlands would be SMALL since the stream flow is intermittent and the applicant would avoid installing wells in the main channels of ephemeral drainages and implement best management practices.	Normal construction activities within the wellfields, at the central plant, along pipelines and access roads have the potential to result in increased sediment yield in surface water runoff, potentially affecting wetlands. However, given the absence of perennial streams, the small area to be affected, potential impacts to surface water and wetlands during construction and decommissioning would primarily be limited to uncommon precipitation or runoff events. These impacts would be further mitigated by implementing best management practices.	There would be no long-term impact to surface water and wetlands following license termination.

8-6

Summary of Environmental Consequences

Table 8-1. Summary of Environmental Consequences

Impact Category (*as applicable*)	Unavoidable Adverse Environmental Impacts	Irreversible and Irretrievable Commitment of Resources	Short-term Impacts and Uses of the Environment	Long-Term Impacts and the Maintenance and Enhancement of Productivity
The Proposed Action (Alternative 1)				
Water Resources (Groundwater) 4.5	There would be a SMALL impact on groundwater. Groundwater would be impacted from *in-situ* recovery (ISR) by consumption of groundwater and degradation of water quality in the 70 Sand production zone.	About 99 percent of the groundwater used during the ISR process would be treated and re-injected into the subsurface; however, about one percent of the groundwater used in the process would be consumed.	Short-term impacts to groundwater would include degradation of water quality within the 70 Sand production zone during operations and the potential drawdown in private wells completed in the same aquifer as the production zone. The potential drawdown would be nominal; so the yield in private wells would not be affected. Private wells would be monitored to mitigate potential impacts.	Both the State of Wyoming and the U.S. Nuclear Regulatory Commission (NRC) require restoration of affected groundwater following operations. The groundwater quality would be restored to ensure that adjacent aquifers would not be affected.
Ecology 4.6.1	There would be a SMALL impact. Construction and decommissioning of the proposed Moore Ranch Project would result in the short-term loss of vegetation of approximately 61 ha [150 ac] and could stimulate the introduction and spread of undesirable and invasive, nonnative species.	Vegetative communities directly impacted by earthmoving activities and wildlife injuries and mortalities would be irreversible. However, the implementation of mitigative measures such as the use of fences to limit wildlife movement and enforcing speed limits would reduce potential impacts to	During any of the ISR phases, direct impacts to ecological resources could include injuries and mortalities caused by either collisions with project-related traffic or habitat removal actions such as the removal of topsoil. Most habitat disruption would consist of scattered, confined drill sites for wells and would not	Some of the vegetative communities that exist within the proposed Moore Ranch Project could be difficult to reestablish through artificial plantings and natural seeding could take many years. Species associated with those communities could be reduced in number or replaced by generalist species.

8-7

Summary of Environmental Consequences

Table 8-1. Summary of Environmental Consequences

The Proposed Action (Alternative 1)

Impact Category (as applicable)	Unavoidable Adverse Environmental Impacts	Irreversible and Irretrievable Commitment of Resources	Short-term Impacts and Uses of the Environment	Long-Term Impacts and the Maintenance and Enhancement of Productivity
			result in large transformation of the existing habitat. Wildlife could be displaced by increased noise and traffic.	
Meteorology, Climatology, and Air Quality 4.7.1	There would be a SMALL impact. During implementation of the proposed action, there would be increased amounts of dust (particulates) from the earthmoving activities to construct the central plant, drill wells and develop the wellfields, lay pipeline, and build access roads to the wellfields. SMALL impacts would also result from vehicular traffic on unpaved roads and from diesel emissions from construction equipment.	There would be no irreversible or irretrievable commitment of air resources from implementing the proposed action.	There would be a temporary, short-term impact on air quality primarily during the construction and decommissioning phases due to earthmoving activities and from vehicle emissions. The effect would be highly localized, temporary. Use of mitigative measure such as applying water to unpaved roads would limit fugitive dust emissions.	No impact. There would be no long-term effect on air quality either from implementing the proposed action or following license termination.
Noise 4.8.1	There would be a SMALL impact. The nearest resident is located about	No impact.	No impact.	No impact. There would be no noise impact from implementing the proposed

8-8

Summary of Environmental Consequences

Table 8-1. Summary of Environmental Consequences

The Proposed Action (Alternative 1)

Impact Category (as applicable)	Unavoidable Adverse Environmental Impacts	Irreversible and Irretrievable Commitment of Resources	Short-term Impacts and Uses of the Environment	Long-Term Impacts and the Maintenance and Enhancement of Productivity
	4.5 km [2.8 mi] east of the center point of the proposed Moore Ranch Project and the site is in a remote location.			action following license termination.
Historic and Cultural 4.9.1	There would be a SMALL impact. There would be no impact to historical and cultural resources recommended eligible for listing on the *National Register of Historic Places*.	No impact.	No short-term impact to historic or cultural resources would be expected. If any unidentified historic or cultural resources are encountered, work would stop and appropriate federal and state officials be notified. Therefore, the potential impact to historic and cultural resources during operation of the facility would be SMALL.	No impact. There would be no impact to historic or cultural resources from the proposed action following license termination.
Visual and Scenic 4.10.1	There would be a SMALL impact on the visual landscape. The area surrounding the proposed Moore Ranch Project contains wellfields, pipelines, and utility lines associated with CBM development.	No impact.	There would be a SMALL short-term impact to the visual landscape from implementing the proposed action. The activities would be consistent with the BLM visual resource classification of the area and the existing natural resource extraction	No impact. There would be no impact to the visual landscape from the proposed action following license termination.

Summary of Environmental Consequences

Table 8-1. Summary of Environmental Consequences

Impact Category (as applicable)	Unavoidable Adverse Environmental Impacts	Irreversible and Irretrievable Commitment of Resources	Short-term Impacts and Uses of the Environment	Long-Term Impacts and the Maintenance and Enhancement of Productivity
The Proposed Action (Alternative 1)				
Socioeconomic 4.11.1	For each phase of the proposed Moore Ranch Project, the socioeconomic impact would be SMALL except during the operations phase when the potential impact to housing could range from SMALL to MODERATE.	Not applicable.	activities. Implementing the proposed action would predominantly have a SMALL impact on the local communities except for housing availability. Although jobs would be created and the purchase of goods and services would contribute to the Campbell County tax base, implementation of the proposed action could affect housing availability and result in a SMALL to MODERATE impact.	Following license termination, individuals who supported activities at the proposed Moore Ranch Project would need to find other employment and there would be a loss of revenue to Campbell County.
Environmental Justice 4.12.1	There would be no disproportionately high and adverse impacts to minority or low-income populations from the construction, operation, aquifer restoration, and decommissioning of the proposed Moore Ranch Project.	Not applicable	There would be no disproportionately high and adverse impacts to minority or low-income populations from the construction, operation, aquifer restoration, and decommissioning of the proposed Moore Ranch Project.	None.
Public and	There would be a SMALL	Not applicable.	There would be a SMALL	No impact. There would be no

8-10

Summary of Environmental Consequences

Table 8-1. Summary of Environmental Consequences

The Proposed Action (Alternative 1)

Impact Category (as applicable)	Unavoidable Adverse Environmental Impacts	Irreversible and Irretrievable Commitment of Resources	Short-term Impacts and Uses of the Environment	Long-Term Impacts and the Maintenance and Enhancement of Productivity
Occupational Health and Safety 4.13.1	impact on public and occupational health from implementing the proposed action. Construction and decommissioning would generate fugitive dust emissions that could result in a dose comparable to that from natural background exposure.		impact from radiological exposure comparable to that from natural background. The radiological impacts from accidents would be SMALL for workers if procedures to deal with accident scenarios were followed, and SMALL for the public because of the facility's remote location. The nonradiological public and occupational health impacts from normal operations, accidents, and chemical exposures would be SMALL if handling and storage procedures were followed.	long-term impact to public and occupational health following license termination.
Waste Management 4.14.1	Waste generation and disposal from all phases of the proposed Moore Ranch ISR activities would result in SMALL impacts on available disposal capacity, since permitted facilities are available to accept the wastes. Construction wastes would be mostly	The energy consumed during the ISR phases, the construction materials used that could not be reused or recycled, and the space used to properly handle and dispose of all waste types (i.e., wells for liquid wastes and permitted disposal space for solid	During all phases, hazards associated with handling and transport of wastes would represent a short-term and SMALL impact.	During all phases, the permanent disposal of wastes in on-site injection wells would represent a SMALL impact on the long-term productivity of the land allocated for these wells.

8-11

Summary of Environmental Consequences

Table 8-1. Summary of Environmental Consequences

The Proposed Action (Alternative 1)

Impact Category (as applicable)	Unavoidable Adverse Environmental Impacts	Irreversible and Irretrievable Commitment of Resources	Short-term Impacts and Uses of the Environment	Long-Term Impacts and the Maintenance and Enhancement of Productivity
	solids, operations wastes would include solids (primarily municipal waste) and liquids (brine, plant washdown water, and others), and decommissioning wastes would include a range of solid wastes (nonhazardous, hazardous, and solid byproduct materials).	wastes) would represent an irretrievable commitment of resources, resulting in a SMALL impact.		

8-12

9 LIST OF PREPARERS

This section documents all individuals who were involved with the preparation of this Supplemental Environmental Impact Statement (SEIS). Contributors include staff from the U.S. Nuclear Regulatory Commission (NRC) and consultants. Each individual's role, education, and experience are outlined below.

9.1 U.S. Nuclear Regulatory Commission Contributors

Behram Shroff: Supplemental Environmental Impact Statement (SEIS) Lead Project Manager
    M.C.P., City/Regional Planning, Georgia Institute of Technology, 1973
    M. S., Engineering, University of California, Berkeley, 1966
    Years of Experience: 29

Jennifer Davis: Socioeconomics and Cultural Resources
    B.A., Historic Preservation and Classical Civilization (Archaeology), Mary Washington College, 1996
    Years of Experience: 11

Nathan Goodman: Ecological Resources
    M.S., Environmental Science, Johns Hopkins University, 2000
    B.S., Biology, Muhlenberg College, 1998
    Years of Experience: 6

Asimios Malliakos: Cost-Benefit Analysis
    Ph.D., Nuclear Engineering, University of Missouri-Columbia, 1980
    M.S., Nuclear Engineering, Polytechnic Institute of New York, 1977
    B.S., Physics, University of Thessaloniki, Greece, 1975
    Years of Experience: 29

Douglas Mandeville: In-Situ Recovery Process
    M.S., Civil Engineering, Clarkson University, 1999
    B.S., Civil Engineering, Clarkson University, 1997
    Years of Experience: 10

Johari Moore: Public and Occupational Health
    M.S., Nuclear Engineering and Radiological Science, University of Michigan, 2005
    B.S., Physics, Florida A&M University, 2003
    Years of Experience: 4

James Park: Reviewer
    M. Ed., Childhood Education, Marymount University, 1999
    M.S., Structural Geology & Rock Mechanics, Imperial College, University of London, England, 1988
    B.S., Geology, Virginia Polytechnic Institute and State University, 1986
    Years of Experience: 15

Ashley Riffle: Reviewer
    B.S., Biology, Frostburg State University, 2009
    Years of Experience: 1

List of Preparers

Elise Striz: Groundwater Reviewer
    Ph. D., Petroleum Engineering, University of Oklahoma, 1998
    M.S. Civil Engineering, University of Oklahoma, 1985
    B.S., Environmental Engineering, Purdue University, 1981
    Years of Experience: 17

Patricia Swain: SEIS Project Manager
    M.S., Geology, University of California, Los Angeles, 1981
    B.S., Geology, University of Texas at Austin, 1976
    Years of Experience: 24

Jim Webb: Public and Occupational Health Reviewer
    M.S., Marketing and Communication, Franklin University, 2000
    M.B.A., Business Administration, Lake Erie College, 1983
    B.S., Radiological Health Physics, Lowell University, 1978
    Years of Experience: 31

9.2 Center for Nuclear Waste Regulatory Analyses (CNWRA®) Contractor Contributors

Hakan Basagaoglu: Water Resources, Cumulative Impacts
    Ph.D., Civil and Environmental Engineering, University of California, Davis 2000
    M.S., Geological Engineering, Middle East Technical University, Turkey 1993
    B.S., Geological Engineering, Middle East Technical University, Turkey 1991
    Years of Experience: 17

Philippe Dubreuilh: Land Use
    Ph.D., Geology, University of Bordeaux, France, 1982
    M.S., Geology, University of Bordeaux, France, 1977
    B.S., Geology, University of Bordeaux, France, 1976
    Years of Experience: 34

Amy Glovan: Ecological Resource
    B.A., Environmental Studies, University of Kansas, 1998
    Years of Experience: 11

Lane Howard: Occupational Health and Safety (Nonradiological and Radiological)
    M.S., Nuclear Engineering, Texas A&M University, 1995
    B.S., Civil Engineering, Texas A&M University, 1988
    Years of Experience: 22

Miriam Juckett: Principal Investigator
    M.S., Environmental Sciences, University of Texas at San Antonio, 2006
    B.S., Chemistry, University of Texas at San Antonio, 2003
    Years Experience: 7

Patrick LaPlante: Transportation and Air Quality
    M.S., Biostatistics and Epidemiology, Georgetown University, 1994
    B.S., Environmental Studies, Western Washington University, 1988
    Years of Experience: 22

James Myers: Waste Management
    Ph.D., Environmental Science & Engineering, Clemson University, 2004
    M.S., Geophysical Sciences, Georgia Institute of Technology, 1990
    B.S., Geology, Michigan State University, 1985
    Years of Experience: 25

Jude McMurry: Geology and Soils
    Ph.D., Geosciences, Texas Tech University, 1991
    M.A., Geological Sciences, University of Texas at Austin, 1982
    B.A., English, McMurry University, 1975
    Years Experience: 35

Robert Pauline: Cumulative Impacts
    M.S., Biology, Environmental Science and Policy, George Mason University, 1999
    B.S., Biology, Bates College, 1989
    Years Experience: 21

Marla Roberts: NEPA Reviewer
    M.S., Geology, University of Texas at San Antonio, 2007
    B.A., Geology, Vanderbilt University, 2001
    Years Experience: 9

John Stamatakos: Project Manager
    Ph.D., Geology, Lehigh University, 1990
    M.S., Geology, Lehigh University, 1988
    B.S., Geology, Franklin and Marshall College, 1981
    Years of Experience: 29

David Turner: Cumulative Impacts
    Ph.D., Geology, University of Utah, 1990
    M.S., Geology, University of Utah, 1985
    B.A., Music/Geology, College of William and Mary, 1981
    Years Experience: 29

Deborah Waiting: Geographic Information System (GIS) Analyst
    B.S., Geology, University of Texas at San Antonio, 1999
    Years Experience: 11

Gary Walter: Waste Management Resources
    Ph.D., Hydrology, University of Arizona, 1985
    M.A., Geology, University of Missouri, Columbia, 1974
    B.A., Chinese and Sociology, University of Kansas, 1969
    Years Experience: 41

Bradley Werling: NEPA Reviewer
    M.S., Environmental Science, University of Texas at San Antonio, 2000
    B.S., Chemistry, Southwest Texas State University, 1999
    B.A., Engineering Physics, Westmont College, Santa Barbara, 1985
    Years of Experience: 25

9.3   CNWRA Consultants and Subcontractors

Susan Courage: Socioeconomics and Environmental Justice
    M.S., Environmental Science, University of San Antonio at San Antonio, 2003
    B.S., Biology, University of San Antonio at San Antonio, 1999
    Years of Experience: 11

Eunice Fedors: Cultural and Historic Resources
    B.A., Historic Preservation, Mary Washington College, 1985
    Years of Experience: 25

<u>Raba-Kistner Inc.</u>

Samuel Blanco: Cumulative Impacts
    M.S., Environmental Science, Texas A&M University – Corpus Christi, 2002
    B.S., Biology, Texas A&M University, Kingsville, 1998
    Years of Experience: 10

10 DISTRIBUTION LIST

The U.S. Nuclear Regulatory Commission (NRC) is providing copies of this Supplemental Environmental Impact Statement (SEIS) to the organizations and individuals listed below. The NRC will provide copies to other interested organizations and individuals upon request.

10.1 Federal Agency Officials

James Hanley
    U.S. Environmental Protection Agency
    Region 8
    Denver, CO

Carol Rushin
    U.S. Environmental Protection Agency
    Region 8
    Denver, CO

Larry Svoboda
    U.S. Environmental Protection Agency
    Region 8
    Denver, CO

Brian Kelly
    U.S. Fish and Wildlife Service
    Mountain-Prairie Region
    Wyoming Field Office
    Cheyenne, WY

Ramon Nation
    Bureau of Indian Affairs
    Wind River Agency
    Ft. Washakie, WY

Gerald Queen
    Bureau of Land Management
    Buffalo Field Office
    Buffalo, WY

Pete Sokolosky
    Bureau of Land Management
    State Office
    Cheyenne, WY

Robert Stewart
    U.S. Department of the Interior
    Office of the Secretary
    Office of Environmental Policy and Compliance
    Denver, CO

Distribution List

10.2 Tribal Government Officials

Perry Brady
    Three Affiliated Tribes
    Tribal Historic Preservation Office
    New Town, ND

Conrad Fisher
    Northern Cheyenne
    Tribal Historic Preservation Office
    Lame Deer, MT

John Murray
    Blackfeet
    Tribal Historic Preservation Office
    Browning, MT

Dale Old Horn
    Crow
    Tribal Historic Preservation Office
    Crow Agency, MT

Donna Rae Peterson
    Cheyenne River Sioux
    Tribal Historic Preservation Office
    Eagle Butte, SD

Arlen Shoyo
    Eastern Shoshone
    Tribal Historic Preservation Office
    Ft. Washakie, WY

Joanne White
    Arapaho
    Tribal Historic Preservation Office
    Ft. Washakie, WY

Joyce Whiting
    Oglala Sioux
    Tribal Historic Preservation Office
    Pine Ridge, SD

Darrel Youpee
    Ft. Peck
    Tribal Historic Preservation Office
    Poplar, MT

10.3 State Agency Officials

Richard Currit
> Wyoming State Historic Preservation Office
> Cheyenne, WY

Don McKenzie
> Wyoming Department of Environmental Quality
> Cheyenne, WY

Mark Rogaczewski
> Wyoming Department of Environmental Quality
> Sheridan District
> Sheridan, WY

10.4 Local Agency Officials

Craig Collins
> Casper Planning Department
> Casper, WY

Joe Coyne
> Converse Area New Development Organization
> Douglas, WY

David Haney
> Wyoming Community Development Authority
> Casper, WY

Lyle Murdock
> Building Department
> Wright, WY

Robert Palmer
> Campbell County Commissioners
> Gillette, WY

Forrest Neuerberg
> Converse County Planning Department
> Douglas, WY

Michael Surface
> Campbell County Economic Development Commission
> Gillette, WY

Distribution List

10.5 Other Organizations and Individuals

Shannon Anderson
    Powder River Basin Resource Council
    Sheridan, WY

Geoffrey Fettus
    Natural Resources Defense Council
    Washington, DC

Cori Lombard
    Natural Resources Defense Council
    Washington, DC

Sarah Fields
    Sierra Club – Glen Canyon Group
    Salt Lake City, UT

Michael Griffin
    Uranium One Inc.
    Casper, WY

Eric Jantz
    New Mexico Environmental Law Center
    Santa Fe, NM

Jim Jones
    Hulett, WY

Steve Jones
    Wyoming Outdoor Council
    Lander, WY

Marion Loomis
    Wyoming Mining Association
    Cheyenne, WY

Wayne Prindle
    Biodiversity Conservation Alliance
    Laramie, WY

Jonathan B. Rattner
    Western Watersheds Project
    Pinedale, WY

Christopher S. Pugsley, Esq. (on behalf of National Mining Association)
    Thompson & Simmons, PLLC
    Washington, DC

Pam Viviano
> Ranchers & Neighbors Protecting Our Water
> Sheridan, WY

Jon Winters
> Uranium One Americas
> Casper, WY

Tony Thompson
> National Mining Association
> Washington, DC

APPENDIX A

CONSULTATION CORRESPONDENCE

A Consultation Correspondence

The Endangered Species Act of 1973, as amended, and the National Historic Preservation Act of 1966 require that Federal agencies consult with applicable state and federal agencies and groups prior to taking action that may affect threatened and endangered species, essential fish habitat, or historic and archaeological resources, respectively. This appendix contains consultation documentation related to these federal acts.

Table A1–1. Chronology of Consultation Correspondence		
Author	**Recipient**	**Date of Letter**
U.S. Nuclear Regulatory Commission (G. Suber)	Wyoming State Historic Preservation Office (M. Hopkins)	April 9, 2008
U.S. Nuclear Regulatory Commission (G. Suber)	U.S. Fish and Wildlife Service (B. Kelly)	April 9, 2008
U.S. Fish and Wildlife Service (B. Kelly)	U.S. Nuclear Regulatory Commission (G. Suber)	May 7, 2008
Wyoming State Parks and Cultural Resources (R. Currit)	U.S. Nuclear Regulatory Commission (G. Suber)	June 5, 2008
U.S. Nuclear Regulatory Commission (A. Kock)	Shoshone Business Council (I. Posey)	February 23, 2009
U.S. Nuclear Regulatory Commission (I. Yu, B. Shroff, and A. Bjornsen)	U.S. Nuclear Regulatory Commission (A. Kock)	March 2, 2009
U.S. Nuclear Regulatory Commission (A. Kock)	Wyoming Game and Fish Department (T. Christiansen)	August 5, 2009
Wyoming Game and Fish Department (T. Christiansen)	U.S. Nuclear Regulatory Commission (A. Kock)	September 3, 2009
U.S. Nuclear Regulatory Commission (A. Kock)	Wyoming State Historic Preservation Office (R. Currit)	October 22, 2009
Wyoming State Parks & Cultural Resources (R. Currit)	U.S. Nuclear Regulatory Commission (A. Kock)	November 3, 2009
Wyoming State Parks and Cultural Resources (R. Currit)	U.S. Nuclear Regulatory Commission (J. Davis)	June 15, 2010

Appendix A

April 9, 2008

Ms. Mary Hopkins
State Historic Preservation Officer
Wyoming State Historic Preservation Office
Department of State Parks
& Cultural Resources
2301 Central Avenue, Barrett Building
3rd Floor
Cheyenne, Wyoming 82002

SUBJECT: INITIATION OF SECTION 106 PROCESS FOR ENERGY METALS CORPORATION'S MOORE RANCH URANIUM RECOVERY PROJECT LICENSE REQUEST (Docket 40-9073)

Dear Ms. Hopkins:

The U.S. Nuclear Regulatory Commission (NRC) has received an application from Energy Metals Corporation for a new radioactive source materials license to develop and operate the Moore Ranch Uranium Recovery Project (an in-situ uranium recovery operation) located in Campbell County, WY. The proposed project will consist of injection/production wellfields, a central plant with ion exchange, resin unloading, elution, precipitation, and yellowcake drying capabilities, and deep injection disposal wells. The Moore Ranch Uranium Recovery Project is located in Township 42 North, Range 75 West, Sections 26, 27, 33, 34, 35, and 36, and Township 41 North, Range 75 West, Sections 1, 2, 3, and 4, and Township 42 North, Range 74 West, Section 31. Maps showing the boundaries of the Moore Ranch Uranium Recovery Project area are enclosed.

In-situ leach mining involves injecting a carbonate/bicarbonate leaching solution and oxidant into a subsurface uranium ore body to release the uranium and then pumping the uranium bearing solution to the surface for further processing to remove and concentrate the uranium.

Surface-disturbing activities associated with the proposed mining in the Moore Ranch Uranium Recovery Project area would involve, at a minimum (1) construction of a 11 acre central processing plant facility, (2) the laying of about 8000 linear feet of pipeline five feet below the ground surface to transport the leaching solution and the uranium-bearing solution; (3) drilling of multiple injection, production, and monitoring wells; and (4) construction of access roads as needed. The proposed total wellfield area to be used for injection and recovery of solution over the ten-year life of the project is approximately 150 acres.

As established in Title 10 Code of Federal Regulations Part 51 (10 CFR 51), the NRC regulation that implements the National Environmental Policy Act of 1969, as amended, the NRC is preparing an environmental assessment (EA) for the proposed action that will tier off a Generic Environmental Impact Statement currently under development. In accordance with Section 106 of the National Historic Preservation Act, the EA will include an analysis of potential impacts to historic and cultural resources. To support the environmental review, the NRC is requesting information from the State Historical Preservation Officer to facilitate the identification of historic and cultural resources that may be affected by the Moore Ranch Uranium Recovery Project

Appendix A

M. Hopkins 2

license application. Any information you provide will be used to document effects in accordance with 36 CFR 800.8(c). After reviewing all the information collected, the NRC will follow up with your office regarding compliance with the National Historic Preservation Act of 1966, as amended, and the Section 106 consultation process.

The Moore Ranch Uranium Recovery Project license application is publicly available in the NRC Public Document Room (PDR) located at One White Flint North, 11555 Rockville Pike, Rockville, Maryland 20852, or from the NRC's Agency wide Documents Access and Management System (ADAMS). The ADAMS Public Electronic Reading Room is accessible at http://www.nrc.gov/reading-rm/adams.html. The accession number for the application is ML072851222.

Please submit any comments concerning this environmental review by mail to the NRC Commission Attn: Mr. Gregory Suber, Mail Stop T-8F05, Washington, DC 20555 within 30 days of the receipt of this letter. If you have any questions, please contact Mr. Behram Shroff of my staff by telephone at 301-415-0666 or by email at bps2@nrc.gov. Thank you for your assistance.

Sincerely,

/RA, by Allen Fetter for/

Gregory F. Suber, Branch Chief
Environmental Review Branch
Environmental Protection
 and Performance Assessment Directorate
Division of Waste Management
 and Environmental Protection
Office of Federal and State Materials
 and Environmental Management Programs

Enclosure: Maps of Moore Ranch
 Uranium Recovery Project

Docket No.: 40-9073

Appendix A

April 9, 2008

Brian T. Kelly, Field Supervisor
U.S. Fish and Wildlife Service
Mountain-Prairie Region
Wyoming Field Office
5353 Yellowstone Road
Cheyenne, WY 82009

SUBJECT: REQUEST FOR INFORMATION REGARDING ENDANGERED OR THREATENED SPECIES AND CRITICAL HABITAT FOR THE PROPOSED LICENSE APPLICATION FOR ENERGY METALS CORPORATION'S MOORE RANCH URANIUM RECOVERY PROJECT (Docket 40-9073)

Dear Mr. Kelly:

The U.S. Nuclear Regulatory Commission (NRC) has received an application from Energy Metals Corporation for a new radioactive source materials license to develop and operate the Moore Ranch Uranium Recovery Project (an in-situ recovery operation) located in Campbell County, WY. The proposed project will consist of injection/production wellfields, a central plant with ion exchange; resin unloading, elution, precipitation, and yellowcake drying capabilities, and deep injection disposal wells. The Moore Ranch Uranium Recovery Project is located in Township 42 North, Range 75 West, Sections 26, 27, 33, 34, 35, and 36, and Township 41 North, Range 75 West, Sections 1, 2, 3, and 4, and Township 42 North, Range 74 West, Section 31. Maps showing the boundaries of the area are enclosed.

As established in Title 10 *Code of Federal Regulations* Part 51 (10 CFR 51), the NRC regulation that implements the National Environmental Policy Act of 1969, as amended, the agency is preparing an environmental assessment (EA) for the proposed action that will tier off a Generic Environmental Impact Statement currently under development. In accordance with Section 7 of the Endangered Species Act, the EA will include an analysis of potential impacts to endangered or threatened species or critical habitat in the action area. To support the environmental review, the NRC is requesting information from the U.S. Fish and Wildlife Service to facilitate the identification of endangered or threatened species or critical habitat that may be affected by the proposed project. Any information you provide will be used to enhance the scope and quality of our review in accordance with 10 CFR 51 and 50 CFR 402. After assessing the information provided by you, the NRC will determine what additional actions are necessary to comply with Section 7 of the Endangered Species Act.

The Energy Metals Moore Ranch Recovery Project license application is publicly available in the NRC Public Document Room (PDR) located at One White Flint North, 11555 Rockville Pike, Rockville, Maryland 20852, or from the NRC's Agency wide Documents Access and Management System (ADAMS). The ADAMS Public Electronic Reading Room is accessible at http://www.nrc.gov/reading-rm/adams.html. The accession number for the application is ML072851222.

Appendix A

B. Kelly   2

Please submit any comments/information that you have regarding this environmental review to me at the U. S. Nuclear Regulatory Commission, Attention: Mr. Gregory Suber, Mail Stop T8F05, Washington, DC 20555 within 30 days of the receipt of this letter. If you have any questions, please contact Mr. Behram Shroff of my staff by telephone at 301-415-0666 or by email at bps2@nrc.gov. Thank you for your assistance.

                            Sincerely,

                            /RA, by Allen Fetter for/

                            Gregory F. Suber, Branch Chief
                            Environmental Review Branch
                            Environmental Protection
                              and Performance Assessment Directorate
                            Division of Waste Management
                              and Environmental Protection
                            Office of Federal and State Materials
                              and Environmental Management Programs

Enclosure: Map of Moore Ranch
                 Uranium Recovery Project

Docket No.: 40-9073

Appendix A

United States Department of the Interior

FISH AND WILDLIFE SERVICE

Ecological Services
5353 Yellowstone Road – Suite 308
Cheyenne, Wyoming 82009

In Reply Refer To:

ES/61411/W.26 /WY08SL0166

MAY - 7 2008

Gregory F. Suber, Branch Chief
Environmental Review Branch
Office of Federal and State Materials and
Environmental Management Programs
U.S. Nuclear Regulatory Commission
Mail Stop T8F05
Washington, D.C. 20555-0001

Dear Mr. Suber:

Thank you for your letter of April 9, 2008, received in our office on April 14, regarding the permit application by Energy Metals Corporation for a uranium in-situ recovery facility in Campbell County, Wyoming. The proposed Moore Ranch Uranium Recovery Project will consist of injection/production wellfields, a central plant with ion exchange; resin unloading, elution, precipitation, and yellowcake drying capabilities; and deep injection disposal wells. Your letter requested that we provide information concerning endangered or threatened species or critical habitat that may be affected by the proposed project.

In response to your letter, the U.S. Fish and Wildlife Service (Service) is providing you with information on (1) federally listed species, (2) migratory birds, (3) wetland and riparian areas, and (4) sensitive species. The Service provides recommendations for protective measures for federally listed species in accordance with the Endangered Species Act (Act) of 1973, as amended (16 U.S.C. 1531 *et seq.*). Protective measures for migratory birds are provided in accordance with the Migratory Bird Treaty Act (MBTA), 16 U.S.C. 703 and the Bald and Golden Eagle Protection Act (BGEPA), 16 U.S.C. 668. Wetlands are afforded protection under Executive Orders 11990 (wetland protection) and 11988 (floodplain management), as well as section 404 of the Clean Water Act. Other fish and wildlife resources are considered under the Fish and Wildlife Coordination Act and the Fish and Wildlife Act (FWCA) of 1956, as amended, 70 Stat. 1119, 16 U.S.C. 742a-742j.

Threatened and Endangered Species

- The following threatened and endangered species may occur in Campbell County, and could also occur on or near this project site. If you determine that the proposed project may affect any of the following listed species, please contact our office to discuss consultation requirements under the Act.

1

Species	Status	Habitat
Black-footed ferret (*Mustela nigripes*)	Endangered	Prairie dog towns
Ute ladies'-tresses (*Spiranthes diluvialis*)	Threatened	Moist soils and wet meadows of drainages below 7000 feet

Black-footed ferret: Black-footed ferrets may be affected if prairie dog towns are impacted. Please be aware that black-footed ferret surveys are no longer recommended in black-tailed prairie dog towns statewide. However, we encourage you to protect all prairie dog towns for their value to the prairie ecosystem and the myriad of species that rely on them. If a field check indicates that prairie dog towns may be affected, you should contact this office for guidance on ferret surveys.

Ute ladies'-tresses: Ute ladies'-tresses is a perennial, terrestrial orchid, 8 to 20 inches tall, with white or ivory flowers clustered into a spike arrangement at the top of the stem. *S. diluvialis* typically blooms from late July through August; however, depending on location and climatic conditions, it may bloom in early July or still be in flower as late as early October. *S. diluvialis* is endemic to moist soils near wetland meadows, springs, lakes, and perennial streams where it colonizes early successional point bars or sandy edges. The elevation range of known occurrences is 4,200 to 7,000 feet (although no known populations in Wyoming occur above 5,500 feet) in alluvial substrates along riparian edges, gravel bars, old oxbows, and moist to wet meadows. Soils where *S. diluvialis* have been found typically include fine silt/sand, gravels and cobbles, and highly organic, peaty soil types. *S. diluvialis* is not found in heavy or tight clay soils or in extremely saline or alkaline soils. *S. diluvialis* seems intolerant of shade and small scattered groups are found primarily in areas where vegetation is relatively open. Surveys should be conducted by knowledgeable botanists trained in conducting rare plant surveys. *S. diluvialis* is difficult to survey for primarily due to its unpredictability of emergence of flowering parts and subsequent rapid desiccation of specimens.

Migratory Birds

Please recognize that consultation on listed species may not remove your obligation to protect the many species of migratory birds, including eagles and other raptors, protected under the MBTA and BGEPA. Of particular focus are the species identified in the Service's *Birds of Conservation Concern 2002*. In accordance with the FWCA (16 USC 2912 (a)(3)), this report identifies "species, subspecies, and populations of all migratory nongame birds that, without additional conservation actions, are likely to become candidates for listing" under the Act. This report is intended to stimulate coordinated and proactive conservation actions among Federal, State, and private partners and is available at http://www.fws.gov/migratorybirds/reports/bcc2002.pdf.

The MBTA, enacted in 1918, prohibits the taking of any migratory birds, their parts, nests, or eggs except as permitted by regulations and does not require intent to be proven. Section 703

Appendix A

of the MBTA states, "Unless and except as permitted by regulations ... it shall be unlawful at any time, by any means or in any manner, to ... take, capture, kill, attempt to take, capture, or kill, or possess ... any migratory bird, any part, nest, or eggs of any such bird..." The BGEPA, prohibits knowingly taking, or taking with wanton disregard for the consequences of an activity, any bald or golden eagles or their body parts, nests, or eggs, which includes collection, molestation, disturbance, or killing.

In order to promote the conservation of migratory bird populations and their habitats, the Service recommends that your agency implement those strategies outlined within the Memorandum of Understanding directed by the President of the U.S. under the Executive Order 13186, where possible. Work that could lead to the take of a migratory bird or eagle, their young, eggs, or nests (for example, if you are going to erect new roads, or power lines in the vicinity of a nest), should be coordinated with our office before any actions are taken.

In situ Uranium Mining

High selenium concentrations can occur in wastewater from in situ mining of uranium ore as uranium-bearing formations are usually associated with seleniferous strata (Boon 1989). The disposal of this wastewater can expose migratory birds to selenium which is known to cause impaired reproduction and mortality in sensitive species of birds such as waterfowl.

The in situ mining wastewater is typically disposed of through deep-well injection or discharge into large evaporation ponds. One mining operation in Converse County disposes of the wastewater through land application using center-pivot irrigation after treatment for removal of uranium and radium.

In 1998, the Service conducted a study of a grassland irrigated with wastewater from an *in situ* uranium mine and found that selenium was mobilized into the food chain and bioaccumulated by grasshoppers and songbirds (Ramirez and Rogers 2002). Disposal of the *in situ* wastewater through irrigation is not recommended by the Service due to the potential for selenium bioaccumulation in the food chain and adverse effects to migratory birds. Additionally, land application may result in the contamination of groundwater and eventually seep out and reach surface waters. Additionally, the selenium-contaminated groundwater could seep into low areas or basins in upland sites and create wetlands which would attract migratory birds and other wildlife.

The Service is also concerned with the potential for elevated selenium in evaporation ponds receiving *in situ* wastewater. Waterborne selenium concentrations ≥ 2 µg/L are considered hazardous to the health and long-term survival of fish and wildlife (Lemly 1996). Additionally, water with more than 20 µg/L is considered hazardous to aquatic birds (Skorupa and Ohlendorf 1991). Chronic effects of selenium manifest themselves in immune suppression to birds (Fairbrother et al. 1994) which can make affected birds more susceptible to disease and predation. Selenium toxicity will also cause embryonic deformities and mortality (See et al. 1992, Skorupa and Ohlendorf 1991, Ohlendorf 2002)

If submerged aquatic vegetation and/or aquatic invertebrates are present in evaporation ponds with high waterborne selenium concentrations, extremely high dietary levels of this contaminant can be available to aquatic migratory birds. Ramirez and Rogers (2000) documented selenium concentrations ranging from 434 to 508 µg/g in pondweed (*Potamogeton vaginatus*) collected from a uranium mine wastewater storage reservoir that had waterborne selenium concentrations ranging from 260 to 350 µg/L.

Wetlands/Riparian Areas

The proposed project area includes tributaries to Ninemile Creek. Wetlands perform significant ecological functions, which include: (1) providing habitat for aquatic and terrestrial wildlife species, (2) aiding in the dispersal of floods, (3) improving water quality through retention and assimilation of pollutants from storm water runoff, and (4) recharging the aquifer. Wetlands also possess aesthetic and recreational values. The Service recommends measures be taken to avoid and minimize wetland losses in accordance with Section 404 of the Clean Water Act, and Executive Order 11988 (floodplain management) as well as the goal of "no net loss of wetlands." If wetlands may be destroyed or degraded by the proposed action, those wetlands in the project area should be inventoried and fully described in terms of their functions and values. Acreage of wetlands, by type, should be disclosed and specific actions should be outlined to avoid, minimize, and compensate for all unavoidable wetland impacts.

Riparian or streamside areas are a valuable natural resource and impacts to these areas should be avoided whenever possible. Riparian areas are the single most productive wildlife habitat type in North America. They support a greater variety of wildlife than any other habitat. Riparian vegetation plays an important role in protecting streams, reducing erosion and sedimentation as well as improving water quality, maintaining the water table, controlling flooding, and providing shade and cover. In view of their importance and relative scarcity, impacts to riparian areas should be avoided. Any potential, unavoidable encroachment into these areas should be further avoided and minimized. Unavoidable impacts to streams should be assessed in terms of their functions and values, linear feet and vegetation type lost, potential effects on wildlife, and potential effects on bank stability and water quality. Measures to compensate for unavoidable losses of riparian areas should be developed and implemented as part of the project.

Plans for mitigating unavoidable impacts to wetland and riparian areas should include mitigation goals and objectives, methodologies, time frames for implementation, success criteria, and monitoring to determine if the mitigation is successful. The mitigation plan should also include a contingency plan to be implemented should the mitigation not be successful. In addition, wetland restoration, creation, enhancement, and/or preservation does not compensate for loss of stream habitat; streams and wetlands have different functions and provide different habitat values for fish and wildlife resources.

Best Management Practices (BMPs) should be implemented within the project area wherever possible. BMPs include, but are not limited to, the following: installation of sediment and erosion control devices (e.g., silt fences, hay bales, temporary sediment control basins,

Appendix A

erosion control matting); adequate and continued maintenance of sediment and erosion control devices to insure their effectiveness; minimization of the construction disturbance area to further avoid streams, wetlands, and riparian areas; location of equipment staging, fueling, and maintenance areas outside of wetlands, streams, riparian areas, and floodplains; and re-seeding and re-planting of riparian vegetation native to Wyoming in order to stabilize shorelines and stream banks.

Sensitive Species

Mountain Plover: Although the Service has withdrawn the proposal to list the mountain plover (*Charadrius montanus*) and we will no longer be reviewing project impacts to this species under the Act, we continue to encourage conservation of this species as it remains protected under the MBTA. Measures to protect the mountain plover from further decline may include (1) avoidance of suitable habitat during the plover nesting season (April 10 through July 10), (2) prohibition of ground disturbing activities in prairie dog towns, and (3) prohibition of any permanent above ground structures that may provide perches for avian predators or deter plovers from using preferred habitat. Suitable habitat for nesting mountain plovers includes grasslands, mixed grassland areas and short-grass prairie, shrub-steppe, plains, alkali flats, agricultural lands, cultivated lands, sod farms, and prairie dog towns. We strongly encourage the development of protective measures with an assurance of implementation should mountain plovers be found within the project area.

Greater Sage-grouse: The Service is currently conducting a review to determine if the greater sage-grouse (*Centrocercus urophasianus*) warrants listing. Greater sage-grouse are dependent on sagebrush habitats year-round. Habitat loss and degradation, as well as loss of population connectivity have been identified as important factors contributing to the decline of greater sage-grouse populations rangewide (Braun 1998, Wisdom *et al.* 2002). Therefore, any activities that result in loss or degradation of sagebrush habitats that are important to this species should be closely evaluated for their impacts to sage-grouse. If important breeding habitat (leks, nesting, or brood rearing habitat) is present in the project area, the Service recommends no project-related disturbance March 1 through June 30, annually. Minimization of disturbance during lek activity, nesting, and brood rearing is critical to sage-grouse persistence within these areas. Likewise, if important winter habitats are present (Doherty *et al.* 2008), we recommend no project-related disturbance November 15 through March 14, annually.

We recommend you contact the Wyoming Game and Fish Department to identify important greater sage-grouse habitats within the project area, and appropriate mitigative measures to minimize potential impacts from the proposed project. The Service recommends surveys and mapping of important greater sage-grouse habitats where local information is not available. The results of these surveys should be used in project planning, to minimize potential impacts to this species. No project activities that may exacerbate habitat loss or degradation should be permitted in important habitats. Additionally, unless site-specific information is available, greater sage-grouse habitat should be managed following the guidelines by Connelly *et al.* 2000 (also known as the WAFWA guidelines).

Appendix A

We appreciate your efforts to ensure the conservation of Wyoming's fish and wildlife resources. If you have questions regarding this letter or your responsibilities under the Act, MBTA or BGEPA, please contact Pedro 'Pete' Ramirez at the letterhead address or phone (307) 772-2374, extension 236.

Sincerely,

Scott Hicks

for Brian T. Kelly
Field Supervisor
Wyoming Field Office

cc: WGFD, Non-game Coordinator, Lander, WY (B. Oakleaf)
WGFD, Statewide Habitat Protection Coordinator, Cheyenne, WY (V. Stelter)

Literature Cited

Boon, D.Y. 1989. Potential selenium problems in Great Plains soils. In L.W. Jacobs, ed. Selenium in agriculture and the environment. American Society of Agronomy, Inc, and Soil Science Society of America. SSSA Special Pub. No. 23. Madison, WI. pp: 107-121.

Braun, C.E. 1998. Sage grouse declines in western North America: What are the problems? Proceedings of the Western Association of Fish and Wildlife Agencies 78:139-156

Connelly J.W., M.A. Schroeder, A.R. Sands, and C.E. Braun. 2000. Guidelines to manage sage-grouse populations and their habitats. Wildlife Society Bulletin 28(4): 967 - 985.

Doherty, K.E., D. E. Naugle, B.L. Walker, and J.M. Graham. 2008. Greater sage-grouse winter habitat selection and energy development. J. Wildl. Manage. 72(1): 187-195.

Fairbrother, A.F., M. Fix, T. O'Hara, and C.A. Ribic. 1994. Impairment of growth and immune function of avocet chicks from sites with elevated selenium, arsenic, and boron. Journal of Wildlife Diseases. 30(2):222-233.

Lemly, A.D. 1996. Selenium in aquatic organisms. Pages 427-445 *in* W.N. Beyer, G.H. Heinz, and A.W. Redmon-Norwood (eds.). Environmental contaminants in wildlife: Interpreting tissue concentrations. Lewis Publishers, Boca Raton, Florida.

Ohlendorf, H.M. 2002. Ecotoxicology of selenium. In *Handbook of Ecotoxicology*, 2nd ed.; Hoffman, D.J., Rattner, B.A., Burton Jr., G.A., Cairns, Jr., J., Eds.; Lewis Publishers, Boca Raton, FL, 2003; pp 465-500.

6

Appendix A

Ramirez, P. and B. Rogers. 2000. Selenium in a Wyoming grassland community receiving wastewater from an in situ uranium mine. U.S. Fish and Wildlife Service Contaminant Report # R6/715C/00. Cheyenne, WY. Sept. 31.

Ramirez, P. Jr. and B.P. Rogers. 2002. Selenium in a Wyoming grassland community receiving wastewater from an *in situ* uranium mine. Arch. Environ. Contam. Toxicol. 42:431-436.

See, R.B., D.L. Naftz, D.A. Peterson, J.G. Crock, J.A. Erdman, R.C. Severson, P. Ramirez, Jr., and J.A. Armstrong. 1992. Detailed study of selenium in soil, representative plants, water, bottom sediment, and biota in the Kendrick Reclamation Project Area, Wyoming, 1988-90. U.S. Geological Survey Water Resources Investigations Report 91-4131. 142 pp.

Skorupa, J.P., and H.M. Ohlendorf. 1991. Contaminants in drainage water and avian risk thresholds. Pages 345-368 *in* A. Dinar and D. Zilberman (eds.). The economics and management of water and drainage in agriculture. Kluwer Academic Publishers, Boston, MA.

U.S. Fish and Wildlife Service. 1989. Black-footed ferret survey guidelines for compliance with the Endangered Species Act, April 1989. U. S. Fish and Wildlife Service, Denver, Colorado and Albuquerque, New Mexico. 15pp.

Wisdom, M.J., B.C. Wales, M.M. Rowland, M.G. Raphael, R.S. Holthausen, T.D. Rich, and V.A. Saab. 2002. Performance of Greater Sage-Grouse models for conservation assessment in the Interior Columbia Basin, USA. Conservation Biology 16: 1232-1242.

Appendix A

ARTS. PARKS. HISTORY.
Wyoming State Parks & Cultural Resources

State Historic Preservation Office
Barrett Building, 3rd Floor
2301 Central Avenue
Cheyenne, WY 82002
Phone: (307) 777-7697
Fax: (307) 777-6421
http://wyoshpo.state.wy.us

June 5, 2008

Gregory F. Suber, Branch Chief
Nuclear Regulatory Commission
Mail Stop T-8F05
Washington, DC 20555

re: Energy Metals Corporation, Initiation of Section 106 Process for the Moore Ranch Uranium Recovery Project License Request (Docket 40-9073) (SHPO File # 0608RLC007)

Dear Mr. Suber:

Thank you for consulting with the Wyoming State Historic Preservation Office (SHPO) regarding the above referenced project.

A search of our records shows that a cultural resource survey has not been conducted for the entire area of potential effect (APE). However, previous surveys of portions of the APE have been completed and have identified numerous historic properties. The proposed project has the potential to adversely affect these properties. Following 36 CFR Part 800, and prior to any ground disturbing activities, we recommend the Nuclear Regulatory Commission carry out appropriate efforts necessary for identification of historic properties, which may include a file search, background research, consultation, consideration of visual effects, sample field investigations or field survey. The identification efforts must be conducted by a consultant meeting the Secretary of the Interior's Professional Qualification Standards (48 FR 22716, Sept. 1983). A report detailing the results of these efforts must be provided to SHPO staff for our review and comment.

We have enclosed a copy of a cultural resource consultants list for your use. Please refer to SHPO project control number #0608RLC007 on any future correspondence dealing with this project. If you have any questions, please contact me at 307-777-5497 or by email at rcurri@state.wy.us.

Sincerely,

Richard L. Currit
Senior Archaeologist

Dave Freudenthal, Governor
Milward Simpson, Director

Appendix A

February 23, 2009

Mr. Ivan Posey
Chairman
Shoshone Business Council
P.O. Box 538
Fort Washakie, WY 82514

SUBJECT: REQUEST FOR INFORMATION REGARDING TRIBAL HISTORIC AND CULTURAL RESOURCES POTENTIALLY AFFECTED BY THE PROPOSED LICENSE APPLICATION FOR URANIUM 1 INC'S MOORE RANCH URANIUM RECOVERY PROJECT IN CAMPBELL COUNTY, WYOMING
(Docket No. 040-09073)

Dear Mr. Posey:

The U.S. Nuclear Regulatory Commission (NRC) has received an application from Uranium 1 Inc. for a new radioactive source materials license to construct and operate the Moore Ranch Uranium Recovery Project (an *in-situ* recovery operation) located in Campbell County, WY. The proposed project will consist of injection/production wellfields, a central plant with ion exchange: resin unloading, elution, precipitation, and yellowcake drying capabilities, and deep injection disposal wells. The Moore Ranch Uranium Recovery Project is located in Township 42 North, Range 75 West, Sections 26, 27, 33, 34, 35, and 36, and Township 41 North, Range 75 West, Sections 1, 2, 3, and 4, and Township 42 North, Range 74 West, Section 31. Maps showing the boundaries of the area are enclosed.

As established in Title 10 Code of Federal Regulations Part 51 (10 CFR 51), the NRC regulation that implements the National Environmental Policy Act of 1969, as amended, the NRC is preparing an Environmental Assessment (EA) for the proposed action that will tier off a Generic Environmental Impact Statement currently under development. The NRC's EA process includes an opportunity for public and inter-governmental participation in the development of the EA. In accordance with Section 106 of the National Historic Preservation Act, the EA will include an analysis of potential impacts to historic and cultural properties. To support the environmental review, the NRC is requesting information to facilitate the identification of tribal historic sites or cultural resources that may be affected by the proposed Moore Ranch Uranium Recovery Project. Specifically, the NRC is interested in learning of any sites that you believe have traditional religious or cultural significance. Any input you provide will be used to enhance the scope and quality of our review in accordance with 10 CFR 51 and 36 CFR 800. After reviewing all of the information collected, the NRC will prepare a draft EA and provide your office with an opportunity to comment.

Uranium 1 Inc.'s Moore Ranch Uranium Recovery Project license application is publicly available in the NRC Public Document Room located at One White Flint North, 11555 Rockville Pike, Rockville, Maryland 20852, or from the NRC's Agencywide Documents Access and Management System (ADAMS). The ADAMS Public Electronic Reading Room is accessible at http://www.nrc.gov/reading-rm/adams.html. The accession number for the application is ML072851222.

I. Posey   - 2 -

Please submit any comments/information that you may have regarding this environmental review within 30 days of the receipt of this letter to the U.S. Nuclear Regulatory Commission Attn: Ms. Andrea Kock, Mail Stop T-8F05, Washington, DC 20555. If you have any questions, please contact Mr. Behram Shroff of my staff by telephone at 301-415-0666 or by email at Behram.Shroff@nrc.gov. Thank you for your assistance.

Sincerely,

/RA/

Andrea Kock, Branch Chief
Environmental Review Branch
Environmental Protection
  and Performance Assessment Directorate
Division of Waste Management
  and Environmental Protection
Office of Federal and State Materials
  and Environmental Management Programs

Enclosures:
1. Map of Moore Ranch Uranium Project
2. Regional Map of Moore Ranch Uranium Project

cc w/enclosures: See next page

Appendix A

March 2, 2009

MEMORANDUM TO: Andrea Kock, Chief
                 Environmental Review Branch
                 EPPAD/DWMEP/FSME

FROM: Irene W. Yu, Project Manager   /RA/
                 Environmental Review Branch
                 EPPAD/DWMEP/FSME

                 Behram Shroff, Project Manager /RA/
                 Environmental Review Branch
                 EPPAD/DWMEP/FSME

                 Alan Bjornsen, Project Manager /RA/
                 Environmental Review Branch
                 EPPAD/DWMEP/FSME

SUBJECT: INFORMAL MEETINGS WITH LOCAL, STATE, AND FEDERAL
                 AGENCIES IN WYOMING REGARDING THE ENVIRONMENTAL
                 REVIEWS BEING CONDUCTED ON THE MOORE RANCH,
                 NICHOLS RANCH, AND LOST CREEK IN-SITU LEACH
                 APPLICATIONS FOR SOURCE MATERIAL LICENSES
                 (DOCKET NOS. 040-09073, 040-09067, 040-09068,
                 RESPECTIVELY)

During the week of January 12, 2008, the U.S. Nuclear Regulatory Commission (NRC) staff and their contractor staff informally met with various local, state, and federal agencies in Wyoming regarding the environmental reviews being conducted on the Moore Ranch, Nichols Ranch, and Lost Creek In-Situ Leach (ISL) applications for Source Material Licenses. The purpose of these meetings was to discuss any comments or concerns they may have on these projects and to better understand the agency's procedures and regulations and how they fit in with NRC's obligations under the National Environmental Policy Act (NEPA). The following is a summary of each meeting and a list of participants.

CONTACT: Irene Yu, DWMEP/FSME
           (301) 415-1951

A. Kock   - 2 -

State Historic Preservation Office (SHPO), Cheyenne, Wyoming – January 12, 2009

Meeting Summary

Regarding the Nichols Ranch Project, we discussed the proximity to the Pumpkin Buttes, which is designated as a Traditional Cultural Property, and the tribal interest in the Pumpkin Buttes. The SHPO is currently working on a programmatic agreement (PA) with the Bureau of Land Management (BLM) pertaining to the Pumpkin Buttes. We discussed potential best management practices (BMPs) and mitigation strategies to be included in the PA such as painting the buildings a certain color to mitigate the visual effect, keeping the buildings a low profile, and adding a public education component. Regarding the Lost Creek Project, we discussed the presence of tribal artifacts with cultural significance in the nearby town of Bairoil. We also discussed the potential presence of paleontological artifacts in the Great Divide Basin because it was at one time covered with water. The mitigation strategies discussed included data recovery (where a discovery plan would be needed) and a public education component. No tribal concerns were discussed for the Moore Ranch Project. For all three projects, we discussed cumulative impacts and the importance of assessing the impacts of ISL in addition to those for coal-bed methane (CBM), oil and gas (O&G), wind, and/or coal, which are all actively underway in Wyoming. We also discussed the Section 106 process and verified NRC's responsibilities and process to submit the cultural resources information to the SHPO.

Meeting Participants

Irene Yu, NRC
Nancy Barker, VHB
Richard Currit, SHPO

Follow-up Items

NRC to talk to BLM about how they want to comment when BLM lands are involved in the Section 106 process. NRC spoke to BLM following the trip about how they want to comment when BLM lands are involved in the Section 106 process. NRC will provide BLM with a copy of the complete cultural resources section of the application for discussion and concurrence prior to submitting the information to the SHPO.

State Engineer's Office (SEO), Cheyenne, Wyoming – January 12, 2009

Meeting Summary

We discussed the importance of the ISL wells being constructed well to prevent cross-contamination between aquifers and that the applicant's provide adequate means for the closure of these wells once the facilities are decommissioned so as not to leave a conduit for cross-contamination. We discussed the differences in the roles and responsibilities of the SEO (focused on water quantity) and of the Department of Environmental Quality (DEQ, focused on water quality). The SEO is responsible for well permitting, which is typically done in permit blocks which allow for a certain number of wells to be constructed within a certain tract of acres. The SEO also issues permits for stormwater management impoundments.

Appendix A

A. Kock   - 3 -

Meeting Participants

Irene Yu, NRC
Nancy Barker, VHB
John Harju, SEO
Harry Labonde, SEO

Follow-up Items

None

<u>Bureau of Land Management State Office, Cheyenne, Wyoming – January 12, 2009</u>

Meeting Summary

NRC staff provided an overview of how and why the draft Memorandum of Understanding (MOU) between NRC headquarters and BLM headquarters was developed and the current status of the draft MOU. Having not reviewed the draft MOU, BLM staff expressed their interest in reviewing the MOU and having the MOU signed at the state level instead of at the headquarter level. BLM has an MOU in place with the DEQ and briefly explained how the MOU specifies the roles and responsibilities of each agency and the points of contact. BLM staff provided NRC staff with a copy of their MOU with DEQ and a copy of the new Department of Interior regulations on implementing NEPA to help NRC in their development of an MOU with BLM. BLM staff also stressed the importance of increased communication between them and the NRC. We discussed both BLM and NRC's NEPA responsibilities for the three ISL projects and whether an environmental assessment (EA) or an environmental impact statement (EIS) is more appropriate. BLM staff sees the main issues with ISL to be related to groundwater quality and cumulative impacts. Specifically, they raised the concern of the possible conflict between the reducing nature of CBM and the oxidizing nature of ISL.

Meeting Participants

Patrice Bubar, NRC (via phone)
Irene Yu, NRC
Nancy Barker, VHB
Larry Claypool, BLM
Ed Heffern, BLM
Larry Jensen, BLM
Bob Janssen, BLM
Janet Kurman, BLM
Pam Stiles, BLM

Follow-up Items

NRC to continue to pursue an MOU with BLM.

A. Kock   - 4 -

Department of Environmental Quality, Cheyenne, Wyoming – January 12, 2009

Meeting Summary

DEQ staff stressed the importance of increased communication between them and the NRC and requested the development of an MOU with the NRC. Since the DEQ issues the permits for the underground injection wells and the aquifer exemption related to ISL, we discussed in great detail DEQ's requirements from the applicant and the issues they have seen thus far in their review of the three project applications. DEQ Land Quality Division staff will coordinate the comments from all other DEQ divisions for their review of NRC's environmental documents. DEQ Water Quality Division staff provided background on the stormwater and groundwater concerns. Specifically, we discussed the different classes of injection wells and which ones apply to ISL facilities, the construction of wells and how important the construction is to minimizing cross-contamination between aquifers, the viability of ISL in an unconfined aquifer, and groundwater restoration. DEQ Air Quality Division staff provided information on air quality issues in the state. DEQ Industrial Siting Division staff provided information related to the sage grouse core areas and provided NRC with a map showing those areas. DEQ Solid and Hazardous Waste Division staff provided background on radioactive/hazardous waste disposal in the state. Regarding the Lost Creek Project, we discussed the need for increased federal and state agency interaction because the site consists primarily of federal lands. Also, DEQ staff raised some wildlife concerns as the Lost Creek Project site is located near a sage grouse core area.

Meeting Participants

Irene Yu, NRC
Nancy Barker, VHB
Carl Anderson, DEQ Solid & Hazardous Waste Division
Mark Conrad, DEQ Water Quality Division
John Corra, DEQ Administration Division
Kevin Frederick, DEQ Water Quality Division
Andrew Keyfaurer, DEQ Air Quality Division
Brian Lovett, DEQ Water Quality Division
Don McKenzie, DEQ Land Quality Division
Darla Potter, DEQ Air Quality Division
Barb Sahl, DEQ Water Quality Division
Chad Schlichtemeier, DEQ Air Quality Division
Tom Schroeder, DEQ Industrial Siting Division
Paige Smith, DEQ Air Quality Division
Lowell Spackman, DEQ Land Quality Division
Ed Heffern, BLM

Follow-up Items

NRC to discuss internally on possible MOU with DEQ. Internal discussions have been held and a call is scheduled with DEQ to discuss this request.

Appendix A

A. Kock   - 5 -

<u>Governor's Planning Office (GPO), Cheyenne, Wyoming – January 13, 2009</u>

Meeting Summary

GPO staff provided an overview of their assistance to several BLM field offices in updating their Resource Management Plans. In addition, we discussed the location of sage grouse core areas and sage grouse conservation initiatives that are being developed or are already underway.

Meeting Participants

Irene Yu, NRC
Nancy Barker, VHB
Tom Blickensderfer, GPO

Follow-up Items

None

<u>Bureau of Land Management Field Office, Rawlins, Wyoming – January 13, 2009</u>

Meeting Summary

The status of the Draft Generic EIS for environmental reviews for ISL facilities (GEIS) and the MOU were discussed. It was explained that the NRC would be the lead agency because of their regulation over milling (not mining) operations. The BLM inquired whether the DEQ should be a cooperating agency. The BLM indicated the state has created an MOU format for federal agencies. Typically, an MOU is made with the state and separate agencies are assigned, as applicable. Shirley Basin & Red Desert, where the Lost Creek site is located, has been extensively explored. The effects of ISLs on freshwater aquifers are critical and applicants need to show that leaching will not occur between aquifers. The Cheyenne Office of the DEQ (Steve Engle-hydrologist) will scrutinize the Lost Creek EA for groundwater issues. The Battle Springs aquifer is a major aquifer in the area. ISLs operate under BLM mining laws and these laws address land use issues. A Plan of Operations will be required by BLM for the Lost Creek site. Currently, they are functioning (exploring) under a Notice (<5 acres of disturbance). An issue of concern is fencing. If fencing of the site is proposed, there are public access issues and wild horse routes that may be impacted. In addition, applicants (ISL operators) need to address effects of their ISL operation on grazing leases. The U.S. Fish & Wildlife Service (FWS) recommends that standard BMPs be used. Their principal concerns are for cattle and raptors. Netting would be required over waste ponds, and over mud pits. The BLM plans on meeting with UR-Energy (applicant) on January 27th on the Lost Creek site.

Meeting Participants

Alan Bjornsen, NRC
Stephanie Davis, Environet
Mark Newman, BLM
Clare Miller, BLM
Patrick Madigan, BLM
Travis Sanderson, FWS

A. Kock   - 6 -

Follow-up Items

NRC to keep BLM Field Offices up to date on status of MOU; BLM to send Environet a copy of the Land Status Map for Wyoming.

Bureau of Land Management Casper Field Office, Casper, Wyoming – January 13, 2009

Meeting Summary

Topics discussed included cumulative impacts, existing coal-related analyses, and hydrology at ISL sites. Specifically, with regards to cumulative impacts, BLM, U.S. Environmental Protection Agency (EPA), and DEQ cooperated on a study of the effects of coal, O&G, CBM, uranium, and wind development in the Powder River Basin. There are several existing coal-related analyses: five coal-related EISs either final or in progress (West Antelope, Wright, and three physical groupings: North, Middle, and South Pods). Chapter 4 in these EISs was recommended as a good resource for NRC's cumulative impacts analysis. Another EIS with good information on cumulative impacts was for Pacific Corporation/Rocky Mountain Corporation's Wind Farm in the northeastern part of the state. BLM's concerns with respect to ISL impacts were about the cross-contamination of groundwater between CBM and ISL and whether NRC was going to require groundwater monitoring. BLM is working on a reliable groundwater model for ISL projects.

Meeting Participants

Behram Shroff, NRC
Stewart Bland, Chesapeake Nuclear
Tracy Hamm, VHB
Patrick Moore, BLM
Tom Foertsch, BLM
Mike Karbs, BLM

Follow-up Items

None

Sweetwater County (SC), Green River, Wyoming – January 13, 2009

Meeting Summary

Safety and emergency issues were the top concerns raised by Sweetwater County (SC). Site access, particularly on the narrow county roads, was of concern with the Bairoil representatives (trucks, dust, noise, etc.). The proposed routes were of concern, along with road improvements, maintenance, and signage. Of special interest was the amount of radiation that could be expected from trucks carrying product from the facility to the next processing facility. The Sweetwater County Fire Department (SCFD) and emergency personnel were concerned with radiation and potential exposure, construction of the facility, access, materials and waste storage, and emergency plans that the applicant would prepare. The SCFD specifically requested that plans of the facility be available to them in case of an actual emergency. Waste

Appendix A

A. Kock - 7 -

disposal was an issue of great importance: what types of waste would be generated; how much would be generated; where would the waste be disposed; and what routes would be used to get there. There is also a limited workforce that is available in the SC area. Even unskilled workers are hard to come by. Other issues that were raised included: impacts to Bairoil's municipal water supply well, potential storm water discharges, waste water ponds, utilities, and air quality (dust).

Meeting Participants

Alan Bjornsen, NRC
Stephanie Davis, Environet
John Radosevich, SC
Steve Horton, SC
John Barton, SC
Dennis Washam, SC
Wayne Silvers, SC
Judy Valentine, SC
Dennis Claman, SC
Robert Robinson, SC
Tony Riga, Bairoil
Sue Ann Riganco, Bairoil

Follow-up Items

NRC to find out what roads are being proposed for access to the facility. NRC to find out the levels of radiation at various locations throughout the facility, as well as during transportation. NRC to inform applicant that the SCFD would like a hazardous materials inventory.

Fremont County Planning Department, Fremont County, Wyoming – January 13, 2009

Meeting Summary

The county has no zoning laws in effect. Reviews are performed for residential subdivisions. Regarding solid waste disposal, the county operates a transfer station and landfill in Riverton. Regarding highway maintenance in the vicinity of Lost Creek (SC), the county only maintains about ten miles of the Crooks Gap-Wamsutter Road south of Jeffrey City. Beyond that point, the road is poorly maintained.

Meeting Participants

Alan Bjornsen, NRC
Stephanie Davis, Environet
Ray Price, Fremont County Planning Department

Follow-up Items

None

Casper Planning Department, Casper, Wyoming – January 13, 2009

Meeting Summary

The main points drawn from this discussion were that rental housing is very scarce, especially affordable housing, and that less expensive housing would be available to ISL workers and families in Glenrock, Douglas, and Wright. However, those cities also have a shortage of affordable housing. Also, the Powder River Basin has good roads and the school capacity and retail establishments are sufficient for the present. Fire and police departments are adequately staffed. Medical and hospital facilities are able to provide good service. Additionally, the industry boom-bust cycles are typical, making it hard to maintain available and affordable housing. The population of Casper is about 53,000 (75,000 including suburbs) and the current economic downturn will likely make housing more affordable. Developers are currently building housing for both upper and lower income families.

Meeting Participants

Behram Shroff, NRC
Stewart Bland, Chesapeake Nuclear
Tracy Hamm, VHB
Craig Collins, Casper Planning Department
Robin Mundell, Casper Planning Department

Follow-up Items

None

Wyoming Community Development Authority (WCDA), Casper, Wyoming – January 13, 2009

Meeting Summary

Discussions centered around the impact of resource extraction, including ISL, on housing. The WCDA was able to provide extensive data on existing housing statewide, and future projections. The main points raised were that rental housing is scarce in the Powder River Basin and Great Divide Basin; single family housing tends to be out of the affordable range; those seeking to move to Wyoming from economically hard-hit areas have a difficult time selling their homes; and the Wyoming economy is doing very well compared to the nation as a whole. Most Moore Ranch and Nichols Ranch workers are expected to live in Casper, Gillette, and other smaller communities such as Wright. The level of healthcare, education, and commercial facilities is generally good. Rawlins would likely be the main base for Lost Creek employees (possibly Wamsutter). There is no office of state planning.

Meeting Participants

Behram Shroff, NRC
Stewart Bland, Chesapeake Nuclear
Tracy Hamm, VHB
David Haney, WCDA
Cheryl Gillam, WCDA

Appendix A

A. Kock — - 9 -

Follow-up Items

None

<u>Lander Chamber of Commerce, Lander, Wyoming – January 13, 2009</u>

Meeting Summary

Inquiries were made regarding housing and workforce. There is some limited housing available in Fremont County (Lander Area), but it's pricey. Jeffrey City may be a better bet as there are still houses there from the oil boom in the late 80s/early 90s. The thought was that there would be sufficient skilled labor available due to the slowdown in the oil industry.

Meeting Participants

Alan Bjornsen, NRC
Stephanie Davis, Environet
Chamber of Commerce Director
Chamber of Commerce Receptionist

Follow-up Items

None

<u>Bureau of Land Management Field Office, Buffalo, Wyoming – January 14, 2009</u>

Meeting Summary

BLM staff explained their responsibilities under NEPA and their review and approval process of Plans of Operations submitted by ISL applicants. BLM staff also provided details on the update to the Buffalo Resource Management Plan, in which they just completed the scoping process. Since no BLM lands are present on the Moore Ranch Project site, BLM staff is not likely to review that application. Regarding the Nichols Ranch Project, BLM staff will provide comments on NRC's environmental documents and request frequent communication with the NRC throughout the environmental review process. BLM staff sees the main issues for the Nichols Ranch Project to be related to cultural resources and tribal concerns since the Pumpkin Buttes was designated a Traditional Cultural Property (TCP) in June 2007. The BLM is in the process of developing a PA for the TCP. BLM staff emphasized the importance of good construction of injection wells and did not seem concerned with CBM operations and ISL operations occurring simultaneously in the same area because of the large distances between CBM wells. BLM has prepared Plan of Development (POD) EISs and a 2003 EIS on CBM and natural gas, which have solid cumulative impacts analyses for the Powder River Basin. FWS staff discussed the locations of sage grouse core areas in the Powder River Basin, the possible need for avoidance of these areas, and the candidate conservation assurances program. FWS staff stated that additional information on sage grouse is present in the Northeast Wyoming Management Plan. FWS staff raised a concern over migratory birds, specifically related to the electrocution of raptors on power poles and they recommended buried power lines or aboveground lines conforming to the requirements set by the Avian Power Line Interaction Committee.

A. Kock   - 10 -

Meeting Participants

Behram Shroff, NRC
Irene Yu, NRC
Nancy Barker, VHB
Tracy Hamm, VHB
Stewart Bland, Chesapeake Nuclear
Brian Kuehl, Clark Group
Lori VanBuggenum, Clark Group
Buck Dumone, BLM
Jerry Queen, BLM
Clint Crago, BLM
Tom Bills, BLM
Paul Beels, BLM
Brad Rogers, FWS
Pete Ramirez, FWS

Follow-up Items

NRC to review BLM's POD EISs and 2003 EIS on CBM and natural gas to see if the cumulative impacts analyses can be incorporated into the NRC documents. NRC to also review the Northeast Wyoming Management Plan for sage grouse.

Department of Environmental Quality District 3 Office, Sheridan, Wyoming – January 14, 2009

Meeting Summary

DEQ staff explained their two tier review process of applications, which consists first of a completeness review and then a technical review (150 days to complete). Both the Moore Ranch and Nichols Ranch ISL applications have been through the completeness review and are undergoing the technical review with Moore Ranch to be completed first. DEQ staff's initial assessment of both applications is that additional information is necessary from the applicant and inconsistencies arise in both applications. DEQ staff's main concerns with both projects are cumulative impacts (whether ISL, CBM, and O&G can all occur simultaneously), groundwater quality resulting from unconfined aquifer conditions (effects on drawdown, ability to limit excursions, restoration), and underground injection well viability (which formation to drill into).

Meeting Participants

Behram Shroff, NRC
Irene Yu, NRC
Nancy Barker, VHB
Tracy Hamm, VHB
Stewart Bland, Chesapeake Nuclear
Brian Kuehl, Clark Group
Lori VanBuggenum, Clark Group
Mark Rogaczewski, DEQ Land Quality Division District 3
Don Fischer, DEQ Water Quality Division
Glenn Mooney, DEQ Land Quality Division District 3

Appendix A

A. Kock   - 11 -

Follow-up Items

None

Department of Environmental Quality District Office, Lander, Wyoming – January 14, 2009

Meeting Summary

A brief update was presented on the status of the GEIS and the EA for Lost Creek. The topic of requests for additional information (RAIs) was discussed. It was found that the DEQ, in addition to the list of RAIs submitted last summer on UR-Energy's application, was currently preparing a much larger list (200 in addition to the initial 45). The DEQ's primary concern is groundwater impact. The Water Quality Division (WQD) determines the class of use of an aquifer, but the EPA determines the exemption boundary. For deep well injection of wastes, the contact at the WQD identified was John Passehl. The DEQ is the agency that issues the actual mining permit, with the BLM concurring. DEQ, however, is also concerned with surface disturbance. If the total amount of disturbance is less than 5 acres, the DEQ issues a Drilling Notification (similar to the BLM's Notice). If the disturbance exceeds 5 acres, a License to Explore is issued (similar to the BLM's Plan of Operation). Bonding is also required by the DEQ and, in fact, the DEQ is the bond holder, even when BLM land is involved. For bond release, 2 years of successive growing seasons must occur after reclamation. Issues, besides groundwater that were raised during the meeting included the need to address solid waste disposal. This includes a complete characterization of the various waste streams, the disposal facilities intended to be used, and if there is to be any hazardous waste generated. The U.S. Department of Game & Fish (DGF) is concerned with the potential impacts to sage grouse. In particular, there appears to be a lek within the boundary of the Lost Creek site. There is a 1/4-mi exclusion area, as well as a 2-mi limited activity area surrounding each lek. The DGF also has an issue with the installation of overhead utility lines (as roosts for raptors). In addition to groundwater quality, groundwater drawdown is an issue. DEQ is asking the applicant to address potential drawdown outside the boundary of the site (up to 3 mi), and to identify users. The DEQ is also concerned with the fault running through the site, and if the potentiometric surface differs either side of it. Regional (outside the permit boundary) well data is also being asked of the applicant by the DEQ. The DEQ questions the need for such a large permit boundary if the ore body only occupies a portion of the site. A new requirement of the DEQ is the need for the applicant to submit data (including well, and GW data) for the first mine unit to operate at an ISL. This seems to be problematic, in that this information is not normally available until after the NRC issues its license. The DEQ is also requesting the applicant to submit additional cross-sections for the Lost Creek site. DEQ is also requesting a more detailed description of the hydrogeology of the site: thicknesses of the confining units, the multiple sands within the primary production zone in the Battle Spring Formation HJ unit, and deep well injection. Stability monitoring is required after uranium recovery is complete (quarterly monitoring for 12 months, then annually, thereafter).

Meeting Participants

Alan Bjornsen, NRC
Stephanie Davis, Environet
Amy Boyle, DEQ
Melissa Bautz, DEQ

A. Kock   - 12 -

Carrie Dobey, DGF

Follow-up Items

DEQ WQD to determine the class of aquifer for the HJ unit, as well as the appropriate monitoring well distribution.

Bureau of Indian Affairs (BIA) Wind River Agency, Fort Washakie, Wyoming – January 15, 2009

Meeting Summary

NRC provided a status of the GEIS, the environmental review process the NRC is undertaking, and proposed ISLs in Wyoming. There was more concern over legacy sites than the proposed new uranium recovery facilities in Wyoming. In particular, the conventional mill near Riverton was discussed because of the groundwater plume. While there are no ISL facilities proposed for the Wind River Reservation, it was told us that anytime a new facility is proposed, all the tribes in Wyoming should be notified. The names of two cultural resource contacts were given to us: Amanda White (Northern Arapaho) and Reed Tidzump (Eastern Shoshone). The counties within the state generally send letters to the tribes for concurrence on cultural matters. It was suggested that when cultural resource studies are performed, tribal elders be contacted so that items other than physical features (e.g., spiritual/sacred views) may be identified. The Wind River Reservation has its own environmental commissions (air, water, etc.).

Meeting Participants

Alan Bjornsen, NRC
Stephanie Davis, Environet
Ray Nation, BIA
Tony Pingree, BIA
Kelly Ferris, BIA
Trisha Cachelin, BIA
John Enos, Shoshone
Steve Babbitts, BIA
Kassel Weeks, WREQC
Don Aragon, WREQC

Follow-up Items

NRC to send copies of draft GEIS (CD) to BIA and Wind River Agency. NRC to send letters to Northern Arapaho and Eastern Shoshone tribes regarding the licensing of the Lost Creek project. The CD and letters were sent in February 2009.

Bureau of Land Management Field Office, Casper, Wyoming – January 15, 2009

Meeting Summary

NRC gave a status of the GEIS and BLM MOU. BLM explained the difference in the types of BLM land. Leasable land, also known as acquired land, is land that the US has bought back the mineral rights. This represents only a small portion of BLM lands. Locatable land is land that

A. Kock   - 13 -

was originally federal-owned, and represents most of BLM lands. BLM was concerned that the GEIS does not distinguish between the two types of land. BLM was pleased to hear that there is progress being made on the MOU, but has a concern about how field office personnel working jointly on a NEPA document with the NRC would be reimbursed for their effort. BLM was also questioning whether the state or field office would participate in the development of the MOU.

Meeting Participants

Alan Bjornsen, NRC
Stephanie Davis, Environet
Tom Foertsch, BLM
Patrick Moore, BLM

Follow-up Items

NRC to send copies of the proposed ISL Wyoming site map to the Casper Field Office and the State BLM Office. NRC sent the copies of the map in February 2009.

Buffalo Chamber of Commerce (COC), Buffalo, Wyoming – January 15, 2009

Meeting Summary

The COC Board raised the issues of impacts to wildlife (specifically to sage grouse) and socioeconomics (specifically housing capacity) in regards to the potential Nichols Ranch Project. The COC Board stated that Kaycee does not have the housing capacity and services that Buffalo has. The COC Board stated that the County school system has the capacity to handle additional students. RV parks and motels still have ample space in the county for workers who choose not to permanently relocate into the County. The COC Board emphasized that like most of the state, the county's population fluctuates with the industry cycles of booms and busts.

Meeting Participants

Irene Yu, NRC
Nancy Barker, VHB
Margaret Dunfee, COC
Various members from the COC Board

Follow-up Items

None

Johnson County Commissioners, Buffalo, Wyoming – January 15, 2009

Meeting Summary

The County Commissioners raised the issues of impacts to socioeconomics, both positive and negative, in regards to the potential Nichols Ranch Project. Specifically, the County

A. Kock   - 14 -

Commissioners mentioned the shortage of housing in Kaycee, the shortage of housing for low-moderate income families in the County, and the poor conditions of Trabing Road (also known as Iragary Road), which is a likely commuter path from Buffalo and Kaycee to the Nichols Ranch Project site. Trabing Road has been heavily utilized by CBM operators and although it is a county-maintained road, the County does not have enough funding currently to upgrade the road. The County Commissioners requested that the path of transport for the yellowcake be described in the NRC's environmental document. We also discussed positive economic impacts from new ISL projects such as the creation of new jobs and the addition to tax base. The County Commissioners stated that emergency response services needed for the Nichols Ranch Project would come from either Buffalo or Kaycee. The County Planner stated that the only local permitting required of the applicant would be for a septic system leach field up to 2,000 gallons in size.

Meeting Participants

Irene Yu, NRC
Nancy Barker, VHB
Smokey Wildeman, Commissioner
Gerald Fink, Chairman
Rob Yingling, County Planner

Follow-up Items

None

Campbell County Economic Development Corporation (CCEDC), Gillette, Wyoming – January 15, 2009

Meeting Summary

The discussion focused on the impact of resource extraction, including ISL, on housing, schools and other community facilities, and socioeconomics. The vacancy rate for rental housing has been close to zero for the last four years; 850 rental units in Gillette have recently been built and fully occupied. The local economy is mineral-based and has gone through boom and bust cycles which have discouraged investment in housing. Local government has extended water and sewer lines well beyond city limits to encourage development. Land is being annexed aggressively by the city as a spur to foster residential development. Two new elementary schools have been built and two more are planned.

Meeting Participants

Behram Shroff, NRC
Stewart Bland, Chesapeake Nuclear
Tracy Hamm, VHB
Michael Surface, CCEDC
Susan Yerke, CCEDC
Brandi Beecher, CCEDC

A. Kock   - 15 -

Follow-up Items

None

<u>Town of Wright, Wright, Wyoming – January 15, 2009</u>

Meeting Summary

A new power plant is being built nearby. O&G extraction and coal mining are active in the vicinity. Almost 200 single family houses have just been built and the town has purchased 113 acres, some of which will be for housing; the land includes water service. It is hard to get developers to come out to a small town of under 5,000 people, although tax credits exist for rural development. There are several private apartments in the community and many employers are building motels and renting rooms to their workers. A new shopping center has been built and the town has one medical clinic. The junior and senior high schools have been combined and capacity is adequate.

Meeting Participants

Behram Shroff, NRC
Stewart Bland, Chesapeake Nuclear
Tracy Hamm, VHB
Lyle Murdock, Wright Building Official

Follow-up Items

None

<u>Converse County Planning Department Douglas, Wyoming – January 15, 2009</u>

Meeting Summary

In discussing where workers from the Powder River Basin may live, Converse County Planning Department Douglas, Wyoming staff thought that the cities of Glenrock and Douglas would likely home bases for workers for the Nichols Ranch and Moore Ranch projects; Midwest and Wright were also mention as possibilities. Some trailer parks might have vacancies but rental apartments are scarce and expensive. There is the potential for new hotels/motels to be built. There are 130 zoned and platted lots for housing but they are without services. The state has a loan program for first-time home buyers. The current population is about 6,000 people, but the city could accommodate a total of 10,000. Schools are close to capacity in Douglas but Glenrock may have some room.

Meeting Participants

Behram Shroff, NRC
Stewart Bland, Chesapeake Nuclear
Tracy Hamm, VHB
Forrest Neuerberg, CCPD
Paul Musselman, CCPD

A. Kock   - 16 -

Follow-up Items

None

Converse Area New Development Organization (CANDO) – January 15, 2009

Meeting Summary

CANDO deals primarily with workforce concerns, local economic development, business recruitment and training, and housing. Ranchers are seeking information about energy companies looking for leases on their property. There is a shortage of workers with uranium mining experience. Locally, there is limited housing and Nichols Ranch and Moore Ranch workers would likely face a 1.5 hour commute, which is typical for the area.

Meeting Participants

Behram Shroff, NRC
Stewart Bland, Chesapeake Nuclear
Tracy Hamm, VHB
Joe Coyne, CANDO
Ed Werner, Consultant to CANDO

Follow-up Items

None

Appendix A

August 5, 2009

Tom Christiansen
Sage Grouse Coordinator
Wyoming Game and Fish Department
Green River Field Office
351 Astle Avenue
Green River, WY 82935

SUBJECT: REQUEST FOR INFORMATION REGARDING SAGE GROUSE HABITATS FOR THE PROPOSED LICENSE APPLICATION FOR URANIUM ONE INCORPORATED'S MOORE RANCH URANIUM RECOVERY PROJECT (Docket 040-09073)

Dear Mr. Christiansen:

The U.S. Nuclear Regulatory Commission (NRC) has received an application from Uranium One Inc. for a new radioactive source materials license to develop and operate the Moore Ranch Uranium Recovery Project (an *in-situ* recovery operation) located in Campbell County WY. The proposed project will consist of the following area located in Township 42 North, Range 75 West, Sections 26, 27, 33, 34, 35, and 36 and Township 41 North, Range 75 West, Range 75 West, Sections 1, 2, 3, and 4, and Township 42 North, Range 74 West, Section 31.

The location of the Moore Ranch project is close to two currently licensed *in-situ* recovery projects, the AREVA (COGEMA) Christensen Ranch Project and the Power Resources Inc. North Butte License Area. Maps showing the proposed project location are enclosed.

As established in Title 10 *Code of Federal Regulations* Part 51 (10 CFR 51), the NRC regulation that implements the National Environmental Policy Act of 1969, as amended, the agency is preparing a Supplemental Environmental Impact Statement (SEIS) for the proposed action that will tier off a Generic Environmental Impact Statement published in June 2009 (NUREG-1910). In accordance with Section 7 of the Endangered Species Act, the SEIS will include an analysis of potential impacts to endangered or threatened species or critical habitat in the proposed project area.

To support the environmental review, the NRC requested information from the U.S. Fish and Wildlife Service (FWS) to facilitate the identification of endangered or threatened species or critical habitat that may be affected by the proposed project. According to a letter sent to the NRC from the FWS dated May 7, 2008, the FWS indicated that they are currently conducting a review to determine if the greater sage-grouse warrants listing and that you may have more information on the greater sage-grouse habitats within the project area and appropriate mitigative measures to minimize potential impacts to the species. Any information you provide will be used to enhance the scope and quality of our review in accordance with 10 CFR 51 and 50 CFR 402. After assessing the information provided by you, the NRC will determine what additional actions are necessary to comply with Section 7 of the Endangered Species Act.

Appendix A

T. Christiansen 2

The Uranium One Inc.'s Moore Ranch Uranium Recovery Project license application is publicly available in the NRC Public Document Room located at One White Flint North, 11555 Rockville Pike, Rockville, Maryland 20852, or from the NRC's Agency Wide Documents Access and Management System (ADAMS). The ADAMS Public Electronic Reading Room is accessible at http://www.nrc.gov/reading-rm/adams.html. The accession number for the application is ML072851229.

Please submit any comments/information that you may have regarding this environmental review within 30 days of the receipt of this letter to the U.S. Nuclear Regulatory Commission, Attention: Mr. Andrea L. Kock, Mail Stop T8F05, Washington, DC 20555. If you have any questions, please contact Mr. Behram Shroff of my staff by telephone at 301-415-0666 or by email at Behram.Shroff@nrc.gov. Thank you for your assistance.

Sincerely,

/RA/

Andrea L. Kock, Branch Chief
Environmental Review Branch
Environmental Protection
 and Performance Assessment Directorate
Division of Waste Management
 and Environmental Protection
Office of Federal and State Materials
 and Environmental Management Programs

Docket No.: 040-09073

Enclosure: Moore Ranch location maps

Appendix A

WYOMING GAME AND FISH DEPARTMENT

5400 Bishop Blvd. Cheyenne, WY 82006

Phone: (307) 777-4600 Fax: (307) 777-4610

Web site: http://gf.state.wy.us

GOVERNOR
DAVE FREUDENTHAL

DIRECTOR
STEVE K. FERRELL

COMMISSIONERS
CLIFFORD KIRK – President
ED MIGNERY – Vice President
CLARK ALLAN
AARON CLARK
JERRY GALLES
MIKE HEALY
FRED LINDZEY

September 3, 2009

WER 329
Nuclear Regulatory Commission
Request for Information Regarding Sage Grouse Habitats
for the Proposed License Application for Uranium One Inc.
Moore Ranch Uranium Recovery Project
Docket No. 040-09073

U. S. Nuclear Regulatory Commission
Attention: Ms. Andrea L. Kock
Mail Stop T8F05
Washington, DC 20555

Dear Ms. Kock:

The staff of the Wyoming Game and Fish Department has reviewed the request for information regarding Sage Grouse Habitats for the proposed license application for Uranium One Incorporated's Moore Ranch Uranium Recovery Project. We offer the following comments for your consideration.

Terrestrial Considerations:

The proposed project area lies within the Pumpkin Buttes and North Converse Pronghorn Herd Units, as well as the and Pumpkin Buttes and North Converse Mule Deer Herd Units. Pronghorn use the areas in question for yearlong and winter/year-long habitat. No crucial winter range for pronghorn is contained within the project area. Mule deer in the region also utilize local habitats as yearlong habitat. Any removal of sagebrush habitats will reduce overall forage for both pronghorn and deer. If sagebrush habitats are disturbed during the uranium extraction process, restoration projects that strive to restore sagebrush and associated native plant species are recommended.

The area also provides winter, breeding, nesting, and brood-rearing habitat for sage grouse, sharp-tailed grouse, and a variety of other sage-dependent non-game birds and small mammals. The project area does not lie a sage grouse Core Area. Currently there are no known leks within the proposed project area. WGFD encourages Uranium One, Inc. to conduct annual sage grouse lek surveys to identify any unkown leks that may occur. Should sage grouse leks be discovered within the project area, WGFD recommends proper steps be taken to avoid disturbance within a

"Conserving Wildlife - Serving People"

Mr. Andrea L. Kock
September 3, 2009
Page 2 – WER 329

2-mile buffer of any active breeding areas (leks) used by sage grouse from March 15 to July 1. In addition, permanent structures should not be placed within ¼ mile of a sage grouse lek.

We recommend that additional wildlife surveys be performed to detect the presence of sensitive or threatened species prior to mining activities that disturb new areas. A winter bald eagle survey should be conducted to document any local roost locations. Surveys should also be conducted to document active nests of other raptor species present on the mine. Listed below are recommended dates that raptor nest sites should be free of disturbance. A 1/2 mile buffer around each nest should be maintained. Exceptions may be granted based on topography or other site-specific factors.

Osprey: April 15-August 1
Bald Eagle: February 15-August 15
Northern Harrier: April 1-July 31
Sharp shinned hawk: May 1 -August 31
Cooper's hawk: April 15 -August 15
Northern goshawk: April 1-August 15
Swainson's hawk: May 1-August 31
Red tailed hawk: March 15-July 31
Ferruginous hawk: April 1- July 31
Golden eagle: February 1-July 31
American kestrel: April 1-August 15
Merlin: April 1-August 15
Peregrine falcon: March 15-August 15
Prairie falcon: March 1-August 15

We recommend contacting and coordinating with the Bureau of Land Management and/or the U.S. Fish and Wildlife Service regarding locations of active raptor nest sites they may be aware of within the project area.

The WGFD recommends that all topsoil be saved and spread over disturbed areas as soon as possible after disturbance to accelerate reclamation. Native plants suitable for wildlife most dependent upon the disturbed site should be planted.

Finally, we encourage the Moore Ranch and Uranium One, Inc. to allow access to properties for the purpose of hunting big game species. Allowing such access would contribute greatly to the successful management of both deer and pronghorn to meet population objectives.

Aquatic Considerations:

This project will not have direct impacts to the aquatic resources of Ninemile Creek. However, we are concerned with the indirect and cumulative impacts to aquatic resources associated with this project. The construction of roads and pads will change how water will run off the landscape. This change will affect the infiltration rate of water, increase the velocity and quantity of water running across the landscape, and potentially could increase erosion and sediment deposition into nearby waterways. Roads have the potential for having the most

Appendix A

Mr. Andrea L. Kock
September 3, 2009
Page 3 – WER 329

profound impact on hydrology. Changes in hydrology across the landscape will then be reflected in changes in the geomorphology of perennial streams downstream of the project area. Ultimately, changes in geomorphology will directly influence aquatic habitat which may impact fish populations.

Currently, we do not have information regarding the effects this in-situ mining on aquatic habitats. Much is known, however, about the effects of increased sediment in streams. Stream channels respond to increased sediment supply by adjusting their pattern (sinuosity) and dimensions. These changes may result in decreased pool depths, decreased riffle area, less diversity in channel substrate and increased lateral instability marked by eroding banks. These changes along with direct effects from increased sediment loading can affect macroinvertebrate populations and diversity and decrease fish habitat. A common impact is a decrease in gravel and cobble used by spawning fish.

Additional information is needed regarding the effects of this project on aquatic habitat. It is recommended the monitoring of cumulative impacts from culverts and roads with 5% slope or greater be conducted.

The following is a protocol that we have developed for the monitoring of culverts and roads with 5% slope or greater. We are more than willing to work Uranium One, Inc. to discuss this protocol and adapt the protocol if needed.

Culverts

The purpose of monitoring culverts is to determine the cumulative impacts of changing the upland surface hydrology, erosion and deposition, and to ensure that they are functioning as designed and they are being maintained.

All culverts installed as part of this project will be monitored by a minimum of the following practices:
- Collect GPS coordinates for each culvert site
- Collect pre-construction photographs of the culvert site; upstream and downstream
- We recommend that several preconstruction photographs be taken overtime/ to record the relative change pre-construction. We recommend that photographs be taken three times between April and November.
- Collect post-construction photographs of the culvert site; upstream and downstream.
- Place a graduated fence post upstream and downstream of each culvert. The posts should have visible markings every 2" to provide a visual reference within each photograph. Fence posts should be placed within 50 feet of the culvert openings. Posts should be placed outside of the mainflow channel so they are not directly affected by storm flow events. Each fence post location will be referenced by GPS.
- GPS the site where photographs will be taken for the upstream and downstream view.
- Culverts and accompanying fence posts will be monitored/photographed three times a year (spring after snow melt, summer, and fall) and after rainfall events accumulating

Mr. Andrea L. Kock
September 3, 2009
Page 4 – WER 329

greater than ½ inch of precipitation as measured at the nearest National Weather Service Monitoring point (if within 10 miles of the site) or at the Facility operations and maintenance building. The summer monitoring period can include a thunderstorm event as long as the monitoring occurs within seven days after the thunderstorm.
- Photographs will be provided to WGFD for review within 45 days. We recommend that a website or ftp site be developed.

If the photographs reveal observable changes from erosion or deposition, consultation between WGFD and industry will occur.

Roads with 5% or greater slope

The purpose of monitoring roads with 5% or greater slope is to determine the cumulative impacts of changing the upland surface hydrology, erosion and deposition, and to ensure that the long-term BMPs that were installed are still functioning and are being maintained.

- Place a graduated fence post midway down the slope and at the bottom of the slope. If a drainage ditch occurs on both sides of the road, post will also needs to be placed on both sides of the road. The posts should have visible markings every 2" to provide a visual reference within each photograph. Posts should be placed outside of the mainflow channel so they are not directly affected by storm flow events.
- Each fence post location will be referenced by GPS.
- GPS the site where photographs will be taken.
- Fence posts will be monitored/photographed three times a year (spring after snow melt, summer, and fall) and after rainfall events accumulating greater than ½ inch of precipitation as measured at the nearest National Weather Service Monitoring point (if within 10 miles of the site) or at the Facility operations and maintenance building. The summer monitoring period can include a thunderstorm event as long as the monitoring occurs within seven days after the thunderstorm.
- Photographs will be provided to WGFD for review within 45 days. We recommend that a website or ftp site be developed.

If the photographs reveal observable changes from erosion or deposition, consultation between WGFD and industry will occur.

If you have any questions or concerns, please contact Mr. Rick Huber, Staff Aquatic Biologist, at 307-777-4558.

Thank you for the opportunity to comment. If you have any questions or concerns, please contact Heather Obrien, Wildlife Biologist at 307-682-1579.

Appendix A

Mr. Andrea L. Kock
September 3, 2009
Page 5 – WER 329

Sincerely,

John Emmerich
Deputy Director

JE: MF: gfb

cc: USFWS
H. Obrien, L. Jahnke- WGFD, Sheridan
Paul Mavrakis, Sheridan Region Fisheries Supervisor

Appendix A

**UNITED STATES
NUCLEAR REGULATORY COMMISSION**
WASHINGTON, D.C. 20555-0001

October 22, 2009

Mr. Richard L. Currit
Senior Archaeologist
Wyoming State Historic
Preservation Office
2301 Central Avenue
Barrett Building, Third Floor
Cheyenne, WY 82002

SUBJECT: URANIUM ONE INC. MOORE RANCH IN-SITU URANIUM RECOVERY PROJECT – SECTION 106 CONSULTATION

Dear Mr. Currit:

In accordance with the provisions in 10 CFR Part 51, the U.S. Nuclear Regulatory Commission (NRC) regulations that implement the National Environmental Policy Act of 1969 and 36 CFR 800.8(c) of the National Historic Preservation Act, NRC is developing a Supplemental Environmental Impact Statement for Uranium One Inc.'s (previously Energy Metals Corporation) request to construct and operate the Moore Ranch In-Situ Recovery (ISR) facility in Campbell County, Wyoming. The facility includes a central processing plant, support buildings, wellfields, and access roads.

By letter dated April 9, 2008, the NRC staff initiated the Section 106 consultation process with the Wyoming State Historic Preservation Office (SHPO) concerning the proposed Moore Ranch ISR. In your office's response of June 5, 2008, it was stated that records showed that a cultural resources survey had not been conducted for the entire Area of Potential Effect (APE). On August 8, 2008, Uranium One provided your office with a Class I literature search, Class II Inventory, and Class III cultural resource survey for the APE, with the intent of completing the Section 106 consultation process. Your office notified Uranium One that the federal agency which issues the license, NRC, was responsible for this process.

On October 1, 2009, Behram Shroff of my staff verified, via an email to you, that your office was in possession of Uranium One's cultural resources surveys and related documentation. By way of this letter, supported by documentation in the SHPO's possession, NRC is seeking to continue and complete the Section 106 consultation process and seeks your concurrence with a finding of no effect.

As discussed in the Class II Inventory, site investigation identified seven sites, 48CA6691-48CA6697, and 25 Isolate Resources/Artifacts, including artifacts from the Paleo-Indian, Middle Archaic, and Late Archaic and Historic periods. Two sites, 48CA6694 and 48CA6696, are considered eligible for nomination to the National Register of Historic Places (NRHP). However, only two sites, 48CA965 and 48CA966, which are listed as not eligible for nomination to the NRHP, are at or near any proposed development areas for the ISR project.

Appendix A

R.L. Currit 2

NRC staff has also corresponded with several Native American tribes concerning the cultural resources in the vicinity of the proposed project. To date, no response has been received.

NRC requests your comments and recommendations within 30 days of receipt of this letter, including the issue of which cultural resources are deemed eligible/ineligible for the NRHP. The agency also seeks your position on the effect of the proposed project on the cultural resources identified. If you have any questions or require additional information, please contact the Environmental Project Manager, Mr. Behram Shroff at (301) 415-0006, or at Behram.Shroff@nrc.gov.

Sincerely

Andrea L. Kock, Chief
Environmental Review Branch
Environmental Protection
 and Performance Assessment Directorate
Division of Waste Management
 and Environmental Protection
Office of Federal and State Materials
 and Environmental Management Programs

Docket No.: 40-9073

cc: D. McKenzie, DEQ, Cheyenne
G. Mooney, DEQ, Sheridan
C. Crago, BLM, Buffalo
P. Beels, BLM, Buffalo
M. Griffin, Uranium One
C. Fisher, Northern Cheyenne THPO

Appendix A

ARTS. PARKS. HISTORY.
Wyoming State Parks & Cultural Resources

State Historic Preservation Office
Barrett Building, 3rd Floor
2301 Central Avenue
Cheyenne, WY 82002
Phone: (307) 777-7697
Fax: (307) 777-6421
http://wyoshpo.state.wy.us

November 3, 2009

Andrea L. Kock, Chief
Environmental Review Branch
U.S. Nuclear Regulatory Commission
Washington, D.C. 20555-0001

re: Uranium One, Inc., Moore Ranch In-Situ Uranium Recovery Project Cultural Resources Inventory (SHPO File # 0608RLC007)

Dear Ms. Kock:

Thank you for consulting with the Wyoming State Historic Preservation Office (SHPO) regarding the above referenced project. We have reviewed the project report and find the documentation meets the Secretary of the Interior's Standards for Archaeology and Historic Preservation (48 FR 44716-42). We concur with your finding that sites are not eligible for the National Register of Historic Places (NRHP) and no further work or protective measures are necessary.

48CA146	48CA963	48CA6173	48CA6693
48CA952	48CA971	48CA6691	48CA6695
48CA961	48CA3400	48CA6692	48CA6697

The following previously recorded sites have been recommended as ineligible for the NRHP but no eligibility justification has been submitted to our office for review. However, these sites meet the current definition of an Isolated Resources and no further work or protective measures are necessary:

48CA965	48CA966	48CA967	48CA968	48CA969

Previously recorded site 48CA970 has been recommended as ineligible for the NRHP but no eligibility justification has been submitted to our office for review. However, this site is no longer extant and no further work or protective measures are necessary.

Previously recorded site 48CA964 has been recommended as eligible for the NRHP but no eligibility justification has been submitted to our office for review. We recommend that this site remain unevaluated for NRHP eligibility pending evaluative testing. Site 48CA964 will not be affected by the project as planned.

Dave Freudenthal, Governor
Milward Simpson, Director

Appendix A

Andrea L. Kock, Chief
November 3, 2009
Page 2 of 2

Sites 48CA6694 and 48CA6696 have been determined eligible for the NRHP. However, we recommend that these sites remain unevaluated for NRHP eligibility pending evaluative testing. Sites 48CA6694 and 48CA6696 will not be affected by the project as planned.

We recommend the U.S. Nuclear Regulatory Commission allow the project to proceed in accordance with state and federal laws subject to the following stipulation:

> If any cultural materials are discovered during construction, work in the area shall halt immediately, the federal agency and SHPO staff be contacted, and the materials be evaluated by an archaeologist or historian meeting the Secretary of the Interior's Professional Qualification Standards (48 FR 22716, Sept. 1983). Additionally, if any future disturbance is planned at the locations of sites 48CA964, 48CA6694 or 48CA6696 that evaluative testing be completed and submitted to our office with a determination of site eligibility and project effect. If eligible and adversely affected, a Memorandum of Agreement implementing appropriate mitigative measures will be required.

This letter should be retained in your files as documentation of a SHPO concurrence with your finding of no historic properties affected. Please refer to SHPO project #0608RLC007 on any future correspondence regarding this project. If you have any questions, please contact me at 307-777-5497.

Sincerely,

Richard L. Currit
Senior Archaeologist

Cc: Glenn Mooney, Wyoming Department of Environmental Quality

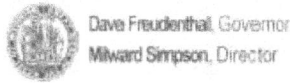
Dave Freudenthal, Governor
Milward Simpson, Director

Appendix A

From:	Richard Currit [RCURRI@state.wy.us]
Sent:	Tuesday, June 15, 2010 12:03 PM
To:	Davis (FSME), Jennifer
Subject:	Status of Site 48CA962

Hi Jennifer,

Our office concurred that site 48CA962 was ineligible for the National Register of Historic Places (NRHP) on 7/26/06, during consultation with the BLM on the "Devon Energy, Cutthroat Coalbed Methane POD".

Since that time, the boundaries of site 48CA6695 have been expanded to incorporate site 48CA962. This resource does not qualify for inclusion on the NRHP.

Sincerely,

Richard L. Currit
Senior Archaeologist
Wyoming State Historic Preservation Office
307-777-5497
rcurri@state.wy.us

APPENDIX B

PUBLIC COMMENTS ON THE DRAFT ENVIRONMENTAL IMPACT STATEMENT FOR THE MOORE RANCH *IN-SITU* RECOVERY PROJECT IN CAMPBELL COUNTY, WYOMING, AND U.S. NUCLEAR REGULATORY COMMISSION RESPONSES

Appendix B

CONTENTS

TABLES .... B–iv
ABBREVIATIONS/ACRONYMS .... B–v

B1	OVERVIEW..		B–1
B2	PUBLIC PARTICIPATION...		B–1
	B2.1	Notice of Intent To Develop the SEIS	B–1
	B2.2	Public Participation Activities	B–1
	B2.3	Issuance and Availability of the SEIS	B–2
	B2.4	Public Comment Period	B–2
B3	COMMENT REVIEW METHODS		B–2
B4	MAJOR ISSUES AND TOPICS OF CONCERN		B–5
B5	COMMENT SUMMARIES AND RESPONSES		B–5
	B5.1	General Opposition...	B–5
		B5.1.1 References..	B–7
	B5.2	General Support	B–7
	B5.3	General Environmental Concerns	B–7
	B5.4	NEPA Process	B–8
		B5.4.1 GEIS/SEIS	B–8
		B5.4.2 Adequacy of Impact Assessment......................................	B–10
		B5.4.3 Range of Reasonable Alternatives	B–15
		B5.4.4 References..	B–16
	B5.5	Purpose, Need, and Scope of the SEIS/GEIS...................................	B–17
		B5.5.1 Description of the SEIS/GEIS Purpose and Need	B–17
		B5.5.2 Use of the GEIS in Site-Specific Environmental Reviews.....	B–19
		B5.5.3 Scope of the SEIS/GEIS	B–20
		B5.5.4 Reference	B–21
	B5.6	Scoping Process and Scoping Report	B–21
		B5.6.1 References..	B–22
	B5.7	SEIS/GEIS Methods and Approach...	B–23
		B5.7.1 Consider Compliance History in Assessing Impacts............	B–23
		B5.7.2 General Comments on SEIS/GEIS Structure, Methods, and Approaches..	B–24
		B.5.7.3 References	B–24
	B5.8	Public Involvement	B–24
		B5.8.1 References..	B–26
	B5.9	Regulatory Issues and Process	B–26
		B5.9.1 NRC as a Regulatory Authority	B–26
		B5.9.2 NRC Authority and Jurisdiction	B–26
		B5.9.3 NRC Policies and Practices	B–27
		B5.9.4 Adequacy of NRC Regulations and Practices	B–28
		B5.9.5 Applicable Rulemaking Efforts	B–30
		B5.9.6 NRC NEPA Process Implementation................................	B–31
		B5.9.7 NRC Licensing Process	B–31
		B5.9.8 Consideration of ISL Facility Safety Record and Compliance History...	B–33
		B5.9.9 Groundwater Restoration Criteria and Methods	B–34

Appendix B

CONTENTS (continued)

	B5.9.10	References	B–37
B5.10	Credibility of NRC		B–39
B5.11	Federal and State Agencies		B–40
	B5.11.1	Roles of Federal, Tribal, State, and Local Agencies	B–40
	B5.11.2	Effects of Changes to Federal or State Regulations on the SEIS	B–42
	B5.11.3	Clarification of Other Federal/State Regulation and Practice	B–43
	B5.11.4	References	B–43
B5.12	Cooperating Agencies and Consultations		B–43
	B5.12.1	References	B–45
B5.13	SEIS Schedule		B–45
	B5.13.1	References	B–45
B.5.14	ISL Process Description		B–46
	B5.14.1	Overview	B–46
	B5.14.2	Preconstruction and Construction	B–46
	B5.14.3	Operations	B–47
	B5.14.4	Aquifer Restoration	B–48
	B5.15.5	Gaseous or Airborne Particulate Emissions	B–48
	B5.14.6	Operational History	B–49
		B5.14.6.1 Historical Operational Experience: Spills and Leaks	B–49
	B5.14.7	Requests for Detailed Information About All ISL Facilities	B–50
	B5.14.8	References	B–50
B5.15	Financial Surety		B–51
	B5.15.1	Reference	B–52
B5.16	Alternatives		B–52
	B5.16.1	References	B–54
B5.17	Land Use		B–55
	B5.17.1	Ownership Issues, Surface, and Mineral Rights	B–55
	B5.17.2	Amount of Land Affected and Type, Degree, and Duration of Potential Impacts	B–55
	B5.17.3	Mitigation and Reclamation Issues	B–58
	B5.17.4	References	B–60
B5.18	Transportation		B–61
	B5.18.1	References	B–63
B5.19	Geology and Soils		B–63
	B5.19.1	Soil Disturbance Concerns	B–63
	B5.19.2	Soil Impacts From Surface Spills	B–65
	B5.19.3	Permanent Change to Rock Formations	B–66
	B5.19.4	References	B–66
B.20	Groundwater Resources		B–67
	B5.20.1	General Concerns About ISL and Groundwater Contamination	B–67
	B5.20.2	Importance of Water and Consumptive Water Use	B–67
		B5.20.2.1 Site Characterization	B–68

Appendix B

CONTENTS (continued)

		B5.20.2.2	Aquifer Exemption and Baseline Water Quality	B–70
		B5.20.2.3	Control of Operational Impacts, Excursion of ISL Solutions, and History	B–72
		B5.20.2.4	Exploratory Drill Wells, Abandoned Wells, and Old Mines	B–75
	B5.20.3	Aquifer Restoration and Decommissioning: Methods and Operational Experience		B–76
	B5.20.4	Miscellaneous Groundwater Comments		B–78
	B5.20.4	References		B–80
B5.21	Surface Water Resources			B–81
	B5.21.1	Impacts to Surface Drainages and Surface Waters		B–81
	B5.21.2	General Water Resource Concerns		B–82
	B5.21.3	References		B–84
B5.22	Wetlands			B–84
	B5.22.1	References		B–85
B5.23	Ecology			B–86
	B5.23.1	General Ecology		B–86
	B5.23.2	Concerns About the Sage-Grouse		B–86
	B5.23.3	General Comments on Threatened and Endangered Species		B–86
	B5.23.4	Concerns About Mitigation and Timing		B–88
	B5.23.5	Habitat Loss and Fragmentation		B–89
	B5.23.6	Comments on Migratory Birds		B–89
	B5.23.7	General Vegetation Comments		B–90
	B5.23.8	Impacts to Terrestrial Ecology and Wildlife Discussion		B–91
	B5.23.9	Inconsistencies Between Sections		B–93
	B5.23.10	References		B–93
B5.24	Meteorology, Climatology, and Air Quality			B–94
	B5.24.1	Permitting and Regulations		B–94
	B5.24.3	Baseline Air Quality		B–95
	B5.24.4	Impact Assessment		B–95
	B5.24.5	References		B–104
B5.25	Historical and Cultural Resources			B–105
	B5.25.5	Potential Impacts to Cultural, Historical, and Sacred Places		B–105
	B5.25.6	License Conditions to Address Potential Impacts to Historical and Cultural Resources		B–105
	B5.25.8	Historic and Cultural General		B–106
	B5.25.9	References		B–108
B5.26	Socioeconomics			B–108
	B5.26.1	Reference		B–109
B5.27	Public and Occupational Health			B–109
	B5.27.2	Impacts to Members of the General Public		B–109
	B5.27.3	Impacts From Off-Normal Operations or Accidents		B–111
	B5.27.4	General		B–111

Appendix B

CONTENTS (continued)

	B5.27.5	References	B–113
B5.28	Waste Management		B–114
	B5.28.1	General Waste Management Comments	B–114
	B5.28.2	Scope of the Assessment of Waste Management Impacts	B–115
	B5.28.3	Characteristics of Wastes Generated by ISL	B–116
	B5.28.4	Waste Treatment and Disposal Methods	B–118
	B5.28.5	Regulation of Wastes and Disposal Methods	B–119
	B5.28.6	References	B–120
B5.29	Decommissioning		B–120
	B5.29.1	Reference	B–121
B5.30	Cumulative Effects		B–121
	B5.30.1	General Comment: The SEIS Does Not Adequately Address Cumulative Effects	B–121
	B5.30.2	Past, Present, and Reasonably Foreseeable Future Actions	B–122
	B5.30.3	Specific Document Changes or Action Requests	B–123
	B5.30.4	Significance	B–123
	B5.30.5	Other	B–126
	B5.30.6	References	B–127
B5.31	Environmental Justice		B–127
B5.32	Best Management Practices		B–128
	B5.32.1	Enforcement of Mitigation	B–128
	B5.32.2	Completeness of the Mitigation Measures and Best Management Practices	B–129
	B5.32.3	General Comments Related to Mitigation Measures and Best Management Practices	B–130
	B5.32.4	References	B–132
B5.33	Monitoring		B–132
B5.34	Editorial		B–134
	B5.34.1	Grammatical Editorial	B–134
	B5.34.2	Technical Editorial	B–135
	B5.34.3	Regulatory Editorial	B–140

TABLE

B3-1 Public Commenter Name With Affiliation and Comment Document Number B–3

Appendix B

ABBREVIATIONS/ACRONYMS

AEA	Atomic Energy Act
ADAMS	Agency-wide Documents Access and Management System
ACL	alternate concentration limit
BLM	U.S. Bureau of Land Management
CBM	coal bed methane
CEQ	Council on Environmental Quality
CFR	Code of Federal Regulations
DOE	U.S. Department of Energy
DOT	U.S. Department of Transportation
EMC	Energy Metal Corporation US
EPA	U.S. Environmental Protection Agency
FR	*Federal Register*
GEIS	Generic Environmental Impact Statement
ISL	*in-situ* leach
ISR	*in-situ* recovery
IX	ion exchange
LQD	Land Quality Division
MCL	maximum contaminant limit
MIT	mechanical integrity test/testing
NAAQS	National Ambient Air Quality Standards
NEPA	National Environmental Policy Act
NOA	Notice of Availability
NPDES	National Pollutant Discharge Elimination System
NRC	U.S. Nuclear Regulatory Commission
NRHP	National Register of Historic Places
NHPA	National Historic Preservation Act of 1966, as amended
RAI	request for additional information
SDWA	Safe Drinking Water Act
SEIS	Supplemental Environmental Impact Statement
SER	safety evaluation report
SGIT	Sage-Grouse Implementation Team
SHPO	State Historic Preservation Office
TDS	total dissolved solids

Appendix B

ABBREVIATIONS/ACRONYMS (continued)

UCL	upper control limit
UIC	Underground Injection Control
USDW	Underground Source Drinking Water
WDEQ	Wyoming Department of Environmental Quality
WGFD	Wyoming Game and Fish Department
WSEO	Wyoming State Engineer's Office

Appendix B

PUBLIC COMMENTS ON THE DRAFT ENVIRONMENTAL IMPACT STATEMENT FOR THE MOORE RANCH *IN-SITU* RECOVERY PROJECT IN CAMPBELL COUNTY, WYOMING, AND U.S. NUCLEAR REGULATORY COMMISSION RESPONSES

B1 OVERVIEW

On December 11, 2009, the U.S. Nuclear Regulatory Commission (NRC) staff published a notice in the *Federal Register* requesting public review and comment on the Draft Environmental Impact Statement for the Moore Ranch *In-Situ* Recovery (ISR) Project in Campbell County, Wyoming, Supplement to the Generic Environmental Impact Statement for *In-Situ* Leach Uranium Milling Facilities (SEIS) [74 *Federal Register* (FR) 65806] in accordance with 10 Code of Federal Regulations (CFR) Part 51, Environmental Protection Regulations for Domestic Licensing and Related Regulatory Functions. The NRC staff initially established February 1, 2010, as the deadline for submitting public comments on the draft SEIS. The NRC staff subsequently extended this deadline to March 3, 2010 (75 FR 6065). Twenty documents (i.e., email, mail, and facsimiles) were submitted to NRC containing comments on the proposed Moore Ranch project. In addition to the public comment period, the public also had the opportunity to request a hearing {January 25, 2008, [73 FR 4642]}. The deadline to request a hearing expired on March 25, 2008. No requests for a hearing were submitted.

B2 PUBLIC PARTICIPATION

Public participation is an essential part of the NRC environmental review process. This section describes the process for public participation during the NRC staff development of the SEIS.

NRC conducted an open, public SEIS development process consistent with the requirements of the National Environmental Policy Act of 1969 (NEPA) and NRC regulations. The NRC staff met with Federal, State, and local agencies and authorities, as well as public organizations, as part of a site visit to gather site-specific information. Including an extension, NRC provided an 81-day public comment period for agencies, organizations, and the general public to review the draft SEIS and provide comments.

B2.1 Notice of Intent To Develop the SEIS

The NRC staff published a Notice of Intent to prepare the SEIS in the *Federal Register* (74 FR 42332) on August 21, 2009, in accordance with NRC regulations.

B2.2 Public Participation Activities

As described in SEIS Section 1.4.2, the NRC staff met with Federal, State, and local agencies and authorities during the course of an expanded visit to the proposed Moore Ranch ISR Project site and vicinity. The purpose of this visit and these meetings was to gather additional site-specific information to assist in the preparation of the Moore Ranch ISR Project environmental review. As part of this effort to gather additional site-specific information, the NRC staff also contacted potentially interested Native American tribes and local authorities, entities, and public interest groups in person and via e-mail and telephone. Additional

Appendix B

opportunities for public participation in the licensing process for the proposed Moore Ranch Project are described in Section B5.8 of this comment response report.

B2.3 Issuance and Availability of the SEIS

On December 11, 2009, in accordance with NRC regulations, the NRC staff published a Notice of Availability (NOA) of the draft SEIS in the *Federal Register* (74 FR 65806). In this notice, the NRC staff provided information on how to access or obtain a copy of the draft SEIS. Electronic versions of the draft SEIS and supporting information were made accessible through the NRC Agencywide Documents Access and Management System (ADAMS) database on the NRC website (http://www.nrc.gov/reading-rm/adams.html). The public may examine and have copied, for a fee, the draft SEIS and other related publicly available documents from the NRC Public Documents Room. Copies of the draft SEIS were also publicly available at the Campbell County public libraries.

B2.4 Public Comment Period

In the publication of the NOA of the draft SEIS on December 11, 2009 (74 FR 65806), the NRC staff stated that public comments on the draft SEIS should be submitted by February 1, 2010. Members of the public were invited and encouraged to submit related comments through any one of the following means. Electronically, comments could be submitted to the Federal rulemaking or the NRC websites. Written comments could be submitted by mail or facsimile. On February 5, 2010, the NRC staff extended the public comment period to March 3, 2010 (75 FR 6065), in response to public requests for extension submitted in comment letters and emails. The 81-day period for public comments (i.e., from December 11, 2009 to March 3, 2010) exceeds the minimum 45-day comment period required under NRC regulations. The NRC staff identified 691 comments from the 20 documents commenting on the Moore Ranch draft SEIS.

B3 COMMENT REVIEW METHODS

As previously discussed, the NRC staff received 691 comments from 20 documents (i.e., email, mail, and facsimiles) during the comment period. Each of these comments has been included in the following comment summaries and addressed in the responses provided. Each comment was individually identified and responded to using a systematic approach. This approach involved indentifying individual comments from the source documents, consolidating comment information into a database, sorting all comments by topic, and distributing and reviewing all comments by the appropriate staff.

NRC conducted the Moore Ranch draft SEIS comment period simultaneously with the comment period for two other draft SEISs for proposed ISR facilities: Lost Creek and Nichols Ranch. Some commenters provided a single document that included comments for two or three of the proposed projects. NRC screened each document to determine if it applied only to one project or to multiple projects. For documents that commented on multiple projects, copies of the document were provided to each individual project and were treated independently from that point forward. Each document was given a unique number based on the order in which the documents were received. The prefix "MR" was attached to the identification number to indicate that this document, or document copy if originally addressing multiple projects, was for Moore Ranch. For documents addressing multiple projects, commenters had specified which

Appendix B

comments applied to the Moore Ranch Project. Sometimes comments were specifically directed only to the Moore Ranch Project. In other cases, the commenter stated that the same comment applied to multiple projects. Only comments concerning Moore Ranch, either uniquely or jointly, were identified and processed within the Moore Ranch version of the document. Lost Creek and Nichols Ranch specific comments were not identified and processed within the Moore Ranch version of the document.

The NRC staff reviewed all comment documents and identified, marked, and consecutively numbered individual, unique comments in each document. Comment numbers are followed by a two-part numbering system separated by a hyphen. The part of the comment number to the left of the hyphen is the document number. The number to the right of the hyphen is a consecutive, unique- count number for each comment identified in a specific comment document. Table B3–1 lists all commenter names, their affiliations, the comment document number assigned to their commenter letter, and the ADAMS Accession Number for the commenter letter. This table can be used by readers to electronically search the report to locate comments submitted by specific individuals or to find individuals associated with comments described in Section B.5.

In addition to the numbering, each unique comment was also assigned a topic category to facilitate sorting and reviewing comments on similar topics. Topic categories aligned with the topics addressed in Section B.5 of this appendix. Following the initial comment identification review, the identified comments were entered into a database that allowed individual comments to be sorted by topic and distributed to staff for further consideration. Staff then continued sorting and reviewing all comments within specific topic categories, developed comment

Table B3–1. Public Commenter Name With Affiliation and Comment Document Number			
Name	Affiliation	Comment Document Number	ADAMS Accession Number
Anderson, S.	Powder River Basin Resource Council	MR003, MR014	ML100271048 ML100690176
Currit, R.	Wyoming State Historic Preservation Office	MR007	ML100341192
Fettus, G.	Natural Resources Defense Council	MR001	ML100270995
Fettus, G. Lombard, C.	Natural Resources	MR017	ML100850378
Jantz, E.	New Mexico Environmental Law Center	MR005, MR012	ML100270996, ML100890375
Jones, J.	Public	MR010	ML100890223
Jones, S.	Wyoming Outdoor Council	MR006, MR016	ML100271689, ML100740240
Loomis, M.	Wyoming Mining Association	MR019	ML100640056

Appendix B

| Table B3–1. Public Commenter Name With Affiliation and Comment Document Number ||||
Name	Affiliation	Comment Document Number	ADAMS Accession Number
McKenzie, D.	Wyoming Department of Environmental Quality (WDEQ), Land Quality Division	MR018	ML100621314
Bott, K.	WDEQ, Air Quality Division		
Schroeder, T.	WDEQ, Industrial Siting Division		
Conrad, M.	WDEQ, Water Quality Division		
Anderson, C.	WDEQ, Solid and Hazardous Waste Division		
Emmerich, J.	Wyoming Game and Fish Department		
Pugsley, C.	National Mining Association	MR013	ML100690165
Ratner, J.	Western Watersheds Project	MR004	ML100270999
Rushin, C.	U.S. Environmental Protection Agency, Region 8	MR015	ML100680712
Stewart, R.	U.S. Department of the Interior	MR002	ML100341191
Svoboda, L.	U.S. Environmental Protection Agency, Region 8	MR020	ML100890218
Thompson, T.	National Mining Association	MR008	ML102080471
Viviano, P.	Public	MR008	ML100890221
Winter, J.	Uranium One Americas	MR009	ML100570072

summaries and responses for this appendix, and changed the draft SEIS, as necessary, to address the public comments.

Based on the similarity of comments related to a specific topic, as appropriate, staff consolidated same or similar comments within each topic to facilitate developing responses. This approach allowed addressing of multiple, similar comments with a single response, avoiding duplication of effort and enhancing readability of this report. For each comment, or

Appendix B

group of comments, a response has been provided. Each response indicates whether or not the SEIS was modified as a result of the comment.

B4 MAJOR ISSUES AND TOPICS OF CONCERN

The majority of comments received specifically addressed items within scope of the draft SEIS. Topics raised included, and were limited to, a variety of concerns about the purpose, need, and scope of the draft SEIS; regulatory issues; NEPA-related concerns; the description of the ISR process; land use; groundwater; surface water; ecology; air; historic, cultural, and Native American concerns; socioeconomic concerns; public health concerns; waste management; and cumulative effects. Other comments addressed topics and issues that are not applicable to the SEIS, including general support or opposition for uranium milling, legacy of past uranium mining and milling, evaluation of the NRC regulatory program or licensing process, comparison of the Moore Ranch financial assurance to previous restoration funding, compensation requests for loss of private water supplies, environmental impacts at disposal facilities for radioactive byproduct material other than Moore Ranch, and comments not specifically directed towards the draft SEIS [e.g., comments exclusively directed towards the generic environmental impact statement (GEIS)].

B5 COMMENT SUMMARIES AND RESPONSES

Detailed responses to comments are provided in this section. The structure of this section is based on the comment topics provided. Within each topic-specific subsection, the detailed presentation of comment and response information includes the applicable comment identification numbers, comment summaries, and the NRC staff response.

B5.1 General Opposition

Comments: MR001-002; MR002-005; MR010-006; MR012-079
Some commenters found the GEIS "wanting" and noted that the GEIS environmental analysis was deficient in several respects; however, they did not provide examples or citations. Another commenter was opposed to the proposed project because the long-term effects from ISR sites are unknown. Other commenters stated that NRC should "scrap" the current draft SEIS, conduct public scoping meetings, and issue another draft SEIS for public comment without reliance on the GEIS for site-specific analysis.

Response: The GEIS provides criteria for each environmental resource area to help determine the significance level for potential impacts (e.g., SMALL, MODERATE, or LARGE). The NRC staff applied these criteria to the site-specific conditions at the proposed Moore Ranch Project. The NRC recognizes that some commenters are not supportive of the development of either the GEIS or the Moore Ranch SEIS. NRC held a 103-day public comment period for the draft GEIS from July 28, 2008, through November 7, 2008, at which time members of the public were invited to provide comments, including eight public scoping meetings. The NRC considered and responded to comments received on the draft GEIS in Appendix G of the final GEIS [NOA published in the Federal Register on June 5, 2009 (74 FR 27052)]. Therefore, comments on the GEIS are beyond the scope of the Moore Ranch SEIS.

The Moore Ranch draft SEIS public participation process is described in Section 1.4.2 of the final SEIS. The Moore Ranch draft SEIS was prepared to fulfill the requirement under

Appendix B

10 CFR 51.20(b)(8) to prepare an EIS, or supplement to an EIS for the issuance of a source material license. The GEIS provided a starting point for the NRC NEPA analysis at the Moore Ranch site. The Moore Ranch site environmental review was initiated by the applicant submittal and NRC acceptance of the license application for detailed technical review, as discussed in the final SEIS Section 1.6.1. No further changes were made to the Moore Ranch SEIS beyond the information provided in this response.

Comments: MR010-001; MR011-001
Some commenters were adamantly opposed to granting a project permit. Another commenter stated that the ISR process was an injustice to the people of Wyoming.

Response: *The NRC recognizes that some commenters are not supportive of uranium milling. These comments are beyond the scope of the GEIS.*

Comment: MR017-003
A commenter was opposed to *in-situ* milling, and described it as "not a benign substitute for past uranium recovery" and suggested that the NRC "...examine and present to the public a precise history of conditions at in-situ leach (ISL) uranium mining operations..." both pre and postoperation.

Response: *The NRC recognizes that some commenters are not supportive of uranium milling. Chapter 2 of the GEIS provides information on the uranium recovery using the ISR process. Information regarding operational experience at ISR facilities is discussed in Section 2.11 of the GEIS. These comments are beyond the scope of the SEIS.*

Comment: MR017-033
A commenter urged the withdrawal of both the draft SEIS and the final GEIS, for failing to meet the requirements of NEPA. The commenter stated the document was legally deficient because it failed to address a number of substantive matters, take a hard look at the proposed action, consider a reasonable range of alternatives, and analyze cumulative impacts in the region of the proposed action.

Response: *The NRC recognizes that some commenters are not supportive of the development of either the SEIS or GEIS. As previously noted, NRC held a 103-day public comment period for the draft GEIS from July 28, 2008, through November 7, 2008, at which time members of the public were invited to provide comments, including eight public scoping meetings. The NRC considered and responded to comments received on the draft GEIS in Appendix G of the final GEIS (see NOA published in the Federal Register on June 5, 2009 (74 FR 27052). Therefore, comments on the GEIS are beyond the scope of the Moore Ranch SEIS.*

The Moore Ranch SEIS was prepared in accordance with NRC guidance in NUREG–1748 (NRC, 2003) and is consistent with NRC implementing regulations in 10 CFR Part 51. Chapter 2 of the Moore Ranch SEIS describes the proposed action and alternatives and Chapter 5 analyzes the cumulative impact from licensing the proposed Moore Ranch Project. No further changes were made to the Moore Ranch SEIS beyond the information provided in this response.

For detailed comments and responses on topics related to those expressed in some of the general opposition comments, see the following sections of this comment response appendix:

NEPA Process (B5.4); Purpose, Need, and Scope of the SEIS/GEIS(B5.5); Public Involvement (B5.8); and History and Legacy of Uranium Mining (B5.17).

B5.1.1 References

10 CFR Part 40. *Code of Federal Regulations*, Title 10, *Energy*, Part 40, "Domestic Licensing of Source Material." Appendix A, "Criteria Relating to the Operation of Uranium Mills and the Disposition of Tailings or Wastes Produced by the Extraction or Concentration of Source Material From Ores Processed Primarily for Their Source Material Content."

74 FR 27052, NRC, June 5, 2009. "Notice of Availability of Final Generic Environmental Impact Statement for In-Situ Leach Uranium Milling Facilities." *Federal Register*. Vol. 74, No. 107. pp. 27052–27054.

NRC (U.S. Nuclear Regulatory Commission), 2009. NUREG–1910, "Generic Environmental Impact Statement for In-Situ Leach Uranium Milling Facilities." Washington, DC: NRC. May.

NRC, 2003. NUREG–1748, "Environmental Review Guidance for Licensing Actions Associated with NMSS Programs." Washington, DC: NRC. August.

B5.2 General Support

Comment: MR018-096
A commenter was generally supportive of the development of the SEIS, stating it was a "great improvement from prior versions."

Response: *The NRC recognizes that some commenters are supportive of the development of the environmental review for the proposed Moore Ranch Project. However, these comments are beyond the scope of the SEIS.*

Comment: MR019-001
A commenter was supportive of the benefits that the proposed Moore Ranch Project would bring to the surrounding area and to the State of Wyoming to include jobs, tax revenues, and domestically produced uranium to fuel current U.S. nuclear reactors.

Response: *The NRC recognizes that some commenters are supportive of the development of the Moore Ranch in-situ uranium milling facility. However, these comments are beyond the scope of the SEIS.*

B5.3 General Environmental Concerns

Comments: MR003-002, MR006-002
Two commenting organizations stated that uranium exploration and production impacts many of their members since these individuals live, work, or recreate in areas where such activities are conducted.

Response: *The NRC acknowledges that uranium milling activities may impact individuals who live, work, or recreate in and around the proposed Moore Ranch Project. The environmental review documented in this final SEIS addresses potential environmental impacts covering a*

Appendix B

variety of resource areas that can affect impacted individuals. Because the comment was general in nature, no changes were made to the final SEIS.

B5.4 NEPA Process

B5.4.1 GEIS/SEIS

Comment: MR017-006
One commenter stated that NRC, in its final GEIS, had provided little more than a cursory response to comments submitted by the commenter and others on the draft GEIS. For example, the commenter stated that the U.S. Bureau of Land Management (BLM) and the U.S. Environmental Protection Agency (EPA) submitted approximately 77 comments, only 16 of which resulted in changes to the GEIS. The commenter considered that this minimal response meant that NRC did not fulfill its responsibility under NEPA, which requires that agencies respond to comments submitted by the public or cooperating agencies. The commenter stated that NRC responses to comments on the draft GEIS were conclusory and non-responsive, thereby failing a basic requirement of NEPA and the agency duty to supplement, modify, or improve its analyses in response to comment.

Response: *NRC disagrees with the commenter that the final GEIS response to comments was inadequate. NEPA requires an agency to have a reasonable response to comments but does not require an agency to accept recommendations or suggestions of other agencies or commenters. An agency is not obligated to conduct new studies in response to issues raised in comments, nor is it obligated to resolve conflicts raised by opposing viewpoints. The standard requires that agencies identify opposing views found in the comments such that differences in opinion are readily apparent and there is a good faith, reasoned analysis in the response. The NRC published the final GEIS on June 5, 2009 (74 FR 27052). The final GEIS included Appendix G, which was dedicated to identifying and summarizing comments submitted on the draft GEIS and NRC responses to those comments. Pursuant to NRC regulations under 10 CFR Part 51 that implement NEPA, and specifically 10 CFR 51.91(a), NRC responses took one of the following forms:*

(i)   *Modification of alternatives, including the proposed action*
(ii) *Development and evaluation of alternatives not previously given serious consideration*
(iii) *Supplementation or modification of analyses*
(iv) *Factual corrections*
(v)   *Explanation of why comments do not warrant further response, source citing, authorities, or reasons that support this conclusion*

The NRC staff consider its response to comments on the draft GEIS, as documented in the main text and appendices of the final GEIS, to be consistent with NRC responsibilities under its NEPA implementing regulations, 10 CFR Part 51. No further modification to the Moore Ranch SEIS was made beyond the information provided in this response.

Comments: MR012-011; MR017-016
Two commenters stated that there were many figures in the Moore Ranch draft SEIS that were either illegible or barely legible. Both commenters stated that the failure to present legible information within the draft SEIS violated NEPA and its implementing regulations. One

Appendix B

commenter additionally stated that it was difficult to review the data provided from such poor quality reproductions.

Response: NRC regrets that many of the figures in the Moore Ranch draft SEIS did not reproduce well in the printed versions of the document that were mailed to the commenters, among others. Additionally, the electronic version of the Moore Ranch draft SEIS, which NRC made available through its public website, contained the same poor quality figures because the electronic version was simply a scanned copy of the printed version. However, as one of the commenters noted, the NRC published revised figures and posted them on its public website as quickly as possible.

The NRC staff revised 17 figures in this final SEIS.

Comments: MR012-033; MR012-035

One commenter stated that neither the GEIS nor the Moore Ranch draft SEIS were based on accurate data related to the impacts of spills and leaks on water resources and that, therefore, neither document was sufficient under NEPA. As background to this comment, the commenter characterized the NRC staff evaluation in the GEIS and the Moore Ranch draft SEIS of the impacts of spills and leaks on water resources as relying on incomplete and inaccurate data, thus, resulting in a misleading impact evaluation in both documents. Additionally, the commenter considered that the NRC had not conducted the requisite site-specific analysis of impacts from spills and leaks at the Moore Ranch project, but instead the NRC had simply stated that the site-specific conditions at Moore Ranch were consistent with the GEIS affected environment description to find the significance of such impacts to be SMALL.

Response: The NRC site-specific analysis of the potential environmental impacts to water resources from spills and leaks is found in Section 4.5 of the Moore Ranch draft SEIS. This section includes the evaluation of potential surface water and wetlands resources impacts and to near-surface groundwater resources from spills and leaks during operations proposed for the Moore Ranch Project. This site-specific analysis determined that, for the Moore Ranch Project, the significance of potential impacts is expected to be SMALL. This site-specific determination draws on the evaluation found in Sections 4.2.4 and 4.3.4 of the GEIS, wherein evaluation criteria for considering the significance of impacts is provided. For impacts to surface water and wetlands resources the criteria include: size of a spill, success of remediation, use of the surface water for domestic or agricultural purposes, proximity of the spill to surface water, and compliance with storm water and National Pollutant Discharge Elimination System (NPDES) permits issued by the State of Wyoming. In the GEIS, the NRC staff determined that such impacts could be SMALL to MODERATE, depending on site-specific conditions. For potential impacts to near-surface groundwater resources, the criteria included: proximity of the shallow aquifer to the surface, use of the shallow aquifer for domestic or agricultural purposes, and connection of the shallow aquifer to other locally or regionally important aquifers. In the GEIS, the NRC staff determined that impacts to near-surface aquifers could be SMALL to LARGE, depending on site-specific conditions.

As discussed in this response, the NRC staff conducted a site-specific evaluation of impacts to water resources from spills and leaks at the proposed Moore Ranch Project. That evaluation determined that such impacts are expected to be SMALL, given the proposed operations and site-specific conditions. No modification has been made to the SEIS beyond the information provided in this response.

B5.4.2 Adequacy of Impact Assessment

Comments: MR015-038; MR012-061
A commenter rated the Moore Ranch draft SEIS as inadequate pursuant to commenter responsibilities under NEPA and authority under Section 309 of the Clean Air Act. The commenter further indicated belief that the SEIS did not meet the purposes of NEPA and, therefore, should be formally revised and made available for public comment in a supplemental or revised SEIS. The commenter indicated that if its concerns were not addressed, then it would consider elevating the issue for referral to the Council on Environmental Quality (CEQ) for resolution.

Response: The NRC disagrees with the commenter and believes that the Moore Ranch final SEIS adequately addresses all public comments and does not need to be reissued for public comment. For further information on how the SEIS tiers from the GEIS and the process of determining impacts, refer to Section B5.5.2 of this appendix. The NRC recognizes EPA authorities and responsibilities under NEPA and the Clean Air Act to rate draft environmental impact statements. The NRC staff consider that it has prepared the Moore Ranch draft SEIS consistent with its regulations under 10 CFR Part 51 that implement NEPA and its guidance for conducting environmental reviews as found in NUREG-1748 (NRC, 2003). Pursuant to 10 CFR 51.73, the NRC staff issued the Moore Ranch draft SEIS for public comment on December 11, 2009 (74 FR 65806), and the comment period on the document closed on March 3, 2010 (75 FR 6065). As discussed previously in this appendix, 691 comments were received on the Moore Ranch draft SEIS, among which were additional comments raised by this commenter. Consistent with 10 CFR 51.91(a), the NRC staff have considered and responded to all comments received.

Comment: MR016-005
A commenter stated that the Moore Ranch draft SEIS was deficient because the NRC had not balanced the need for the project against the potential impacts to humans and the natural environment.

Response: NRC does not analyze the market conditions or business decision of a private entity to submit a license request as part of its licensing decision. NRC licensing decision is based on the safety evaluation review and environmental review of an applicant submitted license application. The NRC staff prepared the Moore Ranch draft SEIS consistent with regulations under 10 CFR Part 51 that implement NEPA and guidance for conducting environmental reviews found in NUREG-1748 (NRC, 2003). NRC regulations at 10 CFR 51.10(b) state that NRC "recognizes a continuing obligation to conduct its domestic licensing and related regulatory functions in a manner which is both receptive to environmental concerns and consistent with the [NRC's] responsibility as an independent regulatory agency for protecting the radiological health and safety of the public."

As stated in Section 1.3 of this SEIS, the purpose and need of the proposed action is to either grant or deny Uranium One's license application to use ISR technology to recover uranium and produce yellowcake at the proposed Moore Ranch project. As a regulatory agency, NRC has no role in a company's business decision to submit a license application to operate an ISR facility at a particular location. No further changes have been made to the SEIS beyond the information provided in this response.

Appendix B

Comments: MR016-021; MR016-022
A commenter stated that the Moore Ranch draft SEIS and the GEIS were inadequate for the purposes of NEPA, and that the Moore Ranch draft SEIS should be withdrawn, a scoping process begun, and the document subsequently reissued for public comment.

Response: The NRC staff prepared the Moore Ranch draft SEIS consistent with its regulations under 10 CFR Part 51 that implement NEPA and its guidance for conducting environmental reviews as found in NUREG-1748 (NRC, 2003). Additionally, the GEIS, which this final SEIS supplements (see Section 1.4.1), provides a starting point for NRC NEPA analyses for site-specific license applications for new ISR facilities, such as the Uranium One application for the Moore Ranch Project. Section 2.11 of the GEIS discusses historic ISR facility operations licensed by NRC. The NRC staff used this historical information in its generic evaluation of potential environmental impacts from the construction, operation, aquifer restoration, and decommissioning of ISR facility in specific geographic regions of the western United States. This final SEIS tiers and incorporates by reference from the GEIS relevant information, findings, and conclusions, depending upon the similarity between the Uranium One (the applicant) proposed facility, activities, and conditions at the Moore Ranch site with those for the reference facility evaluated in the GEIS.

The scope of the Moore Ranch SEIS is discussed in Section 1.4 of the final SEIS, and opportunities for public participation are discussed in Section 1.4.2. In accordance with NRC regulations in 10 CFR 51.26(d), the NRC staff need not conduct a public scoping process for a SEIS. Further information regarding scoping is addressed in Section B5.6 of this appendix.

Comment: MR017-005
A commenter expressed concern that the GEIS and the Moore Ranch draft SEIS gave short attention to the recurring issues with uranium solution mines in the United States and other countries and ignored long historical evidence that environmental harms, such as the escape of extraction fluids from the ore recovery zone, do occur. The commenter provided examples of harms that the commenter considered NRC failed to analyze in the draft SEIS. The commenter felt it was incumbent upon the NRC to comprehensively address the environmental risks inherent in an expansion of the domestic uranium mining and milling industry and to have sufficient protections in the licensing process to prevent a recurrence of previous environmental harms to the environment and public health.

Response: The NRC evaluated historical information on ISR operations licensed by the NRC (see GEIS, Section 2.11) and considered this historical information to assess the potential environmental impacts from the construction, operation, aquifer restoration, and decommissioning of an ISR facility in specific geographic regions of the western United States. The Wyoming East Uranium Milling Region, where the Moore Ranch Project would be located, is in one of these geographic regions (see GEIS, Section 4.3). The final SEIS tiers and incorporates by reference from the GEIS relevant information, findings, and conclusions to the extent that the applicant proposed facility, activities, and conditions are consistent with the reference facility activities, information, and impact conclusions described in the GEIS. As discussed in Section 1.7.1 of the GEIS, if a new license (e.g., for ISR operations at the Moore Ranch Project) was issued following NRC environmental and safety reviews, NRC would ensure that the applicant complied with the conditions in its NRC license and the applicable regulations through an inspection program managed out of the NRC Region IV office in Arlington, Texas.

Appendix B

No further changes have been made to the SEIS beyond the information provided in this response.

Comments: MR017-007; MR012-026; MR005-005
A commenter asserted that NRC had not taken a "hard look" at the environmental impacts, and that the draft SEIS added little, if anything, to the sufficiency of NRC analysis of environmental impacts from the ISR process. The commenter stated that the draft SEIS analysis mirrored the analysis in the final GEIS and that the lack of a discernable difference between the GEIS environmental impact findings and those contained in the draft SEIS was a clear indication that no searching analyses were performed, as required under NEPA. Another commenter stated that the analysis was deficient, but the commenter did not provide specific examples.

Response: As discussed in Section 1.4.1, the Moore Ranch SEIS supplements the GEIS, which provided a starting point for NRC site-specific analysis of potential environmental impacts from the construction, operation, aquifer restoration, and decommissioning of the proposed Moore Ranch ISR Project facility. The proposed Moore Ranch Project would be located in the Wyoming East Uranium Milling Region, one of the four specific geographic regions evaluated in the GEIS. Table 1-1 of the Moore Ranch SEIS shows the range of environmental impacts expected in the Wyoming East Uranium Milling Region, based on the GEIS analyses. Table 1-1 shows those resource areas (e.g., transportation, groundwater, noise), for which the GEIS concluded there was a range of potential impacts possible, depending on site-specific conditions; therefore, these resource areas were evaluated in detail in the SEIS.

The SEIS evaluated potential impacts in Chapter 4 and relied on a description of the proposed Moore Ranch Project facility and associated activities (SEIS Chapter 2), a description of the affected environment at and in the vicinity of the proposed Moore Ranch Project (SEIS Chapter 3), and the resource criteria identified in the GEIS to assess the significance of the environmental impacts. Each resource area was assessed by comparing the site-specific conditions at the proposed Moore Ranch Project with the conditions documented in the GEIS, in addition to identifying any new or significant information that could result in an environmental impact inconsistent with those identified in the GEIS. Table 2-3 summarizes the potential environmental impacts from implementing the proposed action at the proposed Moore Ranch Project.

No further changes have been made to the SEIS beyond the information provided in this response.

Comments: MR017-008; MR017-012; MR012-035
A commenter stated that NRC should evaluate the potential impact of spills and that the failure to consider spills in the draft SEIS, and more notably in the GEIS, was disappointing. The commenter also stated that the potential impacts to groundwater, surface water, and land from spills should be considered, with respect to a reasonable range of the possible severity of spills.

Response: Sections 4.4.1.2, 4.5.1.1.2, and 4.5.2.1.2 of the Moore Ranch SEIS present the analysis of the potential environmental impacts on land use, surface water, and groundwater from spills associated with the ISR operations at the proposed Moore Ranch Project. Sections 4.3.3.2, 4.3.4.1.2, and 4.3.4.2.2.1 of the GEIS discuss these impacts. Table 1-1 in the Moore Ranch SEIS summarizes the results of the GEIS analyses of impacts to these resources. The SEIS tiered and incorporated by reference from the GEIS, and the analysis demonstrated that the proposed facility, activities, and the site-specific conditions at the proposed Moore Ranch Project were comparable to that considered in the GEIS, which included:

- *Engineering controls to detect pressure changes in the wellfield piping system;*
- *Mechanical integrity testing (MIT) of completed wells prior to their placement in service and subsequent retesting every 5 years;*
- *Alarm systems for individual wells and in header houses;*
- *Daily visual inspection of wellfield monitoring;*
- *A spill response plan to address accidental spills;*
- *Requirements to remediate affected areas and dispose of contaminated soils;*
- *The near-surface aquifer at the proposed site is not an important source for local domestic or agricultural water supplies; and*
- *The near-surface aquifer is not hydraulically connected to other locally or regionally important aquifers.*

Furthermore, should a new license be issued for ISR operations at the Moore Ranch Project, the NRC staff would take necessary actions to respond to reported incidents at the facility, including spills. NRC would also conduct periodic inspections to determine compliance with applicable regulatory requirements, license conditions, and approved procedures. Potential violations and allegations would be evaluated and addressed through either the appropriate NRC enforcement or allegation programs.

No further changes have been made to the SEIS beyond the information provided in this response.

Comment: MR017-028
A commenter stated that NRC has the duty to fully explain the science, technology, and techniques used in the ISR process and to fully analyze and assess the environmental impacts of each aspect of the process. The commenter believes that an environmental review is impossible without historical data on the success rates of the technologies used in the ISR process. Only by doing so, the commenter states, can NEPA's "hard look" requirement be fulfilled.

Response: *Chapter 2 of the GEIS describes the ISR process used to evaluate the potential impacts from an ISR facility. All phases of the ISR facility lifecycle are described and information on the historical operating experience at ISR facilities is provided, with respect to safety significance and issues of public concern such as spills, leaks, excursions, and aquifer restoration. Chapter 2 of the GEIS discussed key aspects of the ISR process common to NRC-licensed ISR facilities to build the foundation for GEIS impact analyses. The focus of Chapter 2 was to discuss significant issues for ISR proposals and their potential environmental impacts, rather than to provide a detailed description of all aspects of every facility that NRC has licensed.*

Detailed information regarding the specific technologies, equipment, and operational practices and parameters applicable to the proposed Moore Ranch Project are provided in the applicant license application and are summarized in Chapters 2 and 6 of the SEIS. The NRC staff evaluated the adequacy of the license application with respect to operational safety and potential environmental impacts and determined that key aspects of the ISR process proposed for implementation at the Moore Ranch Project were consistent with those identified in the GEIS before incorporating by reference the relevant sections of the GEIS into the Moore Ranch SEIS. While NRC guidance discusses methods that are considered acceptable to staff, NRC does not prescribe technology or methods that must be used by an applicant nor is it necessary for NRC to proactively evaluate all available options in the GEIS or elsewhere before applications are received. Past experience suggests that ISR facilities use similar technology; by focusing on what is common, the GEIS provides a reasonable basis for supporting future ISR license application reviews. If an applicant application that includes unproven technology or methods not analyzed in the GEIS, the NRC review may require additional details and performance data and additional environmental impact analysis to verify that safety would be maintained. However, as discussed previously, the NRC staff have determined that the key aspects of the ISR process of Moore Ranch ISR Project were consistent with those identified in the GEIS. No additional changes were made to the Moore Ranch SEIS beyond the information provided in this response.

Comment: MR012-028

A commenter stated that NRC lacked a coherent framework for regulating ISR operations. Absent such a framework, the commenter stated that the NRC analysis in the Moore Ranch draft SEIS was arbitrary and ad hoc, and the public was unable to rely on any objective, consistent standards by which to judge the NRC site-specific environmental analysis. The commenter asserted that this approach confounded public participation in the NEPA process and is neither supported by NEPA nor its implementing regulations.

Response*: NRC disagrees with the commenter. NRC regulations addressing ISR facility licensing include 10 CFR Part 20, 10 CFR Part 40, Appendix A, and 10 CFR Part 51, which provides a coherent framework for regulating ISR facilities. NRC also has approximately 30 years of experience regulating ISR facilities, which has been considered in developing its environmental reviews and safety evaluation reports (SERs) and in promulgating the applicable regulations, guidance, and license conditions that have been used to regulate ISR facilities throughout that timeframe to protect public health and safety and the environment.*
The public participation activities associated with the preparation of the Moore Ranch draft SEIS are described in Section B2 of this comment-response appendix and in Section 1.4.2 of the final SEIS. The public participation process included commenting on the draft SEIS, which was issued for public comment on December 11, 2009. Approximately 700 comments were received.

No further changes have been made to the SEIS beyond the information provided in this response.

Comment: MR012-049

A commenter stated that the NRC analysis of groundwater impacts from aquifer restoration was insufficient and relied entirely on the GEIS framework for analyzing those impacts. As a result, the commenter noted that the NRC analysis was limited to impacts from consumptive use (i.e., water quantity).

Response*: Groundwater impacts from aquifer restoration are discussed in Section 4.5.2.1.3 of the SEIS. As stated there, the potential environmental impacts to groundwater resources during aquifer restoration are related to groundwater consumptive use, waste management practices, and groundwater quality. NRC analysis of impacts from consumptive use determined that such impacts would be SMALL, given that only nominal drawdown would be experienced in private*

Appendix B

wells within 3.2 km [2 mi] of the proposed license area. Similarly, impacts from waste management practices were expected to be SMALL due to the characteristics of the proposed host formations for deep well injection and the rigor of the Wyoming Department of Environmental Quality (WDEQ) permitting process for authorizing deep well injection. Finally, the NRC staff determined that the impact of aquifer restoration on groundwater quality would be SMALL, because groundwater quality in the impacted aquifers would be restored to water quality standards that protect human health and the environment, and that surrounding aquifers would not be impacted.

No further changes have been made to the SEIS beyond the information provided in this response.

Comment: MR015-002
One commenter stressed the need for site-specific information in the analysis of all potential impacts associated with ISR projects, and that the Moore Ranch draft SEIS did not provide adequate information to effectively address key issues.

Response: *As discussed in Section 1.4.1, this SEIS was prepared to fulfill the requirement at 10 CFR 51.20(b)(8) to prepare either an EIS or a supplement to an EIS for the issuance of a source material license for an ISR uranium recovery facility. This final SEIS supplements the GEIS, which provided a starting point for the NRC NEPA analysis (documented here) of the Uranium One license application for the proposed Moore Ranch Project. The NRC site-specific NEPA analysis used detailed information and descriptions of the proposed ISR facility and activities. For example, the characterization of the subsurface stratigraphy and groundwater hydrology at the proposed Moore Ranch Project was based on data collected by the applicant from 422 boreholes and 34 monitor wells. In addition, approximately 2,700 rotary drill holes and approximately 130 core holes were completed by Conoco Minerals Corporation from the 1970s through the mid-1980s when the site was being evaluated for conventional milling. These data were used by the applicant and reviewed by NRC to characterize the site-specific stratigraphy and groundwater hydrology. The hydrologic data collected from these wells were used to evaluate how the ore zone would behave, assuming the applicant injected fluids at a certain rate. For each of the resource areas evaluated in the SEIS, the NRC staff reviewed the information provided by the applicant, validated the information as appropriate, and evaluated the potential impact to the environment in the SEIS.*

No further changes have been made to the SEIS beyond the information provided in this response.

B5.4.3   Range of Reasonable Alternatives

Comments: MR012-007; MR016-002; MR016-004
A commenter noted that while NEPA does not require the NRC to consider every possible alternative to the proposed action, it does require that NRC consider all reasonable alternatives. The commenter stated that NRC failed to consider reasonable alternatives by limiting the analysis to the proposed action and "no action" alternatives. Another commenter disagreed that the SEIS eliminated conventional milling as a reasonable alternative to consider for the Moore Ranch site, and noted that with good topsoil preservation and appropriate reclamation, it may be better from an environmental perspective. Both commenters stated the failure to consider reasonable alternatives was a violation of NEPA. One commenter called for the NRC to reevaluate the alternatives within the GEIS and the Moore Ranch SEIS.

Appendix B

Response: The range of reasonable alternatives is defined by the proposed federal action and the purpose and need for the proposed federal action. As a regulatory agency, the proposed federal action for the Moore Ranch site is a NRC decision of whether to grant or deny the license application of a private party. The purpose and need for the proposed federal action does consider the applicant goals and objectives to extract uranium from a particular location, which helps define what are reasonable alternatives to the proposed federal action.

Reasonable alternatives considered in a site-specific environmental review depend on the proposed action and site conditions. As discussed in Section 2.1 of the SEIS, NRC considered all reasonable alternatives, including the No-Action Alternative not approving the license application. Section 2.2 of the SEIS provides a discussion of alternatives that were considered but eliminated from detailed study and the reasons for the elimination.

As noted in Section 1.4.5 of this final SEIS, NRC evaluated the potential environmental impact from issuing a license to Conoco, Inc., to construct and operate a uranium mill associated with an open-pit mine in the same geographic location now being considered for ISR of uranium (NRC, 1982). Further, as noted in NUREG–1508, underground mining would have more significant environmental impacts than ISR extraction, and the ore from underground mining would require processing at a conventional uranium mill to produce the final product. Significant quantities of tailings (the residual rock materials after uranium removal) would be produced by conventional milling, which are normally disposed of on-site at the conclusion of mill operating life (NRC, 1997). NUREG–0706, Final Generic Environmental Impact Statement on Uranium Milling (NRC, 1980), provides a detailed evaluation of the impacts associated with tailings disposal from conventional uranium milling. The environmental impacts of underground mining and conventional milling would be more significant than those from ISR milling. Therefore, underground mining and conventional milling are not evaluated in the Moore Ranch SEIS.

While the NRC staff consider reasonable alternatives to the proposed action in the environmental review, the only alternative within NRC decision-making authority is to approve or not approve the license application. The NRC has no authority or regulatory control over an applicant's selection of uranium recovery technology to be used at the site. NRC regulatory authority is limited to evaluating an applicant license request to use ISR technology at the site. If the NRC decides to grant the license request, the applicant must comply with the license, NRC regulatory requirements, and any other relevant local, State or Federal requirements to operate their facility.

B5.4.4 References

10 CFR Part 20. *Code of Federal Regulations*, Title 10, *Energy*, Part 20, "Standards for Protection Against Radiation."

10 CFR Part 40. Appendix A. *Code of Federal Regulations*, Title 10, *Energy*, Part 40, Appendix A, "Criteria Relating to the Operations of Uranium Mills and to the Disposition of Tailings or Wastes Produced by the Extraction or Concentration of Source Material from Ores Processed Primarily from their Source Material Content."

10 CFR Part 51. *Code of Federal Regulations*, Title 10, *Energy*, Part 51, "Environmental Protection Regulations for Domestic Licensing and Related Regulatory Functions."

74 FR 27052, U.S. Nuclear Regulatory Commission (NRC). "Notice of Availability of Final Generic Environmental Impact Statement for *In-Situ* Leach Uranium Milling Facilities." *Federal Register*. Vol. 74, No. 107, pp. 27052-27054. June 5, 2009.

74 FR 65806, NRC. "Notice of Availability of Draft Environmental Impact Statement for the Moore Ranch *In-Situ* Recovery (ISR) Project in Campbell County, WY; Supplement to the Generic Environmental Impact Statement for *In-Situ* Leach Uranium Milling Facilities." *Federal Register*, Vol. 74, No. 237, pp. 65806-65810. December 11, 2009.

75 FR 6065, NRC. "Extension of Public Comment Period on the Draft Environmental Impact Statement for the Moore Ranch *In-Situ* Recovery Project in Campbell County, WY; Supplement to the Generic Environmental Impact Statement for *In-Situ* Leach Uranium Milling Facilities." *Federal Register*, Vol. 75, No. 24, pp. 6065-6066. February 5, 2010.

NRC (U.S. Nuclear Regulatory Commission), 2009. NUREG–1910, "Generic Environmental Impact Statement for In-Situ Leach Uranium Milling Facilities." Washington, DC: NRC. May.

NRC, 2003a. NUREG–1748, "Environmental Review Guidance for Licensing Actions Associated with NMSS Programs." Washington, DC: NRC. August.

NRC, 2003b. NUREG–1569, "Standard Review Plan for In-Situ Leach Uranium Extraction License Applications. Final Report." Washington, DC: NRC. June.

NRC, 1997. NUREG–1508, Final Environmental Impact Statement to Construct and Operate the Crownpoint Uranium Solution Mining Project, Crownpoint, New Mexico." Washington, DC: NRC.

NRC, 1980. NUREG–0706, "Final Generic Environmental Impact Statement on Uranium Milling," Project M-25. Washington, DC: NRC.

B5.5 Purpose, Need, and Scope of the SEIS/GEIS

B5.5.1 Description of the SEIS/GEIS Purpose and Need

Comments: MR012-001; MR012-003; MR012-004; MR012-005; MR017-025; MR016-001; MR016-003; MR016-006
Three commenters noted that the statement of purpose and need in the GEIS was too limited, which resulted in a limited analysis of reasonable alternatives in the Moore Ranch SEIS. One commenter stated that because of the construct of the purpose and need in the GEIS, the subsequent Moore Ranch SEIS was too narrow and limited the range of reasonable alternatives either to granting or denying the applicant licensing request, and thus failed to satisfy the fundamental requirements of the NEPA. This commenter stated further that by limiting the purpose and need scope, only two alternatives in the Moore Ranch draft SEIS were evaluated in detail, which the commenter concluded meant that only one alternative, licensing the project, was given serious consideration. The commenter also stated that the alternatives analysis violated both the letter and spirit of NEPA and that if the NRC had articulated a reasonable and legitimate purpose and need, the range of alternatives considered would likewise have been reasonable.

Appendix B

A second commenter stated that NRC should craft a statement of purpose and need in consultation with other involved Federal and State agencies that related the uranium recovery program to broad national objectives within NRC purview, such as "improving remediation of land and water impacts from the recovery of source or byproduct materials" or "ensuring the long-term isolation from the human and natural environment of harmful radionuclides and chemical toxins produced in the nuclear fuel cycle."

A third commenter stated that because of the construct of the purpose and need statement, the applicant had not been required to identify a customer for its product, but rather assumed that such a customer would exist to buy the uranium and that this did not satisfy the NEPA "hard look requirements." This commenter also disagreed with the NRC statement in the Moore Ranch SEIS that NRC does not have a role in a company business decision to submit a license application. The commenter argued that the purpose and need of the project could not be determined without NRC evaluating whether or not the project is economically viable at the Moore Ranch site for the particular product.

Response: The statement of the purpose and need is found in Section 1.3 of this final SEIS and is derived from the proposed federal action. Under the AEA, NRC's has statutory authority to issue licenses for the possession and use of AEA regulated radioactive materials and particular activities involving this material. Based on NRC statutory authority, the proposed federal action is NRC's decision whether to grant or deny a private party's licensing application to conduct ISR operations to extract uranium and produce yellowcake at a particular site. The purpose and need for the proposed federal action does consider the applicant goals and objectives to extract uranium from a particular location, which helps define reasonable alternatives to the proposed federal action. As a result, NRC limits its analysis of alternatives to accomplishing the objective of extracting uranium from the applicant site.

The alternatives to the proposed action are discussed in Sections 2.1 and 2.2 of this final SEIS. As discussed there, the No-Action Alternative (i.e., denial of the license application) was considered in detail in the draft SEIS analysis, while alternative mining and milling methods (conventional and heap leach), alternate sites, alternate lixiviants, and alternate wastewater treatment methods were considered in Section 2.2 of the Moore Ranch SEIS, but were not analyzed in detail. Alternate sites analysis is limited to the occurrence of the subsurface ore body and could consider the placement of the wellfields. Section 2.1.1.2 was added to this final SEIS to discuss alternative wastewater disposal options. Section 4.14.1.2 discusses the impacts from alternative wastewater disposal options. NRC does not analyze the market conditions or business decision of a private entity to submit a license request as part of its licensing decision. An NRC licensing decision is based on the safety evaluation review and environmental review of the license application.

NRC does recognize that NRC performs an analysis of alternative energy production methods and alternative sites in its environmental reviews of nuclear power plant licensing actions. In that case, the proposed action involves the decision of whether to grant or deny the license of an energy production facility, and the facility could perform this function at other locations. Even in these environmental reviews, NRC notes that the decision regarding energy policy and energy planning, including whether to implement energy options like solar power, conservation, or even nuclear power, are also made by the utility and State and Federal (non-NRC) decisionmakers, and NRC does not have authority to make these decisions. If NRC decides to renew or grant an operating license to a nuclear power plant the decision of whether to operate

Appendix B

the nuclear power plant, or an alternative is left up to the appropriate State, utility, and Federal entities.

In comparison, an ISR facility does not generate energy and is a fixed site based on the location of the ore body. As a result, alternative energy production methods and alternative site locations are not related to the proposed federal action to decide whether to grant or deny an applicant license request to extract uranium from a particular site. NRC has not included an analysis of alternate geographic locations, alternative energy production methods, or market conditions in this final SEIS.

Sections 2.1.1.2 and 4.14.1.2 were added to the final SEIS to discuss alternative waste disposal options in response to these comments.

B5.5.2   Use of the GEIS in Site-Specific Environmental Reviews

Comments: MR005-004, MR012-036
Commenters expressed concerns about how information from the GEIS was incorporated into the Moore Ranch draft SEIS. One commenter stated that a regional description of the affected environment could not substitute for a meaningful description and analysis of the impacts on the environment from the proposed Moore Ranch ISR Project. Another commenter stated that because the Moore Ranch SEIS was one of three tiered from the GEIS, that analysis of the relationship between the three SEISs and the GEIS was warranted and expressed concern about how information was incorporated from the GEIS.

Response: *The relationship of the Moore Ranch SEIS to the GEIS is discussed in Sections 1.1 and 1.4, and as noted, the SEIS is a supplement to the GEIS, wherein the GEIS provided a starting point for NRC NEPA analyses for site-specific license applications for new ISR facilities and for applications to amend or renew existing ISR licenses. The Moore Ranch SEIS tiers and incorporates by reference from the GEIS relevant information, findings, and conclusions concerning potential environmental impacts. The structure of Chapter 3 of the SEIS was first to describe each resource area (e.g., land use, geology and soils, water resources) at a regional level and then to provide local and site-specific characteristics. The extent to which NRC incorporated GEIS impact conclusions depended on the consistency among the applicant's proposed facility and activities and conditions at the proposed Moore Ranch site and the reference facility description, activities, regional conditions, and information or conclusions in the GEIS. NRC determinations regarding potential environmental impacts and the extent to which GEIS impact conclusions were incorporated by reference are discussed in Chapter 4 of this SEIS.*

Sections 1.7.1 and 1.8 of the GEIS provided a general discussion of the NRC process for reviewing license applications for proposed new ISR uranium recovery projects. An NRC site specific environmental review is conducted for each license application. As discussed in GEIS Section 1.8, each site-specific environmental review will evaluate information provided on all resource areas to ensure sufficient information to assess environmental impacts has been provided in a license applicant environmental report. The applicant environmental report includes a detailed description and assessment of the proposed action, alternatives, site characterization information, and potential environmental impacts. If sufficient information were not provided, NRC would request additional information (RAI) to ensure the information is complete. The GEIS does not relieve the applicant of the need to adequately document site-specific information in its application.

Appendix B

The NRC staff initially rely on applicant information provided by the applicant as well as information and conclusions from a separate detailed safety review conducted by NRC staff in documenting the staff environmental review. NRC staff confirm important attributes of the license application and environmental report through visits to the proposed site location and vicinity, independent research activities, and consultations with appropriate Federal, tribal, State, and/or local agencies. If, after reviewing the detailed information on the site-specific proposal provided by the applicant, the NRC staff find commonality between site conditions and those evaluated in the GEIS, the staff may incorporate by reference into the documentation of the site-specific environmental review the applicable portions or conclusions from the GEIS. Whether or not the staff use information from the GEIS in completing their site-specific environmental review, the conclusions in the site-specific environmental review documentation would be required to have sufficient technical basis.

Section 1.8.3 of the GEIS describes the process by which the NRC staff use the GEIS to help determine the significance of site-specific environmental impacts in the Moore Ranch SEIS (see Chapter 4). As discussed in the GEIS, the GEIS provides criteria for each environmental resource area to help determine the significance level of potential impacts (e.g., SMALL, MODERATE, or LARGE). The NRC staff applied these criteria to site-specific conditions at the Moore Ranch Project to determine the significance of potential impacts. Finally, the NRC staff compared the conditions of the proposed Moore Ranch site and activities under review to the conditions and aspects identified and discussed in the GEIS to see whether the environmental impact conclusions for a particular resource area could be adopted in the Moore Ranch SEIS. The NRC staff compared whether the GEIS impact significance conclusions for a specific resource area could be adopted in full, only in part, or not at all. Chapter 4 of the SEIS discusses the extent to which the GEIS conclusions could be adopted, including the supporting information and data that form the basis for that determination. Additionally, where the GEIS conclusions could be adopted only in part or not at all, the NRC staff also determined the significance of environmental impacts for those resource areas and provided the basis for that determination. For each resource area in Chapter 4 of the Moore Ranch SEIS, the NRC staff provided a conclusory statement (i.e., the one identified by the commenter), which followed the site-specific information and analysis to indicate the extent to which new and significant information affected the ability to adopt impact conclusions from the GEIS. No changes were made to the SEIS beyond the content of this response.

B5.5.3   Scope of the SEIS/GEIS

Comments: MR001-005; MR012-002; MR012-055
Several commenters expressed concern over the scope of the SEIS and relatedly, the GEIS. One commenter stated that certain aspects of the GEIS, including its scope, appeared to be binding upon the SEIS. The commenter further noted that, by improperly limiting the scope of the SEIS, the NRC failed to analyze a number of impact areas. Another commenter stated that additional comments on the GEIS were appropriate,given that the GEIS did not apply to any federal plan or project and did not represent any final NRC regulatory or policy decision, which, therefore, made it impossible for any member of the public to meaningfully comment on the GEIS in a concrete context. The same commenter noted that despite many public comments on the GEIS urging NRC to consider the impacts of previous uranium mining and milling, the NRC deemed that contamination from past uranium mining and milling to be outside the GEIS scope. One commenter requested that the public have an opportunity to review NRC proposed rulemaking on groundwater protection at ISL facilities and urged NRC to extend the draft SEIS

Appendix B

comment period to allow NRC to promptly release its associated draft groundwater protection rule so it could be reviewed concurrently with the draft SEISs.

Response: The scope of this SEIS is discussed in Section 1.4. As discussed there, the NRC staff consider the scope of the GEIS to be sufficient for the purposes of defining the scope of this SEIS. In so stating, the NRC considers that topics determined to be within scope for the GEIS were also within scope for the SEISs. NRC made this determination based on its review of the information provided by the applicant and as a result of meetings with Federal, State, and local agencies and contact with potentially interested Native American tribes and local authorities, entities, and public interest groups in person and via e-mail and telephone (see Section 1.4.2 of this SEIS).

Concerning public involvement in the GEIS, NRC accepted public comments on the scope of the GEIS from July 24 to November 30, 2007, and held three public scoping meetings to aid in this effort. Additionally, NRC held eight public meetings to receive comments on the draft GEIS, published in July 2008. Comments on the draft GEIS were accepted between July 28 and November 8, 2008. Comments received during scoping and on the draft GEIS are available through NRC ADAMS database on the NRC website (http://www.nrc.gov/reading-rm/adams.html). Transcripts of the scoping meeting and draft GEIS comment meetings are available at http://www.nrc.gov/materials/ uranium-recovery/geis/pub-involve-process.html. A scoping summary report is provided as Appendix A to the GEIS (NRC, 2009a). As is evident in the public meeting transcripts and the written comments received during the scoping and public comment period for the GEIS, the NRC staff consider that meaningful and extensive public comments were received on the GEIS.

With respect to the specific comment that contamination from past conventional mining and milling was outside the scope of the GEIS, the NRC noted in Appendix A to the GEIS that such contamination could be assessed as part of a site-specific cumulative impacts evaluation. Chapter 5 of this final SEIS provides the NRC site-specific cumulative effects analysis. In Table 5-1, past uranium recovery operations, including conventional mills within the Wyoming East Uranium Milling Region (where the proposed site is located) are identified. The cumulative impacts evaluation in Chapter 5 of this final SEIS has been revised to clarify and improve the transparency of the analysis.

Regarding the comment concerning the proposed rulemaking on groundwater protection, this SEIS is based on the regulations in effect at the time of writing. This has been clarified in Section 1.5 of this final SEIS.

B5.5.4 Reference

NRC (U.S. Nuclear Regulatory Commission), 2009. NUREG–1910, "Generic Environmental Impact Statement for In-Situ Leach Uranium Milling Facilities." Washington, DC: NRC. May.

B5.6 Scoping Process and Scoping Report

Comments: MR009-044; MR012-057; MR012-059; MR012-60
A commenter stated that NRC did not conduct any public meetings regarding the scope of the Moore Ranch SEIS in contrast to what was done for the GEIS. The commenter stated that instead of public scoping meetings, the NRC met with government agencies and groups it

Appendix B

considered "interested" in the SEIS to determine scope. The commenter stated that the NRC failure to conduct public scoping prevented the public from raising issues, including the cumulative impact of past uranium mining and milling that the commenter stated should have been considered in the Moore Ranch SEIS. Another commenter stated that the NRC failure to conduct public scoping meetings on the Moore Ranch SEIS was a violation of NEPA. Another commenter stated that NRC should describe in more details its "targeted scoping" and emphasize that, while not mandatory under Part 51, it was conducted to provide interested stakeholders with an opportunity to provide public comments.

Response: *NRC conducted a public scoping process for the ISR GEIS, from which the Moore Ranch ISR SEIS is tiered. The scoping process included three public scoping meetings, one of which was in Casper, Wyoming. NRC considered public comments, along with information on ISR technology and regional information to identify the scope of the GEIS for ISR facilities. The process included identifying significant issues to be studied in depth in the GEIS to help evaluate potential environmental impacts to various resource areas and identify other regulatory and consultation requirements for ISR facilities.*

NRC considers the ISR GEIS to be a final environmental impact statement (EIS) and that the environmental reviews for a specific license application to be a supplement to the ISR GEIS. According to NRC regulation 10 CFR 51.92(d), the NRC staff is required to prepare a supplement to a final EIS in the "same manner as the final EIS except that a scoping process need not be used." Furthermore, even if a scoping process is conducted, NRC regulations do not require the scoping process to include public scoping meetings (see 10 CFR 51.26(b).

The NRC staff interacted with multiple Federal, tribal, State, and local agencies and/or entities during the preparation of the Moore Ranch SEIS for consultation purposes and to gather information on potential issues, concerns, and environmental impacts related to the proposed ISR facility at the Moore Ranch site, as described in Section 1.7.3. The NRC staff used information from these interactions and other site-specific information to evaluate whether issues identified during the scoping process for the GEIS were adequate for the Moore Ranch environmental review and whether specific GEIS conclusions or findings were applicable to the Moore Ranch Project. NRC used this information to prepare a draft supplemental EIS which was issued for public comment.

Comments received on the draft SEIS were considered in the development of this final SEIS. In particular, the cumulative impact analysis in Chapter 5 of the Moore Ranch SEIS has been revised in response to public comments received on the draft SEIS and considers past uranium mining and milling.

B5.6.1   References

10 CFR Part 51. *Code of Federal Regulations*, Title 10, *Energy*, Part 51, "Environmental Protection Regulations for Domestic Licensing and Related Regulatory Functions."

NRC (U.S. Nuclear Regulatory Commission), 2009. NUREG–1910, "Generic Environmental Impact Statement for In-Situ Leach Uranium Milling Facilities." Washington, DC: NRC. May.

Appendix B

B5.7 SEIS/GEIS Methods and Approach

B5.7.1 Consider Compliance History in Assessing Impacts

Comment: MR012-039
One commenter stated that the NRC conclusion regarding groundwater impacts disregarded the operational history of other ISR operations that have used the same or similar leak detection and well integrity programs as proposed for the Moore Ranch Project. The commenter provided the example of the Smith Ranch Project as support for their concern.

Response: The NRC conclusions regarding the potential environmental impacts to groundwater for the Moore Ranch Project are provided in Section 4.5.2.1. These impact conclusions are based on facility-specific process descriptions for the Moore Ranch Project and site-specific characteristics at the proposed site. In determining impact conclusions, the NRC staff reviewed information provided by the applicant in its license application as amended (including the technical and environmental reports), information and data independently collected by the staff, and information and data provided in the GEIS. Section 2.11 of the GEIS presents a historical discussion of ISL operations (including the Smith Ranch Project) and Section 2.14 provides reference to specific facilities in Wyoming, Nebraska, and New Mexico. The intent of the information in these sections of the GEIS was to inform the reader regarding which issues have historically resulted in potential impacts at ISL facilities and to provide a range of conditions that may be expected for each of the four ISL phases of ISL. No changes were made to the Moore Ranch SEIS beyond the information provided in this response.

Comment: MR012-050
One commenter stated that the NRC conclusion that impacts to groundwater from groundwater restoration would be small was arbitrary and unreasonable. Additionally, the commenter requested that NRC fully disclose ISR industry groundwater restoration history and then reconsider impacts to groundwater, regionally and locally, based on that history.

Response: NRC conclusions regarding the potential environmental impacts to groundwater from groundwater restoration for the Moore Ranch Project are provided in Section 4.5.2.1.3. As discussed there, NRC analyzed impacts that could result from drawdown, leaks and spills from buried piping, and disposal of waste fluids via deep well injection, and determined that such potential impacts would be SMALL. These impact conclusions are based on facility-specific process descriptions for the Moore Ranch Project and site-specific characteristics at the proposed site. In determining these impact conclusions, the NRC staff reviewed applicant license application information as amended (including the technical and environmental reports), information and data independently collected by the staff, and considered information and data from the GEIS.

NRC published a summary of groundwater impacts from ISR operations at operating facilities that is available through the NRC ADAMS using the Accession Number ML091770402. ADAMS is available on the internet at http://www.nrc.gov/reading-rm/adams.html. Since this information is already publicly available and the NRC analysis of potential impacts to groundwater is based on site-specific information, the SEIS was not revised in response to this comment.

Appendix B

B5.7.2 General Comments on SEIS/GEIS Structure, Methods, and Approaches

Comment: MR012-048
A commenter stated that the GEIS's brief discussion of ISL restoration history implies that while restoration may be difficult, there have been some successful restoration projects. The commenter stated the GEIS's discussion is conclusory and misleading.

Response: *The draft GEIS was published with a NOA on July 28, 2008 (73 FR 43795). NRC held a 103-day public comment period for the draft GEIS from July 28, 2008, through November 7, 2008. During this comment period, members of the public were invited and encouraged to submit related comments online, via e-mail, via regular mail, or orally at one of eight public meetings held on the draft GEIS. The NRC considered and responded to comments received on the GEIS and included these responses in Appendix G of the final GEIS, the NOA, of which was published on June 5, 2009 (74 FR 27052).*

The comments referenced previously were submitted during the public comment period for the draft SEIS for the Moore Ranch ISR Project. Because the comment referenced previous comments solely on the GEIS and does not directly comment on the draft SEIS or provide any site-specific information related to the Moore Ranch ISR Project, it will not be considered further. No changes were made to the SEIS as a result of this comment.

B5.7.3 References

73 FR 43795, NRC (U.S. Nuclear Regulatory Commission). "Notice of Availability of Draft Generic Environmental Impact Statement for *In-Situ* Leach Uranium Milling Facilities." Vol. 73, No. 145, pp. 43795-43798. July 28, 2008.

74 FR 27052, NRC. "Notice of Availability of Final Generic Environmental Impact Statement for *In-Situ* Leach Uranium Milling Facilities." *Federal Register.* Vol. 74, No. 107, pp. 27052-27054. June 5, 2009.

NRC (U.S. Nuclear Regulatory Commission), 2009. NUREG–1910, "Generic Environmental Impact Statement for In-Situ Leach Uranium Milling Facilities." Washington, DC: NRC. May.

B5.8 Public Involvement

Comments: MR009-045; MR013-015
Commenters requested that NRC clarify the additional opportunities for public participation during preparation of the Moore Ranch draft SEIS including public meetings and teleconferences. Another commenter stated that if the SEIS provided a more detailed description of the licensing process, to include scoping and public comment meetings on the GEIS, completion of the SER, and license applicant meetings with NRC staff that this would provide members of the public and interested stakeholders with a better understanding of how focused the NRC licensing process is on transparency and public participation and how extensive the process is on the issues of protecting public health and safety and the environment on a site-specific basis.

Response: NRC provides multiple avenues for public involvement in its licensing process. In the NRC license review process, once an application is received, reviewed for completeness, and accepted for detailed review, NRC formally dockets the application and publishes a notice in the Federal Register. The Federal Register notice announces the availability of the application and provides an opportunity for affected individuals or entities to request a hearing under the NRC formal hearing process. The NOA published in the Federal Register includes the relevant identifying information for the license application so that an interested member of the public can view the application either electronically through the NRC ADAMS [at www.nrc.gov/reading-rm/adams.html] or in person by visiting NRC's public document room.

In the case of the proposed Moore Ranch Project, there have been eight opportunities for public involvement, in addition to opportunities that have been afforded through the NEPA process for the site-specific Moore Ranch supplemental EIS. These include a pre-application meeting, and five publicly-noticed meetings or teleconferences with the applicant to discuss various technical issues or issues associated with the application such as coal bed methane (CBM) operations. The publication of the Notice of Intent to prepare the GEIS also provided an opportunity for public involvement and subsequent notices that extended the public comment period, and the Notice of Opportunity to request a hearing, published in the Federal Register on January 25, 2008 (73 FR 4642). No requests for a hearing were received.

As discussed in Section 1.4.1 of the Moore Ranch SEIS, NRC accepted public comments on the scope of the GEIS, from which the Moore Ranch SEIS is tiered, from July 24 to November 30, 2007, and held three public scoping meetings, one of which was in the State of Wyoming. During the public comment period on the draft GEIS, NRC held eight public meetings to receive comments on the draft GEIS: three of these meetings occurred in the State of Wyoming. The public participation activities associated with the development of the Moore Ranch SEIS are discussed in Section 1.4.2 of the SEIS, and discussion of the NRC licensing process are described in Section 1.6.1 of the SEIS. No changes were made to the final SEIS beyond the information provided in this response.

Comments: MR015-026; MR015-035; MR020-042
A commenter requested that interested stakeholders be involved in the review of any modeling protocol for assessing air quality impacts prior to supplemental work being performed. The same commenter asked if there is a public participation process associated with the establishment and NRC decision to approve, alternate concentration limits (ACLs).

Response: NRC provides multiple avenues for public involvement in its licensing process in the review of an individual ISR facility. For new ISR license applications, NRC will publish a Notice of Intent to prepare a site-specific SEIS and provide details on the scoping process for the SEIS, if applicable. NRC will also publish for public comment a draft SEIS and address stakeholder comments received in its final SEIS. NRC may also make a draft environmental assessment and accompanying draft finding of no significant impact available for public comment.

A licensee must apply for a license amendment for an ACL; a notice is published in the Federal Register for all licensing actions made under the Atomic Energy Act, including amendments. Under NRC regulations, an opportunity is provided for any person whose interest may be affected by an NRC licensing action (see 10 CFR 2.309). Further, NRC performs a safety and environmental review (typically an environmental assessment) as part of evaluating the adequacy of an ACL.

Appendix B

B5.8.1 References

NRC (U.S. Nuclear Regulatory Commission), 2009. NUREG–1910, "Generic Environmental Impact Statement for *In-Situ* Leach Uranium Milling Facilities." Washington, DC: NRC. May.

B5.9 Regulatory Issues and Process

B5.9.1 NRC as a Regulatory Authority

Comments: MR013-010; MR009-003
Two commenters asked for clarification about the NRC and its statutory mission under the AEA and its approach to licensing as an independent regulatory agency. The commenter suggested that all references to the NRC statutory mission in the SEISs be revised with the following language:

"NRC must license facilities, including ISR operations, in accordance with the AEA and the Commission's implementing regulations to protect public health and safety from potential radiological and non-radiological hazards associated with AEA materials and operations."

Response: *The NRC was created after Congress passed the Energy Reorganization Act in 1974. This Act, along with the AEA of 1954, provides the foundation of the NRC regulatory authority. As an independent regulatory agency, NRC reports directly to Congress. Independent agencies can be distinguished from regular executive agencies by their structural and functional characteristics. The NRC has the responsibility in licensing and regulating uranium ISR facilities through the statutory requirements of the Uranium Mill Tailings Radiation Control Act of 1978 and the AEA, as amended. These statutes require that NRC ensure source material, as defined in Section 11(z) of the AEA and byproduct material, as defined in Section 11e.(2) of the AEA, is managed to conform with applicable regulatory requirements. The text within the SEIS is correct. No changes were made to the SEIS beyond the information provided in this response.*

B5.9.2 NRC Authority and Jurisdiction

Comment: MR007-001
One commenter noted an error in the text, which stated that the Wyoming State Historic Preservation Office (SHPO) had declared all sites ineligible for listing on the National Register of Historic Places (NRHP). The commenter further stated that Wyoming SHPO does not necessarily determine the eligibility of sites. The determination of site eligibility is made by the lead agency, which in the case of the Moore Ranch Project is the NRC.

Response: *NRC acknowledges this comment and agrees that the Wyoming SHPO does not determine the NRHP eligibility of historic properties. The responsibility of the SHPO is to concur or not concur with the status of resource eligibility determined by the federal agency assessing the resources (36 CFR 800.4(c)(2)). As such, the text in Section 3.9 Historical and Cultural Resources, Section 4.9, and Section 1.7.2 National Historic Preservation Act of 1966 (NHPA) Consultation was revised in response to this comment.*

Appendix B

B5.9.3 NRC Policies and Practices

Comment: MR008-003
A commenter noted that the NRC staff were not listed as reviewers and asked if the NRC staff were involved with the development of the SEISs.

Response: *Chapter 9 of the final SEIS lists all of the contributors to the SEIS. No change was made to the SEIS.*

Comments: MR009-194; MR009-199
A commenter stated that the NRC should not rely on the Wyoming Game and Fish Department (WGFD) document, "Recommendations for Development of Oil and Gas Resource within Crucial and Important Wildlife Habitats" because this draft document provided recommendations specific to impacts to wildlife from the oil and gas industry and the impacts from the oil and gas industry bear no relation to those posed by ISR uranium mining. The commenter further questioned why NRC referenced BLM's proposed Resource Management Plan since the report had not been finalized and was prepared for the Rawlins Field Office.

Response: *The NRC acknowledges that the WGFD's "Recommendations for Development of Oil and Gas Resources within Important Habitats" (WGFD, 2010) are recommendations, not regulations, and NRC does not have the statutory authority to require the Moore Ranch ISR Project to abide by these recommendations. This final SEIS references these recommendations because they provide a useful basis for assessing impacts and determining reasonable mitigation strategies for wildlife species. Concerning the applicability of the WGFD recommendations to the ISR uranium milling process, the WGFD recommendations for seasonal wildlife stipulations guidelines are reasonable to use for analysis of impacts because the stipulations mainly serve the function of identifying times of the year that particular species are more sensitive to human disturbance. Though specific types of disturbance and magnitude of disturbance varies between oil and gas development and ISR uranium milling, the WGFD guidelines that provide seasonal distance buffers for noise, vehicular traffic, and human proximity provide a valuable gauge for determining impacts on wildlife from the proposed Moore Ranch ISR Project. Additionally, the most recent version of the WGFD recommendations published in April 2010, specifically directs in-situ uranium development to follow stipulations specified for oil and gas development for the greater sage-grouse* (Centrocercus urophasianus) *that were reviewed, even though the proposed Moore Ranch Project is not located in a greater sage-grouse core area.*

Comments: MR009-282; MR013-018
A commenter stated that NRC needed to clarify its approach to wellfield package review and approval; the commenter stated that the new policies are inconsistent with performance-based licensing and the manner in which ISR operations were licensed in the past. The commenter noted that NRC has stated that new licenses would be required to submit some initial wellfield Hydrologic Data Packages until the staff developed a level of comfort with the applicant. Another commenter stated their opinion that it appeared from the language in the SEIS that NRC would be reviewing and approving all wellfield packages rather than just initial wellfield packages. The commenter went on to note that, in their opinion if NRC reviewed and approved all wellfield packages, that would be contrary to and detrimental to the Commission's policy supporting performance based licensing, and that NRC staff should continue to allow Safety and Environmental Review Panels to review and approve wellfield packages under traditional performance-based licensing as has been done in the past.

Appendix B

Response: The NRC agrees with the need to clarify its position on review and approval of wellfield hydrologic data packages. Historically, the NRC reviewed and approved all wellfield packages. During the mid-1990s, the Commission adopted a performance-based approach to licensing. A performance-based regulatory approach is one that establishes performance and results as the primary basis for regulatory decisionmaking, and this approach incorporates the following attributes: (1) measurable (or calculable) parameters (i.e., direct measurement of the physical parameter of interest or of related parameters that can be used to calculate the parameter of interest) exist to monitor system, including both facility and applicant performance, (2) objective criteria to assess performance are established based on risk insights, deterministic analyses, performance history, or both, (3) applicants flexibility in determining how to meet the established performance criteria in ways that would encourage and reward improved outcomes; and (4) a framework in which the failure to meet a performance criterion, while undesirable, will not in and of itself constitute or result in an immediate safety concern.

Current Commission policy allows applicant Safety and Environmental Review Panels (SERP) to review and approve wellfield packages under performance-based license conditions. However, in some wellfields there are particular geologic features (e.g. faults, thin/missing aquitards) or groundwater flow behavior (e.g., unconfined aquifer, leakage across aquitards) that require local field data and testing to be characterized to determine if ISR operations can meet regulatory requirements, the staff may review and approve wellfield packages. Therefore, based on the outcome of the safety review of the technical report, NRC may request by license condition to review and approve the Wellfield Hydrologic Data Package for an individual wellfield if NRC determines that a safety review cannot be adequately completed without the information provided in the package. The discussion in Section 6.3.1.2 of the final Moore Ranch SEIS was revised to clarify this issue.

B5.9.4 Adequacy of NRC Regulations and Practices

Comments: MR009-005; MR009-048; MR009-049; MR013-012; MR013-013
Commenters noted that there are many applicable regulatory provisions other than 10 CFR Part 51 (10 CFR Parts 20 and 40) and that the NRC should mention other applicable guidance and regulatory guides used for completing environmental reviews. Another commenter noted that the NRC discussion of regulatory programs applicable to ISR operations outside the context of the AEA should be expanded to demonstrate how highly regulated the ISR industry is in the United States. The same commenter further noted that the NRC should specify all of the regulatory programs that apply to ISR operations and not limit the discussion in the final SEIS to only 10 CFR Part 51 regulations. Another commenter noted that the SEIS states that ISR operatons are subject to the AEA and NEPA with no mention of other statutory programs such as the Safe Drinking Water Act (SDWA), NHPA, and the Endangered Species Act as implemented in accordance with various state programs. Commenters stated that the final SEIS should make clear how extensive the regulatory oversight is for ISR operations. Commenters noted that multiple agencies oversee ISR operations, often resulting in two or even three layers of financial assurance for each ISR project; a commenter stated this more than assures that adequate site-specific decommissioning and decontamination would be performed.

Response: NRC has to comply with all applicable Federal environmental laws, regulations, and Executive Orders, including its own regulations (in Title 10 of the CFR) and those promulgated by other Federal agencies, so long as compliance would not be inconsistent with other statutory requirements. Section 1.6 of the GEIS identifies agencies involved in a uranium ISR facility, and Section 1.7 discusses the licensing permitting process for an ISR facility. Section 1.6 of the

Moore Ranch SEIS discusses the status of licensing and permitting and associated consultations that pertain to the ISR licensing review at the proposed Moore Ranch Project. The SEIS was prepared in accordance with NEPA requirements and NRC implementing regulations at 10 CFR Part 51.

Furthermore, Appendix B of the GEIS summarizes other Federal statutes, implementing regulations, and Executive Orders potentially applicable to the Moore Ranch licensing review. The description of regulatory programs applicable to ISR operations is provided by the agencies responsible for implementing those programs, and readers should consult the responsible agencies for clarification of their regulations and programs. ISR applicants are ultimately responsible for understanding and complying with all Federal, State, and local permits and regulations, whether described in the GEIS or not. No changes were made to the SEIS beyond the information provided in this response.

Comments: MR012-027; MR012-053
A commenter stated that the NRC analysis of groundwater restoration and excursions shows that the NRC has no coherent framework for regulating ISL operations. They also noted that the NRC staff recognizes that "class of use" is an inappropriate restoration goal and referred to the NRC Regulatory Issue Summary issued in 2009, which concluded that Criterion 5B did not provide for restoration to "class of use" standards (RIS 2009-05). The commenter noted that since "class of use" is a restoration standard that is not legally cognizable, it should not be the basis for analysis of groundwater impacts.

Response: *The NRC has announced its intent to issue a proposed rulemaking specific to groundwater protection at ISR facilities; but to date, this rulemaking has not yet been published in the Federal Register. NRC regulations require that the groundwater quality be returned to the standards identified in Criterion 5B(5) of Appendix A to 10 CFR Part 40. Those standards are background, the values in the table in Criterion 5C of Appendix A to 10 CFR Part 40, or an ACL established by the NRC in accordance with Criterion 5B(6) of Appendix A to 10 CFR Part 40. Criterion 5(B)(5) allows the NRC to approve an ACL for groundwater restoration. A rigorous regulatory process is used for a applicant to receive approval for ACLs as discussed in Appendix C of this SEIS. The applicant must demonstrate that it has attempted to restore hazardous constituents in groundwater to either background concentrations or to the maximum contaminant level (MCL)—whichever level is higher. A license amendment application must be submitted by the applicant to request to establish ACLs. The regulations and criteria used for review of ACL applications is found in 10 CFR Part 40, Appendix A, Criterion 5B(6).*

Within the State of Wyoming, the Wyoming "class of use" standard comes within the 10 CFR Part 40, Appendix A, Criterion 5B(6)(a)(v-vi) and (b)(vi-vii) factors, and thus may be considered as one factor in evaluating ACL requests for ISR facilities located within the state.

In considering ACL requests, particular importance is placed on protecting underground sources of drinking water (USDW). The use of modeling and additional groundwater monitoring may be necessary to show that ACLs in ISR wellfields would not adversely impact USDW.

Before an ISR applicant is allowed to extract uranium, the EPA under 40 CFR Part 146.4 and in accordance with the SDWA must issue an aquifer exemption covering the portion of the aquifer in which the ore zone is located. The EPA cannot exempt the portion of the aquifer unless it is found that "it does not currently serve as a source of drinking water" and "cannot now and would not in the future serve as a source of drinking water." Because of these criteria, only impacts

Appendix B

outside of the exempted aquifer are evaluated. In most cases, the water in aquifers adjacent to the uranium ore zones does not meet drinking water standards. The NRC would not approve an ACL if it would impact adjacent USDW.

Further guidance for the review of ACLs for ISR facilities is being developed in a revision of NUREG–1569, Standard Review Plan for the In-Situ Leach [Recovery] Uranium Extraction License Applications. Existing guidance for the review of ACLs is contained in NUREG–1620, Standard Review Plan for the Review of a Reclamation Plan for Mill Tailings Sites Under Title II of the Uranium Mill Tailings Radiation Control Act of 1978, Section 4.3.

No changes were made to the SEIS in response to these comments beyond the information provided in this response.

Comment: MR017-026
A commenter stated that for the NRC to craft an appropriate "Purpose and Need for Agency Action," the agency must work with its Federal colleagues at EPA, the U.S. Department of Energy (DOE), and the U.S. Department of the Interior to develop a regulatory framework for uranium recovery cleanup and licensing that protects public health and the environment. The commenter asserted that the NRC refuses to issue a draft groundwater protection rule for nearly five years, and that it is past time for NRC to develop a coherent set of protective environmental requirements for ISR uranium mining; the commenter also stated that developing a draft groundwater rule would be a start.

Response: *The NRC has announced its intent to issue a proposed rulemaking specific to groundwater protection at ISR facilities; but to date, this rulemaking has not yet been published in the Federal Register. COMJSM–06–001 (2006) directed the staff to focus on eliminating dual regulation of groundwater by NRC and EPA. The Commission stated that the NRC should retain its jurisdiction over the wellfield and groundwater under its AEA authority, but should defer active regulation of groundwater protection programs to either EPA or to the EPA-authorized state through the EPA Underground Injection Control (UIC) program. The analysis in the Moore Ranch SEIS is based on existing regulations at the time the final SEIS is published. The status of ongoing rulemaking activities is provided on the NRC public website at www.nrc.gov. Because no proposed rule is available to discuss, no changes were made to the Moore Ranch SEIS beyond the information provided in this response.*

B5.9.5   Applicable Rulemaking Efforts

Comments: MR012-054; MR001-007, MR012-030
A commenter stated that NRC has an ad hoc approach to ISL regulation. A commenter asserted that the "class of use" restoration standard used in both the GEIS and the Moore Ranch SEIS indicated a fundamental problem with the NRC regulatory framework and was concerned that the GEIS would become the proxy for ISL regulations. The commenter stated that NRC does not have regulations specifically relevant to ISL operations but rather has adapted some of the conventional milling regulations to apply to ISL operations and "filled in the remaining gaps with license conditions," the standard review plan for ISL facilities, and the GEIS. Another commenter stated that because the public did not have a sense of the timing, scope, and coverage of the proposed groundwater rule and any associated NEPA process, the initial close of the public comment period on the Moore Ranch draft SEIS in February 2010 was

Appendix B

unreasonable. One commenter also stated that the SEIS and GEIS should be withdrawn and reissued after regulations are complete.

Response: *NRC is currently working on a proposed rulemaking specific to groundwater protection at ISR facilities. The analysis in the Moore Ranch SEIS is based upon the current regulations in 10 CFR Part 40. Until and if the above proposed rulemaking is made final, license applications will continue to be reviewed and licensed in accordance with current regulations.*

As background, COMJSM–06–001 (2006 directed the staff to focus on eliminating dual regulation of groundwater by NRC and EPA. The Commission stated that the NRC should retain its jurisdiction over the wellfield and groundwater under its AEA authority, but should defer active regulation of groundwater protection programs to either EPA or to the EPA-authorized state through the EPA UIC program. The status of ongoing rulemaking activities is provided on the NRC public website at www.nrc.gov.

B5.9.6 NRC NEPA Process Implementation

Comment: MR017-002
A commenter noted that without vigorous compliance with NEPA requirements and adherence to strict environmental protections that the environmental history of uranium mining could be repeated.

Response: NRC understands and recognizes there are serious legacy issues resulting from the decades of uranium mining activities from the 1940s through the 1970s when waste from uranium mines was not cleaned up after mines were shut down. NRC regulation of ISR facilities includes ensuring the necessary measures are taken by ISR operators to confine mobilized uranium and other constituents within the wellfield where the facility is operating, ensuring monitoring programs are in place to provide early detection of any migration of process fluids away from the wellfield, and enforcing necessary corrective actions to prevent uranium from contaminating adjacent water sources to ensure the public is protected.

The Moore Ranch SEIS was prepared consistent with NRC regulations at 10 CFR Part 51 that implement NEPA and in accordance with NRC guidance in NUREG–1748 (NRC, 2003). Because the comment was general in nature, no changes were made to the Moore Ranch SEIS beyond the information provided in this response.

B5.9.7 NRC Licensing Process

Comments: MR009-006; MR013-009
Two commenters requested that a more complete description of the NRC licensing process be provided for those not familiar with that process. They stated that the process included NRC safety and environmental reviews of the entire license application (including the technical and environmental reports), NRC RAIs, and a public participation process. One of the commenters stated that the SEIS should clarify the link between the NRC's environmental and safety reviews.

Response: SEIS Sections 1.6 and 1.6.1 discuss in general the NRC licensing process for the Moore Ranch Project. These sections refer to Section 1.7.1 of the GEIS for a more complete discussion of the NRC licensing process. Further, as stated in SEIS Section 1.6.1, NRC

Appendix B

detailed technical review of the Moore Ranch license application is comprised of two parallel reviews: a safety review and an environmental review. The safety review focuses on assessing compliance with the applicable requirements of 10 CFR Parts 20 and 40 and Appendix A to Part 40, while the environmental review is conducted in accordance with the regulations in 10 CFR Part 51. The results of these two detailed reviews support NRC licensing decision. Figure 1.7-1 of the GEIS provides a general flow diagram for the NRC licensing process, including safety and environmental reviews.

It is common during the detailed technical review of a license application for NRC to request additional applicant information to ensure the application is complete. In some cases, multiple rounds of RAIs are possible. For applications that are not complete, this iterative process is designed to provide the applicant the necessary feedback to supplement the application so it is complete. The public participation process for this SEIS is discussed in Section 1.4.2 of the Moore Ranch SEIS.

As indicated by the commenters, there is some overlap between safety and environmental reviews. This is most clearly seen in topics such as groundwater resources and protection and radiological dose to workers and members of the public. The NRC staff conducting the environmental and safety reviews collaborate, as necessary, during the conduct of these parallel reviews.

Throughout the SEIS, NRC has used the term "license application" to be inclusive of all aspects of the application, including the applicant's technical report, environmental report, and responses to NRC RAIs. The reference sections following SEIS Chapters 2, 3, 4, 5, and 6 are reflective of the reliance on all aspects of the application as described previously.
No further changes were made to the SEIS beyond the information provided in this response.

Comment: MR009-050
One commenter requested that language should be added to SEIS Section 1.6.1 to mention that NRC issued RAIs and that the applicant responded.

Response: *The focus of SEIS Section 1.6.1 is to provide only the initial acceptance of the Moore Ranch license application and to discuss the NRC licensing process in general. NRC did make use of the applicant responses to RAIs in its description of the proposed action (Chapter 2), in the description of the affected environment (Chapter 3), and in the analysis of potential environmental impacts (Chapter 4). No further changes were made to the SEIS beyond the information provided in this response.*

Comment: MR009-061
One commenter requested that in Section 2.1 (Alternatives Considered for Detailed Analysis), NRC should include mention of the applicant's technical report and of consultations with the NRC staff conducting the safety review.

Response: *The resources mentioned by the commenter were not primary sources used by NRC in determining the range of reasonable alternatives to be considered for detailed analysis in the SEIS. Therefore, they were not identified in Section 2.1. No further changes to the SEIS were made beyond the information in this response.*

Appendix B

B5.9.8 Consideration of ISL Facility Safety Record and Compliance History

Comments: MR010-002; MR011-002; MR017-001; MR012-040
Two commenters stated that there are several ISL uranium sites in Wyoming, and that all of these facilities have a history of leaks, spills and excursions. Another commenter wanted to remind the NRC of the dreadful uranium mining environmental history which is likely to repeat itself without meaningful oversight, vigorous compliance with the requirements of NEPA, and adherence to strict environmental protections. The commenter further added that NRC is seemingly on a path that will doom us to repeat this chapter of American history. Another commenter stated that there was a fundamental contradiction between actual operational data and NRC conclusions regarding the magnitude of impacts in both the GEIS and the Moore Ranch SEIS were contrary to NEPA.

Response: Section 2.11.2 of the GEIS discusses leaks, spills, and excursions that have occurred at existing ISR facilities. Excursions and mechanical integrity failures have been reported for past and current ISR facilities, but in most cases they have been controlled and did not pose a threat to human health or the environment. Three ISRs are currently operating: two in Wyoming [Uranium One Irigaray and Christensen Ranch facility (formally owned by Cogema Mining, Inc.)], and the PRI Smith Ranch—Highland Uranium Project), and one in Nebraska (the Cameco Crowe Butte Project). Excursion history and corrective action for all of these sites can be found in annual reports and correspondence between the NRC and the applicants in NRC public document room.

All ISR facilities have the potential to have leaks, spills, and excursions, and the purpose of the oversight program is to help ensure that leaks, spills, and excursions are minimized. Oversight activities, including inspection activities, verify that ISR facility operations, aquifer restoration, and decommissioning activities are being conducted according to NRC regulations. NRC enforcement programs and policies are effective at verifying if applicants are in compliance with NRC regulations, and takes appropriate enforcement action if a licensed facility is not in compliance.

Surface containment of leaks and spills is required for all storage tanks (EPA 2006). In addition, spill prevention plans are required of each ISR facility (EPA 1994). For chemicals stored at ISR facilities, concrete berms with containment equivalent to at least the volume of the tank are required. Spill reporting varies from state to state. NRC requires that a applicant report a spill within 24 hours if it meets the criteria in 10 CFR Parts 20 and 40.60. Otherwise, NRC typically requires, by license condition, that if a leak or spill meets state reporting requirements, it must also be reported to the NRC. Leaks and spills must be characterized and cleaned up to regulatory requirements [see 10 CFR Part 40 Appendix A, Criterion 6(6)].

The following is a summary of the regulatory requirements that apply to ISR facilities that are designed to address leaks, spills, or excursions from these facilities.

Surface impoundments (including ponds) are designed with a capacity to hold the anticipated volume of liquids and are operated and maintained to prevent overtopping from normal operations, rainfall, run-on from upstream areas, wind and wave action, and equipment malfunctions. Monitoring wells (both up- and down-gradient) are installed, in addition to requiring a liner under each surface pond (NRC, 2008). Liner materials require a construct tha has sufficient physical properties and strength to withstand the anticipated physical stresses and environmental conditions. Liners are typically constructed with leak detection systems that

Appendix B

allow for leak identification and repair. The leak detection systems are checked for the presence of liquids on a regular basis.

To detect and prevent excursions to the overlying and underlying aquifers, NRC issues a license condition that requires operators to perform mechanical integrity testing for all injection and production wells (NRC, 2003). This test is conducted every five years to ensure that the wells do not develop leaks. To ensure that excursions are identified early, excursion monitoring wells are installed. Horizontal excursion monitoring wells are placed in a perimeter ring surrounding the wellfield in the production aquifer. In addition, vertical excursion monitoring wells are installed in the overlying and underlying aquifers (NRC, 2003). The monitoring wells are usually sampled twice a month for excursion indicators. When excursion indicators exceed predetermined upper control limits (UCLs), it may signal that production fluids are moving out of the wellfield boundary. If an excursion is confirmed, the applicant must begin corrective actions to control the excursion and must continue corrective action until the excursion is controlled. The location of the excursion monitoring wells, the choice of excursion indicators, and the process for determining the UCLs are all reviewed before a license is approved.

Prior to the recovery of uranium from an aquifer, the EPA must declare the portion of the aquifer where production would occur, exempt as a USDW (see 40 CFR 146.4). In addition, if liquid byproduct material is to be disposed of via deep well injection, the EPA must also declare the deep-well receiving aquifer exempt, as well. The production aquifer exemption area would be restored to the standards in 10 CFR Part 40, Appendix A, whereas the injected byproduct material fluid would remain in the exempted aquifer.

The NRC performs an environmental review of an applicant license application to determine the environmental effects of operating the proposed ISR facility. The Commission determined it would prepare a SEIS for each license application to fulfill its responsibilities under the NEPA (see 10 CFR 51.20). NEPA requires that all Federal agencies consider environmental values in the conduct of their work. No changes were made to the SEIS beyond the information provided in this response.

B5.9.9   Groundwater Restoration Criteria and Methods

Comments: MR012-016; MR012-044
A commenter asserted that the NRC practice of averaging poor groundwater quality with good groundwater quality to characterize preextraction groundwater quality misrepresents the impacts from groundwater restoration. The commenter stated that NRC tied groundwater restoration in the Moore Ranch SEIS to the average of poor groundwater quality in the immediate ore zone with good groundwater quality outside the ore zone but within a mine area. The commenter stated that Table 3-2 of the draft SEIS gave the impression that the groundwater in the aquifer within the proposed mine boundary exceeded EPA and Wyoming water quality standards for several constituents but elsewhere in the SEIS, the NRC disclosed that there were wells with good quality water. For example, the SEIS disclosed that there were either active domestic or stock wells within a 3.2 km [2-mi] radius. The commenter asserted that the practice of averaging good groundwater quality with poor groundwater quality is incomplete and misleading and skewed the impact analysis to minimize potential groundwater impacts from ISL mining in general and at the Moore Ranch site in particular, and that the use of this "mathematical artifice" inflated the preextraction contaminant levels within project boundaries to create an impression that pre-extraction groundwater quality is poor and

Appendix B

restoration is possible. Finally, the commenter stated that if groundwater quality within and outside of an ore zone were analyzed separately and not averaged, then the adverse impact on groundwater outside of the ore zone would be substantially larger.

Response: The commenter is referring to the need to establish a baseline for groundwater quality in the proposed license area before ISR operations begin. As part of the site characterization to obtain a license, the applicant is required to determine the average preoperational water quality for all aquifers in, above, below, and outside the proposed wellfield(s) to generally characterize the water quality of each aquifer across the entire license area. However, this general preoperational average is not the same as the average baseline water quality of the uranium-bearing aquifer for a specific wellfield. The average baseline water quality for a specific wellfield is determined only from water quality measured in wells installed within the production ore zone aquifer in each licensed wellfield, and it is this specific average that is used to determine groundwater restoration target values in individual wellfields. Contrary to the comment, this average baseline water quality does not include wells "outside the ore zone."

Comments: MR012-051; MR012-052

A commenter stated that it appears that NRC evaluates groundwater restoration impacts assuming that if baseline groundwater quality is not achieved, "class of use" quality would be achievable and that this analysis ignores NRC regulations governing ISL groundwater restoration that make no mention of "class of use" as a restoration standard, and mandates that groundwater must be restored to background or the MCLs listed in Criterion 5D.

Response: The commenter is correct that NRC has used "class of use," a state designation under the SDWA, as a restoration goal. The "class of use" standard for restored groundwater quality was based on restoration standards provided in NUREG–1569. NRC has determined that the primary and secondary restoration standards in NUREG–1569 are inconsistent with the restoration standards in 10 CFR Part 40, Appendix A, Criterion 5B(5). NRC has notified licensees and applicants in Regulatory Information Summary, RIS 09-05, dated April 29, 2009, that the restoration standards listed in NUREG-1569, Section 6.1.3(4) are not consistent with those listed in 10 CFR Part 40, Appendix A. NRC requires that licensees commit to achieve the restoration standards in Criterion 5B(5). A licensee can apply for a license amendment for an ACL only after showing that restoration to the background level or MCL is not practically achievable for a particular constituent. NRC reviews the ACL request using the criteria articulated in 10 CFR 40, Appendix A Criterion 5B(6). The State designation of "class of use" for an aquifer can be one of the factors that are considered during NRC review of the ACL request. A discussion of the additional Criterion 5B(6) requirements for ACLs is presented in Appendix C.

Comments: MR012-016; MR015-005; MR015-028; MR015-029

Several commenters were concerned with the potential establishment of ACLs as groundwater restoration targets prior to the completion of adequate restoration, and that the draft SEIS did not fully assess the operational requirements and constraints associated with restoration activities. A commenter noted that although the SEIS acknowledged that the water quality goal in the portion of the aquifer where extraction occurs is pre-ISR baseline conditions, the discussion concluded by stating that the demonstration of restoration must comply with the requirements in 10 CFR Part 40, Appendix A, which allows for restoration target values that do not meet the pre-ISR baseline. The commenter noted that although EPA standards in 40 CFR Part 192 allow NRC to use this practice, ACLs are above baseline or MCL values.

Appendix B

Response: *For any given groundwater constituent, licensees and applicants are subject to the three groundwater quality standards listed in 10 CFR Part 40, Appendix A, Criterion 5B(5) – background, MCL, or ACL. Specifically, under Criterion 5B(5), the concentration of a hazardous constituent must not exceed (a) the NRC-approved background concentration of that constituent in groundwater; (b) the respective MCL value in the table in paragraph 5C if the constituent is listed in the table and if the background level of the constituent is below the value listed or; (c) an ACL established by the NRC. Under Criterion 5B(6),requests for ACLs would only be considered after a applicant has demonstrated that restoring the constituent at issue to background or MCL values is not practically achievable at a specific site. Only ACLs that present no significant hazard may be proposed by applicants for NRC consideration. The NRC may establish a site specific ACL for a hazardous constituent if it finds that the proposed limit is as low as reasonably achievable, after considering practicable corrective actions, and that the constituent would not pose a substantial present or potential hazard to human health or the environment as long as the ACL is not exceeded. A discussion of the additional Criterion 5B(6) requirements for ACLs is presented in Appendix C of the final SEIS.*

Comments: MR010-005; MR011-005; MR017-021; MR017-023; MR012-042; MR012-043
A commenter noted that NRC has not been able to accomplish restoration of groundwater to baseline values for all groundwater constituents in any ISL wellfield to date. Another commenter stated that NRC had mischaracterized the ISR mining's groundwater restoration efficacy history and unreasonably minimized the impacts from groundwater restoration. A commenter stated that restoration has only been accomplished by lowering the standards. Another commenter similarly stated that restoration to either background levels or MCL standards has been aspirational rather than a reality, and that regulators—whether NRC or Agreement States—have allowed ACLs to be established for some constituents. The commenter asserted that restoration standards have been a moving target for all ISR mining sites and that NRC has made it nearly impossible for a reader to analyze environmental impacts because of the lack of a detailed and comprehensive history of ISR restoration operations.

Response: *The commenters are correct that, to date, restoration to backgroundwater quality for all constituents has proven to be not practically achievable at licensed NRC ISR sites (NRC, 2005; NRC, 2004; NRC, 2003). In the past, NRC has applied "class of use," a state designation under the SDWA, as a secondary restoration goal to approve these restorations. The "class of use" standard for restored groundwater quality was based on restoration standards provided in NUREG–1569. The "class of use" standard was neither treated nor approved as an ACL. The NRC has since determined that the primary and secondary restoration standards in NUREG–1569 are inconsistent with the restoration standards in 10 CFR Part 40, Appendix A, Criterion 5B(5). The NRC notified licensees and applicants in Regulatory Information Summary, RIS 09-05, dated April 29, 2009, that the restoration standards listed in NUREG–1569, Section 6.1.3(4) are not consistent with those listed in 10 CFR Part 40, Appendix A, applicant must commit to achieve the restoration standards in Criterion 5B(5).*

For all licensees, NRC would require an ACL for any constituents that do not meet the primary baseline standards. NRC would perform an SER and environmental review when a licensee applies for a license amendment to establish ACLs for a particular constituent after it demonstrates it is not practically achievable to restore the wellfield to either background or MCL levels.

Comment: MR012-047
A commenter asserted that the issuance of waivers and (aquifer) exemptions should be part of the analysis and that NRC must analyze the impact of the waivers and exemptions. The commenter stated that by not doing this both the GEIS and SEIS fail the NEPA "hard look" standard.

Response: EPA granting of an aquifer exemption is an administrative action that allows the licensee to use ISR technology to extract uranium from the ore body. NRC does not do an environmental impact analysis of the EPA decision of whether or not to grant an aquifer exemption. NRC environmental review analyzes the potential environmental impact of NRC licensing the proposed action to use ISR technology to extract uranium from the ore body if it receives an aquifer exemption from the EPA. However, the NRC No-Action Alternative analysis is equivalent to the environmental impact analysis of not receiving an aquifer exemption from the EPA because the license would not be able to use ISR technology to extract uranium from the ore body.

The commenter appears to consider the ACL as a "waiver" from the primary goal of restoring constituents to background levels. Neither the terms of Criterion 5B nor the AEA provision on which it is based—AEA Section 84c—support the waiver concept. As stated in Criterion 5B(6), ACLs reflect site-specific limits that are found to be "as low as reasonably achievable." NRC performs an SER and environmental review when a licensee applies for a license amendment to establish ACLs for a particular constituent after it demonstrates is not practically achievable to restore the wellfield to background or MCL levels. A discussion of the additional Criterion 5B(6) requirements for ACLs is presented in Appendix C of the final SEIS.

Comments: MR009-081; MR009-180
A commenter stated that the SEIS needed to provide a more precise description of the Appendix A Criterion 5(B)(5) standards for groundwater restoration and clearly state that the standards are baseline or an MCL, whichever is higher, or an ACL and then provide a description of an ACL.

Response: For any given groundwater constituent, licensees and applicants are subject to the three groundwater quality standards listed in 10 CFR Part 40, Appendix A, Criterion 5B(5)— background, MCL or ACL. A licensee must try to restore to background or MCL, whichever is higher. After demonstrating that background or MCL is not practically achievable, a licensee may request a license amendment for approval of an ACL for a particular constituent. A discussion of the NRC requirements for application, review and establishment of a site-specific ACL is presented in Appendix C of the final SEIS.

NRC has revised the SEIS to state that licensees and applicants must commit to achieve the groundwater quality standards in 10 CFR Part 40, Appendix A, Criterion 5B(5) for all restored aquifers.

B5.9.10 References

10 CFR Part 20. *Code of Federal Regulations*, Title 10, *Energy*, Part 20, "Standards for Protection Against Radiation."

10 CFR Part 40. Appendix A. *Code of Federal Regulations*, Title 10, *Energy*, Part 40, Appendix A, "Criteria Relating to the Operations of Uranium Mills and to the Disposition of Tailings or

Appendix B

Wastes Produced by the Extraction or Concentration of Source Material from Ores Processed Primarily from their Source Material Content."

10 CFR Part 51. *Code of Federal Regulations*, Title 10, *Energy*, Part 51, "Environmental Protection Regulations for Domestic Licensing and Related Regulatory Functions."

36 CFR Part 800. *Code of Federal Regulations*, Title 36, *Parks, Forests, and Public Property*, Part 800, "Protection of Historic Properties."

40 CFR Part 146. *Code of Federal Regulations*, Title 40, *Protection of Environment.* Part 146, "Underground Injection Control Program: Criteria and Standards."
71 FR 77266, U.S. Nuclear Regulatory Commission. "EPA proposed amendments to the SPCC rule." December 26, 2006.

EPA (U.S. Environmental Protection Agency), 1994. National Oil and Hazardous Substance Pollution Contingency Plan (as amended). Final Rule, 59 FR 47384. September.

NRC (U.S. Nuclear Regulatory Commission), 2009a. NUREG–1910, "Generic Environmental Impact Statement for *In-Situ* Leach Uranium Milling Facilities." Washington, DC: NRC. May.

NRC, 2009b. "Uranium Recovery Policy Regarding (1) the Process for Scheduling Licensing Reviews and (2) the Restoration of Groundwater at Licensed Uranium *In-Situ* Recovery Facilities," Regulatory Issue Summary 2009-05. April 29. ADAMS Accession No. ML083510622.

NRC, 2008. Regulatory Guide 3.11, Revision 3, "Design, Construction, and Inspection of Embankment Retention Systems at Uranium Recovery Facilities." Washington, DC: NRC. November.

NRC, 2006. "Regulation of Groundwater Protection at In Sit Leach Uranium Extraction Facilities." SRM to COMJSM–006–001. March 24. ADAMS Accession No. ML060830525.
NRC, 2005, "Technical Evaluation Report: Review of Cogema Mining Inc.'s Irigaray Mine Restoration Report, Production Units 1 Through 9." Source Materials License SUA–1341. ADAMS Accession No. ML062570181.

NRC, 2004. "Review of Power Resources, Inc's A- Wellfield Groundwater Restoration Report for the Smith Ranch –Highland Uranium Project.'" June 29. ADAMS Accession No. ML041840700.

NRC, 2003a. License Amendment 15, Crow Butte Resources *In-Situ* Leach facility, License No. SUA-1534, Wellfield #1 Restoration Acceptance. Letter and Attachments from D. Gillen to M.L. Griffin dated February 12, 2003. ADAMS Accession No. ML03044055.

NRC, 2003b. NUREG–1748, "Environmental Review Guidance for Licensing Actions Associated with NMSS Programs." Washington, DC: NRC. August.

NRC, 2003c. NUREG–1620 "Standard Review Plan for the Review of a Reclamation Plan for Mill Tailings Sites under Title II of the Uranium Mill Tailings Radiation Control Act of 1978." Final Report. Washington, DC: NRC.

NRC. 2003d. NUREG–1569, *Standard Review Plan for In-Situ Leach Uranium Extraction License Applications. Final Report.* Washington, DC: NRC. June 2003.

NRC, 2000. NUREG–1724 "Standard Review Plan for the Review of DOE Plans for Achieving Regulatory Compliance at Sites With Contaminated Groundwater Under Title I of the Uranium Mill Tailings Control Act." Draft Report. Washington, DC: NRC.

WGFD (Wyoming Game and Fish Department), 2010. "Recommendations for Development of Oil and Gas Resources Within Important Wildlife Habitats." Version 6.0. Cheyenne, Wyoming: WGFD. April.

B5.10   Credibility of NRC

Comments: MR001-006, MR012-009; MR012-010; MR012-024; MR012-028; MR012-032; MR012-035; MR012-037; MR012-064.
Two commenters questioned NRC credibility in their submitted comments. The commenter asserted that NRC turned a "blind eye" to the analysis of cumulative impacts and stated the analysis of potential groundwater excursions masked the potential effect to the underlying aquifer and thus evaded meaningful analysis. The commenter concluded the analysis of groundwater impacts was arbitrary and ad hoc, which the commenter stated was not supported by NEPA or its implementing regulations. The commenter stated that NRC evaded performing meaningful analysis of impacts on surface waters at the Moore Ranch site because the analysis disregarded the close proximity of mining operations at the Moore Ranch site. Another commenter stated that the groundwater protection rule had "fallen off the table" for the benefit of an industry that wished to proceed with materials licensing under a less- than-protective regulatory framework.

Response: With regard to the general comment made regarding NRC credibility, the commenter is reminded that NRC is an independent federal agency that has no ownership of any nuclear or ISL facility. NRC regulates licensees by conducting a thorough and independent review of each application for a license consistent with its congressional mandate and NRC regulations for safety and environmental review. Once a license is granted, NRC enforces its regulations and license conditions by conducting regular inspections of operating facilities. If inspections detect noncompliance, fines and other punitive measures can be taken depending on the severity of the infraction. Because the comments were general in nature, no specific changes in the Moore Ranch SEIS were made in response to these comments.

With regard to the specific comments that are related to the commenter views of NRC credibility, it can be noted that (a) the analysis of cumulative impacts in Chapter 5 of the Moore Ranch final SEIS was revised in response to public comments; (b) the commenter concerns regarding the analysis of groundwater impacts are addressed in Section 5.9.12 of this comment response appendix; (c) in contrast to the assertion that NRC evaded performing meaningful analysis of impacts on surface water, Section 3.5.1.4 of the SEIS discusses the seasonal variability in surface water quality that is largely influenced by the CBM operations in the area; (d) Section 5.5.1 of the SEIS discusses the cumulative effect on surface water from both the proposed licensing of the Moore Ranch Project and the other activities occurring in the area including CBM operations; and (e) the status of the proposed rule with new standards applicable to ISL facilities is discussed in Section B5.9.6 of this comment response appendix.

Appendix B

B5.11 Federal and State Agencies

B5.11.1 Roles of Federal, Tribal, State, and Local Agencies

Comments: MR002-003; MR013-008; MR009-054
Three commenters stated that the SEIS should reflect finalization of the MOU between NRC and BLM (NRC, 2010). One commenter further noted that although NRC did not recognize BLM as a Cooperating Agency on the Moore Ranch draft SEIS, the MOU will allow the two agencies to work more closely on ISR uranium recovery projects in states where NRC has the licensing authority and BLM has administrative responsibilities for surface management an/or minerals. The same commenter also stated that the intent of the MOU is to improve interagency communications; facilitate the sharing of special expertise and information; and coordinate the preparation of studies, reports, and environmental documents. A second commenter requested that NRC update references to the MOU to indicate that it had been finalized. The third commenter requested that the SEIS clarify what MOU provisions are applicable to the NRC Moore Ranch Project review.

Response: SEIS Section 1.7.3.1 discusses NRC coordination with the BLM during the preparation of the SEIS. Included in that discussion is mention of the MOU, which was in a draft form at the time the draft SEIS was published. As indicated by the commenter, the MOU has since been finalized. Section 1.7.3.1 has been revised to reflect this fact and the general goals of the MOU.

For the Moore Ranch Project specifically, the MOU formalizes the types of interactions and coordination already occurring between NRC and the BLM. No further changes were made to the SEIS in response to this comment beyond those changes identified in this response.

Comments: MR002-001; MR009-200
Two commenters provided comments on the BLM role with respect to the Moore Ranch Project. One commenter stated that BLM has no authority, as the project is located on surface lands under private ownership. The other commenter stated that the Moore Ranch Project is not located on lands administered by the Rawlins, Wyoming, BLM office.

Response: SEIS Section 3.2 (Land Use) details the ownership of surface and mineral rights at the Moore Ranch Project. As discussed there, the Moore Ranch Project would be located on lands with surface rights owned by private entities (more than 85 percent of the lands within the proposed project boundaries) and by the State of Wyoming (approximately 14 percent). The mineral rights are owned by private entities, the State of Wyoming, and the BLM, although BLM rights have been leased to Uranium One. Therefore, where appropriate in this SEIS, NRC has clarified discussion to clearly indicate that BLM does not have a role at the Moore Ranch Project with respect to surface lands and rights.

Comment: MR009-013
One commenter requested that NRC, when discussing the packaging and transportation of dried yellowcake in the Executive Summary, state that the SEIS notes that packaging and transportation of yellowcake is completed in compliance with NRC and U.S. Department of Transportation (DOT) regulations.

Appendix B

Response: SEIS Section 4.3.1.2 (Transportation Impacts—Operation Impacts) states that transportation of yellowcake is performed in accordance with NRC and DOT regulations. NRC has clarified this elsewhere in this final SEIS where appropriate.

Comments: MR009-056; MR009-057; MR009-059
One commenter expressed concern over comments attributed to staff of the WDEQ in public meetings with the NRC. The commenter stated that the appropriate groundwater restoration standard under Wyoming statute and regulation is "class of use" and not "baseline conditions," as apparently stated by the WDEQ staff. Additionally, the commenter noted that the WDEQ staff comments about groundwater restoration standards do not appear in the reference cited in the text.

Response: Final SEIS Section 1.7.3.4 has been modified to reflect the level of discussion cited in the reference. Additionally, NRC notes that for NRC licensing purposes, the groundwater restoration standard will be baseline conditions, MCLs in 10 CFR Part 40 Appendix A Table 5C, or ACLs approved by NRC.

Comment: MR009-149
One commenter noted that the SEIS is incorrect in stating that industrial safety aspects associated with the use of hazardous chemicals at the proposed Moore Ranch ISR Project would be regulated by the Wyoming Division of Mine Inspection and Safety. The commenter stated that the Occupational Safety and Health Administration regulates industrial safety (including chemical safety) at ISR mines in Wyoming.

Response: Final SEIS Sections 3.12.3 and 4.13.1.2.3 have been modified to reflect this change in regulatory oversight.

Comment: MR009-212
One commenter stated that approval of the site decommissioning plan by Federal and State agencies other than NRC is not required.

Response: NRC agrees with the commenter. SEIS Section 4.6.1.4 has been revised to clearly indicate that the site decommissioning plan is approved by NRC.

Comment: MR013-008
One commenter requested that tables in the draft SEIS detailing applications or requests that have been or will be filed by the applicant be updated in the final SEIS.

Response: NRC agrees with the comment. Table 1-2 in Section 1.6.2 has been updated to reflect the current status of the applicant's permitting with other federal, tribal, and state agencies.

Comment: MR013-019
One commenter stated that the WDEQ conducts detailed reviews of all ISR wellfield packages in Wyoming. Additionally, the commenter considers the NRC review of one or more multiple wellfield packages unnecessary and duplicative.

Response: Because these comments address details about the NRC licensing and the State of Wyoming permitting processes, NRC considers the comments to be beyond the scope of the Moore Ranch SEIS.

Appendix B

Comment: MR020-020
One commenter stated that the SEIS contained inaccurate statements about the UIC program. Specifically, the commenter stated that the process for exempting an aquifer in Wyoming first involves State approval of the exemption request to be followed by EPA approval of the State action. The commenter notes that both the State process and the EPA final approval are required.

Response: *NRC agrees with the comment. Discussion of the aquifer exemption process has been clarified throughout the SEIS, where appropriate, to reflect the State process and the EPA authority for final approval.*

B5.11.2 Effects of Changes to Federal or State Regulations on the SEIS

Comments: MR018-002; MR018-003; MR018-004; MR018-005; MR018-007
One commenter noted that EPA has made revisions to the National Ambient Air Quality Standards (NAAQS) for lead and nitrogen dioxide, and that EPA has proposed revisions to the primary sulfur dioxide NAAQS and to the 8-hour standard for ozone.

Response: *SEIS Section 3.7.2 has been revised to reflect the EPA revisions to the lead and nitrogen oxide NAAQS. Additionally, NRC has modified this section to reflect that the State of Wyoming has developed stricter standards for annual and 24-hour sulfur dioxide. Finally, as necessary, NRC has reanalyzed air quality impacts in Chapter 4 to reflect these EPA and Wyoming standards.*

NRC has decided not to reflect proposed standards in the SEIS, as the SEIS is written with the regulations currently in effect at the time of its writing. Should the proposed standards be finalized in the future, NRC will reflect the final air quality standards in future environmental reviews.

Comment: MR018-006
One commenter noted that, although Table 1-2 in Section 1.6.2 suggests that an air quality permit may not be needed for the Moore Ranch Project, the project is still subject to the permitting requirements of the Wyoming Air Quality Standards and Regulations.

Response: *NRC has revised Table 1-2 to reflect that an air quality permit may be needed pursuant to the standards and regulations identified by the commenter.*

Comment: MR018-095
One commenter stated that the State of Wyoming has not yet entered into rulemaking to revise the standards for annual PM_{10} or 24-hour PM_{10}. However, the commenter requests that the SEIS still note these standards.

Response: *NRC has decided not to reflect proposed standards in the SEIS, as the SEIS is written with the regulations currently in effect at the time of its writing. Should the proposed standards be finalized in the future, NRC will reflect the final air quality standards in future environmental reviews.*

Appendix B

B5.11.3 Clarification of Other Federal/State Regulations and Practices

Comment: MR009-055
One commenter requested that the discussions with BLM described in Chapter 1 of the draft SEIS be made specific to the proposed Moore Ranch project.

Response: SEIS Section 1.7.3.1 has been revised to provide specific reference to the proposed Moore Ranch ISR facility.

B5.11.4 References

NRC (U.S. Nuclear Regulatory Commission), 2010. "Notice of Availability of a Memorandum of Understanding Between the Nuclear Regulatory Commission and the Bureau of Land Management." *Federal Register.* Vol. 74. p. 1088. January 8.

10 CFR Part 40. Appendix A. *Code of Federal Regulations*, Title 10, *Energy*, Part 40, Appendix A, "Criteria Relating to the Operations of Uranium Mills and to the Disposition of Tailings or Wastes Produced by the Extraction or Concentration of Source Material from Ores Processed Primarily from their Source Material Content."

B5.12 Cooperating Agencies and Consultations

Comment: MR020-005
One commenter suggested it would be appropriate to include, under "1.7.2 National Historic Preservation Act of 1966 Consultation," a reference to coordination with tribes.

Response: NRC acknowledges this comment. A complete list of the tribes can be found in SEIS Section 1.7.3.3 Interactions with Tribal Governments. Additionally a copy of the letter sent to each tribe is included in Appendix A. Although each tribe was sent a separate letter on February 23, 2009 only a single letter is included. Clarification text was added to Appendix A in response to this comment.

Comment: MR020-006
One commenter states the SEIS is inconsistent is the status and agreement of the Wyoming SHPO of the eligible status of sites within the Moore Ranch Project area.

Response: The comment is noted, and the text has been revised to correct inconsistencies. The Wyoming SHPO, in a letter dated November 3, 2009, to the NRC, concurred that the sites located within the Moore Ranch Project area are ineligible for listing on the NRHP. In addition, the letter states, "We recommend the U.S. Nuclear Regulatory Commission allow the project to proceed in accordance with State and Federal laws subject to the following stipulation: If any cultural materials are discovered during construction, work in the area shall halt immediately, the federal agency and SHPO staff be contacted, and the materials be evaluated by an archaeologist or historian meeting the Secretary of the Interior's Professional Qualification Standards (48 FR 22716, September 1983). Additionally, if any future disturbance is planned at the locations of sites 48CA964, 48CA6694, or 48CA6696 that evaluative testing be completed and submitted to our office with a determination of site eligibility and project effect. If eligible and adversely affected, a Memorandum of Agreement implementing appropriate mitigative

Appendix B

measures will be required." Section 1.7.2 of the SEIS has been revised to include this statement and a copy of the letter has been appended to the document.

Comment: MR020-007
One commenter noted the NHPA discussion does not identify which parties have been determined as consulting parties under 36 CFR 800.2(c).

Response: NRC agrees with the comment, and Section 1.7.2 was revised to further indicate the consulting parties.

Comment: MR020-008
One commenter noted there is no documentation of communication of the finding of no effect to any parties other than the six cc recipients of the letter. Under 36 CFR 800.4(d)(1), all consulting parties must be informed of a finding of no effect.

Response: NRC acknowledges this comment. The final SEIS Section 1.7.2 National Historic Preservation Act of 1966 Consultations was revised to include that all consulting parties were informed of a finding of no effect.

Comment: MR020-009
One commenter stated that no rationale is provided for the selection of these nine tribes for consultation.

Response: Section 3.9.2.1.1 Ethnology—Identification and Evaluation states that tribes that have heritage interest in the proposed Moore Ranch project area were consulted. Section 1.7.3.3 Interactions with Tribal Governments has been revised to include information on Tribal selection for consultation.

Comments: MR020-010, MR020-025
One commenter stated that the list of enclosures documenting that tribes other than the Eastern Shoshone were sent the February 23, 2009, consultation letter is not included in the Appendix. Therefore, it is not possible to confirm whether Tribal Historic Preservation Officers were included in the distribution.

Response: NRC acknowledges this comment. All Tribal Historic Preservation Officers were included on the distribution list on the letters dated February 23, 2009. A footnote has been added to the Table A-1 to clarify the distribution list. Note however, that there is only one ADAMS Accession number for all letters sent to the tribes.

Comments: MR020-011
One commenter noted that the consultation letter included at A-39 appears to be a draft (it is dated October 22, 2009, not October 23, 2009 as in the citation at 1-12, and includes formatting marks. The commenter requests the appendix include a copy of the final version.

Response: NRC acknowledges this comment. A copy of the finalized and signed letter to the Wyoming SHPO (October 23, 2009) was included in the References for Chapter 4, and the text has been revised to reflect the most current letter date.

Appendix B

B5.12.1 References

36 CFR Part 800. *Code of Federal Regulations*, Title 36, *Parks, Forests, and Public Property*, Part 800. "Protection of Historic Properties."

48 FR 22716. "Archeology and Historic Preservation: Secretary of the Interior's Standards and Guidelines." *Federal Register*. September 1983.

B5.13 SEIS Schedule

Comments: MR001-001; MR003-001; MR004-001; MR005-001; MR006-001
Commenters requested that the comment period on the Moore Ranch SEIS be extended to provide interested stakeholders sufficient time to adequately review the SEIS. Some commenters referred to the large size of the Moore Ranch SEIS and the need for more time to read and collect referenced information to better contribute to the process through the submission of substantive comments. Commenters also noted the comment period overlapped with seasonal holidays in December, thus reducing the time to review the document.

Response: On December 11, 2009 (74 FR 65806), the NRC staff published a notice in the Federal Register requesting public review of and comment on the "Draft Environmental Impact Statement for the Moore Ranch ISR Project in Campbell County, Wyoming, Supplement 1 to the Generic Environmental Impact Statement for In-Situ Leach Uranium Milling Facilities." In the publication of the NOA of the draft SEIS, the staff stated that the close of the public comment period closed on February 1, 2010. On February 5, 2010, the NRC staff extended the public comment period to March 3, 2010, in response to public requests for extension received via comment letters and email. The 82-day period for public comment (i.e., from December 11, 2009 to March 3, 2010) exceeds the minimum 45-day comment period required under NRC regulations. By letter and email, 20 individuals submitted 691 comments on the Moore Ranch draft SEIS.

Comment: MR005-002
A commenter stated that the public comment period on the Moore Ranch draft SEIS was insufficient to allow for meaningful public participation and was inconsistent with the purpose and intent of the NEPA, which encourages meaningful public participation in the environmental decision-making process.

Response: The NRC staff initially established February 1, 2010, as the deadline for submitting public comments on the Moore Ranch draft SEIS. In response to public comments, the NRC staff extended the public comment period to March 3, 2010 (75 FR 6065).

B5.13.1 References

74 FR 65806, NRC (U.S. Nuclear Regulatory Commission). "Notice of Availability of Draft Environmental Impact Statement for the Moore Ranch *In-Situ* Recovery (ISR) Project in Campbell County, WY; Supplement to the Generic Environmental Impact Statement for *In-Situ* Leach Uranium Milling Facilities." *Federal Register*. Vol. 74, No. 237. pp. 65806–65810. December 11, 2009.

Appendix B

75 FR 6065, NRC. "Extension of Public Comment Period on the Draft Environmental Impact Statement for the Moore Ranch *In-Situ* Recovery Project in Campbell County, WY; Supplement to the Generic Environmental Impact Statement for *In-Situ* Leach Uranium Milling Facilities." *Federal Register*, Vol. 75, No. 24, pp. 6065-6066. February 5, 2010.

B5.14 ISL Process Description

B5.14.1 Overview

Comment: MR009-011
One commenter stated that uranium is not technically leached from solution, but removed.

Response: *NRC agrees with this comment. SEIS text in the Executive Summary was revised to accurately describe this uranium milling process.*

Comment: MR012-022
One commenter expressed concern that NRC conducted an analysis of impacts from excursions related to ISR uranium recovery from unconfined aquifers. The commenter believes that this is contradictory to an NRC position in the GEIS that ISR technology can only be used in confined aquifer systems.

Response: *The GEIS is not an NRC requirement, nor does it constitute an agency policy for or against a particular type of aquifer or uranium ore formation geometry. As stated in the text box in GEIS Section 2.1, the "[h]ydrogeologic (formation) geometry must prevent uranium-bearing fluids (i.e., lixiviant) from vertically migrating." Although this type of isolation may best be achieved by the presence of confining, low permeability layers such as shales or clays both above and below the uranium-bearing horizon, the GEIS does not identify a specific type of geometry that must exist for a license to be granted. Instead, the applicant defines the wellfield boundaries "...based on the geometry of the specific uranium mineralization." (GEIS Section 2.1), and NRC evaluates the potential impacts from the proposed approach. The impact analysis presented in Section 4.5.2 of this final SEIS is based on the hydrogeologic information for the Moore Ranch site. In addition, the NRC staff are conducting a detailed technical review of the applicant license application to evaluate whether the proposed Moore Ranch ISR facility can be operated in accordance with the applicable NRC regulations in 10 CFR Part 40.*

B5.14.2 Preconstruction and Construction

Comment: MR004-002
Western Watershed Project staff commented that their organization had not received all of the reports and analyses they requested.

Response: *NRC staff do not consider this request to be a comment on the draft SEIS. This comment is beyond the scope of the SEIS.*

Comment: MR009-012
One commenter noted that the well pattern for Moore Ranch was not limited to a five-spot pattern and that alternate patterns were possible.

Response: NRC acknowledges the possibility of a pattern other than five-spot, and a modification was made to the SEIS to reflect this.

Comment: MR009-072
One commenter noted that Section 2.1.1.2, Construction Activities, does not include a description of all construction activities similar to Section 2.1.1.2 in the Uranerz Nichols Ranch SEIS. Specifically, there is no discussion of well construction and testing methods and other structures and systems in this section of the Moore Ranch SEIS. This additional information would help to better describe the Proposed Action.

Response: The NRC staff have reviewed the discussion of construction activities in the SEIS. The text has been modified to include discussions of well construction and testing.

Comment: MR009-168
One commenter noted that the only potential spills during construction could be petroleum product leaks from drill rigs, vehicles, and heavy equipment, as there are no chemical or radioactive materials present that could be spilled.

Response: The NRC staff acknowledge this comment. However, no changes have been made to the SEIS beyond this comment response.

Comment: MR020-043
One commenter was concerned that the discussion in the SEIS indicated that fugitive dust and combustion emissions during construction would be short term, however, well construction was projected to last a number of years.

Response: The NRC staff have reviewed the discussion of meteorology, climatology, and air quality impacts in the SEIS. The short-term impact due to dust and combustion emissions during construction was evaluated in terms of the duration of the emissions and the lasting impacts beyond emission-causing construction activities themselves. While construction activities would be ongoing for a number of years, the impacts from the emission-causing activities are not expected to have a lasting impact on air quality when activities cease. No modifications have been made to the SEIS beyond this response.

B5.14.3   Operations

Comment: MR009-067
One commenter was concerned that the discussion of chemical storage was inconsistent with the measures proposed by the applicant for the Moore Ranch facility.

Response: The NRC staff have reviewed the discussion of chemical storage in the SEIS and modified the text to be consistent with the information provided by the applicant in their technical report.

Comment: MR009-073
One commenter suggested adding a discussion of the management of production bleed and other liquid effluents to the SEIS.

Response: The NRC acknowledges the request, and a discussion about management of production bleed and other liquid effluents has been added to the final SEIS.

Appendix B

Comment: MR009-074
One commenter noted that the stated five steps identified as part uranium mobilization in the SEIS were not part of mobilization, but uranium processing.

Response: The NRC staff have reviewed the discussion of uranium processing in the SEIS. The text has been modified to address the inconsistencies identified by the commenter.

Comment: MR009-206
One commenter indicated that the discussion of impacts to wildlife in the SEIS stated that leaks from pipelines could expose wildlife to toxic chemicals; however, there are no toxic chemicals present in lixiviant.

Response: The NRC staff have reviewed the discussion of impacts to wildlife in the SEIS. The text in the SEIS has been modified to clarify the potential exposures to wildlife.

Comment: MR018-036
One commenter noted that the SEIS did not appear to address the hydrostatic testing of pipelines or testing/purging of wells.

Response: As noted in SEIS Section 2.1.1.2.4, the applicant has stated that all pipelines would be pressure tested before being placed into service. In addition, the applicant has indicated its intent to use engineering controls to monitor flow rates and pressure changes in the piping system, and alarms to alert the operator of losses in pressure that could indicate problems with piping or well integrity (SEIS Section 6.3.2). The commenter is correct that integrity testing of wells was not addressed in the draft SEIS, and Section 2.1.1.2.3 has been revised to reflect the planned well development and testing program described by the applicant in Section 3.1.2 of its technical report for the proposed Moore Ranch ISR facility in Energy Metal Corporation US (EMC, 2007; Uranium One, 2008).

B5.14.4 Aquifer Restoration

B5.14.5 Gaseous or Airborne Particulate Emissions

Comments: MR015-015; MR015-016; MR015-023; MR020-039; MR020-045
One commenter, EPA, stated that the air quality analysis is not adequate because detailed emission inventories for drill rig engines, fugitive dust, and facility operations are not presented. They indicated the emission inventories are needed so that NRC can decide on the extent of the air impact analysis. An additional comment from the same commenter noted on pages 4-45 to 4-47 that while the number of vehicles and equipment used for the proposed action are less than evaluated in the GEIS, the emissions are not included in the air quality impact analysis.

Response: The draft SEIS included emissions estimates for fugitive dust but did not provide detailed drilling rig or operational emissions estimates. In response to public comments, the staff updated the SEIS with additional information on the fugitive dust calculations and provided emissions estimates for diesel powered drilling and construction equipment. Details of the diesel emissions calculations are summarized in a new Appendix D in the final SEIS. The staff also provided more details regarding emissions from facility operations in SEIS Section 2.1.1.1.6.1. Section 4.7 of the final SEIS, the air quality impact analysis, was revised to incorporate by reference the revised emissions information that was added to

Appendix B

Section 2.1.1.1.6.1. The more detailed emissions estimates support the conclusions in the GEIS and SEIS that ISR facilities are not major sources of airborne emissions and draft SEIS impact conclusions were not changed by providing the more detailed emissions information.

B5.14.6   Operational History

Comment: MR009-112
One commenter noted that the discussion of evaporation ponds as an alternate waste disposal method in draft SEIS Section 2.2.5 was based on an older NRC evaluation of an earlier (1982) application for a conventional uranium recovery operation at the Moore Ranch site. The commenter noted that although the text in the SEIS did provide an historical context, additionally, a more recent basis is available for evaluating evaporation.

Response*: The NRC staff agrees that the discussion of evaporation ponds as an alternate wastewater disposal option could be strengthened. The text in SEIS Section 2.1.1.2.1 has been modified to improve the clarity of the discussion in response to this comment.*

B5.14.6.1   Historical Operational Experience: Spills and Leaks

Comment: MR012-031
One commenter stated that the water resource impact analysis relies heavily on the leak and spill surveys presented in two documents that are incomplete and inaccurate. The two documents identified by the commenter are the GEIS and the NRC memorandum "Staff Assessment of Groundwater Impacts from Previously Licensed *In-Situ* Uranium Recovery Facilities (NRC, 2009).

Response*: NRC conclusions regarding the potential environmental impacts to water resources for the Moore Ranch Project are provided in SEIS Sections 4.5.1 and 4.5.2. In determining impact conclusions, the NRC staff reviewed information provided by the applicant in its license application as amended (including the technical and environmental reports), information and data independently collected by the staff, and information and data from the GEIS and the NRC memorandum (NRC, 2009). The intent of the GEIS and NRC memorandum is not to provide an exhaustive listing of site-specific information, but rather an accurate understanding of the types and magnitudes of impacts that have been encountered at NRC-licensed facilities. No changes were made to the Moore Ranch SEIS in response to this comment.*

Comment: MR012-034
One commenter stated that for water resource impacts, NRC acknowledges the record of ISL operations spills and leaks to a certain extent. The commenter provided the example of the Smith Ranch Project as support for their concern over the limited explanation.

Response*: NRC conclusions regarding the potential environmental impacts to water resources for the Moore Ranch Project are provided in SEIS Sections 4.5.1 and 4.5.2. These impact conclusions are based on facility-specific process descriptions for the Moore Ranch Project and site-specific characteristics at the proposed site. In determining impact conclusions, the NRC staff reviewed information provided by the applicant in its license application as amended (including the technical and environmental reports), information and data independently collected by the staff, and information and data provided in the GEIS. Section 2.11 of the GEIS presents an historical discussion of ISL operations (including the Smith Ranch Project) and*

Appendix B

Section 2.14 provides reference to specific facilities in Wyoming, Nebraska, and New Mexico. The intent of the information in these sections of the GEIS was to inform the reader about which issues have historically resulted in potential impacts at ISR facilities and to provide a range of conditions that may be expected for each of the four ISR phases. Because the SEIS discussion is considered appropriate, no changes were made to the SEIS in response to this comment.

B5.14.7 Requests for Detailed Information About All ISL Facilities

Comment: MR012-046
One commenter stated that instead of disclosing the average groundwater concentrations, the SEIS should provide all groundwater sampling data and written lab reports, including details like constituent concentrations and sampling date and locations. The commenter stated that if this information was not available, the SEIS should disclose that fact.

Response: *As described in NRC guidance (NRC, 2003), an applicant, in support of its license application, is asked to provide site baseline information including groundwater quality at and in the vicinity of the site. An NRC-accepted list of constituents to be sampled for determining baseline water quality is provided in this guidance as well as a method for the applicant to propose a list of constituents that is tailored to a particular location. NRC guidance states that to determine background groundwater quality conditions, at least four sets of samples, spaced sufficiently in time, should be collected and analyzed for each constituent. The applicant provided this summary groundwater quality information and it is summarized in SEIS Section 3.5.2.3.3. Detailed information, such as the type requested by the commenter, is contained in the applicant environmental and technical reports (EMC, 2007a,b). No changes were made to the SEIS beyond the information provided in this response.*

B5.14.8 References

10 CFR Part 40. Appendix A. *Code of Federal Regulations*, Title 10, *Energy*, Part 40, Appendix A, "Criteria Relating to the Operations of Uranium Mills and to the Disposition of Tailings or Wastes Produced by the Extraction or Concentration of Source Material from Ores Processed Primarily from their Source Material Content."

EMC (Energy Metals Corporation US), 2007a "Application for USNRC Source Material License, Moore Ranch Uranium Project, Campbell County, Wyoming, Environmental Report." Casper, Wyoming: Uranium 1 Americas Corporation. ADAMS Accession Nos. ML072851222, ML072851229, ML072851239, ML07285249, ML07285253, and ML07285255. October 12.

EMC, 2007b. "Application for USNRC Source Material License, Moore Ranch Uranium Project, Campbell County, Wyoming, Technical Report." Casper, Wyoming: Uranium 1 Americas Corporation. ADAMS Accession Nos. ML072851222, ML072851258, ML072851259, ML072851260, ML072851268, ML072851350, ML072900446. October 12.

NRC (U.S. Nuclear Regulatory Commission), 2009a. NUREG–1910, "Generic Environmental Impact Statement for In-Situ Leach Uranium Milling Facilities." Washington, DC: NRC. May.

NRC, 2009b. "Staff Assessment of Groundwater Impacts from Previously Licensed *In-Situ* Uranium Recovery Facilities." Memorandum from C. Miller to Chairman Jaczko, et al. July 10. ADAMS No. ML091770402.

NRC, 2003. NUREG–1569, "Standard Review Plan for *In-Situ* Leach Uranium Extraction Licensee Applications—Final Report." Washington, DC: NRC. June.

B5.15   Financial Surety

Comment: MR009-089

A commenter noted that, while the financial assurance cost estimate for the decommissioning plan must be approved by NRC, the cost estimate need not be in place until operations are ready to commence.

Response: 10 CFR Part 40, Appendix A, Criterion 9 states that a financial surety arrangement must be established prior to the commencement of uranium recovery operations. NRC revised the Moore Ranch SEIS Section 2.1.1.1.8 in response to this comment to state that an initial surety estimate is required to cover the first year of operation.

Comments: MR009-099; MR013-021

Commenters requested that the SEIS financial assurance discussion be more descriptive. One commenter noted that financial assurance is a key component of ISR facility licensing and has been a contentious issue in the past. Additionally, the commenter considered financial assurance as an excellent example of a mitigation measure to protect against a licensee's potential financial difficulties. The commenters provided examples of additional topics to be discussed, such as (1) the types of financial assurance instruments available to licensees, (2) how financial assurance cost estimates are developed, (3) when a financial assurance cost estimate needs to be approved and posted with the agency, and (4) when the cost estimate is to be updated. One commenter also requested that reference be made to the requirements in 10 CFR Part 40, Appendix A, Criterion 9 and that the term financial "assurance" should be used rather than financial "surety."

Response: The Moore Ranch SEIS discusses financial assurance in Section 2.1.1.1.8, which references NRC financial assurance requirements, both 10 CFR Part 40, Appendix A, Criterion 9 and Section 2.10 of the GEIS that provide the detail requested by the commenters. Furthermore, the NRC staff review financial surety in detail as part of its review for the SER, which is conducted in parallel with the environmental review. Section 2.1.1.8 of the Moore Ranch SEIS has been modified to direct the reader to Part 40 Appendix A to and to Section 2.10 of the GEIS for further details about financial assurance.

NRC uses the term "financial surety" in the Moore Ranch SEIS to be consistent with its use in 10 CFR Part 40, Appendix A, Criterion 9. The term "surety instruments" is used to refer to mechanisms (e.g., surety bonds, cash deposits, and irrevocable letters of credit) for holding the funds deemed sufficient to cover the costs of site decommissioning and restoration. No changes were made to the SEIS in response to this comment beyond the information provided in this response.

Appendix B

Comment: MR009-100
One commenter stated that NRC should modify the SEIS text to indicate that surety instruments other than an irrevocable letter of credit are acceptable to the NRC and, therefore, available for use by the applicant.

Response: NRC agrees with the comment and revised the text in Section 2.1.1.8 of the Moore Ranch SEIS to indicate other forms of financial surety instruments acceptable to the NRC.

Comment: MR017-024
One commenter stated that the Moore Ranch draft SEIS failed to analyze the applicant financial assurance and decommissioning plan and did not compare the current applicant financial assurance and decommissioning plans with previous restoration funding in terms of dollars, plan, and likely results.

Response: The Moore Ranch SEIS discusses financial assurance in Section 2.1.1.8, which explains that an initial surety estimate is required to cover the first year of operation and that annual revisions to the surety estimate would be required by NRC and WDEQ to reflect existing operations and planned construction or operation the following year. This discussion also notes that the NRC staff review financial surety in detail as part of the NRC SER, which is prepared in parallel with the environmental review. The commenter request for a comparison to previous restoration funding is beyond the purpose and scope of the SEIS. Therefore, no changes were made to the SEIS in response to this comment beyond the information provided in this response.

B5.15.1   Reference

10 CFR Part 40. Appendix A. *Code of Federal Regulations*, Title 10, *Energy*, Part 40, Appendix A, "Criteria Relating to the Operations of Uranium Mills and to the Disposition of Tailings or Wastes Produced by the Extraction or Concentration of Source Material from Ores Processed Primarily from their Source Material Content."

B5.16   Alternatives

Comments: MR012-006; MR017-027
Comments were received stating that NRC should reevaluate the alternatives analyses in the GEIS and Moore Ranch SEIS. A commenter stated that the scope of the SEIS (and the GEIS) forestalled viable alternatives by assuming that uranium mining will occur, the ISL process will be used, and that the proposed Moore Ranch site is appropriate. The commenter further noted that the SEIS did not explore a range of options within ISL mining.

Response: NEPA requires Federal agencies to consider alternatives to their proposed Federal actions as well as to their environmental impacts. Alternatives can be divided into two classes: primary alternatives which are alternatives that can substitute for the agency-proposed action to accomplish the action in another manner and secondary alternatives that allow the proposed action to be carried out in a different manner. The previous comments concern both primary and secondary alternatives to the proposed Federal action.

Reasonable alternatives for a particular Federal action are defined by the proposed Federal action and the purpose and need for the proposed Federal action. As a regulatory agency, the

Appendix B

proposed Federal action for the site is an NRC decision to grant or deny the license application of a private party. The purpose and need for the proposed federal action does consider the applicant goals and objectives to extract uranium from a particular location, which helps define what are reasonable alternatives to the proposed federal action.

Reasonable alternatives considered in a site-specific environmental review depend on the proposed action and site conditions. As discussed in Section 2.1 of the final SEIS, NRC considered reasonable alternatives, including the No-Action Alternative, not approving the license application and the alternative of approving the application. Section 2.2 of the final SEIS provides a discussion of alternatives that were considered but were eliminated from detailed study and the reasons for their having been eliminated. These alternatives included conventional mining and milling, conventional mining and heap leaching, siting another ISR facility location, alternate lixiviants, and alternate wastewater treatment methods. These alternatives were eliminated from detailed study because they would cause greater environmental impacts than the proposed action. Section 2.1.1.2 of the final SEIS discusses alternative wastewater disposal methods.

While the NRC staff consider reasonable alternatives to the proposed action in the environmental review, the only alternative within NRC's decision-making authority is to approve or not approve the license application. The NRC has no authority or regulatory control over the applicant selection of uranium recovery technology to be used at the site. NRC regulatory authority is limited to evaluating the applicant's request for a license to use ISR technology at the site. If NRC decides to grant the license request, the applicant must comply with the license. NRC regulatory requirements and any other relevant local, State or Federal requirements to operate their facility.

Comment: MR012-008
A commenter noted that NRC failed to consider the reasonable alternative of altering proposed project boundaries to reduce the potential environmental impact to surface water, and NRC also did not eliminate Wellfield 2 from consideration because of the hydraulic connection between groundwater in the proposed target aquifer [the 70 Sand aquifer and an overlying aquifer (the 68 Sand aquifer)].

Response: *NRC staff note that the footprint of the subsurface ore zone cannot be altered; therefore, development of the wellfield is constrained by the occurrence of the ore zone. However, since publication of the draft SEIS, the applicant has proposed to avoid well installation in the main channels of ephemeral drainages. The commenter is also correct that the production zone in Wellfield 2 would include the 70 and 68 Sand in the regions where the "70 underlying shale" is thin or absent. The decision to combine these aquifers into one production zone was made by the applicant, not NRC. NRC notes the groundwater quality of the "68 Sand," as measured by the applicant, exceeded the EPA MCL drinking water standards for U, Ra 226, gross alpha and selenium. NRC also notes that the applicant would be required to obtain exempt aquifer status for the "68 Sand" from EPA before it may be treated as part of the production zone. No changes were made to the SEIS beyond the information provided in this response.*

Comments: MR015-001; MR015-010
A commenter noted that the consideration of only Class I UIC injection wells as the waste disposal method was inadequate. The commenter noted that other waste disposal alternatives, such as (1) treatment and disposal via a Class V injection well; (2) treatment and discharge to

Appendix B

surface waters under an NPDES permit; and (3) other potential wastewater disposal methods, such as land disposal and evaporation ponds, should have been considered in the SEIS.

Response: *In response to public comments, the final Moore Ranch SEIS was revised to expand the discussion of alternative wastewater disposal options that were not proposed by the license applicant. Section 2.1.1.2 of the final SEIS discusses the previously referenced waste disposal options and Table 2-3 compares the options, and Section 4.14.1.2 discusses the potential impacts from implementing the alternative wastewater disposal options. If licensed, the licensee would have to request a license amendment before using one of these alternative wastewater disposal options. NRC would perform an environmental and safety review on the proposed wastewater disposal method before deciding whether to grant the request.*

Comment: MR018-033
A commenter stated that the SEIS needed to analyze alternatives that minimize the amount of surface disturbance and topsoil removal, and that pipelines should be colocated with roads to minimize surface disturbance.

Response: *Table 2-3 of the final SEIS compares the potential impact to land area from implementing alternative wastewater disposal options and shows that the ISR uranium recovery method results in significantly less land disturbance than either evaporation ponds or land application.*

Comment: MR020-001
A commenter stated that chemical precipitation and reverse osmosis were dismissed for cost rationales rather than environmental or human health effects, and that evaporation ponds were eliminated as a waste disposal mechanism without adequate analysis.

Response: *The SEIS considered chemical precipitation and reverse osmosis and eliminated them from detailed analysis because they would result in a brine residual and formation of sludge being formed. As noted in Section 2.2.5, these alternatives would require onsite storage (resulting in additional land disturbance and potential impacts to cultural and biotic resources). they create waste that could potentially be characterized as hazardous or mixed waste, increasing the regulatory threshold. They also require additional transportation to ship wastes offsite, therefore, increasing the potential for occupational exposure and transportation accidents. Evaporation ponds at the Moore Ranch site were previously evaluated when Conoco Inc. proposed to use conventional milling at the proposed Moore Ranch site (NRC, 1982), and the use of evaporation ponds as an alternative wastewater disposal option is considered in Section 4.14.1.2.1 of the Moore Ranch final SEIS. No changes were made to the SEIS in response to these comments beyond the information provided in this response.*

B5.16.1 Reference

NRC (U.S. Nuclear Regulatory Commission), 2009. NUREG–1910, "Generic Environmental Impact Statement for In-Situ Leach Uranium Milling Facilities." Washington, DC: NRC. May.

Appendix B

B5.17 Land Use

B5.17.1 Ownership Issues, Surface, and Mineral Rights

Comment: MR002-002
One commenter noted that the BLM administers all minerals under the Stock Raising Homestead Act patented lands and that multiple mineral development conflicts should be avoided on split-estate land. The BLM is managing all the mineral rights on approximately half of the Moore Ranch land that is split-estate land.

Response: The NRC staff acknowledged in GEIS comment response G5.19.2 on the Ownership Issues, Surface, and Mineral Rights, that ISL operators need to not only lease mineral rights but also obtain the consent of surface owners to access the land; explore, construct, and operate their ISL facilities. Staff also recognized in GEIS comment response G5.19.3 on the Amount of Land Affected and Type; Degree and Duration of Potential Impacts, that a lease for uranium extraction on a permitted area an ISL operator would secure, would likely take precedence over future mineral rights for oil and gas, CBM, or other mineral resources in the same area. For this project, however, where 465 CBM wells and related infrastructure are present either on or near the proposed Moore Ranch permitted area, the uranium extraction mineral lease would likely need to be intermixed with other and different mineral leases, such as for gas extraction by CBM. The NRC staff acknowledge that these multiple mineral development leases should be intermixed without potential conflict under the BLM administration. But consistent with the GEIS, the NRC staff can only acknowledge, in general terms, that the potential Moore Ranch ISL operator would need to reach agreements with preexisting mineral rights lease holders under the BLM administration. Because such agreements between mineral rights lease holders and the BLM are private and, therefore, beyond the scope of this SEIS, no changes to the Moore Ranch SEIS were made beyond this response.

Comment: MR018-099
A staff member of the Industrial Siting Division of the WDEQ indicated that the Industrial Siting Council may require a permit for the ISR facility.

Response: As indicated in SEIS Section 1.6.2, a new uranium ISR milling site would need several permits from the WDEQ. NRC acknowledges that the Moore Ranch ISR may require a permit from the Industrial Siting Division of the WDEQ. The Council works with the Industrial Siting Division to determine if a Section 109 Permit pursuant to Wyoming Statute (W.S.) § 35-12-109 of the Industrial Development Information and Siting Act would be required. According to this Division, permits are required of all projects with a construction cost of $175.5 million or more, and some business types need such a permit regardless of the cost of construction. Because it has not been determined at this time whether such a permit would be required, no changes were made to the SEIS beyond the information provided in this response.

B5.17.2 Amount of Land Affected and Type, Degree, and Duration of Potential Impacts

Comment: MR009-063
A commenter suggested that NRC indicates that construction of a uranium ISR facility is small compared to other industrial facilities, as was indicated in the Nichols Ranch SEIS.

Appendix B

Response: The Nichols Ranch SEIS indicated that, comparatively, trucking activities during construction would be minor compared to other industrial activities because the Nichols Ranch project construction was deemed to be a fairly small construction project. This statement provides a relative comparison because no specific information is provided on other industrial construction projects. Thus, this statement does not provide important or critical information needed to support decisionmaking.

Because the information and clarification provided in this response rely on information already in the SEIS and, comparatively, on information in the Nichols Ranch SEIS, no changes to the Moore Ranch SEIS were made beyond this response.

Comment: MR009-140
A commenter indicated there are no coal extraction operations in the Moore Ranch Project area.

Response: The discussion in Section 3.10, Visual and Scenic Resources, was revised to be consistent with the discussion of land use in Section 3.2 of the final Moore Ranch SEIS indicating there are no active coal operations in the proposed license area.

Comment: MR009-152
A commenter indicated that it is incorrect to state that operating two wellfields sequentially would limit impact to the land and open one decommissioned wellfield to other uses like grazing.

Response: While the NRC staff indicated, consistent with the GEIS, that land impact could diminish in sequential operation of wellfields because initial decommissioning and reclamation could start at a wellfield where aquifer restoration is completed while operations would continue at other wellfields, the NRC staff agree that complete restoration of the land uses, such as grazing, to preconstruction conditions and lifting of all access restrictions would only occur after license termination. To further stress flexibility in operation, decommissioning and reclamation, the NRC staff noted that the applicant could decommission and reclaim the site in stages. Nevertheless, NRC revised the discussion in Section 4.2.1.2 of the final SEIS in response to the comment.

Comment: MR009-171
A commenter noted that in the discussion of the No-Action Alternative, the SEIS stated that CBM production in the proposed license area would continue, and the 61 ha [150 ac] would be restricted from CBM production. The commenter noted that Section 4.14.1.3 of the license application states that the applicant believes that uranium recovery development and CBM activities can be coordinated.

Response: The comment is noted. Section 4.5.1.2 of the Moore Ranch final SEIS has been revised to clarify CBM production in the proposed license area.

Comment: MR009-287
A commenter indicated that either CBM or oil and gas exploration would not be limited over the lifetime of the Moore Ranch ISR Project because they can coexist.

Response: As noted and corrected for comment MR009–171, the NRC staff acknowledge that either CBM or oil and gas exploration would not be restricted within the approximate 66 ac of the Moore Ranch license area that would be used for the operation of the facility, as the

Appendix B

applicant explained uranium ISR operations and CBM and oil and gas exploration could coexist. The text in Table 8-1 of the SEIS was corrected accordingly.

Comment: MR018-029
A commenter indicated that the construction design should account for natural features of the land (i.e., topography and drainage) so natural drainage would not be disrupted.

***Response**: It is part of standard engineering and best management practice that a site construction be designed and conducted to minimize disruption to and avoid blockage of natural surface water drainage features. For the Moore Ranch ISR Project, the NRC staff find that, based on the information the applicant provided on its limited construction area, its limited and intermittent number of surface water and wetlands features on site, and its implementation of best management practices, the potential impacts to natural surface water drainage features and to wetlands associated with the construction of roads, the installation of power lines, the construction of wells and pipelines, building of the plant, and the related vehicular traffic are expected to be small, consistent with the GEIS findings.*

For example, during construction of the proposed Moore Ranch ISR Project, construction of Wellfield 2 would have the potential to impact Upper Wash #2 to Simmons Draw because this intermittent channel crosses proposed Wellfield 2. One culvert, located on the access road connecting the central plant to the main access road, would be constructed to maintain the current surface drainage conditions. Power line poles would be not be installed in or near intermittent streams.

Because of the configuration of the site topography and that of the ore bodies, an estimated 13 wells from Wellfield 2 may be installed in ephemeral channels or delineated wetlands, only if unavoidable. No wells would be installed in existing ponds. If some wells would be installed in intermittent streams, they would be drilled during the dry season, and erosion and sedimentation control measures would be implemented during the drilling period to minimize impacts. Upon completion of the well construction, reseeding and mulching would occur to stabilize loose soil.

A PVC pipeline that would need to bisect wetlands and ephemeral channels would be buried perpendicular to the channels during the dry season in narrow trenches cut with small excavating equipment. Excavated native soil would be used to backfill the trenches to restore the natural channel preexisting grade. Reseeding and mulching would be implemented to restore soil stability. Also, a single trunkline connecting the pipeline network to the header houses would be located in uplands so it would not cross any washes or stream channels.

Finally, the central processing plant and support buildings would be constructed away and landward of all intermittent channels on flat terrain and above the peak flood elevation of the intermittent channels.

These examples illustrate that construction design and planned activities would take into account the natural features of the land to help minimize potential impacts to natural surface water drainage features and to wetlands.

Because this response addresses the comment by providing confirmation and clarification with information already in the SEIS, no changes to the Moore Ranch SEIS were made beyond this response.

Appendix B

Comment: MR018-030
One commenter indicated that concentrated runoff that would cause erosion should be avoided by appropriately designing roads, pipelines, and other structures.

Response: As detailed in the NRC staff response to related comment MR018-029, the NRC staff find that for the Moore Ranch Project, the applicant information on construction design and planned construction activities would account for the presence and characteristics of the natural features of the land so that natural drainage would not be disrupted and surface runoff from roads, pipelines, corridors, and other structures would not be concentrated to potentially cause additional erosion.

Because this response refers to the detailed response provided for related comment MR018-029 and because confirmation and clarification is provided in this response with information already in the SEIS, no changes to the Moore Ranch SEIS were made beyond this response.

Comment: MR018-035
A commenter suggested that during drilling operations, portable tanks should be used to contain drilling mud and other fluids instead of excavating mud pits in the ground. This would reduce surface disturbance and reclamation cost and would also reduce risks of contamination and erosion.

Response: The applicant is proposing the use of mudpits during wellfield construction and drilling activities. Mudpits typically remain open for less than 30 days and the applicant would follow the WDEQ-LQD guidelines on topsoil and subsoil management at uranium ISR facilities. The NRC staff concluded that impacts to soils would be small based on the limited size of the wellfield areas, the applicant's proposed topsoil removal and stockpiling activities, and the short duration of mud pit usage and pipeline trenching. The NRC staff recognizes that alternative methods are available to limit the potential impacts from the use of mud pits during well drilling activities. The text in Section 4.4.1.1 was clarified to acknowledge the availability of alternative methods to mitigate impacts from mudpits.

B5.17.3 Mitigation and Reclamation Issues

Comment: MR009-016
A commenter indicated that land use restrictions would be lifted at the end of decommissioning and reclamation activities and noted that NRC wrote, "The potential impact to land use would diminish as restoration activities were completed and the use of wellfield header houses was complete."

Response: NRC staff believe there is no apparent contradiction or inconsistency as this comment implies. Consistent with the GEIS and the SEIS on this Moore Ranch Project, NRC staff indicated impact to land use during the aquifer restoration phase would diminish because decommissioning and reclamation activities could be initiated at a particular wellfield as the aquifer restoration phase would end. This is the case at uranium ISR projects where the operations and restoration of different wellfields are planned to be staggered over time. For the Moore Ranch Project, the applicant is planning to operate and restore two separate wellfields sequentially. Wellfield 2 would begin to be restored while operations at Wellfield 1 would continue (SEIS Section 2.1.1.2.3). Thus, as aquifer restoration activities end at Wellfield 2,

Appendix B

some decommissioning and restoration activities could be initiated while operations or aquifer restoration continue at Wellfield 1. That is why the NRC staff indicated that land use impact would diminish, as aquifer restoration activities would be completed.

Because clarification is provided in this response with information already in the SEIS, no changes to the Moore Ranch SEIS were made beyond this response.

Comment: MR009-090; MR013-007
Two commenters pointed out that the NRC should use the phrase "unrestricted use" when referring to post-reclamation land use.

Response: The NRC acknowledges the comments, and has added language to SEIS Section 2.1.1.1.5 to clarify terms for "unrestricted use."

Comment: MR013-016
The commenter states that the draft SEIS structure indicates that aquifer restoration is separate from the surface reclamation stage of an ISR facility lifecycle. The commenter seeks clarification on whether 10 CFR 40.42, "Expiration and termination of licenses and decommissioning of sites and separate buildings or outdoor areas," can be applied to groundwater restoration. The commenter also seeks clarification on the timeline in which a decommissioning plan is required to be submitted to the NRC.

Response: According to the Commission Decision regarding Hydro Resources, Inc. (NRC, 2000), the NRC staff are required to review a decommissioning plan prior to issuing a license. Section 6.5 of NUREG–1569 (NRC, 2003) contains staff guidance for reviewing decommissioning plans. Sections 6.1 through 6.4 address the decommissioning/restoration activities to be included in the application including groundwater restoration, soils reclamation, building decommissioning, and post-decommissioning surveys. Therefore, the intent of the aforementioned Commission Decision and NUREG–1569 is to review a decommissioning plan that addressed full facility build-out for the life of the facility.

Unlike other facilities, the precise as-built conditions are unknown prior to operations because continued exploration may result in alterations to proposed wellfields. Such alterations affect the required wellfield infrastructure. Therefore, a more detailed decommissioning plan would be required 12 months prior to decommissioning a facility or a portion thereof. This plan would comply with §40.42. Regarding financial assurance, the Commission stated in the Hydro Resources, Inc. decision that a surety is not required prior to licensing, but one is required prior to operations.

As stated in letters to applicants dated July 7, 2008 (NRC, 2008), the timeliness and decommissioning regulations apply to ISRs; therefore, alternate schedules must be submitted if restoration/decommissioning would require more than 2 years. Because the timeliness in decommissioning rule applies not only to entire facilities but to portions thereof, restoration schedules apply to individual wellfields.

Comment: MR018-033
A commenter indicated that alternatives to minimize surface disturbance and topsoil removal need to be analyzed and pipelines should be located along roads to minimize such disturbance.

Appendix B

Response: *As described in the Moore Ranch SEIS, measures and alternatives were taken into account to minimize surface disturbance and soil removal. First, the minimum amount of land needed for construction of the plant and related construction was considered and the topsoil to be stripped would be stockpiled for reuse following WDEQ Land Quality Division (LQD) guidelines. Topsoil and subsoil to be excavated at other areas (e.g., well pad, mud pit, pipeline trenches and crossing) would be replaced where removed, whenever possible. Where evident and practical, segregation of topsoil and subsoil would be conducted at this site, and the subsoil would be replaced first in excavations, followed by topsoil, then regrading and reseeding activities. Also, the two wellfields construction would be staggered over several years so soil disturbance would be further minimized because it would not occur during the same construction season over the whole area of the two wellfields.*

Most major pipelines from header houses to the processing plant would be installed in trenches that would not be located along access roads at this site. Further inquiry would be needed to determine if cost, more favorable topography, shortest routes, and other factors have prompted the applicant to design its major pipeline routes away from the site access roads and if the applicant has considered, in this particular situation, alternatives for minimizing surface and soil disturbance in locating trenches over undisturbed land away from access roads.

During the decommissioning phase, all roads at the Moore Ranch ISR with exceptions to be determined later by interested parties site, would be reclaimed by removing road surfacing materials and culverts, recontouring, and preparing and reseeding the soil. Other areas within the permit area such as the wellfields, header houses, and pipeline and well areas would be reclaimed by regrading the ground surface to the approximate preconstruction topographic contours, reestablishing natural drainage, replacing salvaged soil, and revegetating those areas.

These examples illustrate that measures and alternatives considered would be implemented along with these and other accepted best management practices during the construction and decommissioning phases to minimize and mitigate impacts to surface and soil disturbances.

Because this response addresses the comment by providing confirmation and clarification with information already in the SEIS, no changes to the Moore Ranch SEIS were made beyond this response.

B5.17.4   References

10 CFR Part 40. *Code of Federal Regulations*, Title 10, *Energy*, Part 40, "Domestic Licensing of Source Material."

NRC (U.S. Nuclear Regulatory Commission), 2008. Letter from K. McConnell, Deputy Director, U.S. Nuclear Regulatory Commission, to M. Collings, President, Power Resources, Inc. Subject: Compliance with 10 CFR 40.42's Timely Decommissioning Requirements. July 7. ADAMS No. ML081480293.

NRC, 2003. NUREG–1569, "Standard Review Plan for In-Situ Leach Uranium Extraction License Applications." Final Report. Washington, DC: NRC. June. ADAMS No. ML032250177.

NRC, 2000. Commission Memorandum and Order in the Matter of Hydro Resources, Inc. CLI–00–08. May 25.

Appendix B

B5.18 Transportation

Comment: MR009-032
In reference to draft SEIS text on page xxi; line 19, a commenter suggested that in addition to less yellowcake shipments during the aquifer restoration phase, NRC should add that fewer chemical supply shipments would also be realized during this period.

Response: The NRC staff agree with the commenter that during aquifer restoration there would be fewer chemical supply shipments because uranium processing activities would diminish during the aquifer restoration phase. The text of the draft SEIS was modified to acknowledge a decrease in chemical shipments during aquifer restoration in response to this comment.

Comments: MR009-159, MR009-160
A commenter noted that Table 4-2 in the draft SEIS contains several values that do not match the applicant response to the NRC RAI (Uranium One, 2009). Specifically, the commenter indicated the percentage increase in truck traffic at mile sign 118.726 should be 0.3, and not 0.2; the percentage increase in auto traffic at mile sign 149.24 should be 4.1, not 4.2; the percentage increase in auto traffic at mile sign 131.793 should be 6.7 not 7.2; and the percentage increase in auto traffic at mile sign 137.12 should be 6.6, and 7.1. The commenter noted that similar corrections should be made to draft SEIS Table 4-3 as well.

Response: The NRC staff checked the information, and found inconsistencies in the information that was submitted in the applicant RAI that were corrected in the draft SEIS table. Therefore, no changes were made to SEIS Tables 4-2 and 4-3.

Specifically, based on applicant provided information the percent of auto traffic should be the projected auto traffic volume divided by auto traffic count (derived from the reported all vehicle count minus the truck count). For the comments regarding mile sign 131.73 and 137.12, the values in the applicant spreadsheet do not follow this approach and no information was provided to explain why these values would be calculated differently, so this was assumed to be an error in the applicant table. Regarding the comment on values for mile sign 118.726, the draft SEIS value is not 0.2 as the commenter suggests (the value is 0.3, the value recommended by the commenter), so no changes to that value were needed. Regarding the comment on values for mile sign 149.24, the value in applicant Table 4.2-1 and the draft SEIS table is 3, not 4.1 or 4.2 as the commenter suggests.

Comment: MR018-011
One commenter, referring to sections 3.2.2 and 3.3.2, stated that transportation routes would be determined by the Wyoming Department of Transportation District Engineer.

Response: While the comment is not specific, the staff have interpreted the comment as referring to sections in the GEIS that describe the affected environment for transportation in the Wyoming milling regions. This is because there are no such discussions in the Moore Ranch draft SEIS Section 3.2.2 and there is no Section 3.3.2 in that report. While the comment period for the GEIS has ended, the staff understand that individual states can specify routes that are acceptable to them for hazardous material transportation. For yellowcake shipments that travel through multiple states, NRC applicants and their carriers are required to comply with all

Appendix B

applicable state laws in addition to the applicable transportation regulations of the NRC in 10 CFR Part 71 and DOT regulations in 49 CFR 171 to 189. Because the comment was not specific to the material in the SEIS no changes to the SEIS were made in response to the comment.

Comment: MR018-052; MR018-066

One commenter suggested the transportation impacts were trivialized because the SEIS did not account for hauling all waste generated from all phases including construction, operations, restoration, and decommissioning. The commenter also expressed the view (referring to executive summary page xvii, in particular) that the volume of contaminated soil that would need to be shipped offsite for disposal could be significant. The commenter also indicated the SEIS did not provide waste volumes for decommissioning wastes.

Response: *Sections 2.1.1.1.7 and 4.3 of the draft SEIS provided a discussion of the estimated magnitude of traffic generated by the proposed action. Section 4.3, in particular, includes tabulated estimates of traffic generated for each phase of the project. The traffic estimates in Section 4.3 include shipments of both and municipal solid wastes. To respond to the comment, the NRC staff reviewed the traffic estimates and found the waste shipment estimates for decommissioning low; therefore; more detailed estimates were developed. To develop the decommissioning waste shipment estimates, the NRC staff calculated annual and cumulative decommissioning waste volumes that were added to Section 2.1.1.1.6.3 (solid waste) and decommissioning waste shipment estimates that were added to Section 2.1.1.1.7 (transportation). The decommissioning waste volumes were based primarily on information submitted in the applicant surety estimate (EMC, 2007a), which includes a detailed accounting of decommissioning costs, including costs to excavate contaminated soil and ship the soil to a licensed facility for disposal as byproduct material. The waste volumes include estimated contaminated soil volumes that could be generated in wellfields from leaks and spills during operations. Because a significant proportion of the decommissioning waste is from wellfields and the wellfields are planned to be decommissioned using a phased approach over a four-year period, the highest expected annual average daily waste transportation is expected to occur during periods when two wellfield decommissioning actions overlap and the plant facilities decommissioning is also underway. The resulting annual volume of wastes equates to about one truck shipment per day. As a result of this revision, information in Table 4-4 was updated, however, because the magnitude of trucking is still low relative to the exiting road traffic, the impact conclusions were not changed.*

Comment: MR018-083

Referring to draft SEIS page 5-8, a commenter noted that section 5.3 states that transportation impacts for solid wastes would be small, however, they requested that impacts be reevaluated because the local landfill is expected to be closing in the near future, and waste would have to be shipped over a greater distance.

Response: *In response to the comment, the staff reevaluated the circumstances regarding the status of the Midwest landfill in Edgerton, Wyoming. While the commenter is correct that the landfill is planned to close in 2010, WDEQ permitted the landfill operator to operate as a transfer facility that would transfer wastes to the balefill in Casper, Wyoming. As a result, the direct transportation impacts would not change, as the solid wastes from the proposed action would be shipped to the location that is indicated in the draft SEIS. The transfer operation from Edgerton to Casper would have some additional indirect transportation impacts (including additional traffic and air emissions), however, at the stated annual waste volumes discussed in*

Appendix B

the SEIS (which, along with related truck shipment estimates, have been updated in response to other comments) the contribution to truck traffic from annual shipments of solid waste for landfill disposal is expected to be approximately one truck per day or less (see revisions to Sections 2.1.1.1.6.3, 2.1.1.1.7, and 4.3 in the final SEIS). This additional amount of traffic on roads that serve over 200 trucks per day (Section 4.3) is not expected to represent a significant change from current conditions and, therefore, the impact conclusions in the SEIS were not changed in response to the comment.

B5.18.1 References

10 CFR Part 71. *Code of Federal Regulations*, Title 10, *Energy*, Part 71. "Packaging and Transportation of Radioactive Material."

EMC. 2007a. "Application for USNRC Source Material License, Moore Ranch Uranium Project, Campbell County, Wyoming, Technical Report." Casper, Wyoming: Uranium 1 Americas Corporation. ADAMS Accession Nos. ML072851222, ML072851258, ML072851259, ML072851260, ML072851268, ML072851350, ML072900446. October 2, 2007.

Uranium One (Uranium One Americas, Inc.), 2009 "Re: Additional Information Requested for the Moore Ranch *In-Situ* Uranium Recovery Project License Application Environmental Report, First Set of Responses." Letter (June 19) from J. Winter to M. Fliegel, NRC. Casper, Wyoming: Uranium One Americas, Inc.

B5.19 Geology and Soils

B5.19.1 Soil Disturbance Concerns

Comment: MR009-017
A commenter noted that topsoil removal in wellfields would affect a much smaller area than indicated by a statement in the SEIS Executive Summary.

Response: *Section 2.1.1.1 of the SEIS described that the proposed license area for the Moore Ranch Project comprises a surface area of about 2,879 ha [7,110 ac], of which less than 61 ha [150 ac] would be affected by the proposed ISR activities. Within the affected area, the main activities resulting in topsoil removal would be construction of the processing plant facility and development of the wellfields (i.e., drilling of wells, excavation of pipelines, and construction of wellfield access roads). In contrast, the Executive Summary of the SEIS stated that these same activities would involve the removal of topsoil covering about 61 ha [150 ac]. The NRC staff acknowledge that the latter statement is misleading because it implies that all topsoil throughout the area would be removed. The text of the Executive Summary has been revised accordingly.*

Comment: MR018-013
A commenter noted that the applicant could significantly reduce the overall amount of surface disturbance of soils by using portable tanks and closed loop mud systems to contain drilling mud and other fluids instead of digging and reclaiming hundreds of mud pits.

Response: *Mud pits are commonly used during drilling activities to control the spread of fluids, minimize the potential area of soil contaminated by used drilling fluids and cuttings, and*

Appendix B

enhance evaporation (SEIS Section 4.5.2.1.1). Section 4.4.1.1 of the SEIS identified that mud pits would be constructed by removing the topsoil from a designated pit area, placing it in a separate location, then excavating the subsoil to the desired depth and depositing it next to the pit area. After drilling was complete and the mud pit was no longer needed (typically within about 30 days from the initial excavation), the excavated subsoil would be used to fill in the pit, and the topsoil would be replaced on top. Given the brief period of time each mud pit is used, the limited size of each construction area, and the implementation of best management practices to restore and revegetate the topsoil in the filled-in pits, the SEIS concluded that the potential environmental impacts to soils from mud pits would be small. The NRC staff acknowledge that, as an alternative to excavated mud pits, the use of portable tanks and closed loop mud systems is a viable construction technique that would further mitigate the environmental impact of soil disturbance in the project area for the large number of wells (approximately 850) that the applicant proposes to drill in developing Wellfields 1 and 2. However, the difference in technology would not affect the conclusion that the proposed drilling activities are expected to have a small impact on soils in the proposed project area. No changes have been made to the SEIS beyond this response.

Comments: MR018-034, MR018-041, MR018-093
Several commenters noted the importance of vegetation in mitigating the environmental impacts caused by disturbed soil. One commenter noted that the amount of disturbed land surface should be limited whenever possible because vegetation is the main factor controlling erosion. Another commenter stated that disturbed areas should be revegetated as soon as possible with saved topsoil and plants suitable for wildlife. A commenter suggested vegetation should be mowed rather than bladed to minimize soil disturbance because soils that remain in place may be reclaimed more easily than soils that have been removed.

Response: *The NRC staff acknowledge that vegetation is an important factor in controlling erosion and maintaining the site ecology. Section 4.4.1 of the SEIS identified that soils within the proposed wellfields have a moderate to severe risk from both wind and water erosion. Although the practice of mowing vegetation rather than blading to minimize soil disturbance is not feasible for the proposed construction activities that require excavation of topsoil, the applicant has identified that soil disturbance at the proposed project area would be limited to less than about 2 percent of the total license area. Moreover, much of the disturbed soil will be localized in the area that includes the construction of the central plant and associated facilities (SEIS 2.1.1.1). Most of the remainder of the disturbed soils would result from well drilling operations, trenching and burial of pipelines, and construction of secondary access roads. The applicant described that disturbed soils would be restored and reclaimed on a scale of weeks or months (e.g., drilling sites and open pipeline trenches) or several years (e.g., roads and buildings), following best management practices. Topsoil would be segregated and salvaged during excavation activities and protected from erosion while it was stockpiled. The applicant identified mitigation measures to minimize soil impacts, including reestablishing temporary or permanent native vegetation as soon as possible after disturbance in accordance with WDEQ guidelines and implementing best management practices to retain sediment within the disturbed areas. Because the information used to determine the significance of disturbed soil impacts is provided in the SEIS, no changes beyond this response have been made to the SEIS.*

B5.19.2 Soil Impacts from Surface Spills

Comment: MR017-010
A commenter was concerned that even small spills may have a cumulatively significant effect on soils and groundwater.

Response: Section 4.4.1.2 of the SEIS described that impacts from surface spills and leaks from the proposed Moore Ranch Project would be limited by (i) engineering controls to detect pressure changes in the piping system, (ii) a containment system for spills in the central processing plant, and (iii) a leak detection system for pond liners. The potential for spills would be further mitigated by implementation of onsite standard operating procedures and by the need to comply with NRC and WDEQ requirements for spill response and reporting of surface releases. For example, Section 4.5.2.1.2.1 of the SEIS noted that contamination surveys would be performed after repairing wellfield leaks, and contaminated soils could be immediately remediated if concentrations exceeded regulatory requirements or might be cleaned up later during decommissioning. Monitoring wells would be in place to provide an extra level of surveillance in the wellfield to detect any impacts to the near-surface aquifers from either surface spills or casing leaks. Based on these considerations, the long-term impact to soils and groundwater from spills would be expected to be SMALL. Because information used to determine the significance of the effect of spills on soils and groundwater is included in the SEIS, no changes to the SEIS were made beyond this response.

Comments: MR009-021; MR017-011
Two commenters noted that spills during construction would not include radioactive materials or hazardous waste, except for fuel. One commenter claimed inconsistent information about the control and management of spills during construction.

Response: Neither the SEIS (2009; Section 4.5.2.1.1) nor NUREG–1910 (Section 4.3.4.2.1) suggests the potential for radioactive or hazardous toxic spills during the construction phase, other than spills of fuels and lubricants from construction equipment.

The applicant is required to establish a detection monitoring program to protect USDW from potential spills and leaks, in compliance with 10 CFR Part 40, Appendix A, Criterion 7A and 40 CFR 144.54. Once a facility is licensed, the applicant is also required to implement corrective action to prevent movement of any spills or leaks into USDW, in compliance with 40 CFR 144.55.

The applicant is required to obtain construction and industrial stormwater NPDES permits from the State (through WDEQ) prior to commencement of ISR activities (SEIS 2009; Table 1-2). As part of this permit, the applicant would implement best management practices, such as implementation of a spill prevention and cleanup plan to minimize soil contamination (NUREG–1910, Sections 7.4 and 4.3.4.2.1).

Under these circumstances, cumulative impacts of small spills on soil and near-surface aquifers would be small.

Comments: MR018-048, MR018-051
A commenter expressed concern about management of contaminated media, such as whether soils and other media contaminated by spills and other releases would be transported to solid waste disposal or treatment facilities in Wyoming.

Response: Contaminated media, such as soils, would be managed in the same manner as any other waste generated at the site. The NRC noted in Section 4.14 of the SEIS that ISR facilities generate radiologically contaminated wastes, including contaminated soils, structures, and liquids that are classified as byproduct material. If soils, construction material, piping, or other media become contaminated with byproduct material, then that media would be handled, stored, and disposed of in the same manner as byproduct material. The NRC requires an ISR facility to have an agreement in place with a licensed disposal facility to accept byproduct material before ISR operations begin. The existing facilities that are licensed by NRC to accept byproduct material for disposal are the Pathfinder-Shirley Basin uranium mill tailings impoundment in Mills, Wyoming, and the Rio Algom Ambrosia Lake uranium mill tailings impoundments near Grants, New Mexico. Additionally, two sites in Utah and one in Texas are licensed by NRC Agreement States to accept byproduct material for disposal. Because the information provided in the SEIS about the disposal of contaminated materials generated by the proposed Moore Ranch Project is sufficient to support the evaluation of environmental impacts, no changes were made to the SEIS in response to this comment.

B5.19.3 Permanent Change to Rock Formations

Comment: MR009-018
A commenter disagreed with the statement in the Executive Summary of the SEIS that ISR operations would permanently change the composition of the uranium-bearing rock.

Response: The ISR operations would dissolve and remove uranium from the target sandstones, thereby lowering the uranium concentration in the rock. In this regard, the change in composition of the rock is permanent, but the uranium mineralization in the sandstones is only a minor component of the total composition of the rock. The NRC staff agree that the statement, as written, could convey to interested stakeholders the impression that ISR operations could result in significant, large-scale impacts on site geology when no such impacts would occur. In response to the comment, the sentence in the Executive Summary of the SEIS was revised accordingly.

B5.19.4 References

10 CFR Part 40. Appendix A. *Code of Federal Regulations*, Title 10, *Energy*, Part 40, Appendix A, "Criteria Relating to the Operations of Uranium Mills and to the Disposition of Tailings or Wastes Produced by the Extraction or Concentration of Source Material from Ores Processed Primarily from their Source Material Content."

40 CFR Part 144. *Code of Federal Regulations*, Title 144, *Protection of Environment*, Part 144, "Underground Injection Control Program."

NRC (U.S. Nuclear Regulatory Commission), 2009. NUREG–1910, "Generic Environmental Impact Statement for In-Situ Leach Uranium Milling Facilities." Washington, DC: NRC. June.

Appendix B

B5.20 Groundwater Resources

B5.20.1 General Concerns About ISL and Groundwater Contamination

Comment: MR016-014
A commenter expressed skepticism about the SEIS's assurances regarding spills and leaks, given ISL history.

Response: An applicant is required to obtain construction and industrial stormwater (NPDES) permits from the State (through WDEQ) prior to commencement of ISR activities. As part of this permit, the applicant would implement best management practices, such as a spill prevention and cleanup plan to minimize potential impacts on soil and groundwater due to leaks and spills on the ground surface (NUREG–1910, Sections 7.4 and 4.3.4.2.1).

The applicant is required to establish a detection monitoring program to protect USDW from potential spills and leaks, in compliance with 10 CFR Part 40, Appendix A, Criterion 7A and 40 CFR 144.54. Once a facility is licensed, the applicant is also required to implement corrective action to prevent movement of any spills or leaks into USDW in compliance with 40 CFR 144.55.

To be compliant with these regulations, the applicant would install a leak detection system on impoundments for early detection of any potential leaks (NUREG–1910, Section 2.3.2). The applicant would conduct mechanical integrity testing of each well to check for leaks or cracks in the casing (NUREG–1910, Section 2.4.1.3), in compliance with 40 CFR 146.8. The applicant would install meters and control valves in individual well lines to monitor and control flow rates and pressures for each well to maintain water balance and to aid in identifying leaks (NUREG–1910, Section 2.3.1.1). The applicant would measure and record pipeline pressure to monitor for potential leaks and spills that might result from the failure of fittings and valves (NUREG–1910; Section 2.4.1.2). The applicant would implement corrective action to prevent movement of any spills or leaks into USDW. Under these circumstances, potential impacts on groundwater due to spills and leaks would be small to moderate. No changes were made to the SEIS beyond the information provided in this response.

B5.20.2 Importance of Water and Consumptive Water Use

Comments: MR009-020; MR009-022; MR020-026
A commenter noted that consumptive water use would be minimal during construction and would be limited to water used for well drilling and dust suppression activities. Another commenter noted that the assessment of groundwater impacts due to consumptive water use during construction by relating the consumptive water use to the existing water supply is not helpful. A commenter noted an incorrect statement in the SEIS about water source for the consumptive water use during construction.

Response: Section 4.5.2.1.1 of the SEIS discusses consumptive water use during the construction phase of the proposed Moore Ranch Project. The applicant has stated that most water used for the proposed Moore Ranch Project would come from a well completed in the 40 and 50 Sand aquifer at depths of 143 to 180 m [470 to 590 ft] below the ground surface, much deeper than the shallower aquifers beneath the site. Consumptive water use during the construction phase of the proposed Moore Ranch Project would generally be limited to dust control, drilling support, and cement mixing. Impacts from groundwater consumptive use during construction would be minor and temporary.

No changes were made to the SEIS beyond the information provided in this response.

Comment: MR009-185
A commenter noted that groundwater consumptive use during aquifer restoration would be reduced by 75-80 percent by using reverse osmosis.

Response: *Groundwater consumptive use during aquifer restoration is generally reported to be greater than during ISR operations. One reason for increased consumptive use during restoration is that no water is reinjected during groundwater sweep. The rate of groundwater consumptive use during aquifer restoration would be lowered during the reverse osmosis phase, because up to 70 percent of the pumped groundwater treated with reverse osmosis can be reinjected into the aquifer. Groundwater consumptive use could be further decreased during the reverse osmosis phase if brine concentration is used, in which case up to 99 percent of the pumped water could be suitable for reinjection (NUREG–1910, Section 4.3.4.2.3). The SEIS text was modified in response to this comment.*

Comments: MR010-003; MR011-003
A commenter claimed that the large volume of waste water alone will have significant impacts on the State.

Response: *During ISR operations, the average production bleed rate at the Moore Ranch Project is estimated at a percentage of the maximum injection rate, which would be 114 L/min [30 gal/min] (SEIS 2009, Section 2.1.1.2.3). The production bleed is removed from the circuit and needs to be disposed. The purpose of the production bleed during ISR operations is to maintain a negative water balance to ensure that there is a net inflow of groundwater into the wellfield to minimize the potential movement of lixiviant and its associated contaminants out of the wellfield (NUREG–1910, Section 2.4.1.2).*

The applicant is required to dispose of production bleed produced during ISR operation and restoration while ensuring public health and safety in compliance with 10 CFR Part 40, Appendix A 5A for groundwater protection; 5B for secondary groundwater protection; and 5C for maximum values for groundwater protection. The applicant proposed deep well disposal of production bleed at Moore Ranch and has applied for a UIC Class I injection permit from WDEQ, as part of the license application. The permit application is currently under WDEQ review. WDEQ would grant the permit only if deep disposal practice is protective of human health and the environment.

B5.20.2.1 Site Characterization

Comments: MR009-023; MR020-022
A commenter pointed out inconsistent information on the existence of production wells within the project area. Another commenter wanted to know how domestic wells within the permit area are to be handled.

Response: *There is one domestic well and four stock wells within the license area (environmental report, 2007, Section 3.4.1.2). The domestic well within the permit boundary is permitted for industrial and domestic uses by Rio Algom Mining Corporation. Although this well* *is permitted for domestic uses, there is currently no occupied residence within the project area (SEIS 2009, Section 3.5.2.3.4). Four stock wells within the licensed area are not licensed through the State Engineers Office. There is no irrigation groundwater well within the permit area (environmental report, 2007, Section 3.4.1.2).*

At Moore Ranch, the uranium-bearing 70 Sand occurs between 30-100 m [100-330 ft] below ground surface (environmental report, 2007, Appendix A). There are three domestic water wells ranging from 41.7 to 134 5 m [137 to 440 ft] in depth within the 3.2-km [2-mi] radius of the Moore Ranch Site (SEIS 2009, Section 3.5.2.3.4), but only one of them lies within the permit boundary.

The applicant is required to obtain an aquifer exemption permit from the State (through WDEQ) first. Then, the applicant is required to obtain the final approval from the EPA on the State action. If the domestic well is completed in the exempted portion of the ore-bearing aquifer, the well cannot be used as a source of drinking water in compliance with 40 CFR Part 146. In this case, the domestic well is required to be properly plugged and abandoned prior to commencement of ISR operations. Similarly, if industrial or livestock wells are completed in the exempted portion of the aquifer, these wells are required to be properly plugged and abandoned to avoid potential negative impacts on targeted bleed rates during ISR operations and also to minimize potential adverse impacts on the environment. Upon completion of ISR operations, the applicant is required to return groundwater quality in the exempted portion of the production aquifer to restoration standards, in compliance with 10 CFR Part 40, Appendix A, Criterion 5B(5).

The other two domestic wells are outside the permit boundary, but within the 3.2-km [2-mi] radius of the Moore Ranch Site. Therefore, these two domestic wells would be protected through the excursion monitoring and remediation in compliance with 40 CFR 144.54 and 40 CFR 144.55 during ISR operation and post-operation periods (NUREG–1910, Sections 2.11.4 and 8.3.1.1).

Comments: MR016-007; MR016-008
A commenter noted that NRC does not consider requiring an alternative site for the facility, altering boundaries of the facility, or extensive testing requirements for detailed hydrological characterization of the site for protection of underground drinking water resources.

Response: *ISR operations under NRC regulations are risk-informed and performance based. An applicant is required to meet State, NRC, and other Federal agency requirements and regulations. This includes, but is not limited to, permit requirements in compliance with 40 CFR 144.51, monitoring and reporting in compliance with 40 CFR 144.54, implementing corrective action to protect USDW in nonexempted portions of the ore-bearing aquifer and surrounding aquifers, and meeting restoration standards, in compliance with 10 CFR Part 40, Appendix A, Criterion 5B(5).*

If the applicant meets these and other related regulatory requirements, the applicant, under NRC regulation, has flexibility in proposing permit boundaries, designing wellfields and monitoring well rings, and selection of restoration techniques. NRC is interested in whether the applicant is in compliance with regulatory requirements, but it is up to the applicant how to meet these regulatory requirements.

Appendix B

Although the applicant has flexibility with wellfield designs, the applicant is still required to seek approval from the State (through WDEQ) when they finalize their locations. At the time the draft SEIS was prepared, the applicant had not yet finalized the wellfield designs. Therefore, the decision on the acceptability of the locations and wellfield designs for Wellfield 2 (applicable also to other wellfields) would be made by WDEQ. The permit boundary is proposed by the applicant, but the applicant needs an approval from the State. The decision on the aquifer exemption is to be made by WDEQ and EPA. The applicant is not required to propose alternative sites, if the applicant is committed to meet NRC, State, and other Federal Agency regulations and requirements for the proposed ISR actions.

The applicant is required to acquire sufficient data to determine hydrogeological characteristics of the production aquifers, confining layers, and other important aquifers, in compliance with 10 CFR 51.45, which requires a description of the affected environment containing sufficient data to aid the Commission in its conduct of an independent analysis for evaluating public health and safety (including potential impacts on underground water sources) associated with the proposed ISR actions.

At the Moore Ranch Site, the 68 Sand and the 70 Sand coalesce at Wellfield 2, and hence, the 68 Sand is expected to be designated as an exempted aquifer. However, the final decision on the aquifer exemption for the 68 Sand is to be made by WDEQ and the EPA.

Comments: MR009-079; MR009-273
A commenter noted that "monitoring well" is not the correct nomenclature for wells from which baseline water quality levels are obtained. The commenter wanted to know when the baseline water quality, wellfield delineation, and restoration standard, UCLs are to be determined with respect to the time for submission of a license application.

Response: *An acceptable set of baseline water quality samples should be taken from all wellfield perimeter monitor wells, all upper and lower aquifer monitor wells, and at least one production/injection well per acre in each wellfield (NUREG–1569, 5.7.8.3). Therefore, the term "monitoring" well is correctly used in the text, but it has been replaced with "baseline monitoring well" to improve the clarity.*

The applicant is required to establish baseline water quality prior to the submission of a license application (NUREG–1910, Section 2.2). The excursion parameters and UCLs are determined based on the baseline water quality sampled from monitoring wells placed in the ore-bearing, underlying, and overlying aquifers, when applicable (NUREG–1910, Section 2.4.1.3).

Therefore, the UCLs should be established prior to ISR operations. The applicant is required to collect enough information to generally locate the ore body and understand the natural system during the prelicensing period (NUREG–1910; Section 2). After the license is issued, the applicant is required to acquire more detailed geologic and hydrologic information as the site to be developed is brought into production. Hence, details for wellfield delineation and restoration standards would be developed after the license issuance.

B5.20.2.2 Aquifer Exemption and Baseline Water Quality

Comments: MR009-025; MR018-016; MR018-020; MR018-023; MR020-014; MR020-032
A commenter noted an incomplete definition for suitability of an aquifer for deep well disposal activities. Another commenter asked to remove the TDS (total dissolved solids) threshold of

Appendix B

3,000 ppm in assessing the suitability of a aquifer for deep well disposal. Several commenters noted that "aquifers with TDS concentration greater than 3,000 ppm are not a potentially useable underground source of drinking water" is not a correct statement.

Response: The applicant is required to obtain a UIC Permit from the State (through WDEQ) for deep well disposal within the permit boundary at the Moore Ranch Site. Suitability criteria for an aquifer for deep well disposal are provided in the WDEQ Water Quality and Regulations for Underground Management of Hazardous Waste (2005, Chapter 8, Section 6). According to these regulations, not only water quality of the targeted deep aquifer (groundwater in the targeted deep aquifer is required to be at least Class IV; the total dissolved content is to be in excess of 10,000 ppm), but also geologic and hydrogeological characteristics of the targeted deep aquifer, economic feasibility of water production from the targeted deep aquifer, and potential impacts of deep disposal of process fluids on the existing water rights and other mineral resources need to be considered for determining suitability of an aquifer for deep well disposal.

As the commenter pointed out, any water with TDS ≤10,000 ppm may be considered as a USDW in accordance with 40 CFR 146.3. WDEQ classified groundwater as Class I (suitable for domestic uses), if it has a TDS ≤500 ppm; Class II (suitable for agricultural uses), if it has a TDS ≤2,000 ppm and Class III (suitable for livestock), if it has a TDS ≤5,000 ppm (WDEQ Water Quality Rules and Regulations, Table 1). For a deep aquifer to be considered for deep disposal, groundwater in this aquifer should meet Class IV use with a TDS in excess of 10,000 ppm (WDEQ's Water Quality Rules and Regulations; Sections 4 and 6). Several updates were made throughout the SEIS text in response to these comments.

Comments: MR009-174; MR009-178; MR020-20; MR020-021; MR020-027; MR020-032
A commenter requested clarification on who would issue an aquifer exemption, the State of Wyoming or EPA. Another commenter stated that the Moore Ranch SEIS did not accurately define an aquifer exemption and that the definition only covered part of the criteria for an aquifer exemption. The commenter noted that exempting an aquifer from being a USDW does not guarantee that it would not, in the future, be used as a drinking water source, unless there is a State law to this effect. The commenter also noted that the draft SEIS incorrectly restated the definition of an USDW, as defined in 40 CFR 144.3.

Response: The EPA issues the aquifer exemption status. The EPA criteria for an aquifer exemption is found in 40 CFR 146.4. The regulation states that an aquifer, or a portion thereof, may be determined to be an "exempted aquifer," if it meets the following criteria: (a) it does not currently serve as a source of drinking water; and (b) it cannot now and would not in the future serve as a source of drinking water; or (c) the TDS content of the groundwater is more than 3,000 and less than 10,000mg/l, and it is not reasonably expected to supply a public water system. A USDW, as defined in 40 CFR 144.3, is an aquifer or its portion which (1) supplies any public water system; or (2) contains a sufficient quantity of groundwater to supply a public water system; and (i) currently supplies drinking water for human consumption; or (ii) contains fewer than 10,000 mg/l TDS; and (iii) is not an exempted aquifer. The Federal regulation of exempt aquifers is enforced by EPA, who has the responsibility to ensure that an exempted aquifer is not used as a source of drinking water. Section 4.5.2.1.2.3 of the final SEIS was revised to reflect this regulation.

Appendix B

B5.20.2.3 Control of Operational Impacts, Excursion of ISL Solutions, and History

Comments: MR009-028; MR009-076; MR009-077; MR009-176

A commenter asked for inclusion of a discussion on the potential for vertical excursions due to well casing failure. The commenter noted that an increase in flow rate is not effective for recovering vertical excursions. Regarding excursion monitoring, the commenter noted that monthly sampling instead of biweekly sampling would be required to detect excursions. The commenter suggested that excursions would be less likely during restorations than during operations.

Response: Poor well integrity involving a cracked well casing or leaking joints between casing sections could cause vertical excursions (NUREG–1910, Section 2.4.1.3). The applicant is required to take preventative measures against vertical excursions prior to operations, including well integrity tests (NUREG–1910, Section 1). The applicant is required to conduct mechanical integrity testing of each well to check for leaks or cracks in the casing, in compliance with 40 CFR 146.8.

An applicant typically recovers horizontal excursion by adjusting the flow rates of nearby injection and production wells to increase process bleed in the area of excursion. On the other hand, a applicant typically recovers vertical excursions by adjusting injection and production flow rates in the area of excursion and pump directly from the affected monitoring wells or from other wells drilled for that purpose (NUREG–1910; Section 2.4.1.4). Therefore, adjusting the flow rates of the nearby injection and production wells to increase in process bleed in the area of excursion is not necessarily effective to recover vertical excursions, and vertical excursion, are often more difficult to recover than horizontal excursions (NUREG–1910, Sections 2.4.1.4 and 2.11.4).

In accordance with NUREG–1910 and NUREG–1569, applicants are required to sample monitoring wells at least biweekly during well operations to detect potential excursions.

Because (i) groundwater restoration takes place when a wellfield is no longer used to produce uranium, and (ii) the applicant is required to restore groundwater quality in the production aquifer to NRC-approved restoration standards during aquifer restoration in compliance with 10 CFR Part 40, Appendix A, Criterion 5B(5), excursions during restoration would be less likely than during ISR operations.

Comments: MR015-036; MR020-018; MR020-031

A commenter noted that the SEIS concluded that drinking water impacts would be small, if all preventive measures are in place. The commenter noted that the SEIS did not address potential impacts to groundwater if there was an excursion during operation, restoration, and postoperation. The commenter questioned the conclusion that no environmental impacts occurred because of historical excursions. Another commenter asked for a thorough analysis of the potential environmental impacts of excursions on groundwater restoration estimates.

Response: Potential impacts to nonexempted aquifers at the ISR site would depend on the frequency and longevity of excursions, if they were to occur. The applicant must establish and maintain groundwater monitoring programs for early detection of vertical and horizontal excursions; have procedures to analyze excursions, determine how to control and remediate excursions to ensure public health and safety, and report it to NRC during operation, restoration, and postrestoration periods (NUREG–1910, Section 2.4.1.4; NUREG–1569, Section 5.7.8.3).

The applicant is required to establish an excursion monitoring system and corrective action plans in compliance with 10 CFR Part 40, Appendix A Criterion 7A; 40 CFR 144.54, and 40 CFR 144.55.

As noted in NUREG–1910, the applicant would acquire more geologic and hydrogeological information during construction and operations for determining the locations of production, injection, and monitoring wells in proposed wellfields at the Moore Ranch Site. Once the exact location of production, injection, and monitoring wells is finalized and additional hydrogeological data are acquired, more site-specific assessments for potential impacts of excursions on groundwater resources can be made. However, at the time the SEIS was prepared, the applicant had not finalized wellfield designs (Uranium One, 2009), which would require approval from WDEQ.

Section 2.11 of the GEIS discusses historical operation of ISR uranium milling facilities which includes a discussion of excursions in Section 2.11.4. In addition, NRC staff evaluated groundwater impacts from three previously licensed in-situ uranium recovery sites in 2009, in response to direction from the Commission. The staff acknowledged that certain parameters can require a long time to reach preextraction concentration levels and that in most cases excursions were reported and controlled. The staff concluded that in all cases there was no threat posed to human health or to the surrounding aquifers.

Excursions from the exempted aquifer that migrate into nonexempted aquifers would result in an environmental impact during ISR operation and restoration. The impacts would be SMALL if the excursions are temporary and recoverable. The applicant would establish and maintain excursion monitoring, control, and remediation (corrective action) programs at the ISR site.

With respect to restoration cost estimates for excursions, the applicant is required to return groundwater quality in the exempted production aquifer to the NRC-approved restoration standards in compliance with 10 CFR Part 40, Appendix A, Table 5C. The applicant is required to provide financial sureties, as part of license application, to cover costs associated with restoration and remediation at the ISR facility, in compliance with 10 CFR Part 40, Appendix A, Criterion 9. NRC reviews financial sureties annually and additional costs associated with potential restoration delays due to excursions are covered by revised financial sureties, which require NRC approval. Additional information on financial sureties are provided in Section B5.15 of this comment-response appendix, and Section 2.1.1.1.8 of the final SEIS, discussing financial surety, was revised in response to public comments.

Comments: MR016-009, MR012-023
Two commenters noted that ISR mining would be in an unconfined aquifer, which is a significant departure from past practice.

Response: *The uranium-bearing proposed production aquifer (the 70 Sand) is not completely saturated in most parts of the Moore Ranch Site (SEIS 2009; Section 3.5.2.3). As the commenter pointed out, the unconfined nature of the uranium-bearing production aquifer at the Moore Ranch ISR Site is different from typical, confined uranium-bearing production aquifers (NUREG-1910, Section 2.1.2). However, permit requirements as part of a license application, implementation of a closed circuit between production and injection wells during ISR operation, enforced groundwater protection and remediation standards, and the requirements for excursion monitoring and control practice are similar for confined and unconfined ore-bearing aquifers. No changes to the SEIS were made beyond the information provided in this response.*

Appendix B

Comments: MR016-011; MR016-012; MR020-028
A commenter noted that potential impacts due to consumptive water use during ISR operation to nearby wells that are in hydraulic connection with the production zone need to be included in the discussion on page. 4-27 of the SEIS. Another commenter questioned "small" potential impacts on groundwater during ISR operations, given that the applicant expects that miles of drawdown be needed to recover excursions.

Response: *Potential impacts to nearby wells completed in the 68-70 Sand (underlying aquifer), the 70 Sand (production aquifer), and 70-72 Sand (overlying aquifer) within 3.2 km [2 mi] of the Moore Ranch ISR facility were evaluated on page 4-27 in the SEIS (2009; pages 4-27 and 4-28). The overlying 72 Sand is separated from the production aquifer through an upper Mudstone, E Coal, and Lower Mudstone confining unit. The underlying 68 Sand is separated from the production zone through a lower confining unit that consists of sequence clays and silts; although, as noted in the previous comments, the 68 Sand and the 70 Sand coalesce (environmental report 2007, 3.4.3.2). In the areas where the 68 and the 70 Sand coalesce, the applicant committed to not install any production wells in the 68 Sand and assumed in their analysis that drawdown in both the 68 and 70 Sand would be identical, which is conservative (in general, if there is no production well in the 68 Sand, drawdown in the 68 Sand would be smaller than in the 70 Sand). According to the applicant numerical simulation results, expected drawdown near the permit boundaries at the end of ISR operation would be on the order of meters, if the bleed production is maintained at less than 3 percent during production. Therefore, based on this conservative analysis, miles of drawdown would not be expected, even if the bleed rate was increased from 3 percent in the event of an excursion. No changes were made to the SEIS beyond the information provided in this response.*

Comments: MR009-182; MR012-023; MR012-024; MR012-025; MR012-029 MR016-010
A commenter stated that the NRC new approach to analyzing impacts from ISR operations in an unconfined aquifer was insufficient. The commenter asserted that NRC did not meaningfully evaluate potential excursions from the 70 Sand to the underlying 68 Sand at the Moore Ranch site and instead chose to include in the production zone where the 68 and 70 Sands coalesce. The commenter stated that by fiat NRC evaded evaluating the potential impacts to the underlying, good quality aquifer and the alternatives and mitigation measures that would follow from such an analysis. Another commenter noted that in the area where the 68 and 70 Sands coalesce, the 68 Sand would also need to be designated as an exempt aquifer.

Response: *The NRC staff initially rely on applicant information as well as information and conclusions from a separate NRC detailed safety review, which is documented in the NRC environmental review. The Moore Ranch SEIS described the expected environmental impacts arising from ISR operations in the "70 Sand" unconfined aquifer. The commenter is correct that the SEIS did not include a comprehensive technical analysis of ISR operations in the "70 Sand" unconfined aquifer. This analysis can instead be found in the NRC SER of the Moore Ranch Technical Application. The SER includes an extensive review of ISR operations in the "70 Sand" unconfined aquifer, which were provided in the technical application. Specifically, NRC reviewed the field tests provided by the applicant to demonstrate the field behavior of the "70 Sand" unconfined aquifer. NRC also reviewed the applicant comprehensive groundwater flow model, which included software specifically designed for unconfined aquifers to simulate flow behavior in the ISR production and restoration phases at Moore Ranch. Finally, NRC reviewed an applicant excursion scenario simulation provided to demonstrate excursion behavior and excursion capture in the "70 Sand" unconfined aquifer. The applicant will also be*

Appendix B

required to provide NRC with its Wellfield Hydrologic Test Data package to verify the field behavior of the 70 Sand aquifer and to support the results of the groundwater flow simulations. Operations in the 70 Sand unconfined aquifer would not commence until NRC has reviewed and approved this wellfield package to ensure the operations in the Moore Ranch 70 Sand unconfined aquifer can be conducted safely and minimize any expected environmental impacts. The commenter is correct that the production zone in Wellfield 2 would include the 70 and 68 Sand in the regions where the "70 underlying shale" is thin or absent. In those regions where the 70 and 68 Sand coalesce, the 68 Sand will be considered part of the production zone and would be exempted. Therefore, any movement or recovery fluids into the exempt zone of the 68 Sand would not be considered an excursion. The applicant decided to combine these aquifers into one production zone, not NRC. NRC notes the groundwater quality of the "68 Sand," as measured by the applicant, exceeded the EPA MCL drinking water standards for uranium Ra 226, gross alpha and selenium. NRC notes that the applicant would be required to obtain exempt aquifer status for the "68 Sand" from EPA before it may be treated as part of the production zone. Therefore, NRC concludes the potential environmental impact to the "68 Sand" where it coalesces with the "70 Sand" in Wellfield 2 has been adequately addressed. No changes were made to the final SEIS beyond the information provided in this response.

Comment: MR020-031
A commenter noted that the Moore Ranch SEIS stated that NRC staff had reviewed reports of 60 excursions at 3 NRC licensed ISR facilities and concluded that none resulted in environmental impacts. The commenter contended that movement of a contaminant into an USDW is an environmental impact and requested that the final SEIS explain how the NRC defines environmental impacts in these cases.

Response: *The commenter is correct that movement of a contaminant into a USDW may constitute an environmental impact. However, NRC does not define an excursion as contamination that moves into a USDW. An excursion is defined as an event where a monitoring well in overlying, underlying, or perimeter well ring detects an increase in specific water quality indicators, usually chloride, alkalinity and conductivity, which may signal that fluids are moving out from the wellfield. These specific water quality parameters are used because they are present in high concentrations in the ISR production fluids and are "conservative" in the sense that they move at roughly the same rate as the groundwater flow and are not significantly attenuated by adsorption or reduced by other factors. Therefore, they serve as early indicators of imbalance in the wellfield flow system to notify operators to take appropriate actions. The perimeter monitoring wells are located in a buffer region surrounding the wellfield within the exempted portion of the aquifer. These wells are specifically located in this buffer zone to detect and correct an excursion before it reaches a USDW. The overlying and underlying monitoring wells are located in aquifers that are separated from the ore zone by aquitards, which NRC has determined have sufficient thickness and integrity to prevent an excursion. However, in all cases, any excursion that lasts longer than 60 days is required to undergo corrective action to meet the drinking water protection standards in 10 CFR Part 40, Appendix A 5(B) 5. To date, no excursions from an NRC-licensed ISR facility has contaminated a USDW. No changes were made to the SEIS beyond the information provided in this response.*

B5.20.2.4 Exploratory Drill Wells, Abandoned Wells, and Old Mines

Comments: MR016-016; MR018-026
Two commenters noted that the applicant would be required to take preventive measures for inadequately plugged oil wells within the radius of influence of the injection activities. Similarly,

Appendix B

another commenter noted that all properly plugged or unplugged drill holes need to be identified; and those that are not properly abandoned need to be adequately plugged and abandoned before ISR operations commence.

Response: *As part of site characterization, the applicant is required to identify locations of abandoned wells, including depth, type of use, condition of closing, plugging procedure, and date of completion within the site and within 0.4 km [0.25 mi] of the wellfield boundary in compliance with 10 CFR 51.45, which requires a description of the affected environment containing sufficient data to aid the Commission in its conduct of an independent analysis [NUREG–1569, Section 2.2.3.1(e)].*

The applicant discussed abandoned drill holes at the Moore Ranch site in the environmental report (2007, Section 3.3.4). From the 1970s to mid-1980s, nearly 2,700 rotary drill holes and 130 core holes were completed by Conoco. Drill holes completed by Conoco were reported to be plugged in accordance with Wyoming Statute WS 35-11-401, in effect at the time, except for several drill holes that required additional abandonment work, according to WDEQ-LQD District III personnel.

The applicant noted that EMC conducted verification drilling in late 2006 totaling 157 holes and 20 monitor wells. All drill holes were plugged in accordance with Wyoming Statute WS35–11–401, as documented. The applicant provided a list of all drill holes known to the applicant in environmental report (2007, Table 3.3-1) and a location map of these known drill holes in the environmental report (2007, Figure 3.3-13).

The applicant did not report any abandoned oil wells within the Moore Ranch site. The NRC staff are not aware of any abandoned oil wells within the Moore Ranch site. The closest oil production to the Moore Ranch Site is in the Pine Tree field within 1.6 km [1 mi] to the west of the Moore Ranch Permit area is primarily from the Shannon Formation at depths of 3,050–3,350 m [10,000 to 11,000 ft].

The applicant is required to submit information about abandoned drill and exploratory wells, dry holes, and wells within the Moore Ranch Site, as part of the UIC permit application in compliance with 40 CFR Parts 144 and 146. Therefore, WDEQ would also evaluate the applicant survey and findings for abandoned drill and exploratory wells and dry holes prior to commencement of ISR operations.

B5.20.3 Aquifer Restoration and Decommissioning: Methods and Operational Experience

Comment: MR009-084
A commenter noted an incorrect restoration process step.

Response: *Groundwater recovered from the restoration field is passed through an ion exchange (IX) unit prior to reverse osmosis and electro dialysis reversal units. Reverse osmosis electro dialysis reversal units, as down-end treatment units, are used to separate clean water (permeate) from brine. A correction has been made in the SEIS.*

Appendix B

Comments: MR009-086; MR015-032
A commenter noted that bioremediation suggested previously as a remediation process has been removed from SEIS (2009). Another commenter noted that the draft SEIS should evaluate alternative methods that could be used to meet restoration goals.

Response: ISR actions under NRC regulations are risk-informed and performance-based. Therefore, the applicant has flexibility in choosing and implementing proper remediation techniques to achieve NRC-approved restoration standards in compliance with 10 CFR Part 40, Appendix A, Criterion 5B(5). No additional changes were made to the SEIS in response to these comments.

Comments: MR009-087; MR015-033
A commenter asked whether a 6-month stability period would be sufficient for long stabilization at the ISR facility. Another commenter noted that the stability monitoring period has been extended to a minimum of 12 months.

Response: The applicant confirmed in their license application revisions that they extended the stability monitoring program to a minimum of 12 month. The SEIS has been revised to include this correction.

Comments: MR009-081; MR010-004; MR011-004; MR012-017; MR012-018; MR012-041; MR012-042; MR012-045; MR013-011; MR015-031; MR016-013; MR016-014; MR016-015; MR017-020; MR017-022; MR017-004
Several commenters raised concerns about the NRC assessments for operational impacts of ISR activities on groundwater by noting that the assessments are not consistent with the existing data. A few commenters asked for clarification on groundwater restoration standards. A commenter noted that groundwater restorations are difficult at best at ISR sites. A commenter pointed out that for cases in which baseline levels are not recovered after restoration, potential effects on surrounding USDW are not evaluated in the SEIS. Another commenter expressed disbelief on proper containment of contamination after aquifer restoration allegedly based on historical data. A commenter asked for additional information on the historical analysis of aquifer restoration at ISR sites. Several commenters disagreed with the NRC assessment that impacts on groundwater would be small and temporary after restoration, and noted that aquifers used for ISR operations have never been restored to baseline/preextraction conditions. A commenter claimed that baseline sampling procedure is biased and inflated and there is no successful groundwater restoration to preoperational baseline levels at ISR sites.

Response: Under the Federal Underground Injection Control (UIC) program, the ISR production aquifer must receive an exemption from EPA, that the aquifer, or part of the exempted aquifer, is not now, and would never be a source of drinking water and, thus, would no longer be protected under the Safe Drinking Water Act (SDWA). Hence, groundwater in exempted aquifers cannot be considered as a source of drinking water supply even after restoration.

NRC licensees are required to return water quality parameters to the standards in 10 CFR Part 40, Appendix A, Criterion 5B(5). As stated in the regulations: "5B(5)—At the point of compliance, the concentration of a hazardous constituent must not exceed—(a) The Commission approved background concentration of that constituent in the groundwater; (b) The

Appendix B

respective value given in the table in paragraph 5C if the constituent is listed in the table and if the background level of the constituent is below the value listed; or (c) An alternate concentration limit is established by the Commission."

To establish the preoperational nonradiological and radiological groundwater baselines within the proposed permit boundaries and adjacent properties, to collect samples over a period of at least 1 year, from at least four sets of groundwater samples with sufficiently spaced time. An acceptable set of samples should include all wellfield perimeter monitor wells, all lower and upper aquifer monitor wells, and at least one production/injection well per acre in each wellfield. Baseline samples are collected with a sampling density of not less than one for [16,187 m^2 [4 ac]. Because the applicant is required to collect baseline water quality before ISR operations begin, the baseline sampling procedure outlined previously would provide adequately unbiased preoperational groundwater quality measures at the proposed ISR site.

These standards are implemented during aquifer restoration to ensure protection of public health and safety in compliance with 10 CFR Part 40, Appendix A, Criterion 5B(5) for protection of groundwater and USDW in surrounding nonexempted aquifers. The applicant is required to provide financial sureties to cover planned and delayed restoration costs in compliance with 10 CFR Part 40, Appendix A, Criterion 9. The NRC reviews financial sureties annually. Although the goal of groundwater restoration is meeting baseline values or MCLs whichever is greater, it is recognized under the EPA standards and NRC regulations that ACLs can be used in circumstances where achieving those standards is impracticable or impossible. ACLs must present no substantial present or potential hazard to human health or the environment. A discussion of the NRC requirements for application, review and establishment of a site-specific ACL is presented in Appendix C.

All ISR restorations completed under the NRC regulations have met restoration standards in compliance with their licenses. Examples of successfully completed groundwater restorations or delayed restoration activities are provided in NUREG–1910, Section 2.11.5. Licensees are also required to establish routine regional aquifer monitoring programs as a license condition. The data from those monitoring programs do not show impacts to USDWs attributable to an ISR Facility (NRC 2009).

B5.20.4 Miscellaneous Groundwater Comments

Comments: MR009-062; MR009-096; MR018-018; MR018-024; MR020-023; MR020-033; MR018-027; MR018-028

A commenter asked that a sentence be included in the SEIS stating two Class I wells would be required in the Lance Formation for deep disposal of process fluids. Another commenter asked to remove the information on (i) depths to targeted deep aquifers for deep well disposal, and (ii) injection capacity to these aquifers. The commenter marked some corrections on the depth to the Lance Formation and the number of injection wells proposed in the Lance Formation for deep well disposal. Another commenter noted that the suitability of Lance Formation for deep well disposal was not supported by evidence in the SEIS. A commenter asked for an explanation of why the Teapot-Tackla Parkman and Lance Formations are considered as a USDW, if Class I wells are to be installed below these formations for deep well disposal. Another commenter noted that deep well disposal activities should not affect coal production near the ISR site. Another commenter noted the existing groundwater wells within the Moore Ranch area are usually shallower than the CBM wells, and they would not likely be impacted by

Appendix B

deep disposal. The commenter also noted that oil field water supply wells are not likely be affected by deep well disposal.

Response: The final SEIS has been revised to acknowledge that from two to four Class I UIC wells could be drilled at the proposed Moore Ranch Project for disposal of liquid effluent as discussed in Section 4.5.2.1.2.3. The WDEQ will evaluate the suitability of the formations proposed for deep well injection and would only grant a UIC permit if the applicant can demonstrate that liquid effluent could be safely isolated in a deep aquifer. The State of Wyoming is currently reviewing a permit application for up to four Class I disposal wells at the proposed Moore Ranch Project as noted in Table 1-2 of this SEIS. The cumulative impacts analysis in the SEIS considers other past, present, and reasonably foreseeable activities affecting groundwater resources and includes discussion of CBM and oil and gas operations as noted by the commenters.

Comments: MR009-180; MR015-030; MR020-029; MR020-030; MR015-034
Several commenters raised concerns about how NRC determines ACLs and how ACLs assure public health and safety. A commenter asked for a proper definition of an ACL. Commenters want to know at what point NRC would make the decision to set ACLs and what public health and safety standards NRC uses to approve or reject ACLs. Other commenters questioned the accuracy of the assessment of a SMALL groundwater impact during aquifer restoration because of potentials for permanent degradation to groundwater quality and lack of information on how often NRC approves ACLs.

Response: NRC revised the draft SEIS to state that licensees and applicants must commit to achieve the groundwater quality standards in 10 CFR Part 40, Appendix A, Criterion 5B (5) for all restored aquifers. These standards state the concentration of a hazardous constituent must not exceed (a) the Commission approved background concentration of that constituent in groundwater; (b) the respective value in Table 5C if the constituent is listed in the table and if the background level of the constituent is below the value listed or; (c) an alternative concentration limit established by the Commission. An ACL is not a primary restoration goal and will only be considered after a license has demonstrated that primary restoration goals are not practically achievable at a specific site. ACLs that present no significant hazard may be proposed by the licensees for Commission consideration. The Commission may establish a site-specific ACL for a hazardous constituent as provided in 5B(5) if it finds that the proposed limit is as low as reasonably achievable, after considering practicable corrective actions, and that the constituent will not pose a substantial present or potential hazard to human health or the environment as long as the ACL is not exceeded. Appendix C of the SEIS discusses the NRC requirements for application, review and establishment of a site-specific ACL. In addition, ACL application review procedures for NRC staff are available in the following documents: January 1996 Staff Technical Position: Alternate Concentration Limits for Title II Uranium Mills, NUREG-1620 and NUREG-1724.

Comments: MR009-181; MR009-184; MR009-275; MR009-288
A commenter claimed that the WDEQ requires a hydraulic connection between monitoring and production wells, but no specific NRC guidance is required. The commenter emphasized that monitoring wells are not within the production zone. The commenter also noted that at Moore Ranch, monitoring wells are to be placed in the 60 Sand only in areas where the 70 Sand and the 68 Sand coalesce.

Response: The requirement for the hydraulic connection between the monitoring wells and the production wells is addressed in NUREG–1569, Section 5.7.8.1(4) of, which is NRC guidance.

Ore zone monitoring wells are placed to detect any lixiviant moving out of the production zone. Thus, ore zone monitoring wells are placed outside the production zone to detect horizontal excursions (NUREG–1910, Sections 2.11.4 and 8.3.1.1, Figure 2.3-1).

Monitoring wells below the ore zone are required within the production zone for early detection of vertical excursions. In the areas of Wellfield 2 where a confining unit exists between the 70 Sand and the 68 Sand, the applicant would place monitoring wells in the 68 sand at a spacing of 1 per 4 acres (RAI Response to 5.12). Although the applicant has not finalized their wellfield designs, which would require WDEQ approval, the applicant noted that monitoring wells would be placed at a spacing of 1 well per 4 acres in the underlying 60 Sand in the areas where the 70 and 68 sand coalesce.

B5.20.5 References

10 CFR Part 40. Appendix A. *Code of Federal Regulations*, Title 10, *Energy*, Part 40, Appendix A, "Criteria Relating to the Operations of Uranium Mills and to the Disposition of Tailings or Wastes Produced by the Extraction or Concentration of Source Material from Ores Processed Primarily from their Source Material Content."

40 CFR Part 144. *Code of Federal Regulations*, Title 144, *Protection of Environment*, Part 144, "Underground Injection Control Program."

40 CFR Part 146. *Code of Federal Regulations*, Title 40, *Protection of Environment*. Part 146, "Underground Injection Control Program: Criteria and Standards."

EMC (Energy Metal Corporation US), 2007a. Environmental Report. Application for USNRC Source Material License Moore Ranch Uranium Project Campbell County, Wyoming.
EMC, 2007b. Technical Report. Application for USNRC Source Material License Moore Ranch Uranium Project Campbell County, Wyoming.

NRC (U.S. Nuclear Regulatory Commission), 2009. NUREG–1910, "Generic Environmental Impact Statement for In-Situ Leach Uranium Milling Facilities." Washington, DC: NRC. May.

NRC, 2009. "Staff Assessment of Groundwater Impacts from Previously Licensed In-Situ Uranium Recovery Facilities." Memorandum from C. Miller to Chairman Jaczko, et al. ADAMS Accession No. ML091770402. July 10.

NRC, 2003. NUREG–1569, "Standard Review Plan for In-Situ Leach Uranium Extraction License Applications." Final Report. Washington, DC: NRC. June. ADAMS No. ML032250177.

Uranium One (Uranium One Americas), 2009. RAI Response to technical report 7.2.9.2.2 and 3.1.3. Surface Water Impacts. ML 092450317. Received on August 27, 2009.

Uranium One, 2008. RAI Response to 5.12. Groundwater and Surface Water Monitoring Programs. ML 082060527. Received on July 11, 2008.

Appendix B

B5.21 Surface Water Resources

B5.21.1 Impacts to Surface Drainages and Surface Waters

Comments: MR009-019; MR009-047; MR009-167; MR012-037; MR018-031; MR018-032; MR018-038; MR018-039; MR018-040; MR018-047

A commenter noted that potential impacts to surface water resources in the Wyoming East Uranium Milling Region are rated as SMALL. The commenter noted that because waste water would be disposed into a deep aquifer during ISR operation, there would not be any surface water impacts during operations and restoration. The commenter also noted that water pumped during the placement of production wells in intermittent channels would not be released directly into the channel.

On the other hand, several commenters raised concerns about surface water impacts due to ISR activities. A commenter noted that NRC disregarded close proximity of mining operations to surface water features at Moore Ranch. A commenter suggested that pipelines should be routed around wetlands or be constructed perpendicular to wetland features to minimize impact. Another commenter noted that wells placed in ephemeral channels may lead to increased erosion, increased risk of breached structures, and potentials for releases of processed fluids to ephemeral channels. The commenter also noted that runoff and erosion from roads, culverts, and ephemeral channels could cause accelerated channel alterations.

Response: The SEIS concluded that the potential impacts of ISR activities on surface waters in the Wyoming East Uranium Milling Region would be SMALL (NRC 2009, Section 4.3.4.1). The applicant analyzed and reported natural and man-made surface water features at and near the proposed Moore Ranch Project and assessed the potential site-specific impacts from ISR activities.

The applicant is required to obtain industrial and construction NPDES permits from WDEQ as part of license application (SEIS 2009, Table 1-2) for the protection of surface water and wetlands).

These permits involve best management practices for spill prevention and control and disposal of water produced during placement of wells at the ISR site during construction. Moreover, the applicant would install meters and control valves in individual well lines to monitor and control flow rates and pressures for each well to maintain water balance and to aid in identifying leaks (NUREG–1910, Section 2.3.1.1). The applicant would also measure and record pipeline pressure to monitor for potential leaks and spills that might result from the failure of fittings and valves (NUREG–191, Sections 2.4.1.2).

If the UIC permit is granted WDEQ, then liquid effluent produced during ISR operations would be disposed into UIC-permitted deep aquifers. Thre would be no impact to surface water from normal operations. Potential impacts to surface waters would occur from accidental spills or leaks. The applicant noted that installation of monitor, injection, and production wells in main ephemeral stream channels would be avoided, if possible. The applicant would seek approval from WDEQ for the final wellfield layouts (SEIS 2009, Section 6.3.1.2). The location, distribution, and alignment of pipeline networks would be associated with the exact position of injection and production wells in each wellfield. The location, design, and distribution of pipeline on and near the wetlands and intermitent channels within the permit boundary comply with the

Appendix B

WDEQ construction permit. If a well were to be installed in an intermittent channel, the applicant would install the well within the high water marks with adequate structural wellhead protection (e.g., concrete berms or reinforced steel/concrete well covers) to protect the well during potential flood conditions. The applicant would reseed disturbed areas during construction soon after wellfields are constructed to minimize the risk of potential for erosion.

The applicant would use diversion ditches and culverts to prevent excessive erosion and to control runoff. In areas where runoff has concentrated, the applicant has committed to use energy dissipaters to slow the flow of runoff to minimize erosion and sediment loading in surface water runoff. The applicant would implement best management practices to monitor and reduce erosion impacts in accordance with storm water management plans developed as part of itsWYPDES permit. The applicant is committed to implement soil erosion mitigation in accordance with WDEQ Rules and Regulations, Chapter 3, Environmental Protection Performance Standards and to construct roads to minimize erosion through practices such as surfacing with a gravel road base, constructing stream crossings at right angles with adequate embankment protection and culvert installation, and providing adequate road drainage with runoff control structures and revegetation.

Comment: MR012-073
A commenter noted that NRC did not address how the impacts of surface water contamination and erosion due to livestock grazing would interact with surface water impacts from the ISR activities.

Response: *Surface water is not a source of consumptive water use at the proposed Moore Ranch Project; hence, surface water impacts due to livestock grazing (erosion or surface water contamination) would be insignificant for the ISR related activities. Any potential spills or discharge to surface water features (which may have direct or indirect impacts on livestock grazing) during the ISR lifecycle would be controlled by the Construction and Industrial WYPDES permits. The applicant would obtain the WYPDES permits from WDEQ before ISR operations began. No further modification to the SEIS was required beyond the information provided in this response.*

B5.21.2 General Water Resource Concerns
Comments: MR018-001; MR018-014; MR018-015

A commenter noted that the draft SEIS did not contain the most recent hydrological information submitted to the WDEQ/LQD because this information was submitted to the WDEQ/LQD after the printing of the draft SEIS.

Response: *The applicant would acquire more geologic and hydrogeological information during construction and operations before finalizing locations of production, injection, and monitoring wells. Once the applicant finalizes wellfield layouts and designs, the applicant would seek approval for the locations and wellfield designs from the WDEQ. This informaiton was unavailable at the time the draft SEIS was issued.*

The final SEIS has been revised to reflect current hydrological data that has become available since the publication of the draft SEIS.

Appendix B

Comment: MR018-012

A commenter noted that the State of Wyoming Constitution gives control of the "waters of the State," both surface and ground, to the State Engineer and that water used by any projects would be required to obtain the necessary permits (GEIS, Sections 3.2.4 and 3.3.4).

Response: This comment relates to GEIS Sections 3.2.4 and 3.3.4. The text in these sections states that water resources are described in terms of surface waters, wetlands, "Waters of the United States," and groundwater. The commenter noted that the State of Wyoming has jurisdictional control over the waters of the State. The comment is noted. The final SEIS, Table 1-2, identifies environmental approvals required for the proposed Moore Ranch Project including permits related to both surface water and groundwater as noted by the commenter. No changes were made to the document in response to this comment.

Comment: MR020-022

A commenter stated that the Moore Ranch SEIS identified three wells permitted as domestic wells that are completed at depths close to that of the 70 Sand production zone that could be impacted by ISR operations. The commenter noted that these statements indicate there is a potential of current use of these wells as a drinking water source, even though there are no residences located in the proposed license area. The commenter stated that even though these wells are not being used primarily for human consumption, it does not rule out the possibility of their use as current drinking water sources. The commenter further noted that the Moore Ranch SEIS did not adequately explain why these wells would not be used in the future as a drinking water source. Finally, the commenter asked if there was a process for the State of Wyoming to rescind the permits for domestic use.

Response: Three domestic water wells are located within the 3.2 km [2-mi] buffer for the proposed Moore Ranch license area. One well, UM 1575 2 33 42 75, is located within the proposed license area. Rio Algom Mining Corporation installed this well in 1972, and the Wyoming State Engineer's Office (WSEO) permit for this well, P12299.0W, states that the well is for domestic and industrial use with a note that it may be used for uranium exploration drilling. The well is located about 1,219 m [4,000 ft] to the southwest of the Wellfield 1 monitoring well ring. It is, therefore, hydrologically upgradient of the wellfield. The well is screened from 106–134 m (348–440 ft) below ground surface. Review of geologic cross sections of this area indicates that the screen depth would place it in the "60 Sand" aquifer. The "60 Sand" aquifer is below an aquitard under the 68 Sand aquifer, which is below another aquitard under the Wellfield 1 "70 Sand" production zone. Given the vertical stratigraphic and horizontal separation of this well from ISR operations in the "70 Sand," it would not likely be impacted by ISR operations. It could be used for domestic purposes in the future. The two other domestic wells, 9 Mile #1 and 9 Mile #2, are located at the limit of the 3.2 km [2-mi] buffer to the southeast of the proposed license area. According to the 9 Mile #1 permit, P9309.0W, the well was installed in 1971 and is screened at 58–76 m [190–250 ft] below ground surface. According to P12240.0W, the 9 Mile #2 well was installed in 1965 to a total depth of 55 m [180 ft] below ground surface with no screen interval information. Both wells are located at least 4.8 km [3 mi] away and hydrologically upgradient from Wellfield 2. No cross sections were available to assess in which aquifer the wells were completed. The Moore Ranch application indicated there is no current use of these wells for domestic purposes. Given their distance from and upgradient location from the ISR operations, they would not likely be impacted by operations. These wells could also be used for domestic purposes in the future.

Appendix B

The WSEO, which issues permits for all groundwater wells, does not have a process to rescind permits. The WSEO can cancel permits if they do not meet well installation and completion specifications; but they do not cancel permits based on water quality. No changes were made to the SEIS in response to this comment beyond the information provided in the response.

B5.21.3 References

40 CFR Part 144. *Code of Federal Regulations*, Title 144, *Protection of Environment*, Part 144, "Underground Injection Control Program."

EMC (Energy Metal Corporation US), 2007a. Environmental Report. Application for USNRC Source Material License Moore Ranch Uranium Project Campbell County, Wyoming.

EMC, 2007b. Technical Report. Application for USNRC Source Material License Moore Ranch Uranium Project Campbell County, Wyoming.

NRC (U.S. Nuclear Regulatory Commission), 2009. NUREG–1910, "Generic Environmental Impact Statement for In-Situ Leach Uranium Milling Facilities." Washington, DC: NRC. May.

Uranium One (Uranium One Americas), 2009. RAI Response to technical report 7.2.9.2.2 and 3.1.3. Surface Water Impacts. ML 092450317. Received on August 27, 2009.

Uranium One, 2008. RAI Response to 5.12. Groundwater and Surface Water Monitoring Programs. ML 082060527. Received on July 11, 2008.

B5.22 Wetlands

Comments: MR018-043; MR018-044; MR018-046; MR020-019
A commenter noted that jurisdictional determination of water at the Moore Ranch Site was not adequately addressed in Section 3.5.1 and 4.5.1 of the SEIS. Another commenter noted that although wetlands within the permit boundary may not be considered jurisdictional waters of the United States, all naturally occurring wetlands are considered waters of the State and they are protected under Wyoming Law; therefore, all naturally occurring wetlands should be protected during the proposed project. The commenter noted that the magnitude of impacts to wetlands would be based on the effect the proposed project had on wetlands and hydrologic function, not whether a permit is required.

Response: Table 2.8-14 of applicant's technical report (2007) includes a list of wetlands within the Moore Ranch project area and the applicant jurisdictional recommendation.. The applicant recommended all wetlands in the project area to be non-jurisdictional under Section 404 of the Clean Water Act due to lack of connection to navigable waters, and they do not support interstate commerce. Since publication of the draft SEIS, the proposed Moore Ranch Project has received a determination from the U.S. Army Corps of Engineers (USACE, 2010). The proposed activities are consistent with activities authorized under Nation Wide Permit #12.

NRC recognizes that surface waters and wetlands, whether they are jurisdictional or not, are protected by the State. In case the applicant conducts ISR related activities on or near wetlands within the permit boundary, the applicant is required to monitor, control, mitigate, and remediate any potential spills in accordance with 40 CFR 144.54; 40 CFR 144.55; and 40 CFR Part 40,

Appendix B

Appendix A, Criterion 7A. The applicant is required to obtain industrial and construction WYPDES permits from WDEQ as part of the license application (SEIS 2009, Table 1-2) prior to commencement of ISR activities. As part of these permits, the applicant would implement best management practices, such as implementation of a spill prevention and cleanup plan, to minimize soil contamination. The applicant is also committed to implementing soil erosion mitigation in accordance with WDEQ Rules and Regulations, Chapter 3, Environmental Protection Performance Standards (environmental report 2007, Section 5.3).

Potential impacts to surface waters and wetlands are discussed in SEIS (2009, Section 4.5.1). NRC analyzes potential impacts to surface waters and wetlands based on independent review of the site characteristics. The GEIS concluded that most construction impacts to surface water would be SMALL, but could potentially be MODERATE if a U.S. Army Corps of Engineers permit is required. All fragmented wetlands and ponds within the permit boundary are formed by discharges from CBM and livestock wells (SEIS 2009, Section 3.5.1.2.1)

B5.22.1 References

10 CFR Part 40. Appendix A. *Code of Federal Regulations*, Title 10, *Energy*, Part 40, Appendix A, "Criteria Relating to the Operations of Uranium Mills and to the Disposition of Tailings or Wastes Produced by the Extraction or Concentration of Source Material from Ores Processed Primarily from their Source Material Content."

40 CFR Part 144. *Code of Federal Regulations*, Title 144, *Protection of Environment*, Part 144, "Underground Injection Control Program."

EMC (Energy Metal Corporation US), 2007a. Environmental Report. Application for USNRC Source Material License Moore Ranch Uranium Project Campbell County, Wyoming.

EMC, 2007b. Technical Report. Application for USNRC Source Material License Moore Ranch Uranium Project Campbell County, Wyoming.
NRC, 2009. NUREG–1910, "Generic Environmental Impact Statement for *In-Situ* Leach Uranium Milling Facilities." Washington, DC: NRC. May.

Uranium One (Uranium One Americas), 2009. RAI Response to technical report 7.2.9.2.2 and 3.1.3. Surface Water Impacts. ML 092450317. Received on August 27, 2009.

Uranium One, 2008. RAI Response to 5.12. Groundwater and Surface Water Monitoring Programs. ML 082060527. Received on July 11, 2008.

USACE (U.S. Army Corps of Engineers), 2010. "Subject: Response to a Preconstruction Notification (PCN)." Letter to J. Winter from M.A. Bilodeau, Program Manager, Department of the Army Corps of Engineers. May 10.

Appendix B

B5.23 Ecology

B5.23.1 General Ecology

Comment: MR018-094
One commenter noted that based on operational and mitigative practices described in the SEIS, they had no concerns regarding aquatic ecology.

Response: The NRC acknowledges the comment and notes that support of the practices described within the SEIS is outside the scope of responses. No changes have been made to the SEIS in response to this comment.

B5.23.2 Concerns About the Sage-Grouse

Comment: MR017-014; MR017-017
One commenter suggests that extra care should be taken to protect sage-grouse and its habitat and minimize potential impacts.

Response*: NRC recognizes that sage-grouse are a species of great concern in Wyoming and has consulted with stakeholders as described in section 1.7. Section 3.6.1.2.2 describes the limited habitat found on and in the vicinity of the proposed Moore Ranch site to support sage-grouse existence. No large expanses of contiguous sagebrush occur within several kilometers of the proposed Moore Ranch project and few sage-grouse have been documented in the area. No sage-grouse leks have been discovered either on or near the proposed Moore Ranch Project; the nearest known sage-grouse lek is located approximately 4.0 km [2.5 mi] to the northwest of the Moore Ranch Project area (BLM, 2009). Changes were also made in Section 3.6.3 to include the U.S. Fish and Wildlife Service's (FWS) rule listing the sage-grouse as a candidate species, the revised WGFD Recommendations for Development of Oil and Gas Resources Within Important Wildlife Habitats, and the BLM revised National Sage-Grouse Habitat Conservation Strategy (75 FR13909; WGFD, 2010; BLM, 2010). However, the NRC is not bound by the WGFD recommendations or BLM guidelines and does not have the statutory authority to enforce wildlife mitigation measures upon a licensee. Mitigative measures would be negotiated by the applicant and the agency with statutory authority. NRC believes that the sage-grouse analyses are supported by sufficient technical bases, whether tiered from the GEIS or based on supplemental staff analyses. Based on this information, no changes were made to the SEIS.*

B5.23.3 General Comments on Threatened and Endangered Species

Comment: MR002-006
One commenter noted that the survey conducted to determine threatened or endangered species was not included in the draft SEIS, so it could not be determined if it was accurate. They recommended that the final appendix should include these surveys.

Response*: The NRC summarizes the baseline ecological survey results submitted by the applicant for inclusion in the SEIS. The applicant submitted an environmental report for the application for source material license at the proposed Moore Ranch ISR project in October 2007. The environmental report can be located using the ADAMS accession numbers ML72851222, ML072851229, ML072851239, ML07285249, ML07285253, and ML07285255.*

NRC does not have the obligation to ensure, or the regulatory authority to enforce, that surveys are conducted according to those standards established by other agencies with regulatory authority. NRC is not obligated to provide supporting documentation as an attachment to the SEIS and has made the documents provided by the applicant available to the public through the NRC website. Because the requested information is available, no changes were made to the SEIS.

Comment: MR002-009
One commenter noted that while the draft SEIS indicates that no bald eagles roost within the Moore Ranch boundaries, recent information shows that they roost nearby, and that the Final SEIS should be updated and expanded accordingly.

Response*: Information cited by the commenter indicates the presence of bald eagle roosts within the proposed Nichols Ranch ISR project area, which is approximately 13.7 km [8.5 mi] from the proposed Moore Ranch ISR project. NRC staff reviewed the cited BLM Environmental Assessment for Yates Petroleum Corporation, All Day Plan of Development (BLM, 2008). Because of this response, Sections 3.6.3 and 4.6.1.1.4 were amended to reference the nearby eagle roost sites.*

Comments: MR009-196; MR009-197
One commenter suggested that the NRC misapplied conclusions made in the GEIS (NUREG–1910) to the proposed Moore Ranch ISR project.

Response*: NRC recognizes that the proposed Moore Ranch ISR project is not located within a core population area for sage-grouse and does not support active sage-grouse leks. The NRC reference to the WGFD Recommendations for Development of Oil and Gas Resources Within Important Wildlife Habitats in the SEIS suggests examples of mitigation measures that could be observed to reduce potential impacts to wildlife, including sage-grouse, but the recommendations are not limited to sage-grouse mitigation measures. The inclusion of the WGDF recommendations as examples of mitigation measures do not imply that the applicant is bound to them. The WGFD has no specific authority to require adoption or implementation of the recommendations, and NRC cannot enforce mitigation measures on an applicant. NRC believes the inclusion of this document as a reference in the SEIS is warranted since the sage-grouse is listed as a federal candidate species, sage-grouse occur in the project county and region, and NRC has referenced this document for mitigation examples in other SEIS projects for proposed ISR facilities. Based on the aforementioned reasons, no changes were made to the SEIS.*

Comment: MR009-208
One commenter stated that a conflicting statement is made regarding potential impacts to threatened and endangered species as a result of the proposed Moore Ranch Project.

Response*: NRC states in section 4.6.1.2.4 that "No impacts to federally-listed threatened and endangered species would be expected to occur…". Stating that no impacts are expected does not equate to "there would be no impacts." Section 4.6.1.2.4 has been amended to state that "Continued mitigation would be implemented to ensure potential impacts to threatened and endangered species remain SMALL." This change is to clarify that NRC does not expect impacts to threatened and endangered species; however, any potential impacts that could occur would be SMALL.*

Appendix B

Comment: MR002-004
One commenter noted that FWS plans to reopen the comment period on the proposed rule to list the mountain plover as a threatened species, and that the Endangered Species Act requires Federal agencies to confer with the FWS on any action that is likely to jeopardize the continued existence of any species proposed for listing.

Response: *The mountain plover (Charadrius montanus) is a small ground bird that is currently listed as a species of greatest conservation need, as designated by the WGFD. The mountain plover is known to occur throughout the State of Wyoming; however, it was not observed by private and agency biologists within the project area during repeated surveys over multiple, consecutive years between 2003 and 2007. Therefore, the Moore Ranch project is not expected to have impacts to mountain plovers. As stated in Section 5.5.4 of the Environmental Report, the applicant would conduct annual wildlife monitoring at the project site during the lifespan of the project including, annual raptor surveys between late April and early May, or as required, which is also the breeding season for the mountain plover. Should mountain plovers or plover nests be observed during monitoring events, the applicant would consult with the FWS. At this time, any applicable permits would be obtained from the appropriate agencies. No changes were made to the SEIS beyond the information provided in this response.*

B5.23.4   Concerns About Mitigation and Timing

Comment: MR002-005
One commenter encouraged NRC and project planners to develop and implement protective measures, should mountain plovers occur within the project area, and provided a list of potential protective measures.

Response: *As discussed in comment MR002-004, the mountain plover is known to occur throughout the State of Wyoming but has not been observed within the Moore Ranch project area. The applicant would obtain WGFD or BLM approval before beginning operations. Should mountain plovers or plover nests be observed within the Moore Ranch project area during surveys, Uranium One would consult with appropriate agencies to develop and implement protective measures, as directed. Since mountain plovers are not a concern for the Moore Ranch project at this time, no changes were made to the SEIS.*

Comments: MR009-198; MR018-089; MR018-090; MR018-091; MR018-092; MR018-098;
One commenter recommends the applicant conduct annual sage-grouse lek surveys, conduct additional wildlife surveys prior to new disturbance, conduct winter bald eagle and raptor nest surveys, review BLM and/or FWS raptor nest records, and avoid raptor nests during restriction time periods. Another commenter believes that annual sage-grouse surveys are not necessary.

Response: *Sections 3.5.5.3.1 of the Uranium One environmental report states that supplemental information to the wildlife surveys conducted by the Uranium One consultant for the project area was obtained from several sources, including WGFD, FWS, and BLM records from surveys conducted by their respective agency biologists in and near the vicinity of the proposed Moore Ranch project. The environmental report further explains that "because much of the project area has been included in wildlife monitoring efforts annually since 2003, the WGFD reduced the study area for raptors and other migratory birds to the portions of the proposed Moore Ranch License area and 1-mile perimeter not already encompassed by overlapping studies in recent years." (EMC, 2007a). Sage-grouse and threatened and endangered species survey areas were not reduced. Wildlife surveys conducted by the*

Appendix B

Uranium One consultant targeted bald eagle winter roost sites, sage-grouse leks, nesting raptors and eagles, mountain plovers, and other avian species of concern. As stated in Section 6.4.2 of the SEIS, Uranium One would conduct annual wildlife monitoring at the project site during the lifespan of the project, including annual raptor surveys between late April and early May, or as required. The survey would cover all areas of planned activity and a 1-mile area around the activity for the life of the project. Section 4.6.1 of the SEIS explains that if threatened or endangered species were identified in the project site during surveys, mitigation plans to avoid and reduce impacts to potentially affected species would be developed.

NRC does not have the statutory authority to request that an applicant conduct additional surveys, commit to conduct surveys during the project, enforce mitigation measures, or modify the information presented in the application when NRC has completed an acceptance review. If the applicant agrees to conduct additional studies, monitoring, or adhere to threatened and endangered species guidelines, then the licensee would implement those commitments.

B5.23.5   Habitat Loss and Fragmentation

Comment: MR018-087

One commenter suggested that removal of sagebrush would reduce forage for pronghorn and deer, and restoration projects should strive to restore sagebrush and native plant species.

Response: *NRC explains in Section 4.6 that the proposed project construction and operation may result in the disturbance of 61 ha [150 ac] of land, incrementally, for up to 12 years through the life of the ISR facility. Section 4.6 of the SEIS also discusses the potential impacts to big game and increased the potential for non-native plant species*

NRC acknowledges that in arid environments, natural revegetation could take years, and certain vegetative communities could be difficult to reestablish through artificial plantings. However, temporary and permanent revegetation planned by the applicant in a phased (sequential) schedul, would increase the rate at which a disturbed area is returned to a state similar to preconstruction and restore the wildlife forage lost during the project. Section 2.1.1.5.3 describes that revegetation practices would be conducted in accordance with WDEQ-LQD regulations and the ISR permit, including an extended reference area, which would ensure the disturbed area is reclaimed with the same vegetation type as adjacent undisturbed areas. Since the commenter concerns regarding big game and vegetation have been addressed in the SEIS, no changes were made in response to this comment.

B5.23.6   Comments on Migratory Birds

Comment: MR002-007

One commenter suggested that a migratory bird conservation plan be developed for the project, and that the FWS be consulted regarding potential impacts to eagles. The commenter pointed out that an eagle permit is required if a project "takes" eagles or their nests (active or inactive). The commenter wanted information regarding the status of the bald eagle included in the final SEIS.

Response: *SEIS Section 4.6.1.1.2.3 states that raptor nest surveys would be conducted annually. If nests were discovered during these surveys, the applicant would take appropriate mitigation measures, such as moving the nest, to ensure the protection of the species.*

Appendix B

The NRC acknowledges that consultation with the FWS concerning the eagle take permit rule is appropriate for the Moore Ranch ISR Project. NRC contacted the FWS on March 15, 2010, to discuss whether an eagle permit would be appropriate for the proposed Moore Ranch ISR project (NRC, 2010). The FWS concluded that the NRC would not need to further pursue consultation with the FWS regarding bald eagles and would not need to obtain an eagle take permit at this time. Accordingly, Section 4.6.1.1.4 has been updated to reflect this new information. Because NRC did not need to enter into consultation with the FWS regarding an eagle take permit, this consultation would not be added to the description of agency consultations in Section 1.7. However, the memorandum summarizing the teleconference with FWS (NRC, 2010) has been added to Appendix A of the final SEIS and Section 4.6.1.1.4 has been updated to reflect the information described in this comment response.

Regarding the status of the bald eagle (Haliaeetus leucocephalus), page 3-31, line 4 of the draft SEIS explains that the bald eagle was delisted from threatened status in 2007, but is still protected under the Bald and Golden Eagle Protection Act and the Migratory Bird Treaty Act. Suitable habitat is present on the Moore Ranch site for golden eagles; however, suitable habitat for bald eagles is limited within the proposed license area because of the lack of trees. No bald or golden eagles were observed during surveys conducted by the applicant between 2003 and 2007. Since the eagle status information is provided in the SEIS, no changes were made to the SEIS.

Comment: MR017-009
One commenter expressed a concern of potential impacts to wildlife from selenium contamination.

Response: *NRC acknowledges that wildlife may be temporarily exposed to contamination from spills and leaks in the SEIS Section 4.6.1.2.2. The license application for the Moore Ranch Project did not propose use of land application. The commenter references a document that uses the Highland Uranium in-situ uranium mine located in Converse County, Wyoming, as a study area and reports elevated concentrations of selenium in food sources, soil, and water within the study area (Ramirez, 2000). Note that the Highland mine used land application through irrigation for wastewater disposal, which is not planned at the proposed Moore Ranch site. In addition, the report does not discuss leaks or spills known to occur at the study area. SEIS Section 4.6.1.1.3 explains that there are no aquatic habitats on the proposed Moore Ranch site that support fish or macroinvertebrates. Therefore, there would be no impact to aquatic wildlife.*

The potential of toxic chemical impacts is discussed in the SEIS, and NRC has determined that the potential impacts to wildlife from evaporation ponds or chemical spills and leaks would be SMALL. The commenter does not provide new information that should be considered; therefore, no changes were made to the SEIS.

B5.23.7   General Vegetation Comments

Comment: MR009-195
One commenter suggests that the WGFD *Recommendations for Development of Oil and Gas Resources Within Important Wildlife Habitats* are not appropriate to include as general potential mitigation measures for the proposed project because an SEIS should consider site-specific characteristics.

Appendix B

Response: *NRC has provided a comprehensive site-specific analysis of the proposed Moore Ranch ISR project. NRC references the WGFD Recommendations for Development of Oil and Gas Resources Within Important Wildlife Habitats in the SEIS to provide examples of mitigation measures that could be implemented to reduce potential impacts to wildlife. The inclusion of the WGDF recommendations as examples of mitigation measures do not imply that the applicant is bound to them. The WGFD has no specific authority to require adoption or implementation of the recommendations, and NRC cannot enforce applicant implementation of mitigation measures. NRC believes the inclusion of this document as a reference in the SEIS is warranted since sage-grouse is listed as a federal candidate species, sage-grouse occur in the project county and region, and NRC has referenced this document for mitigation examples in other SEIS projects for proposed ISR facilities. Based on the aforementioned reasons, no changes were made to the SEIS.*

Comment: MR009-291
One commenter asserted that not all vegetation would be lost within the wellfield footprint, and that vegetation loss with the entire wellfield should not be considered an unavoidable impact.

Response: *NRC has provided Table 8-1 in the SEIS to summarize environmental consequences per resource area. The description of unavoidable consequences for ecological impacts from the proposed project includes the short-term loss of vegetation covering approximately 23 ha [57 ac], which is the footprint of the proposed wellfields. NRC agrees with the commenter that not all of the vegetation within the wellfield would be lost during the life of the proposed project; however, the expected impact to vegetation also includes the disturbance from the central plant and from developing the infrastructure that includes laying pipeline and constructing access roads. As described in Section 4.6.1.1.1, an estimated 61 ha [150 ac] of upland grassland would be affected by construction disturbance, and similarly for decommissioning, under current development plans. Table 8-1 of the SEIS has been modified to clarify that direct and short-term impacts of an estimated 61 ha [150 ac] of vegetation would be impacted.*

B5.23.8   Impacts to Terrestrial Ecology and Wildlife Discussion

Comment: MR009-193
One commenter questioned the applicability of referring to the document, "Recommendations for Development of Oil and Gas Resources Within Important Wildlife Habitats" published by the WGFD within the SEIS because it was specifically developed for the oil and gas industry.
Response: *NRC would establish site-specific license conditions for the Moore Ranch ISR Project, but only within the limits of the legislative authority granted by Congress. State and other Federal agencies would also establish permit conditions for the proposed Moore Ranch Project based upon their statutory and regulatory authority. The WGFD has the lead for the protection of sage-grouse. Although there are no regulations regarding the protection of the sage-grouse, the WGFD, in cooperation with the Wyoming Governor Sage-Grouse Implementation Team (SGIT), has developed guidelines for various industries operating in different locations within Wyoming. The NRC staff have been working with the SGIT and its sub-committees to better define the State agency roles and to develop guidelines for the ISR uranium industry. In addition, the WGFD recently issued an update to "Recommendations for Development of Oil and Gas Resources Within Important Habitats" (April 2010), which contains revised guidelines for sage-grouse protection that would be applied to the uranium extraction industry. These guidelines address (1) standard mitigation practices (for all wildlife); (2) specific best management practices for sage-grouse; and (3) stipulations for development in*

Appendix B

sage-grouse core areas that would be monitored by the WGFD. If a license were to be granted, the Moore Ranch ISR facility would be routinely inspected by WGFD for compliance with the requirements and conditions of the sage-grouse guidelines. For more information on the sage-grouse issue, the reader is referred to Section B5.25 of this comment-response appendix.

Chapter 7 of the GEIS provides a general overview of the types of best management practices, mitigation measures, and management actions that have historically been used at ISR facilities to avoid or reduce potential environmental impacts.

No change was made to the Moore Ranch SEIS beyond the information provided in this response.

Comment: MR009-207

A commenter noted that the NRC states (in Section 4.6.1.2.2.4) that impacts to reptiles and amphibians would be small. However, in Section 4.1.1.1.2, the SEIS states that there would be no impacts to amphibians and reptiles.

Response: *NRC states in Section 4.6.1.1.2.4 that "no impacts to reptiles or amphibian populations would be expected." Stating that no impacts are expected does not equate to "there would be no impacts." Section 4.6.1.2.2.4 has been revised to state that the potential impact (to reptiles and amphibians) would be SMALL. This change is to clarify that NRC does not expect impacts to reptiles or amphibians; however, any potential impacts that could occur would be small.*

Comments: MR017-013; MR017-019

One commenter stated that the NRC should take a hard look at wildlife impacts and explore alternatives or requirements that would reduce or eliminate adverse environmental impacts.

Response: *The SEIS was prepared in accordance with NRC guidance in NUREG–1748 (NRC, 2003) and is consistent with NRC regulations at 10 CFR Part 51 that implement NEPA. The draft SEIS was published for public comment in December 2009. In March 2010, the FWS listed the sage-grouse as a candidate species. Subsequently, the Wyoming BLM made amendments to the National Sage-Grouse Habitat Conservation Strategy and the Wyoming Governor SGIT continues to discuss an evaluation process for impacts to sage-grouse. NRC has included a discussion regarding those recommendations in the SEIS and has changed some of the significance levels for wildlife impacts. It should be noted that the proposed project is not located within a core population area for sage-grouse. NRC has made a reasonable effort to provide a discussion of potential mitigation measures that would limit impacts to wildlife; however, NRC does not have the statutory authority to enforce wildlife mitigation measures at a licensed facility. NRC believes that the wildlife analyses are supported by sufficient technical bases, whether tiered from the GEIS or based on supplemental staff analyses. Because the comment does not provide any additional information to incorporate into the SEIS, no changes were made to the SEIS beyond the information provided in this response.*

Comment: MR018-088

One commenter suggests that allowing hunting activities for big game would help in the management of pronghorn and deer.

Response: *NRC understands that mule deer and pronghorn are abundant in the proposed Moore Ranch Project area. No crucial big game habitat or migration corridors occur on or within several kilometers of the proposed license area. Section 1.7.3.5 explains that the WGFD is responsible for controlling all game in Wyoming. Section 3.2 of the SEIS explains that over 85 percent of the land is owned by private entities, and about 14 percent of the land is owned by the State of Wyoming. NRC does not have regulatory authority to require that the applicant allows hunting activities during operations. The applicant can make arrangements with the private land owners or engage in consultation with the WGFD regarding hunting arrangements. No changes were made to the SEIS in response to this comment.*

B5.23.9   Inconsistencies Between Sections

Comments: MR009-029; MR009-201, MR009-202

One commenter suggested that the BLM and WGFD recommendations to reduce impacts to ecology are inconsistently applied.

Response: *NRC has explained in the response to MR009-193 that the BLM and WGFD recommendations referenced by the commenter are not required mitigation measures; but serve as mitigation examples that can reduce impacts to ecology resources. The documents are referenced where some, but not all, potential applicable mitigations measures are relevant. The SEIS has been modified to clarify that these recommendations are examples.*

B5.25.10   References

10 CFR Part 51. *Code of Federal Regulations*, Title 10, *Energy*, Part 51, "Environmental Protection Regulations for Domestic Licensing and Related Regulatory Functions."

74 FR 46836, FWS. "Eagle Permits; Take Necessary To Protect Interests in Particular Localities." *Federal Register*. Volume 74, No. 175, pp. 46836-46879. September 11, 2009.

75 FR 13909, FWS. "Endangered and Threatened Wildlife and Plants; 12-Month Findings for Petitions to List the Greater Sage-Grouse *Centrocercus urophasianus*) as Threatened or Endangered." Federal Register: Volume 75, No. 55, pp. 13909-13959. March 23, 2010.

BLM (U.S. Bureau of Land Management), 2010. Instruction Memorandum No. 2010-071. Subject: Gunnison and Greater Sage-grouse Management Considerations for Energy Development (Supplement to National Sage-Grouse Habitat Conservation Strategy). Washington, D.C. U.S. Department of Interior, BLM. March 5.

BLM, 2009. "Sage-Grouse Management." <http://www.blm.gov/wy/st/en/field_offices/Buffalo/wildlife/data.html#SG>. (14 April 2009).

BLM, 2008. "Bureau of Land Management Buffalo Field Office Environmental Assessment for Yates Petroleum Corporation All Day POD Plan of Development WY–070–08–026." Cheyenne, Wyoming: BLM. August 28.

Appendix B

EMC (Energy Metal Corporation US), 2007. "Application for USNRC Source Material License, Moore Ranch Uranium Project, Campbell County, Wyoming, Environmental Report." Casper, Wyoming: Uranium 1 Americas Corporation. ADAMS Accession Nos. ML072851222, ML072851229, ML072851239, ML07285249, ML07285253, ML07285255. October 2.

FWS (U.S. Fish and Wildlife Service), 2009. "50 CFR Parts 13 and 22 Eagle Permits; Take Necessary To Protect Interests in Particular Localities; Final Rules." Washington, DC: U.S. Department of the Interior. September 11, 2009.

NRC (U.S. Nuclear Regulatory Commission), 2010. "Summary of Teleconference with Pedro Ramirez, Wyoming Field Office, U.S. Fish and Wildlife Service, Regarding Eagle Take Rule for the Proposed Nichols Ranch ISR Project (Docket No. 040-09067) and the Proposed Moore Ranch ISR Project (Docket No. 040-09073). Washington, DC: NRC. March 25. ADAMS No. ML100760621.

NRC, 2003. NUREG–1748,"Environmental Review Guidance for Licensing Actions Associated With NMSS Programs—Final Report." Washington, DC: NRC. August.

Ramirez, Jr., P. and B. Rogers, 2000. "Selenium in a Wyoming Grassland Community Receiving Wastewater from an *In-Situ* Uranium Mine." Contaminant Report Number: R6/715C/00. Cheyenne, Wyoming: FWS, Region 6. September.

WGFD (Wyoming Game and Fish Department), 2010. Recommendations for Development of Oil and Gas Resources Within Important Wildlife Habitats. Version 6.0. Cheyenne, Wyoming: WGFD. April.

B5.24   Meterology, Climatology, and Air Quality

B5.24.1   Permitting and Regulations

Comment: MR020-002
One commenter, referring to page xxi of the Executive Summary (environmental impacts from air quality) of the draft SEIS, noted the sections state the proposed project would not be subject to Title V of the Clean Air Act without providing a basis for either statement. The commenter also requested a detailed air emission inventory should be developed and used to evaluate Clean Air Act programs that may apply, including Prevention of Significant Deterioration New Source Review, Maximum Achievable Control Technology, National Emission Standards for Hazardous Air Pollutants, and Title V (Permits).

Response: While the NRC staff analysis of emissions within the context of Clean Air Act regulations supports the assessment of potential environmental impacts that is required by the NEPA, as amended, the authority to enforce Clean Air Act Regulations in Wyoming rests with the WDEQ. In that role, the WDEQ, at the time of this writing, is currently evaluating the applicant's air quality construction permit application and deciding whether to issue a permit to the applicant in the near future. Construction permits recently issued to other proposed ISR facilities (WDEQ, 2009; WDEQ, 2010) include a condition that the applicant must obtain a permit to operate in accordance with Chapter 6, Section 2(a)(iii) of the Wyoming Air Quality Standards and Regulations. According to the language of that cited requirement, such a permit applies to facilities that are not subject to the provisions of Chapter 6, Section 3 of the Wyoming

Appendix B

Air Quality Standards and Regulations. Section 1 of the Wyoming Air Quality Standards and Regulations refer to Section 3 as the state operating permit program required under Title V of the Clean Air Act. Section 1 also refers to the required Section 2 operational permit as a minor source permit to operate. This information indicates the WDEQ has concluded that each of the other proposed ISR facilities is not considered a major source of emissions nor are they subject to Title V operating permit requirements. In the review of the proposed action, the NRC staff have not identified any emissions information that would suggest the proposed Moore Ranch facility would be permitted differently; however, this would not be known for certain until a permitting decision has been made by the WDEQ. The Executive Summary has been modified to clarify the air impacts information.

Regarding the other Clean Air Act programs mentioned by the commenter, it is the NRC staff's understanding that these programs apply to major stationary sources of emissions and, based on the discussion in the preceding paragraph, do not need to be evaluated further in the SEIS. Should the WDEQ determine that some or even all of the aforementioned regulatory programs apply to the proposed action, then the applicant would need to comply with any applicable permitting requirements the WDEQ has the authority to enforce.

In response to these and other comments, the staff also updated the discussion of proposed air emissions in Section 2.1.1.1.6.1 of the SEIS and provided, in a new Appendix D, supporting calculations of mobile non-road diesel emissions from well drilling activities and construction equipment. The staff also updated portions of the air impact analysis (Section 4.7) and the Executive Summary to reflect this additional information and provide additional supporting bases for air impact conclusions. The commenter should be aware that the executive summary is a brief summary of the impact findings and does not normally contain a detailed description of supporting bases. The complete bases for impact conclusions are documented in the impact analysis in Section 4.7.

B5.24.2 Baseline Air Quality

Comment: MR020-024
One commenter, referring to the description of the affected environment for air quality on page 3-39 (Section 3.7.2), noted the proposed project is 117 mi from Wind Cave National Park, which is the nearest Clean Air Act Class 1 Prevention of Significant Deterioration Area and 77 miles from Cloud Peak Wilderness area, which is a Sensitive Class II area. They requested the SEIS identify all nearby Class I and II Areas.

Response: *In response to these comments, the staff verified the commenter information and added the recommended Prevention of Significant Deterioration sites to Section 3.7.2 of the final SEIS.*

B5.24.3 Impact Assessment

Comments: MR009-093; MR009-217; MR020-046
A commenter requested clarification in the air quality impact analysis (SEIS, Section 4.7) of what mitigations would be implemented by the applicant to reduce air emissions. Another commenter suggested the air impact analysis should include a discussion of how down-flow IX columns and vacuum dryers are beneficial to protecting public health and safety. The same commenter noted that the first paragraph in Section 2.1.1.1.6.1 mentions uranium particulate

B–95

Appendix B

emissions; however, the application of vacuum dryer technology would eliminate significant uranium particulate emissions.

Response: In response to the request for clarification of the proposed mitigations, text was added to the description of the proposed action in SEIS Chapter 2, to the air and public and occupational health impact in SEIS Sections 4.7 and 4.13, and to the air impact analysis executive summary. Because the proposed ISR facility is not considered a major source of nonradiological air pollution (see revised Section 4.7 in the final SEIS and responses to other air quality public comments in this section) the most significant mitigation proposed by the applicant to mitigate potential nonradiological air quality impacts is their proposed fugitive dust control measures that call for application of water or other agents to unpaved roads to control fugitive dust emissions (Uranium One, 2009). Regarding the comment about down-flow IX and vacuum dryer technology (both related to potential radiological air impacts), the staff clarified that these technologies are part of the proposed action in SEIS Sections 2.1.1.1.6 and 4.13.1.2.1 regarding public and occupational health impacts where the health and safety impacts of radiological air emissions are evaluated. The staff also corrected the first paragraph of draft SEIS Section 2.1.1.1.6.1 to more accurately reflect the reduced emissions from the proposed application of vacuum dryer technology.

Comments: MR009-213; MR009-215; MR009-216; MR009-218

One commenter suggested corrections be made to statements in the draft SEIS air quality impact section that summarize GEIS analyses. Specifically, regarding statements about emissions that may be associated with suspension of dried spill areas and radiological impacts (page 4-45, line 25 and page 4-47, line 1 of the draft SEIS), the commenter suggested such emissions were not considered in the GEIS. The commenter noted the draft SEIS included no information about site-specific characteristics that suggest this process would impact air quality and they suggested deleting the statement. Regarding a statement (page 4-46, line 40 in the draft SEIS) about other potential sources of nonradiological emissions, including fugitive dust and fuel from equipment, the commenter noted the GEIS does not discuss fuel emissions.

Response: While some of the sections of the GEIS refer to dried spills as an inhalation hazard (e.g., Section 4.2.3.2, Operation Impacts to Geology and Soils), this process is not discussed in the sections of the GEIS that evaluate air impacts or public and occupational health impacts. The statements in the GEIS describe resuspension of dried spill deposits as a potential route for exposure from spills, but the GEIS does not elaborate on the magnitude of the impact other than to reference an earlier bounding analysis (Mackin et al., 2001).

The referenced bounding analysis in Mackin et al., (2001) involved a dose calculation based on a conservative, hypothetical exposure scenario (i.e., a maximally exposed individual) that evaluated annual radiation dose to an individual. In this scenario, the individual was assumed to live in a residence and farm land that was contaminated by a 58,000 L [15,322 gal] spill of pregnant lixiviant. The analysis evaluated radiation doses from exposure to external radiation fields, inhalation of resuspended soil, and consumption of produced crops and livestock. The calculated doses are 1.4 to 2.6 mSv/yr [140 to 260 mrem/yr] at years 1 and 25. While the calculated doses are above the NRC 10 CFR Part 20 public dose limit of 1 mSv/yr [100 mrem/yr], the results are potentially misleading due to the conservative and, in some instances, unrealistic assumptions in the exposure scenario used in this dose calculation (see the following). Some of the conservative assumptions in the referenced analysis include that (i) a large spill would go undetected and unreported, (ii) all spilled radionuclides would remain in the top 15 cm of soil for the initial soil radionuclide inventory (no initial leaching or runoff),

(iii) there would be no attenuation of air concentrations from the effect of downwind mixing and dispersion (e.g., inherent characteristic of the mass loading model used), and (iv) the unreported soil contamination would not be detected by NRC required site decommissioning radiological surveys prior to NRC release of the site for unrestricted use. Actual doses from more plausible scenarios are likely to be far lower than the aforementioned conservative estimates and are, therefore, not evaluated further in this comment response. An operating facility would also have an NRC-approved monitoring system in place and the NRC staff would inspect the facility at least annually to verify compliance with NRC public and worker dose limits.

In response to these and other comments, the staff removed the mostly redundant discussion of potential radiological air impacts from the air quality impacts section of the draft SEIS (Section 4.7), including specific references to resuspension of dried spill areas. Based on the preceding discussion, the statements had placed undue emphasis on resuspension of spill deposits as an air quality emission concern when the supporting risk analysis information suggests otherwise. Corrections were also made because, for practical reasons, the SEIS addresses radiological impacts from air releases in the Public and Occupational Health and Safety Impacts Section (4.13) and not in the Meteorology, Climatology, and Air Quality Impacts Section (4.7) that addresses the potential nonradiological impacts to air. Section 4.13 did not specifically evaluate potential dose impacts from resuspension of dried wellfield spill areas because it focused on more severe radiological accident scenarios that are expected to have greater consequences than a scenario involving resuspension of dried spill areas. The staff also changed discussions of fuel emissions to refer to combustion engine emissions to clarify the original intent of the statements.

Comments: MR012-009; MR012-012; MR015-006
One commenter expressed concern about the greenhouse gas emissions from the proposed Moore Ranch ISR facility and the impact of these emissions on climate change. Another commenter suggested the draft SEIS ignores climate change impacts based on what was stated as the imprecise nature of the science. The commenter noted that the exact extent and timing of climate change is not certain, but that many adverse impacts have already been documented and such impacts will continue into the future. Citing draft guidance from CEQ (2010) (to help Federal agencies improve their consideration of greenhouse gas emissions and climate change in evaluations of proposals for Federal actions), the commenter stated that despite the evolving nature of climate change science, Federal agencies have an obligation to consider both greenhouse gas emissions emitted from proposed projects and the impacts the action has on natural resources that could also be affected by climate change.

Response: *As one commenter noted, the state of the science of climate change is evolving. The NRC staff acknowledge the changing state of the science on climate change and the evolving Federal role in evaluating the potential environmental impacts of Federal actions. The NRC approach to evaluating potential climate change impacts from NRC licensing actions is also evolving and continues to evolve as more information becomes available that NRC staff can use to evaluate potential impacts.*

To address these and other comments regarding the need for the NRC staff to consider and evaluate the potential impact of greenhouse gas emissions on the global climate, the staff have calculated annual and cumulative CO_2 emissions from applicant use of diesel construction equipment during construction and decommission of the production wellfields and facilities. Because operating ISR facilities are not major sources of CO_2 or other greenhouse gas emissions, the NRC staff expect construction equipment emissions (including well drilling rigs)

Appendix B

produced during both construction and decommissioning phases to represent the majority of greenhouse gas emissions from the proposal. The emissions estimates are documented in a new Appendix D and are summarized in Section 2.1.1.1.6. The NRC staff also added an evaluation of potential impacts to climate from the calculated construction equipment emissions from the proposed facility in Section 4.7. The cumulative air impact analysis was also updated to evaluate the impact of the emissions in the context of other past, present, and reasonably foreseeable future actions.

Comments: MR012-013; MR012-014; MR012-015; MR012-019
One commenter provided a number of comments related to climate change and the potential impacts of climate changes on the potential environmental impacts from the proposed Moore Ranch ISR facility. They provided a report from the U.S. Global Change Research Program entitled "Global Climate Change Impacts in the United States" (Karl et al., 2009) as the technical basis for predictions of climate change in the region where the facility is proposed. They suggested the report shows that climate changes in the region have the potential to impact the proposed facility and, therefore, such impacts should be evaluated by NRC in the SEIS. Specifically, they noted the area can expect reduced snowpack and spring runoff and disruption of precipitation over the next decades. Regarding the potential increase in precipitation, they requested NRC disclose and evaluate how increased soil saturation, flooding, and aquifer recharge would interact with project impacts.

Response*: With regard to future changes in climate altering the potential impacts of the proposed action, the staff evaluated the report cited by the commenter, and found that the projected changes in climate over the 10-year time scale of the licensing period for the proposed facility were limited in degree and unlikely to significantly change the intensity of the potential impacts evaluated in the final SEIS. For example, the projected changes in precipitation for a high-emissions scenario were discussed for the latter part of this century (years 2080 through 2090) as 10 to 15 percent above current values for the area of Wyoming where the proposed site would be located. Changes during the next 10 years would be expected to be much less than the values reported for the end of the century. The staff could not identify information in that report to suggest that over the next 10 years there would be the types of changes indicated by the commenter (e.g., soil saturation, flooding, recharge effects). Projected temperature changes are also cited in the report as long-term consequences. The cited report includes projected changes in average temperature for year 2020 as ranging from a slight decrease in the present temperature to a maximum of approximately 2 degrees higher than present temperatures. The resource area that would be expected to be the most sensitive to small changes in ambient temperature would be the local ecology. Potential changes to the regional ecology from a rise in average temperature (including invasive species, fire, erosion, desertification) would occur whether the site were licensed or not, but localized effects could be exacerbated to some degree by proposed site activities and the changes in the ambient temperature. In response to these comments, NRC staff added discussion of the potential impacts from projected changes to climate in Section 5.7 (Cumulative Air Quality Impacts) of the final SEIS.*

In Section 5.7.1.4, the NRC staff determined that the overall effect of projected climate change on the proposed Moore Ranch facility would be SMALL. The small, predicted increases in temperatures and precipitation over the next decade would have no effect on the proposed Moore Ranch Facility in any of the ISR phases.

Appendix B

Regarding the portion of a comment that suggested potable water sources outside the ore zone would be sacrificed, as discussed in the GEIS, the NRC licensees are required to return wellfield water quality parameters to the standards in 10 CFR Part 40, Appendix A, Criterion 5(B)(5) or another standard approved in their license (NRC, 2009). In general, favorable hydrogeological conditions for effective isolation of ore-bearing aquifers and containment of recovery solutions; integrity and continuity of impermeable confining layers;, successful implementation of restoration techniques; and continuous and effective monitoring of wellfields during ISL operations, restorations, and stabilization periods are expected to limit potential environmental impacts. The NRC staff are not aware of any incident in which nonexempt portions of an ore-bearing aquifer have been contaminated by ISL operations under NRC regulations. Additional comments and responses related to potential groundwater resource impacts are provided in Section B5.20. As no specific changes were suggested by the comment, no changes to the draft SEIS were made in response to the comment.

Comment: MR012-021

Referring to the Chapter 5 (cumulative impact analysis) discussion of climate change in the draft SEIS, a commenter suggested the draft SEIS failed to consider the impacts of climate change by not disclosing all greenhouse gas emissions. The commenter noted the emissions for the proposed site discussed in the draft SEIS are incomplete because they do not include the emissions from other nuclear fuel cycle facilities such as facilities involved in uranium conversion, uranium enrichment, and nuclear fuel fabrication.

Response: *Evaluation of environmental impacts from other nuclear fuel cycle facilities is beyond the scope of the current licensing action regarding whether or not to grant a license to the proposed Moore Ranch ISR facility. NRC evaluates the potential safety and environmental impacts of other fuel cycle facilities when those facilities are proposed or their licenses are amended. Because the requested information is beyond the scope of the current licensing action, no changes were made to the SEIS in response to the comment.*

Comments: MR015-004; MR015-014; MR015-017; MR015-019; MR015-024; MR015-025; MR020-040; MR020-041

One commenter stated that the draft SEIS lacked information on air pollutant emissions, and the impact analysis is inadequate to assess the impacts of those emissions. The commenter also stated that ISR projects would likely result in a deterioration of air quality due to emissions from drill rig engines, fugitive road dust, and uranium processing activities. The commenter suggested that projects similar in scope require hundreds of wells and multiple deep injection wells, and without a complete air quality analysis, such activity is likely to have significant adverse local air quality impacts. The commenter was particularly concerned about the air emissions from the truck-mounted diesel drilling rigs and the drilling of hundreds of wells for the Moore Ranch ISR project. The commenter suggested this level of development may have cumulative emission rates in excess of several hundred tons per year of nitrogen oxides, particulates, and other priority air pollutants. They requested a screening analysis be conducted for air emissions to identify far field impacts including visibility parameters for Class I and sensitive Class II air sheds. They also requested that a near field air analysis be conducted to evaluate direct air impacts.

Response: *In response to this and other comments, NRC staff reviewed the applicable sections of the draft SEIS and added more detailed information on emissions from drilling rigs, construction equipment, and unpaved roads (i.e., fugitive road dust) to Section 2.1.1.1.6.1 and Appendix D. The NRC staff also added information to Section 3.7.2 on nearby Class I and*

Appendix B

Class II areas that could potentially be impacted by emissions generated by the proposed action; and added text to Section 4.7 to clarify the NRC staff approach to evaluating impacts and improve the transparency of the NRC bases for impact conclusions.

The NRC staff estimates of annual nitrogen oxide and particulate emissions from drilling rigs and construction equipment are approximately 18.1 t/yr [20 T/yr] and 0.79 t/yr [0.87 T/yr] as discussed in Appendix D. The NRC staff estimated the nitrogen oxide emissions could be as high as 29.9 t/yr [33 T/yr] if the applicant drilled all four deep disposal wells in one year; however, this would be a 1 year maximum, as no additional deep wells would need to be drilled in later years. Applicant estimates of fugitive road dust are 14.5 t/yr [16 T/yr] (EMC, 2007), and their close proximity to State Highway 387 limits the number of miles traveled on unpaved roads and the resulting dust emissions. The magnitude of the annual diesel emissions calculated by the NRC staff and dust emissions provided by the applicant are well below the several hundreds of tons suggested by the commenter.

The differences among the commenter assumed emissions and the NRC staff and the applicant calculations might be the result of differing levels of understanding about the details of the proposed action. Perhaps the commenter is more familiar with oil and gas drilling that penetrates to far greater depths than the water wells proposed for the Moore Ranch ISR Project. Oil and gas wells can take longer to drill, require greater horsepower drilling rig engines, consume fuel at a higher rate, and produce greater hourly emissions. Another factor that may not have been considered by the commenter is the phased approach the applicant plans for wellfield development, constructing approximately one wellfield per year rather than all wellfields at once. The difference between the emissions assumed by the commenter and those calculated by NRC staff is significant because the commenter assumes emissions would be at levels that are well above the current stationary source threshold for major emitters, whereas the NRC staff estimates are well below this threshold. Had the NRC staff estimates been at the same levels assumed by the commenter, the level of the NRC staff concern for potential air quality impacts would be similar to that of the commenter. However, given that the calculated values are much lower than the values assumed by the commenter and well below the major source threshold, NRC staff conclude that the emissions would be unlikely to change the current attainment status of the region surrounding the site nor would the emissions be likely to destabilize the local air quality. Therefore, additional detailed quantitative air analyses are not warranted to support the evaluation of nonradiological air impacts. Short-term and intermittent, visible air emissions are possible to the local area surrounding the site when vehicles travel on unpaved roads. Such impacts would be reduced, but not eliminated, by road treatments proposed by the applicant (Uranium One, 2009).

The scope of the air impact analysis in Section 4.7 is intentionally limited to consideration of non-radiological air quality impacts. This is because, as noted in the draft SEIS, radiological air emissions are regulated by NRC and are addressed in Section 4.13 as a public and occupational health and safety topic, whereas nonradiological emissions are regulated by the State and the EPA and are best evaluated separately in Section 4.7.

Comment: MR015-018
One commenter indicated the proposed project may adversely impact nearby Federal Class I areas, which require special protection of air quality and air quality related values such as visibility.

Appendix B

Response: The mobile nonroad diesel emissions from construction and mobile fugitive road dust emissions from all phases are the emissions from the proposed action that have the greatest potential to impact nearby Prevention of Significant Deterioration areas based on the NRC staff understanding of the types and magnitudes of emissions associated with ISR facilities and the information provided by the applicant on this specific proposal. As discussed in Section 2.1.1.1.6.1, the applicant estimated fugitive road dust emissions to be approximately 13.6 t/yr [15 T/yr] if not controlled; however, the applicant proposes to control these emissions by water application or other means of road treatment.

All other emissions information reviewed by the NRC staff support the conclusion that NRC staff expect the proposed action would not be comparable to nor considered a major source of emissions (e.g., a stationary source that emits or has the potential to emit 90.7 t/yr [100 T/yr] of an air pollutant to 9.1 t/yr [10 T/yr] of any individual hazardous air pollutant, or 22.7 t/yr [25 T/yr] of any combination of hazardous air pollutants as defined in Sections 501 and 112 of the Clean Air Act). While NRC staff recognize the stationary source requirements, by definition, do not apply to mobile sources of emissions, these requirements apply to the same types of air pollutants that are emitted by the mobile sources proposed by the applicant and the threshold values are the levels of emissions that trigger a substantial increase in the requirements that must be met to ensure the protection of air quality. NRC staff conclude that such emissions (i.e., well below the major source thresholds) in an area with meteorology favorable for dispersion would be unlikely to impact air quality in the nearest Class I area to the proposed action. The Class I area, Wind Cave National Park, is located about 188 km [117 mi] to the east of the Moore Ranch site. Cloud Peak Wilderness Area, the closest Class II area to the proposed action located about 124 km [77 mi] to the northwest of the Moore Ranch site is also unlikely to be impacted by the magnitude of proposed emissions-generating activities. In addition to the magnitude of emissions and distance, the predominant wind direction at the proposed site is from the southwest and, therefore, would carry emissions to the northeast, away from the Class II area.

While the NRC staff analysis of emissions within the context of Clean Air Act regulations supports the assessment of potential environmental impacts that is required by the NEPA, as amended, the authority to enforce Clean Air Act Regulations in Wyoming rests with the WDEQ, and they are responsible for making applicability and compliance decisions regarding the regulations that implement the Clean Air Act. In that role, the WDEQ is currently evaluating the applicant air quality construction permit. If a permit is granted, it is expected to clarify what additional air quality permits would be required for the proposed action and could specify additional controls to limit emissions (e.g., radon, fugitive road dust controls) based on the NRC staff review of other ISR facility construction air permits (WDEQ, 2009; WDEQ, 2010). Should the air quality in the nearby Class I areas become degraded in the future, the WDEQ has the authority and would be expected to take appropriate corrective actions to reestablish attainment air quality in these protected areas.

In response to this and other comments about the potential impacts of air emissions, NRC staff updated the discussion of proposed air emissions in Section 2.1.1.1.6.1 of the draft SEIS and provided, in a new Appendix D, supporting calculations of mobile nonroad diesel emissions from well drilling activities and construction equipment. The air quality impact analysis discussion in SEIS Section 4.7 was also updated to reflect the updated emissions information.

Appendix B

Comment: MR015-037
A commenter suggested NRC expand the discussion of greenhouse gas emissions and climate change in the draft SEIS. Specifically, they requested NRC staff consider the projected regional climate changes and the project contribution to these changes. They also requested NRC staff quantify the annual and cumulative greenhouse gas emissions and discuss the link between greenhouse gas emissions and climate change. A discussion of mitigation measures for greenhouse gas emissions was also requested.

Response: *To address these and other comments regarding the need for NRC staff to consider and evaluate the potential impact of greenhouse gas emissions on the global climate, NRC staff have calculated annual and cumulative CO_2 emissions from applicant use of diesel construction equipment during construction and decommission of the production wellfields and facilities. Because operating ISR facilities are not major sources of CO_2 or other greenhouse gas emissions, the construction equipment emissions (including well drilling rigs) produced during both construction and decommissioning phases are expected by the NRC staff to represent the majority of greenhouse gas emissions from the proposal. The emissions estimates are documented in a new Appendix D and summarized in Section 2.1.1.1.6. The staff also added an evaluation of potential impacts to climate from the calculated construction equipment emissions from the proposed facility in Section 4.7. The cumulative air impact analysis was also updated to evaluate the impact of the emissions in the context of other past, present, and reasonably foreseeable future actions. The revised impact analyses included discussion of the current understanding of the link between greenhouse gas emissions and global climate change. Based on the nature of the emissions (e.g., construction equipment) and the lack of available CO_2 mitigations for such equipment, Section 5.7.1.5 was added to the SEIS. These general mitigation measures discussed in Section 5.7.1.5 would be implemented to minimize the overall GHG emissions at the proposed Moore Ranch Project.*

Comments: MR020-034; MR020-035
Referring to page 4-43 of the draft SEIS (Section 4.7, Meteorology, Climatology, and Air Quality Impacts), a commenter noted that no project-specific emissions estimates were provided in the draft SEIS. The commenter noticed that the draft SEIS references GEIS Section 2.7.1, which includes emissions estimates for the Crownpoint ISR facility from a 1997 NRC Final Environmental Impact Statement (NRC, 1997). The commenter indicated the draft SEIS did not discuss how that facility, and therefore its emissions estimates, relate to the proposed facility. The commenter also suggested that the referenced emissions estimates from 1997 were not current and should be updated.

Response: *In response to this and other comments about the potential impacts of air emissions, NRC staff updated the discussion of proposed air emissions in Section 2.1.1.1.6.1 of the draft SEIS and provided, in a new Appendix D, supporting calculations of mobile nonroad diesel emissions from well drilling activities and construction equipment. NRC staff also updated the air quality impact analysis discussion in SEIS Section 4.7 to reflect the updated emissions information. Text was also added to SEIS Section 4.7 to compare attributes of the Crownpoint facility and the proposed action to establish a more transparent basis for adopting the GEIS air impact analyses in the SEIS.*

Comment: MR020-036
Referring to page 4-45 of the draft SEIS (Section 4.7, Air Quality Impacts), a commenter expressed that while there is a discussion of air quality impacts, neither the draft SEIS nor GEIS have an air impact analysis.

Appendix B

Response: *While the commenter was not specific about the type of impact analysis that was expected, some additional clarification regarding some of the limitations that affect the content of Section 4.7 may be informative. First, Section 4.7 of the draft SEIS describes the potential impacts to nonradiological air quality based on the NRC staff review of the proposed action that was summarized in draft SEIS, Chapter 2, and the accessible environment that was summarized in SEIS, Chapter 3. The NRC staff approach to documenting the impact analyses in the Chapter 4 impact sections is to avoid repetitive discussions of information that was previously discussed in prior chapters by referencing and summarizing discussed information. This approach may have contributed to an appearance of incompleteness. In response to the comment, the NRC staff have reviewed the section, incorporated additional references and discussion of referenced information to add transparency to the support for the analysis and the bases for conclusions. Another factor that limits the scope of the air impact analysis is that the analysis in Section 4.7 is intentionally limited to consideration of nonradiological air quality impacts. This is because, as noted in the draft SEIS, radiological air emissions are regulated by NRC and are addressed in Section 4.13 as a public and occupational health and safety topic, whereas nonradiological emissions are regulated by the State and the EPA and are best evaluated separately in Section 4.7. The NRC staff evaluation of potentially nonradiological impacts in Section 4.7 of the draft SEIS is further limited because as stated in that section and in the GEIS, ISR facilities are not major emitters of nonradiological air pollutants and, consistent with NRC NEPA implementing regulations at 10 CFR Part 51, Appendix A (Item 7), the level of information considered in detail reflects the depth of analysis required for sound decisionmaking.*

In response to this and other comments, the NRC staff reviewed the applicable sections of the draft SEIS and added more detailed information on emissions to Section 2.1.1.1.6.1 and Appendix D; added information on nearby Class I and Class II areas to Section 3.7.2 that could potentially be impacted by emissions generated by the proposed action; and added text to Section 4.7 to clarify the NRC staff approach to evaluating impacts and improve the transparency of the NRC staff bases for impact conclusions. The additional emissions information confirms the proposed Moore Ranch ISR project would be a minor source of nonradiological emissions that the NRC staff conclude would be unlikely to change the current attainment status of the region surrounding the site nor would the emissions be likely to destabilize the local air quality. Short-term and intermittent, visible air emissions are possible to the local area surrounding the site (for example when vehicles travel on unpaved roads).

Comments: MR020-037; MR020-038; MR020-047
One commenter, referring to Section 4.7, Meteorology, Climatology, and Air Quality Impacts, expressed that the draft SEIS discusses potential local impacts of carbon monoxide and particulate emissions, as well as impacts to particulate matter NAAQS without providing bases for the impact conclusions. The commenter highlighted another statement in the SEIS section that describes the insignificance of pollutant emissions that present a potential for cumulative impacts as unsupported by any emission inventory or modeling.

Response: In response to this and other comments, the NRC staff reviewed draft SEIS Section 4.7 and added text to clarify the NRC staff approach to evaluating impacts and to clarify the NRC staff bases for impact conclusions. The revisions included consideration in Section 4.7 of the detailed information on emissions from drilling rigs, construction equipment, and unpaved roads (i.e., fugitive road dust) that was added to Section 2.1.1.1.6.1 and Appendix D in response to other comments. The specific statement about particulate matter and compliance

Appendix B

with NAAQS emphasized by the commenter was paraphrasing impact conclusions from the GEIS, and the text was clarified to more explicitly associate the statement with the GEIS. The statement in draft SEIS Section 4.7 about cumulative impacts was deleted because cumulative impacts are addressed in SEIS, Chapter 5.

B5.24.4   References

CEQ (Council on Environmental Quality), 2010. "Draft NEPA Guidance on Consideration of the Effects of Climate Change and Greenhouse Gas Emissions." Letter (February 18) from Nancy Sutley to the Heads of Federal Departments and Agencies. Washington DC: CEQ.

EMC (Energy Metal Corporation US), 2007. "Application for USNRC Source Material License, Moore Ranch Uranium Project, Campbell County, Wyoming, Environmental Report." Casper, Wyoming: Uranium 1 Americas Corporation. ADAMS Accession Nos. ML072851222, ML072851229, ML072851239, ML07285249, ML07285253, ML07285255. October 2.

Mackin, P.C., D. Daruwalla, J. Winterle, M. Smith, and D.A. Pickett, 2001. NUREG/CR–6733, "A Baseline Risk-Informed Performance-Based Approach for *In-Situ* Leach Uranium Extraction Licensees." Washington, DC: NRC. September.

NRC (U.S. Nuclear Regulatory Commission), 1997. NUREG–1508, "Final Environmental Impact Statement To Construct and Operate the Crownpoint Uranium Solution Mining Project, Crownpoint, New Mexico." Washington, DC: NRC. February.

NRC, 2009. "Uranium Recovery Policy Regarding: (1) The Process for Scheduling Licensing Reviews of Applications for New uranium Recovery Facilities and (2) The Restoration of Groundwater at Licensed Uranium *In-Situ* Recovery Facilities." Regulatory Information Summary 2009-05. ADAMS Accession No. ML083510622. April 29.

Karl, T.R., J.M. Melillo, and T.C. Peterson,eds., 2009. "Global Climate Change Impacts in the United States". New York City, New York: Cambridge University Press.

Uranium One (Uranium One Americas), 2009. "Response to Request for Additional Information for the Moore Ranch *In-Situ* Uranium Recovery Project License Application (TAC JU011)." ADAMS Accession Number ML092450317. August 31.

WDEQ (Wyoming Department of Environmental Quality), 2010. "Re: Permit No. CT-7896." Letter (January 4) from D.A. Finley to J. Cash, Lost Creek ISR, LLC. Cheyenne, Wyoming: WDEQ, Air Quality Division.

WDEQ, 2009. "Re: Permit No. CT-8644." Letter (October 2) from D.A. Finley to M. P. Thomas, Uranerz Energy Corporation. Cheyenne, Wyoming: WDEQ, Air Quality Division.

Appendix B

B5.25 Historical and Cultural Resources

B5.25.1 Potential Impacts to Cultural, Historical, and Sacred Places

Comment: MR007-005
One commenter requested a text change to indicate that, under conditions in its license, the applicant would likely be required to stop work upon discovery of previously undocumented historical or cultural resources.

Response: NRC acknowledges this comment and has revised text in SEIS Section 4.9.1.1 to address inadvertent discovery.

Comment: MR007-006
One commenter noted that contrary to the Moore Ranch SEIS text the Wyoming SHPO does not possess the legal authority to require that work stop upon discovery of previously undocumented historic or cultural resources, etc.

Response: NRC acknowledges the comment. SEIS Section 4.9.1.2 has been revised in response to this comment.

Comment: MR007-007
One commenter stated that the Moore Ranch SEIS is a federal undertaking. As such, the requirements of Section 106 of the NHPA apply regardless of land ownership, and minimization/mitigation of adverse effects is required.

Response: Any Federal undertaking defined in 36 CFR 800.16(y) is subject to Section 106 of the NHPA regardless of land ownership. Text in SEIS Section 5.9 has been revised to clarify the status of the Moore Ranch as a Federal undertaking.

Comment: MR009-035
One commenter noted that the Executive Summary states, "The identified eligible sites would be avoided and; therefore, there would be no impact." The commenter further states that while technically correct, the nearest eligible site is located over 1 mile away from the current proposed areas of surface disturbance.

Response: NRC acknowledges this comment. Because of the general nature of the comment, no revisions were made to the SEIS.

B5.25.2 License Conditions to Address Potential Impacts to Historical and Cultural Resources

Comments: MR009-036; MR009-037; MR009-038; MR009-292
One commenter stated the applicant did not propose the preparation of an Unidentified Discovery Plan and is not aware of any NRC guidance that discusses the contents or requirements of such a plan. The Executive Summary states, "If any identified historic or cultural resources were encountered during the construction phase of the proposed Moore Ranch Project, they would be evaluated following procedures in an Unidentified Discovery Plan developed prior to initiation of construction." This text is incorrect.

Appendix B

Response: The development of an Unidentified Discovery Plan is an NRC proposed mitigation measure. This proposed mitigation measure was not mentioned within SEIS Section 4.9. Since the issuance of this SEIS, the Wyoming SHPO has concurred with the NRC finding that no historic properties would be affected by the proposed action. The SHPO noted in its November 3, 2009, letter that, "If any cultural materials are discovered during construction, work in the area shall halt immediately, the federal agency and SHPO staff be contacted, and the materials be evaluated by an archaeologist or historian meeting the Secretary of the Interior's Professional Qualification Standards (48 FR 22716, Sept 1983)." The staff concur with this recommendation. Additionally, the applicant has agreed to condition its license (if issued) to include a stop-work provision in case historic and cultural resources are encountered. The SEIS Executive Summary and Chapter 4 were revised in response to this comment.

Comment: MR013-017
One commenter stated that NRC should be more specific in each final SEIS as to when license conditions are imposed on its licensees with respect to control (e.g., elimination or mitigation of a potential impact) so that members of the public and interested stakeholders are aware that NRC is regulating that activity.

Response: NRC acknowledges this comment. With respect to cultural resources, a license condition can address ISR operators regarding an ongoing responsibility to monitor for unidentified historic and cultural resources during site construction and operation and, if and when such resources are identified, to cease operations and seek appropriate consultations with State and Federal agencies. Because of the general nature of this comment, no specific changes were made to the SEIS in response to this comment.

B5.25.3   Historical and Cultural General

Comment: MR007-002
One commenter noted that while the 50-year cut-off date for possible inclusion on the NRHP is a good rule of thumb, it is not a hard and fast rule, which fully excludes sites younger than that for inclusion.

Response: NRC acknowledges this comment. There is a general stipulation that a resource is 50 or more years old, however, there are exceptions to this criteria if the property displays unique, outstanding, or ethnographic characteristics that would deem it eligible for listing on the NRHP. SEIS Section 3.9.1 has been revised in response to this comment.

Comment: MR007-003
One commenter stated the Moore Ranch SEIS incorrectly titled Section 3.9.3 Historic Properties Listed in the National and State Registers. The commenter noted that that the State of Wyoming does not maintain a Register of Historic Places.

Response: NRC agrees with this comment. The title and text in Section 3.9.3 has been revised in response to this comment.

Comment: MR007-004
One commenter requested clarification in Moore Ranch Section 4.9.1.1.1. The text, in part, discusses eligibility for the NRHP under criteria in 36 CFR 60.4(a)-(d) or Traditional Cultural Properties, or both. Per National Register Bulletin 38, "Guidelines for Evaluating and Documenting Traditional Cultural Properties," in order for a property to be eligible as a

Appendix B

Traditional Cultural Properties, it must be eligible under one of the four criteria of eligibility set forth in 36 CFR 60.4. This should be clearly stated in the document.

Response: *NRC acknowledges this comment. The Moore Ranch SEIS Sections 4.91.1 and 1.7.2 have been revised in response to this comment.*

Comments: MR009-107; MR009-151
The applicant notes that in Section 2.2.3, Alternate Site Location, the discussion states that the alternate plant site locations would have resulted in potential impacts to cultural resources. The applicant further notes the response to environmental report RAI 2.5 Number 1, which discusses the plant site alternatives and does not discuss cultural resources because surveys have already been conducted in both areas, and potential sites are avoided. The applicant suggests revisions to Section 2.2.3.

Response: *NRC acknowledges this comment. SEIS Section 2.2.3 has been revised in response to this comment.*

Comment: MR009-220
One commenter noted the current SHPO eligibility of the site was not accurate in the Moore Ranch SEIS.

Response: *NRC acknowledges this comment. Section 4.9.1.1 has been revised to reflect the November 2009, Wyoming SHPO decision that states the sites are ineligible.*

Comment: MR009-286
The applicant requested that NRC reference the fact that ongoing monitoring and protection of historic and cultural resources will be required under a license condition.

Response: *NRC acknowledges this comment. SEIS Section 5 or 6 has been added to address this comment.*

Comments: MR020-048; MR020-049
One commenter stated that the shift from recapping the GEIS to describing the additional findings of the SEIS is abrupt and confusing. Additionally, the commenter requests clarification on the status of eligibility of the sites on the Moore Ranch Project area.

Response: *NRC acknowledges this comment. The SEIS has been revised to clarify the language and to incorporate a November 3, 2009, letter from the Wyoming SHPO. This letter concurs with the NRC finding that the sites located in the project area are ineligible for listing on the NHRP. Appendix A of the final SEIS contains a copy of this letter.*

Comments: MR020-053; MR020-54
One commenter noted a lack of discussion of Tribal impact, specifically stating that the CEQ guidance recommends considerations in the EJ analysis of impacts on Tribal cultural and subsistence resources. Although NRC does consider itself bound by CEQ guidance, the GEIS specifically requires consideration of those resources for sites in the Nebraska-South Dakota-Wyoming Uranium Milling Region. It notes the importance of hunting, gathering, and cultural resources to Tribal populations in the region. The Moore Ranch site is located within this region. Although the SEIS considers cultural resources outside the environmental justice context (and finds them to be small), it does not discuss them in the environmental justice

Appendix B

analysis. An impacts on cultural resources discussion should be added to the environmental justice section. Furthermore, there is no indication that Tribal subsistence resources were considered at all. (The SEIS notes that there are large herds of antelope and deer in the project area, and at least some of the project surface area is not privately owned.) Regardless of whether the SEIS correctly identified the impact area, it is possible that Tribal members from outside the area use cultural and subsistence resources within the area. Therefore, that possibility should either be eliminated by a specific analysis of Tribal use, or impacts on cultural and subsistence resources should be addressed.

Response: *NRC acknowledges this comment. As the GEIS notes the importance of hunting, plant gathering, and cultural resources to Tribal populations in the area, the text has been revised to provide justification for the conclusions that there would be no impacts on EJ issues in regards to Tribal land use within the specific project area.*

B5.25.4 References

36 CFR Part 800. *Code of Federal Regulations*, Title 36, *Parks, Forests, and Public Property*, Part 800. "Protection of Historic Properties."

36 CFR Part 60. *Code of Federal Regulations*, Title 36, *Parks, Forests and Public Property*. Part 60, "Natural Register of Historic Places."

B5.26 Socioeconomics

Comment: MR018-008
One commenter expressed concerns over the two mining districts in Wyoming being split.

Response: *The two mining districts were developed for analysis in the GEIS. Although the SEIS is tiered from the GEIS, the SEIS includes site-specific analysis of socioeconomic factors for the Moore Ranch project site. The SEIS uses a Region of Influence limited to the project site area (Campbell County).*

Comment: MR018-009
One commenter expressed concerns that political subdivisions would cause socioeconomic data to be collected incongruently.

Response: *The SEIS uses a Region of Influence limited to the project site area (Campbell County). Socioeconomic information pertaining to these counties was derived from US Census Bureau information, in addition to different State agency data. Therefore, socioeconomic information in the SEIS is limited to the project area and is taken from resources that typically standardize their collection methods.*

Comment: MR018-010
A commenter expressed concerns about using 10 year old US Census Bureau data in the socioeconomic analyses.

Response: *The GEIS relies on 2000 Census data, which is based on the actual count. The SEIS uses the latest U.S. Census Bureau American Community Survey estimates, which are based on the 2000 Census. The SEIS also uses current State and county estimates.*

Appendix B

Comments: MR009-015; MR009-230
The commenter requested that socioeconomic impacts be divided up front in a positive or negative category.

Response: Although positive and negative descriptions can be used in describing socioeconomic impacts, they are subjective and are not typically used in NRC licensing reviews. As a regulatory agency, the NRC must remain impartial to the socioeconomic negative and beneficial effects associated with the proposed action. This comment does not present any significant new information or arguments that would warrant a change to the final SEIS.

Comment: MR009-231
One commenter requests that the SEIS provide specific examples of industrial activities that are larger in scale than an ISR project.

Response: The commenter refers to Chapter 4, Section 4.11.1.1 that describes the small size of the construction workforce in relation to other construction projects. NRC has revised the final SEIS in response to this comment.

Comments: MR009-232; MR009-235
One commenter stated that the SEIS mentions two different numbers of workers expected for staff operations and during the construction phase (40–60 workers vs. 50 workers).

Response: The comment refers to Chapter 4, Section 4.11.1.1 that describes the number of operations workers for the proposed action (Alternative 1). NRC has revised the final SEIS in response to this comment.

B5.26.1 Reference

NRC (U.S. Nuclear Regulatory Commission), 2009. NUREG–1910, "Generic Environmental Impact Statement for In-Situ Leach Uranium Milling Facilities." Washington, DC: NRC. May.

B5.27 Public and Occupational Health

B5.27.1 Impacts to Members of the General Public

Comment: MR015-027
One commenter who requested the SEIS include an analysis of the potential use of evaporation ponds, further requested that this analysis include radon emission estimates and comparison to applicable Clean Air Act requirements, which could be significant.

Response: The draft SEIS did not evaluate the use of evaporation ponds because evaporation ponds were not included in the applicant proposal, and that proposal was the focus of the NRC staff environmental review. However, in response to this and other comments, additional information was provided in SEIS Sections 2.1.1.2 and 4.14.1.2 to discuss and evaluate potential environmental impacts of options for liquid waste water disposal that were not proposed by the applicant. That evaluation of wastewater disposal options includes consideration of the use of evaporation ponds and how the potential environmental impacts compare with the applicant proposal and other liquid waste management options. The waste

Appendix B

management options are discussed at a general level of detail with regard to radon emissions because there are various implementation options that an applicant could present that would affect the amount of radon emitted from a specific proposal. Additional information is discussed in the following paragraphs to address the commenter concern that radon emissions from evaporation ponds, if used in a modified proposed action, could lead to significant environmental impacts.

The amount of radon that might be emitted if an evaporation pond or ponds were added to the current Moore Ranch ISR proposal can be approximated from radon emissions information provided in the applicant proposal (EMC, 2007). To calculate the emission estimates, the applicant used NRC accepted methods (NRC, 1987) to estimate the annual activity of radon that would be transferred to production fluids from the decay of radium in the ore body. This approach considered variables such as the average production flow rate (i.e., the amount of lixiviant that would be circulated annually through the ore body and pumped to the surface) and the radium content of the ore body. Assuming the radon is in secular equilibrium with the radium in the ore body, the applicant estimated 94.83 TBq/yr [2,563 Ci/yr] of radon would be emitted if all radon in the pumped lixiviant were allowed to escape to the open air.

The amount of this potential total annual radon emission that could be released from an evaporation pond would be proportional to the amount of lixiviant (and, therefore, dissolved radium) that is diverted from the processing circuit as process bleed (1.0 percent of the production flow rate from the applicant proposal) or approximately 0.95 TBq/yr [26 Ci/yr]. This level of emission is well below the applicant annual estimate of 22.37 TBq [604.7 Ci] from all proposed releases.

The applicant evaluated the potential offsite dose impacts of emitting 22.37 TBq/yr [604.7 Ci/yr] of radon using the MILDOSE code (Argonne National Laboratory, 1989). This resulted in a 0.008 mSv/yr [0.8 mrem/yr] dose at the property boundary. Because the calculated dose is proportional to the emission, the dose from the calculated evaporation pond radon emission would be approximately 0.0003 mSv/yr [0.03 mrem/yr] (i.e., 26/604.7 × 0.8), and the combined dose from all proposed radon emissions with the evaporation pond emission dose added would be approximately 0.0083 mSv/yr [0.83 mrem/yr]. This calculated dose is a small fraction of the NRC 10 CFR Part 20 public dose limit of 1 mSv/yr [100 mrem/yr]. The NRC staff consider this calculation sufficient to demonstrate that potential public health impacts from radon releases would be small, and additional analyses or comparisons with other regulatory requirements are not necessary to support this conclusion. A licensed facility would also be required to have an NRC approved environmental monitoring program for radon emissions in place that would report measured radon values to NRC for review on a semi-annual basis. Annual NRC inspections would also verify that applicant safety programs are compliant with NRC regulations in 10 CFR Part 20 and any conditions of their license, thereby providing additional confidence that the facility would be operated safely and within the bounds described in the applicant proposal.

Radon emissions associated with the applicant's proposal are evaluated in SEIS Section 4.13.1.2.1. The use of evaporation ponds is presently not part of the applicant proposal for Moore Ranch. Should the applicant decide in the future to change their proposed approach to wastewater management, they would be required by NRC to amend their proposal, and that amendment would be reviewed for potential environmental impacts as well as for compliance with NRC safety requirements.

Appendix B

B5.27.2 Impacts From Off-Normal Operations or Accidents

Comment: MR009-249
One commenter suggests the SEIS discussion of dose on page 4-73, line 29, should be considered in relation to the 40 CFR Part 190 annual limit of 25 mrem from airborne particulate radioactivity.

Response: This section of the SEIS is addressing potential impacts from possible accident scenarios, in particular a thickener failure and spill of yellowcake slurry. As indicated in 40 CFR Part 190, the EPA standard of 25 mrem applies to any member of the public as the result of exposures to planned discharges of radioactive materials to the general environment from uranium fuel cycle operations. The dose limit regulations discussed in the SEIS apply for the accident analyses and are not planned discharges; therefore, no change to the SEIS has been made in response to this comment.

B5.27.3 General

Comment: MR009-014
One commenter suggested adding the statement that the SEIS evaluates potential public health and safety impacts in addition to the potential environmental impacts mentioned on page xv, line 8.

Response: The NRC acknowledges the comment, and has amended the SEIS text appropriately to include that this draft SEIS evaluates the potential environmental and public health and safety impacts of the proposed action and the No-Action alternative.

Comments: MR009-078; MR009-094; MR009-148; MR009-239; and MR013-020
Two commenters suggested that the SEIS should be more specific about the technologies and processes that are employed at ISL facilities that provide additional protection of public and occupational health and safety. A specific example cited in the comments is downflow IX columns, which provide additional protections by limiting or eliminating potential public and worker exposure to radon gas.

Response: This type of equipment is discussed in the SEIS, as well as the GEIS, and is part of the analysis in which the radiological impacts to the public and workers are evaluated as SMALL. Because these topics are already addressed and the impacts are classified as SMALL, no changes were made to the SEIS in response to this comment.

Comment: MR009-248
One commenter suggested additional descriptions of the systems in place to mitigate accidents such as spills at ISR facilities.

Response: The primary references used in the SEIS, such as the GEIS and Mackin, et al, 2001, go into detail on the accident analysis that is summarized in the SEIS. Therefore, no changes were made to the SEIS in response to this comment.

Appendix B

Comments: MR009-240; MR009-241; MR013-022
One commenter suggested the NRC staff discussion of radiation protection issues should reference comparisons of potential radiation dose to natural background and should not be limited to comparison to NRC dose limits.

Response: Since NRC dose limits are well below natural background levels of radiation, a comparison of public dose from an ISL facility that is generally well below dose limits would be even further below natural background levels of radiation. The public generally perceives a marked difference from radiation exposure from man-made sources than that from natural background radiation levels. NRC requires that worker and public radiation doses be quantified as effective dose equivalent in millirem per year, which is intended to normalize doses by the expected health risk. This is achieved for different types of radiation and different body tissues by using weighting factors for radiation (alpha, beta, gamma, neutrons) and for body tissues (bone marrow, reproductive organs, lens of the eyes) to convert the radiation absorbed by a person to a common scale (in units of millirem) for determining compliance with NRC radiation protection requirements and for assessing the potential for harm or detriment. When this is accomplished, if a person is exposed to the same dose from background radiation or from releases from ISL facilities, there is no difference in the expected health effects. NRC staff understand that members of the public can perceive involuntary man-made risks as more hazardous than voluntary natural risks. Because the SEIS discussion is considered appropriate, no changes were made to the SEIS beyond the information provided in this response.

Comment: MR009-251
One commenter suggested that hydrochloric acid be added to the list of hazardous chemicals proposed for use at this site.

Response: Based on information provided by the applicant in an RAI response (Uranium One, 2009), hydrochloric acid and the associated protective provisions have been added to the SEIS.

Comment: MR009-258
One commenter suggested the SEIS discussion on page 4-81, line 21, should address occupational exposure at such a well site per NRC recent policies and statements on Part 40.32(e) prelicensing site construction authorizations and Part 20.2002 dose assessments.

Response: Occupational exposure from all operations at the Moore Ranch facility is addressed in detail in previous sections of the SEIS and by reference to the GEIS; therefore, no additions have been made to the SEIS beyond this comment response.

Comment: MR009-259
One commenter noted that data from Section 4.14.1.2 regarding calculations of radiation exposure resulting from deep well injections was incorrect and needed to be updated.

Response: The sentence noted by the commenter has been removed from the SEIS (new Section 4.14.1.1.2) because the calculations were irrelevant to the conclusion. By definition, the WDEQ could not issue a permit for Class I injection if a complete exposure pathway existed that could result in public consumption. If deep well disposal is conducted in accordance with applicable UIC regulations, this type of disposal of liquid effluent would be protective of human health and the environment. Radiation doses to the public would be expected to be near

*zero (due to an incomplete exposure pathway) and well below the public limit of 1 mSv
[100 mrem] per year.*

Comment: MR020-055
One commenter recommended the statement on SEIS, page 5-18, that states, "Because hazardous and radioactive wastes are closely monitored throughout the United States, the potential impact from these activities would be expected to be SMALL" should be deleted. This is suggested because the commenter says not all hazardous and radioactive wastes are closely monitored and that monitoring does not necessarily completely guard against the occurrence of accidents.

Response: *The NRC acknowledges the comment, and the SEIS has been modified for clarification.*

B5.27.4   References

40 CFR Part 190. *Code of Federal Regulations*, Title 40, *Protection of Environment*. Part 190, "Environmental Radiation Standards."

Argonne National Laboratory, 1989. "MILDOS-AREA (Computer Code)-Calculation of Radiation Dose from Uranium Recovery Operations for Large-Area Sources." Argonne, Illinois: Argonne National Laboratory.

EMC (Energy Metal Corporation US), 2007 "Application for USNRC Source Material License, Moore Ranch Uranium Project, Campbell County, Wyoming, Technical Report." Casper, Wyoming: Uranium 1 Americas Corporation. ADAMS Accession Nos. ML072851222, ML072851258, ML072851259, ML072851260, ML072851268, ML072851350, ML072900446. October 12, 2007.

Mackin, P.C., D. Daruwalla, J. Winterle, M. Smith, and D.A. Pickett, 2001. NUREG/CR–6733, "A Baseline Risk-Informed Performance-Based Approach for *In-Situ* Leach Uranium Extraction Licensees." Washington, DC: NRC. September.

NRC (U.S. Nuclear Regulatory Commission), 2009. NUREG–1910, "Generic Environmental Impact Statement for *In-Situ* Leach Uranium Milling Facilities." Washington, DC: NRC. May.

NRC, 1987. "Regulatory Guide 3.59, Methods for Estimating Radioactive and Toxic Airborne Source Terms for Uranium Milling Operations." Washington, DC: NRC.

Uranium One (Uranium One Americas), 2009. "Responses to Request for Additional Information for the Moore Ranch *In-Situ* Uranium Recovery Project License Application (TAC JU011)." ADAMS Accession No. ML092450317. August 31.

Appendix B

B5.28 Waste Management

B5.28.1 General Waste Management Comments

Comment: MR009-091
One commenter pointed out that injection and production feed lines are discussed in the SEIS as surface equipment, but feed lines are actually buried.

Response: The NRC acknowledges the comment, and the SEIS has been modified to address this inconsistency.

Comment: MR009-261
One commenter suggested that the applicant has committed to having an agreement for disposal of byproduct material in place before operations, not before construction, as stated in the SEIS.

Response: The NRC staff acknowledge that the suggested change is consistent with NRC requirements with respect to byproduct material disposal, as stated in SEIS Section 4.14, and the SEIS has been modified.

Comments: MR015-003; MR015-013
One commenter was concerned with the narrow range of waste disposal alternatives and limited discussion regarding waste management impacts in the SEIS.

Response: NRC staff have reviewed the discussions of waste disposal alternatives and waste management impacts in the SEIS. The discussions in SEIS Chapters 2, 3, and 4 have been modified to include additional discussion of wastewater disposal options and provide more detailed discussion regarding waste management impacts.

Comments: MR018-049; MR018-050; MR018-053; MR018-054; MR018-067; MR018-074; MR018-076; MR018-078; MR018-081
One commenter indicated that only estimates of solid wastes for the operations phase are provided and that estimates for all phases are needed to determine if adequate landfill capacity exists.

Response: NRC staff has reviewed the discussions of waste management in the SEIS. As discussed in revised SEIS section 3.13.2, proposed activities during operations would annually generate approximately 1,530 m^3 (2,000 yd^3), and decommissioning activities would generate about 45,500 m^3 (59,500 yd^3) of solid waste (i.e., non-radioactive solid waste [general trash], construction and demolition debris, or byproduct material that complies with NRC unrestricted release limits). This estimated range applies to the life of the project. The discussion in SEIS section 3.13.2 has been modified, based on information from the applicant, to address the capacity and projected life of the City of Casper landfill, located in Casper, Wyoming. This landfill has a permitted capacity of 317,000,000 m^3 (414,000,000 yd^3).

Comments: MR018-056; MR018-097
A commenter noted that the Wyoming Solid Waste Program encourages applicants to consider developing on-site recycling plans during the construction, operation, restoration, and decommissioning phases of facilities. The commenter also noted that a solid waste permit may

Appendix B

be required, depending on the volume and location of solid waste accumulated on-site before transportation to a disposal facility.

Response: The NRC staff acknowledge and appreciates the commenter encouragement of applicants to use recycling to reduce waste management impacts and notes that the applicant is required to coordinate waste management activities in accordance with applicable State and Federal laws. No modifications have been made to the SEIS beyond this comment response.

Comment: MR018-068
One commenter requested that Table 1-2, showing the waste management impact as SMALL, be updated because waste volumes were unknown.

Response: Based on the NRC evaluation of the waste management issue, there is sufficient information about the capacity and projected life of landfills in the area to determine that the impact is accurately stated as SMALL. No changes were made to the SEIS in response to this comment.

B5.28.2 Scope of the Assessment of Waste Management Impacts

Comments: MR009-041; MR020-013
Two commenters noted that during decommissioning, the potential exists for some equipment, materials, and buildings to require disposal as byproduct material or hazardous waste.

Response: The NRC acknowledges the comments, and the SEIS has been modified to reflect these potential waste management options.

Comment: MR015-011
One commenter was concerned with the lack of estimates of the amount of wastewater that would be generated by the project.

Response: SEIS, fig. 2-5, shows a water balance for the proposed project. Figure 2-5 contains estimates of 114L/min [30 gal/min] for the maximum flow rate of production bleed and 38 L/min [10 gal/min] of miscellaneous plant waste water that includes plant wash-down water and bleed stream from the elution and precipitation circuits during operations. During groundwater restoration, discharge from IX or reverse osmosis processes, or both, would increase by an estimated 380 L/min [100 gal/min]. SEIS Section 2.1.1.1.6.2 has been revised to include these liquid waste estimates.

Comment: MR020-003
One commenter wanted to know how waste management impacts were quantified or what thresholds were designated for the impacts classifications.

Response: SEIS Section 4.14 discusses waste management impacts. NRC did not designate quantitative thresholds for the various impact classifications. In the SEIS, waste management impacts were often assessed in terms of the capacity or availability of treatment or disposal facilities. SEIS Sections 2.1.1.1.6 and 3.12 discuss the expected amounts of various wastes generated by the proposed action. Because the SEIS discussion is considered appropriate, no changes were made to the SEIS beyond the information provided in this response.

Appendix B

B5.28.3 Characteristics of Wastes Generated by ISL

Comment: MR009-095
One commenter expressed concern that the liquid wastes section of the SEIS does not contain any discussion of brine from the reverse osmosis system, which is one of the most significant sources of liquid waste.

Response: The NRC staff acknowledge this concern. The SEIS has been modified to include a discussion of liquid wastes generated during groundwater restoration.

Comments: MR009-260; MR013-014
One commenter asserts that the discussion of waste classification within the SEIS is unclear. The NRC should adopt the format espoused in the Generic Environmental Report (GER) submitted by the National Mining Association. The GER used a format with radiological and non-radiological byproduct material and liquid and solid byproduct material. Another commenter suggested that the final SEIS be reformatted so that members of the public and interested stakeholders clearly understand the difference between wastes at ISL facilities that are classified as byproduct material and nonradiological wastes. The second commenter also noted that the staff should follow the format presented in the GER issued by National Mining Association.

Response: The NRC staff have revised the text in Chapters 2, 3, and 4 of this SEIS in a manner similar to that proposed by one of the commenters to clarify the description of wastes generated by the proposed action. As indicated in the revised Sections 2.1.1.1.6, 3.13, and 4.14.1, the proposed action would generate liquid and solid byproduct material, as well as other hazardous and nonhazardous liquid and solid wastes. As stated in these sections and in Section 4.2.1 of the NRC SER, all liquid byproduct material, whether radiological or not, is proposed for disposal via state-permitted well injection. Remaining liquid wastes consist of standard sanitary wastewater and uncontaminated well development and well test water. Also, as stated in these sections, solid byproduct material would be disposed of either at a site licensed to receive such waste or at a municipal waste disposal site, if the waste meets NRC unrestricted release criteria. Since liquid byproduct material would not be segregated according to its radiologic content, the NRC staff do not believe it is necessary to further distinguish between radiological and nonradiological byproduct material.

Comment: MR018-019
One commenter indicated that groundwater sweep could generate more liquid waste than production bleed.

Response: The NRC acknowledges that this may be true. The SEIS was modified in response to this comment.

Comments: MR009-098; MR018-055; MR018-057; MR018-058; MR018-070
One commenter was concerned that the SEIS description of handling and disposal of hazardous wastes was not consistent with pertinent local, State, and Federal regulations. Another commenter noted that the SEIS should be more specific about waste disposal locations.

Response: The NRC staff has reviewed the discussions of solid waste management in the SEIS. Sections 2.1.1.1.6, 3.13.2, and 4.14.1 of the SEIS have been modified with more explicit

Appendix B

statements with respect to types of wastes generated, proposed disposal locations, and compliance with pertinent State and Federal regulations governing hazardous waste handling and disposal.

Comments: MR018-060; MR018-061; MR018-062; MR018-064; MR018-065; MR018-069; MR018-072; MR018-077; MR018-080
One commenter indicated that the volume of solid byproduct material generated and the capacity of the solid byproduct material disposal sites being considered were needed to determine waste management impacts.

Response: The NRC staff have reviewed the discussion regarding solid byproduct material management in the SEIS. The discussion in SEIS Section 3.13.2 identifies the Pathfinder Mines Corporation Shirley Basin site as a potential facility for the disposal of byproduct material. In addition, the discussion of waste management impacts in the SEIS indicates that NRC requires that an applicant have a byproduct material disposal agreement in place prior to operations. This agreement would include byproduct material generated throughout the life of the project, including decommissioning. The applicant is currently negotiating an agreement with Pathfinder Mines Corporation. The applicant has also identified the Energy Solutions disposal site in Clive, Utah as an alternate disposal location and is negotiating a draft disposal agreement with Energy Solutions. The environmental impacts of disposing a specified amount of byproduct material at any potential byproduct material disposal facility would be covered in the environmental impact statement or environmental assessment as part of the licensing of that disposal facility. The evaluation of the environmental impacts on the disposal facility is beyond the scope of this document, but is evaluated as part of the licensing process for the disposal facility. As stated throughout the SEIS, the local environmental impacts that result from the disposal of solid byproduct material would be small. Consequently, no modifications have been made to the SEIS beyond this comment response.

Comment: MR018-071
A commenter indicated that uncontaminated solid waste was not clearly defined and required clarification.

Response: The NRC staff have reviewed the discussion of uncontaminated solid waste in the SEIS. The final SEIS revised the discussion in Section 2.1.1.1.6.3 to clarify the different types of solid wastes that would be generated by the proposed action.

Comments: MR018-073; MR020-016
Multiple commenters indicated that used oil storage and spent battery generation and disposal are regulated as both hazardous and solid waste by the State of Wyoming and may require a permit.

Response: NRC staff have reviewed the discussion regarding used oil and spent batteries in the SEIS. The SEIS states that the applicant would develop management plans to meet the WDEQ regulatory requirements. Wyoming has primacy in the management of solid and hazardous wastes at the site, and Wyoming regulations would be the compliance standard. The applicant has proposed to dispose of both its hazardous and nonhazardous waste at the Casper, Wyoming landfill and special waste and diversion facility. The SEIS text in Sections 2.1.1.1.6.3, 3.13.1, and 4.14.1 has been modified to include additional information concerning waste disposal.

Appendix B

Comment: MR020-004
A commenter requested clarification of the term "other solid wastes" and noted that some construction materials, such as organic solvents, paints, used oil, and paint thinners, may be classified as hazardous wastes subject to regulation under *Resource Conservation and Recovery Act*.

Response: *The Executive Summary and Sections 2.1.1.1.6 and 3.13 of the final SEIS have been modified to more clearly describe waste generation and disposal during the construction phase of the proposed Moore Ranch Project.*

B5.28.4   Waste Treatment and Disposal Methods

Comments: MR015-007; MR015-008; MR015-009; MR015-012
One commenter was concerned with the deep well disposal of liquid wastes because of the waste water composition (radioactive and nonradioactive components) and potential impacts to the receiving strata and other USDW.

Response: *Uranium One has identified deep well disposal as its preferred liquid waste disposal option. Under the SDWA, EPA was granted primary authority to regulate underground injection and protect current and future sources of drinking water. EPA implements this responsibility through its UIC program. EPA has authorized the State of Wyoming to administer the UIC programs in accordance with EPA regulations. NRC expects the applicant's compliance with these and any other applicable regulatory requirements. The applicant is responsible for obtaining authorization from the State for a Class I UIC Permit. Unless authorized by rule or by permit, any underground injection is unlawful and violates the SDWA and UIC regulations. Before an NRC-licensed uranium ISL facility can begin operations at any project site, the applicant must obtain the necessary UIC authorizations. The terms of the UIC permit would dictate the concentrations of components (radioactive and nonradioactive), and injection rates allowable for the proposed well. In the event that the applicant is unable to obtain the proposed Class I UIC permit, an amendment to its NRC license application would be required. Any license amendment request would be subject to a safety and environmental review by the NRC and subject to public comment. No modifications have been made to the SEIS beyond this response.*

Comment: MR020-017
One applicant pointed out that evaporation ponds used for storage of byproduct material (not currently considered for waste management) are considered a source of radon and subject to requirements of 40 CFR Part 61, Subpart W, and approval of construction is required under 40 CFR Part 61, Subpart A. These requirements should be included in the SEIS if evaporation ponds are included.

Response: *The NRC staff acknowledges these requirements associated with the use of ponds used for storage of byproduct material. The applicant is not proposing to use evaporation ponds. However, the staff has included a discussion of evaporation ponds and associated requirements as an option under the proposed action in Section 2.1.1.2.1 of the SEIS.*

Appendix B

B5.28.5 Regulation of Wastes and Disposal Methods

Comment: MR018-037
One commenter noted that the discharge of water from wells or pipelines (from hydrostatic testing or from well testing and purging) onto the land surface or into surface water channels requires a permit.

Response: The NRC staff acknowledge this comment. As stated in SEIS Section 4.5.2, well water would be discharged to the surface in accordance with approved permits from the State of Wyoming, which the applicant would obtain prior to any release. No further changes have been made to the SEIS beyond this response.

Comment: MR018-059
One commenter expressed concern that the SEIS contains no references to the role the State of Wyoming plays in authorizing additional byproduct material disposal facilities, if needed, and that the byproduct material is defined as solid waste by Wyoming statute and subject to state regulatory requirements.

Response: The NRC acknowledges the commenter's concern. Section 3.13.3 has been modified to include the permitted and current capacity of the applicant's proposed disposal site for this material (Pathfinder Mines Corporation Shirley Basin site). However, a discussion of construction and authorization of additional byproduct material disposal facilities goes beyond the scope of this document. Concerning the definition of byproduct as "solid waste" and the State of Wyoming regulatory authority thereof, the NRC agrees that such waste would be subject to Wyoming solid waste regulations if it meets NRC criteria for unrestricted release. However, byproduct material is regulated by NRC under 10 CFR Part 40 and is not "solid waste" according to 40 CFR 261.4(a)4. Because Wyoming is a non-agreement state, NRC retains jurisdiction over byproduct materials.

Comment: MR020-012
One commenter stated that the ISR facility may be subject to Emergency Planning and Community Right-to-Know Act and Toxic Substance Control Act and requested that the SEIS discuss the extent to which the ISR facility would comply with these regulations.

Response: SEIS Section 4.13.1.2.3 identifies the primary regulations applicable to the use and storage of chemicals and includes the topics mentioned by the commenter. NRC, though not the regulatory authority for either of these Acts, expects its licensees to comply with these and all other applicable regulations. SEIS Sections 1.5 and 1.6 discuss the role of other Federal, Tribal, State, and local agencies in regulating and permitting an ISR facility. Because the SEIS discussion is considered appropriate, no changes were made to the SEIS beyond the information provided in this response.

Comment: MR020-015
One commenter was concerned that the definition in the SEIS of conditionally exempt small quantity generator did not fully explain the requirement for this exemption, or the consequences if the site fails to meet the requirements.

Response: Section 2.1.1.1.6.4 of the SEIS has been modified to include the requirements for a conditionally exempt small quantity generator, as well as the consequences if the site fails to meet the requirements.

Appendix B

B5.28.6 References

10 CFR Part 40. *Code of Federal Regulations*, Title 10, *Energy*, Part 40, "Domestic Licensing of Source Material."

40 CFR Part 61. *Code of Federal Regulations*, Title 40, *Protection of Environment.* Part 61, "National Emission Standards for Hazardous Air Pollutants (NESHAPS)."

B5.29 Decommissioning

Comment: MR009-088
One commenter noted that the Moore Ranch SEIS discussion of the decommissioning process did not clearly indicate that a detailed decommissioning plan would need to be approved and the NRC license amended before decommissioning could begin.

Response: *NRC agrees with the comment and has revised SEIS Section 2.1.1.1.5 to state that NRC approval of the decommissioning plan is required prior to the start of site decommissioning.*

Comment: MR009-165
A commenter stated that NRC should note that the Moore Ranch license application included a detailed discussion of decommissioning and decontamination planning for review.

Response: *The comment is noted. Details regarding the planning activities for decommissioning and decontamination are addressed in Chapter 6 of the Moore Ranch SER. No new information was added to the final Moore Ranch SEIS beyond the information provided in this response.*

Comment: MR009-211
A commenter noted that the SEIS stated that the applicant should submit a reclamation plan for approval by the appropriate State and Federal agencies, but that the requirement was for the applicant to submit a decommissioning plan for NRC review and approval at least 12 months before final decommissioning commences.

Response: *The comment is noted. Section 4.6.1.4 of the final SEIS was revised in response to the comment.*

Comments: MR009-227; MR009-228
A commenter noted that the Moore Ranch draft SEIS stated, "Once project operations cease, the central plant and support structures would be decommissioned and removed." The commenter noted that the Moore Ranch license application had indicated the possibility that some structures could be decontaminated and released, if desired by the landowner and approved by the NRC. The commenter also noted that the SEIS stated, "Uranium One would submit a site reclamation plan to NRC in accordance with 10 CFR Part 40 before the license was terminated." The commenter noted, per 10 CFR Part 40, that Uranium One would submit a decommissioning plan to the NRC 12 months before final site decommissioning commences.

Response: *The comments are noted. Section 4.10.1.4 of the final SEIS was revised in response to these comments.*

Appendix B

B5.33.1 Reference

10 CFR Part 40. *Code of Federal Regulations*, Title 10, *Energy*, Part 40, "Domestic Licensing of Source Material."

B5.30 Cumulative Effects

B5.30.1 General Comment: SEIS Does Not Adequately Address Cumulative Effects

Comments: MR001-003; MR005-003; MR012-065; MR012-066; MR012-072; MR014-002; MR016-017; MR016-018; MR017-032

Multiple commenters expressed concern that the SEIS does not adequately address cumulative effects. For example, several commenters noted that the SEIS provides a list of other EISs prepared by different agencies, but with no associated meaningful analysis. Other commenters expressed concern that cumulative impacts were presented as conclusory statements with an inadequate basis. Another commenter expressed concern that the SEIS only considered Federal actions in the cumulative effects analysis. One comment noted that the SEIS does not provide an adequate discussion of potential cumulative impacts from nearby uranium and oil and gas activity. Also, another commenter noted that the cumulative effects analysis in the SEIS was not transparent and was not developed with sufficient public input.

Response: The NRC staff believe that the information presented in Chapter 5 of the SEIS is valid and relevant to the assessment of potential cumulative effects. Mitigation measures are described throughout Chapters 4 and 5 of the SEIS, and additional monitoring measures are described in Chapter 6. The cumulative effects analysis presented in Chapter 5 of the SEIS has been revised to improve the transparency and clarity of the analysis and provide a more detailed discussion of potential cumulative effects for critical resource areas, such as Land Use (SEIS Section 5.2), Groundwater (SEIS Section 5.5.2), Ecological Resources (SEIS Section 5.6), Air Quality (SEIS Section 5.7), and Socioeconomics (SEIS Section 5.11).

Comments: MR012-056; MR012-058; MR012-062; MR012-063; MR012-064; MR016-019; MR016-020

Commenters expressed concern that the cumulative effects analysis presented in the GEIS (NRC, 2009) was inadequate and was used to constrain the scope of the cumulative effects analysis in the SEIS. For example, one commenter noted that the SEIS does not consider the cumulative impacts of past uranium mining and milling combined with the current project. One commenter noted that the GEIS deferred conclusions on the potential cumulative impacts to the site-specific SEIS. Because the site-specific cumulative effects analysis presented in the SEIS is based heavily on information presented in the GEIS, the commenter concluded that the SEIS does not address the NEPA requirements with respect to cumulative impacts.

Response: The relationship between the GEIS and the site-specific SEIS is described in SEIS Section 1.4.1. Revisions to the GEIS are beyond the intended scope of the public comment process associated with the SEIS. The NRC staff believe that the cumulative impacts from past uranium mining and milling projects have been adequately considered. All known uranium recovery sites in the vicinity of Moore Ranch have been listed in Table 5.1. Locations of nearby uranium recovery projects are illustrated in Figures 5.1 and 5.2, while Figure 5.3 shows other energy development projects within a 50 mile radius of the proposed project site. The NRC staff

Appendix B

believe that the information presented in SEIS Section 5 is valid and relevant to the assessment of potential cumulative effects. The cumulative effects analysis presented in SEIS Section 5 has been revised to improve the transparency and clarity of the analysis and provide a more detailed discussion of potential cumulative effects for critical resource areas, such as Land Use (SEIS Section 5.2), groundwater (SEIS Section 5.5.2), Ecological Resources (SEIS Section 5.6), Air Quality (SEIS Section 5.7), and Socioeconomics (SEIS Section 5.11). The NRC staff believe that NEPA requirements have been adequately addressed.

B5.30.2 Past, Present, and Reasonably Foreseeable Future Actions

Comments: MR010-007; MR011-006
Two commenters stated that no studies have been conducted to identify the cumulative effects of locating multiple ISR facilities closely together.

Response: The cumulative impacts analysis presented in Chapter 5 of the SEIS includes a discussion of past, present, and reasonably foreseeable future uranium recovery operations, both for conventional mining and milling and ISR technologies. All known uranium recovery sites in the vicinity of Moore Ranch have been listed in Table 5.1, which also includes the distance and direction from the proposed project site. These facilities are regulated by NRC, and the potential environmental impacts from these facilities are (or will be) evaluated in accordance with NRC NEPA requirements in 10 CFR Part 51. In addition, the cumulative effects analysis has been revised to improve the clarity and transparency of how past, present, and reasonably foreseeable future actions relating to uranium recovery were considered.

Comments: MR012-067; MR012-068; MR012-069; MR012-070; MR012-071; MR017-031
Multiple commenters expressed concern over the possible cumulative effects that could result from past, present, and reasonably foreseeable future actions associated with other resource extraction operations in the Powder River Basin, such as CBM production, oil and gas production, and coal mining. For example, several commenters noted that the SEIS should include a disclosure of the types and amounts of contaminants released from CBM operations into aquifers and surface waters, and provide a detailed analysis of the incremental impacts from the Moore Ranch ISR. Other commenters stated that the SEIS should include an analysis of the potential for cross-contamination from wells associated with CBM, oil and gas, and coal mining operations.

Response: Potential impacts to groundwater resources and the effects of waste management practices at the site are described in Sections 4.5.2 and 4.14 of the SEIS. Chapter 5 of the SEIS includes a discussion of other past, present, and reasonably foreseeable future actions associated with resource extraction in the Powder River Basin. In addition, as described in Section 1.7 of the SEIS, NRC has entered into an MOU with BLM to keep current on issues that develop with respect to these operations on public lands. The analysis presented in Section 5.5.2 of the SEIS has been revised to clarify the technical basis for potential cumulative impacts to groundwater resources in the vicinity of Moore Ranch.

Comments: MR012-074; MR012-077; MR017-029
Several commenters expressed concern that the cumulative impacts analysis presented in Chapter 5 of the SEIS did not consider impacts from past uranium mining or milling.
Response: The cumulative impacts analysis presented in Chapter 5 of the SEIS includes a discussion of past, present, and reasonably foreseeable uranium recovery operations, both for conventional mining and milling and ISR technologies. All known past and present uranium

Appendix B

recovery sites in the vicinity of Moore Ranch have been documented and listed in Table 5.1, which also includes the distance and direction from the proposed project site. NRC has regulatory authority for the radiological aspects of these facilities, and the potential environmental impacts from these facilities are (or will be) evaluated in accordance with NRC NEPA requirements in 10 CFR Part 51. In addition, the cumulative effects analysis has been revised to improve the clarity and transparency of how past, present, and reasonably foreseeable future actions relating to uranium recovery were considered.

Comments: MR015-020; MR015-021; MR015-022
Several commenters noted specific cumulative impacts from multiple ISR facilities with respect to the ambient air quality, including effects on NAAQS pollutants such as nitrogen oxides, particulate matter, and ozone. In addition, one commenter noted that the development of multiple ISR facilities could result in air emission levels that could adversely affect the Air Quality Related Values such as visibility in Class I and sensitive Class II areas.

Response*: Ambient air quality is discussed in Section 3.7 of the SEIS, and potential impacts to air quality are discussed in Section 4.7. Visual resources are discussed separately in Sections 3.10 and 4.10 of the SEIS. These sections have been revised to address recent changes in the air quality requirements, and Section 5.7 incorporates these analyses by reference.*

B5.30.3 Specific Document Changes or Action Requests

Comment: MR012-078
One commenter stated that the GEIS is the more appropriate document for conducting a cumulative impacts analysis. The commenter also suggested that NRC reissue the GEIS for public review and comment on its cumulative impacts analysis.

Response*: As described in Section 1.4.1 of the SEIS, the NRC staff believe that the SEIS is the appropriate way to update and supplement the environmental report in the GEIS. Specific revisions to the GEIS are beyond the intended scope of this environmental report. Because the relationship between the SEIS and GEIS and the tiering approach used are described in Section 1.4.1, no changes were made to the SEIS in response to this comment.*

B5.30.4 Significance

Comment: MR012-076
One commenter noted that the cumulative effects analysis presented in the GEIS (NRC, 2009) was inadequate, assuming that most site-specific cumulative impact analyses would require only a Level 1 or Level 2 analyses. The commenter expressed concern that this assumption was used to constrain the scope of the cumulative effects analysis in the SEIS.

Response*: Section 5.4 of the GEIS describes approaches to address cumulative impacts in a site-specific environmental impact statement. These approaches are based on cumulative impacts assessment guidance developed by CEQ, providing examples and assumptions that the NRC staff might use to determine the appropriate level of detail to analyze the potential cumulative effects for a given resource area. The purpose of the information in the GEIS is to outline one methodology that may be used in conducting cumulative effects analysis. It does not to prescribe a particular approach, nor does it presume a particular outcome (e.g., an impact*

Appendix B

significance level) in the site-specific environmental impact statement. In the examples given in the GEIS, a relatively lower level of detail (Level 1) might be, but not necessarily would be, applied to analyze cumulative impacts for a resource area. The relationship between the GEIS and the site-specific SEIS is described in Section 1.4.1 of the SEIS. Revisions to the GEIS are beyond the intended scope of the public comment process associated with the SEIS. The NRC staff believe that the information presented in Chapter 5 of the SEIS is valid and relevant to the assessment of potential cumulative effects. The cumulative effects analysis presented in Chapter 5 of the SEIS has been revised to improve the transparency and clarity of the analysis and provide a more detailed discussion of potential cumulative effects for critical resource areas such as groundwater and land use.

Comments: MR017-010; MR017-011

Two commenters addressed the potential significance of cumulative impacts associated with spills, noting that even small spills might be cumulatively significant for soil and groundwater resources. In addition, the commenter noted that the GEIS identified spills as occurring at ISR facilities but does not assert that these spills are necessarily cleaned up promptly.

Response*: These comments address the discussion of spills contained in the GEIS rather than in the SEIS. GEIS Sections 2.11.2, 4.2.3 and 4.2.4 correctly characterize the NRC approach to regulate and inspect ISR facilities and implement corrective actions to minimize both the likelihood of and the impacts from unplanned spills. Licensees are required to develop and implement a spill control plan prior to beginning uranium recovery operations. NRC ensures these procedures are correctly implemented through its inspection and enforcement process. GEIS Section 2.11.2 also discusses historical information with respect to spills at NRC-regulated ISR facilities. No changes were made to the SEIS in response to the comments.*

Comment: MR017-030

One commenter requested greater detail about how NRC determined the magnitude of cumulative impacts. Specific issues raised include whether cumulative impacts were evaluated on both a geographic and temporal scale, and whether groundwater impacts could be classified as large because the groundwater in the mining areas would never be the same.

Response*: The NRC staff believe that the information presented in SEIS Section 5 is valid and relevant to the assessment of potential cumulative effects. Section 5.1.2 identifies the temporal scale as being from 2007 to 2020. This period represents the time that NRC initially received the license application from the applicant (2007) through expected license termination (2020). The geographic scale varies by resource category and is clearly identified for each resource throughout SEIS Section 5. The cumulative effects analysis presented in SEIS Section 5 has been revised to improve the transparency and clarity of the analysis, including a more detailed discussion of how impact significance was determined for potential cumulative effects for critical resource areas, such as Land Use (SEIS Section 5.2), Groundwater (SEIS Section 5.5.2), Ecological Resources (SEIS Section 5.6), Air Quality (SEIS Section 5.7), and Socioeconomics (SEIS Section 5.11).*

Comments: MR018-063; MR018-084

The commenter (WDEQ) reviewed the draft SEIS and two other pending ISR NRC draft SEISs (i.e., Nichols Ranch, Lost Creek) and estimated the volume of byproduct material that would be generated from the three proposed facilities. The commenter estimate was based on the reported waste volumes for operations and aquifer restoration from the draft SEIS and the decommissioning waste volume reported in the GEIS (6008 yd^3, based on adding information in

Appendix B

Table 2.6-1). NRC would require any solid byproduct material generated from the proposed facilities to be disposed at a licensed facility. The commenter estimate of the cumulative solid byproduct material from the three proposed ISR facilities was 16,067 m3 [21,000 yd3]. The commenter expressed a concern that if this volume of material were disposed in Wyoming, there could be a large impact.

Response: *An important aspect of the NRC staff evaluation of potential waste management impacts is the availability of disposal capacity. As discussed in the GEIS, NRC requires an ISR facility to have an agreement in place with a licensed disposal facility to accept byproduct material that would be associated with facility operations, aquifer restoration, and decommissioning. Such agreements ensure that sufficient disposal capacity for byproduct material would be available throughout the life of the facility.*

As discussed in draft SEIS Section 2.1.1.1.6.3, the applicant does not presently have an agreement in place with a licensed site to accept their solid byproduct material for disposal. The applicant preferred destination for disposal of byproduct material is at the Pathfinder-Shirley Basin site in Mills, Wyoming. If that facility does not have sufficient capacity at the time the request for an agreement is made, then the applicant could engage other low-level radioactive waste disposal facilities that are licensed to accept byproduct material. Another existing facility that is licensed by NRC to accept byproduct material for disposal is the Rio Algom Ambrosia Lake uranium mill tailings impoundments near Grants, New Mexico. Additionally, three sites are licensed by NRC Agreement States to accept byproduct material for disposal (i.e., the EnergySolutions site in Clive, Utah; the White Mesa uranium mill site in Blanding, Utah; and the Waste Controls Specialists site in Andrews, Texas).

At the time of this writing, NRC has received no proposals to expand byproduct material disposal capacity in Wyoming. As discussed in the GEIS (Section G5.32.2), proposals for onsite disposal of byproduct materials at locations without available disposal capacity are uncommon, but if such proposals were received by NRC, they would be evaluated on a case-by-case basis against criteria in 10 CFR Part 40, Appendix A. NRC would evaluate the potential environmental impacts of any such proposals if and when they are received. Based on the disposal options currently available and the disposal agreement that NRC requires prior to operations, the NRC staff continue to conclude that the potential waste management impacts associated with the generation of byproduct material would be SMALL.

In response to this and other comments, the NRC staff reviewed the draft SEIS and, as the commenter has also noted, found the reported facility waste volumes did not include estimates of decommissioning byproduct material. The NRC staff then calculated the amount of solid byproduct material that could be generated from decommissioning activities based primarily on information provided in the applicant surety estimate (EMC, 2007). The calculation results for the Moore Ranch proposal were added to the SEIS Section 2.1.1.1.6.3 discussion of) byproduct material expected to be generated by the proposed action. The estimates were also added to the revised waste management cumulative effects analysis and discussion in Section 5.14.

Appendix B

B5.34.5 Other

Comment: MR012-020
A commenter noted that the cumulative impacts analysis in the SEIS should include consideration of the ISR facility on climate change, and also related effects that climate change might have on the groundwater supply for the region.

Response: *EPA issued regulations for inventorying greenhouse gas emissions on October 20, 2009, and on February 18, 2010. After the draft SEIS was published for comment, CEQ issued draft guidance to agencies on the consideration of the effects of climate change and greenhouse gas emissions in the context of NEPA environmental reports. NRC is currently evaluating the best approaches for how it will address these recent developments with respect to climate change and greenhouse gas emissions while meeting its responsibilities under NEPA. The SEIS has been updated to reflect these recent developments.*

Comment: MR012-075
One commenter expressed concern that the cumulative effects analysis presented in the GEIS (NRC, 2009) effectively pre-determined to what extent cumulative impacts would be analyzed in the SEIS.

Response: *Section 5.4 of the GEIS describes approaches to addressing cumulative impacts in a site-specific environmental impact statement. These approaches are based on cumulative impacts assessment guidance developed by CEQ, providing examples and assumptions that the NRC staff might use to determine the appropriate level of detail to analyze the potential cumulative effects for a given resource area. The purpose of the information in the GEIS is to outline one methodology that may be used to conduct cumulative effects analyses. It does not to prescribe a particular approach, nor does it presume a particular outcome (e.g., an impact significance level) in the site-specific environmental impact statement. The relationship between the GEIS and the site-specific SEIS is described in Section 1.4.1 of the SEIS. Revisions to the GEIS are beyond the intended scope of the public comment process associated with the SEIS. The NRC staff believe that the information presented in Chapter 5 of the SEIS is valid and relevant to the assessment of potential cumulative effects. The cumulative effects analysis presented in Chapter 5 of the SEIS has been revised to improve the transparency and clarity of the analysis and provide a more detailed discussion of potential cumulative effects for critical resource areas such as groundwater and land use.*

Comment: MR014-003
One commenter noted that the SEIS does not provide a discussion of mitigation measures that might reduce potential cumulative impacts associated with nearby uranium and oil and gas production.

Response: *Mitigation measures are described throughout Chapters 4 and 5 of the SEIS, and additional monitoring measures are described in Chapter 6. SEIS Chapter 5 has been revised to provide a more clear and transparent evaluation of the potential cumulative impacts from nearby production of coal, oil and gas, CBM, and uranium.*

Appendix B

B5.30.6 References

10 CFR Part 51. *Code of Federal Regulations*, Title 10, *Energy*, Part 51, "Environmental Protection Regulations for Domestic Licensing and Related Regulatory Functions."

EMC (Energy Metals Corporation US), 2007. "Application for USNRC Source Material License Moore Ranch Uranium Project Campbell County, Wyoming." Technical Report, Volume 3, Appendix D. Casper, Wyoming: EMC. September.

NRC (U.S. Nuclear Regulatory Commission), 2009. NUREG–1910, "Generic Environmental Impact Statement for *In-Situ* Leach Uranium Milling Facilities." Washington, DC: NRC. May.

B5.31 Environmental Justice

Comment: MR020-052
One commenter stated that NRC policy recommends that the minority and low-income proportions in the impact area be the county and state proportions. The commenter further states the SEIS claims that the state was selected as the area for comparison, but that the SEIS later correctly notes the comparison with both the state and counties. The commenter requests that the second paragraph should be modified to reflect this.

Response: *The commenter refers to text in Chapter 4, Section 4.12.1 that describes the methodology used to conduct the EJ impact assessment. NRC has revised the final SEIS in response to this comment.*

Comment: MR020-050
One commenter expressed concern that the SEIS gave an insufficient justification for varying from NRC policy in defining the impact area of the project. The commenter states that the impact area of the project is too broad for the environmental justice analysis and that the data is misleading. The commenter further states that low-income and minority individuals may reside in a community very near a project site and that the analysis would skew this by analyzing a broad impact area.

Response: *The commenter refers to text in Chapter 4, Section 4.12 that describes the deviation from the NRC Policy Statement on Environmental Justice (69 FR 5240). NRC has revised the final SEIS in response to this comment.*

Comment: MR020-051
One commenter states that the SEIS briefly discusses the distribution of minority individuals in the area but insufficiently to conclude that minority communities are not disproportionately impacted by the project.

Response: *The comment refers to text in Chapter 4, Section 4.12.1 that describes the methodology used to conduct the environmental justice impact assessment. NRC has revised the final SEIS in response to this comment.*

Comments: MR020-053; MR020-054
A commenter expressed concern that Tribal cultural resources are not discussed in the environmental justice analysis. The commenter requested that a reference to the cultural

Appendix B

resource analysis be added to the environmental justice analysis. The commenter also stated that there is no indication that Tribal subsistence resources were considered in the cultural resources section, and those resources need to be analyzed in the SEIS.

Response: Subsistence (consumption) hunting or gathering practices were not discussed in the Draft SEIS. NRC has assessed the potential impact to subsistence consumption behavior receptors and has revised the final SEIS in response to these comments.

B5.32   Best Management Practices

B5.32.1   Enforcement of Mitigation

Comment: MR012-038
One commenter stated that the classifying groundwater impacts from leaks and spills as SMALL is unjustified because this relies on the assumption that mitigation measures would be effective.

Response: As noted in SEIS Section 4.5.2., implementation of the required leak detection program and well MITs should mitigate the potential impacts from leaks and spills to shallow (near surface) aquifers and result in SMALL potential impacts. This impact conclusion is based on facility-specific process descriptions for the Moore Ranch Project and site-specific characteristics at the proposed site. In determining impact conclusions, the NRC staff reviewed information provided by the applicant in its license application as amended (including the technical and environmental reports), information and data independently collected by the staff, and information and data provided in the GEIS. Section 2.11 of the GEIS presents an historical discussion of ISL operations, and Section 2.14 provides reference to specific facilities in Wyoming, Nebraska, and New Mexico. The intent of the information in these sections of the GEIS was to inform the reader regarding which issues have historically resulted in potential impacts at ISL facilities and to provide a range of conditions that may be expected for each of the four ISL phases. No changes were made to the SEIS in response to this comment.

Comment: MR017-018
One commenter requested that NRC ensure mitigation measures be implemented by the applicant regarding protection for sage-grouse.

Response: NRC has made a reasonable effort to provide a discussion of potential mitigation measures that would limit impacts to wildlife and sage-grouse; however, NRC does not have the statutory authority to enforce mitigation measures. NRC recognizes that the proposed project may not meet all of the development recommendations established by the Wyoming Governor, BLM, and WGFD. NRC is not bound by the WGFD recommendations or BLM guidelines that the proposed action may not meet the published recommendations. The applicant may be required to consult with appropriate agencies in order to obtain permits or be required to develop a mitigation plan. Mitigation measures may be stipulations of other permitting agencies with statutory authority. NRC believes that the wildlife analyses are supported by sufficient technical bases whether tiered from the GEIS or based on supplemental staff analyses. Because the SEIS discussion is considered appropriate, no changes were made to the SEIS beyond the information provided in this response.

Appendix B

Comment: MR014-001
One commenter questioned why NRC would consider licensing a project in an unconfined [aquifer] location, unless it can demonstrate that the [potential] impacts to water resources can be prevented through "enforceable and effective mitigation measures."

Response: The commenter is correct that the production zone "70 sand" aquifer at Moore Ranch is an "unconfined aquifer." An "unconfined" aquifer is one where the water level is below the overlying aquitard as described in SEIS Section 4.5.2.1.2.2 and shown in Figure 4-1. The term "unconfined aquifer" does not mean that the overlying aquitard, which acts as the top confining layer to the production ore zone is missing. At the proposed Moore Ranch Project, the overlying aquitard that confines the top of the "70 sand" production zone is continuous across the entire site; however, the groundwater flow behavior of an "unconfined aquifer" is different than that of a "confined aquifer," which can impact the ability to prevent and control excursions. The applicant provided field test data and groundwater modeling to demonstrate it could monitor, prevent and capture excursions from the "unconfined aquifer" in the "70 sand" production zone.

Furthermore, the NRC SER will evaluate the potential for excursions and discuss the operational monitoring program that would consist of establishing monitoring well rings around each wellfield to monitor for both horizontal and vertical excursions, as discussed in Section 6.3.1.2 of the final SEIS. In addition, NRC will establish site-specific license conditions for the Moore Ranch ISR facility to address required applicant actions that would further demonstrate the ability to operate an ISR facility in the 70 sand "unconfined aquifer" with either no or minimal impacts to the environment. License conditions are enforceable, therefore, comparable to the "enforceable and effective mitigation measures" to which the commenter refers. No revisions were made to the SEIS beyond the information provided in this response.

Comment: MR018-018
A commenter stated that some of the measurements [depths below ground surface (bgs)] and quantities contradict information provided to the WDEQ in the permit application for the Class I injection well.

Response: NRC acknowledges that reported formation depths to the proposed injection zones may have been revised after NRC received the initial license application from the applicant. Sections have changed from the time the applicant first submitted license application (i.e., environmental and technical reports) to the NRC to the time the Class I disposal application was submitted. Sections 3.5.2.4 and 4.5.2.1.2.3 of the Moore Ranch SEIS have been revised to reflect the depth to potential aquifers based on current information.

B5.32.2 Completeness of the Mitigation Measures and Best Management Practices

Comment: MR009-203
A commenter stated that the mitigation measures presented in the applicant environmental report would be effective in minimizing the potential impacts of the proposed Moore Ranch ISR Project, and that if the NRC determined that additional mitigation measures would be required, then the specific requirements should be disclosed in the SEIS.

Appendix B

Response: Chapter 7 of the GEIS provided a general overview of the types of best management practices, mitigation measures, and management actions that have been used, historically, at ISR facilities to avoid or reduce potential environmental impacts. The NRC staff have been evaluating the adequacy of the proposed safety and environmental monitoring programs at the proposed Moore Ranch Project as part of both the environmental and safety reviews. For example, the seasonal noise guidelines developed by WGFD (WGFD, 2010) were identified as a means to mitigate the potential impact to avian species, as discussed in Section 4.6.1.1.2.3 of the final SEIS. NRC can only establish license conditions within the limits of authority granted by Congress and if a best management practice or mitigation measure became a license condition, they would become subject to NRC inspection and oversight at the Moore Ranch facility. No changes were made to the final SEIS beyond the information provided in this response.

B5.32.3 General Comments Related to Mitigation Measures and Best Management Practices

Comment: MR009-177
A commenter stated that the discussion of excursions and groundwater quality in Section 4.5.2 of the SEIS should have discussed MIT of injection wells for potential well casing failures.

Response: NRC agrees with the commenter and revised the referenced discussion in Section 4.5.2 to acknowledge the performance of MITs. Additionally, operators are required to demonstrate that no significant leaks exist prior to operation of a Class I injection well through an MIT and every 5 years after for the operational life of the well.

Comment: MR009-191
A commenter noted that different fencing methods could be used depending on the purpose of the fence.
Response: NRC agrees with the commenter and clarified the discussion in Section 4.6.1.1.2.1 of the final SEIS in response to the comment.

Comment: MR009-205
A commenter noted that there is no BLM surface ownership at the proposed Moore Ranch Project; therefore, there would not need to be BLM concurrence on seed mixtures used for reseeding.

Response: NRC agrees with the commenter and has revised the discussion in Section 4.6.1.2.1 of the final SEIS in response to the comment.

Comment: MR020-044
A commenter was concerned about air quality impacts from fugitive dust emissions during construction and operation of the proposed Moore Ranch Project and questioned who would develop and enforce a site-specific monitoring plan.

Response: NRC does not have the statutory authority to require compliance with a site-specific monitoring plan regarding fugitive dust emissions. The WDEQ permitting process would be the mechanism used to address air quality. Best management practices, such as the application of water to suppress fugitive dust emissions, have been proposed by the applicant.

Appendix B

Comment: MR009-039
A commenter noted that the applicant would use dust suppression techniques, when necessary, to mitigate air quality impacts but did not propose to use dust suppression as a mitigation technique for visual resource impacts.

Response: In the environmental report, the applicant identified that visual impacts would be mitigated by using harmonizing paint colors for wellfield structures, using existing topographic features, where possible, to screen visual impacts due to constructed features; aligning roads to follow existing topographic contours, where feasible; and removing construction debris from the area as soon as possible (environmental report, Section 5.9). The applicant stated that the fugitive dust arising from construction activities and vehicle traffic on the site access roads would have a negligible effect on air quality, and that dust suppression techniques, such as the application of water on unpaved roads, would further reduce the amount of dust produced. The SEIS was revised to clarify that mitigation measures for visual resource impacts did not specifically include dust suppression techniques.

Comments: MR009-169; MR009-170
A commenter clarified that the mitigation of soil erosion and revegetation of disturbed land areas would follow WDEQ-LQD rules and regulations and associated guidelines, not separate reclamation plans.

Response: The NRC acknowledges that the applicant committed to perform soil erosion mitigation practices in accordance with WDEQ-LDQ rules and regulations, and the applicant committed to conduct revegetation practices in accordance with WDEQ-LDQ regulations and guidelines. The SEIS was revised to clarify applicant commitments.

Comment: MR009-209
One commenter stated that mitigation measures discussed in Section 4.6.1.2.4 of the SEIS for threatened and endangered species do not correlate to those discussed in Section 5.5.5 of the applicant environmental report.

Response: NRC states in Section 4.6.1.2.4 that "Examples of mitigation are ..." related to spill procedures and fencing. NRC references two documents, the applicant environmental report and the NRC summary of consultations with stakeholders. Spill procedures and fencing may have been discussed during the NRC consultations; however, they are not included in the summary document. Therefore, NRC removed the reference to the NRC consultation summary memo from the SEIS. The reference to the environmental report does not specify a particular section of the environmental report where examples of mitigation are located. The applicant discusses mitigation measures in Chapter 5 of the environmental report, and specifically for ecological resources in Section 5.5 of the environmental report. The applicant specifically identifies timing restrictions as a mitigation measure for reducing potential impacts to threatened and endangered species.

As clearly stated in the SEIS, examples of mitigation measures are discussed in Section 4.6.1.2.4, which does not preclude other mitigation measures that may indirectly benefit a resource area. Because spill response procedures and fencing are mitigation measures discussed in the reference environmental report that could benefit threatened and endangered species, NRC has retained the discussion in Section 4.6.1.2.4. Because the environmental report specifies timing restrictions, NRC has included timing restrictions in the SEIS as part of

Appendix B

the discussion. These changes clarify that NRC provides examples of mitigation measures to reduce potential impacts to threatened and endangered species; however, other mitigation measures may be used.

B5.32.4 References

NRC (U.S. Nuclear Regulatory Commission), 2009. NUREG–1910, "Generic Environmental Impact Statement for *In-Situ* Leach Uranium Milling Facilities." Washington, DC: NRC. May.

WGFD (Wyoming Game and Fish Department), 2010. "Recommendations for Development of Oil and Gas Resources within Important Wildlife Habitats." Version 6.0. Cheyenne, Wyoming: WGFD. April.

B5.33 Monitoring

Comment: MR009-034
A commenter stated that no long-term monitoring activities have been proposed for the Moore Ranch Project, and that the site would be released with no restrictions following license termination by the NRC.

Response: *NRC agrees with the commenter. Following completion of license termination activities at ISR facilities, there are no restrictions on site use and there are no long-term monitoring activities. Page xxii of the Moore Ranch final SEIS was revised in response to this comment.*

Comment: MR009-075
A commenter stated that the purpose of the ore zone restoration wells is to establish baseline water quality information and to determine whether restoration standards have been met after restoration. The commenter noted that the ore zone restoration wells are not used to detect excursions.

Response: *NRC agrees with the commenter. The only monitoring wells used to detect horizontal excursions are the perimeter monitoring wells. The monitoring wells drilled within the ore zone are used to collect baseline water quality information and to verify restoration progress, not to detect excursions. Section 2.1.1.1.3.1.3 of the final SEIS has been revised in response to this comment.*

Comments: MR009-267; MR009-268
A commenter stated that the proposed air monitoring and soil monitoring program at the proposed Moore Ranch Project included analysis of U-nat in accordance with Regulatory Guide 4.14, rather than isotopic analysis of U-234, U-235, and U-238; furthermore, the commenter noted there would be no reason to perform isotopic analysis of uranium at an ISR facility.

Response: *The commenter is correct. The applicant has not proposed identifying the different isotopes of uranium. The proposed monitoring includes natural uranium, consistent with Regulatory Guide 4.14. Sections 6.2.1, 6.2.2, and 6.2.4 of the final SEIS has been revised in response to this comment.*

Appendix B

Comment: MR009-269
A commenter stated that the phrase "outside of the license area" is imprecise and that the applicant has committed to monitoring all private wells within 1 km (0.6 mi) of an operating wellfield.

Response: NRC agrees with the commenter. Section 6.2.5 of the final SEIS has been revised to clarify the location of private wells that would be monitored.

Comment: MR009-270
A commenter stated that the groundwater monitoring of all private wells within 1 km [0.6 mi] of an operating wellfield would be performed to provide an early warning of potential impacts from ISR operations.

Response: Section 6.2.5 of the final SEIS includes a statement about the purpose of monitoring within 1 km [0.6 mi] of an operating wellfield to identify potential impacts from ISR operations. No changes were made to the final SEIS in response to this comment.

Comment: MR009-272
A commenter stated that the discussion of physiochemical monitoring in Section 6.3 of the final SEIS should be revised to clarify that ISR processes affect groundwater quality in the production zone during operations, which is exempted under an aquifer exemption, meaning that it cannot now nor ever in the future serve as a source of public drinking water. Restoration is designed to reduce such impacts, and monitoring allows protection of production zone versus adjacent, non-exempt aquifers or portions thereof.

Response: NRC modified the text in Section 6.3 of the SEIS to discuss that the purpose of monitoring is to protect groundwater outside the exempt aquifer from potential contamination.

Comment: MR009-279
A commenter stated that monitoring wells are not placed within the wellfields to detect an excursion.

Response: NRC agrees with the commenter. Perimeter monitoring wells are used to detect potential horizontal excursions. Section 6.3.1.2 of the final SEIS was revised for clarification.

Comment: MR009-281
A commenter objected to a phrase in the SEIS that stated that typical pump tests used for a confined aquifer are ineffective.

Response: In response to the comment, the discussion in Section 6.3.1.2 of the Moore Ranch final SEIS was revised to clarify that more intensive pumping tests are required.

Appendix B

B5.34 Editorial

B5.34.1 Grammatical Editorial

Comment: MR009-040
A commenter noted that the following sentence in the summary appears to be in the wrong section. "The local economy would experience a MODERATE impact from the purchasing of local goods and services and taxes derived from construction equipment and other construction-related activities." The commenter noted that this sentence appears in the section describing impacts from facility operations, not construction.

Response: *The comment is noted, and the SEIS has been revised in response to this comment.*

Comment: MR009-130
One commenter noted a typographical error regarding discussion of a wildlife survey.

Response: *NRC recognizes that the sentence should reference wildlife surveys instead of vegetation surveys and has corrected the sentence accordingly.*

Comments: MR009-082; MR009-108; MR009-115; MR009-119; MR009-120; MR009-128; MR009-129; MR009-145; MR009-155; MR009-156; MR009-175; MR009-266; MR009-278; MR009-070; MR009-118; MR009-154; MR009-157; MR009-162; MR009-163; MR009-210; MR009-234; MR009-238; MR009-245; MR009-252; MR009-277
One commenter noted that in several occurrences throughout the SEIS, references were not properly noted or cited.

Response: *The NRC acknowledges the need for accuracy of references. Each occurrence noted by the commenter was checked for accuracy and updated or modified, as appropriate.*

Comments: MR009-001; MR009-008; MR009-009; MR009-010; MR009-024; MR009-027; MR009-030; MR009-043; MR009-046; MR009-058; MR009-060; MR009-071; MR009-080; MR009-097; MR009-101; MR009-102; MR009-103; MR009-104; MR009-105; MR009-111; MR009-117; MR009-122; MR009-123; MR009-124; MR009-125; MR009-126; MR009-131; MR009-134; MR009-135; MR009-141; MR009-142; MR009-143; MR009-146; MR009-147; MR009-164; MR009-183; MR009-186; MR009-188; MR009-189; MR009-192; MR009-204; MR009-214; MR009-221; MR009-222; MR009-225; MR009-233; MR009-237; MR009-246; MR009-263; MR009-265; MR009-290
Commenters suggested corrections for typographical, format, or grammatical errors in the SEIS.

Response: *Proposed changes were checked for accuracy, determined to be appropriate, and incorporated into the SEIS.*

Comment: MR009-153
One commenter requested that the order of two paragraphs in Section 4.2.1.4 be changed for consistency with other sections.

Response: *The NRC acknowledges the comment and has changed the order of the two paragraphs in 4.2.1.4 for clarity.*

Appendix B

Comment: MR009-284
One commenter suggested a specific wording change to clarify leak detection monitoring of injection and production wells.

Response: *The NRC staff have reviewed the discussion of surface water monitoring in the SEIS. The text regarding leak detection monitoring has been modified to clarify the system proposed for the Moore Ranch project.*

Comment: MR013-002
One commenter stated that as a general observation, the SEIS should be clear and consistent.

Response: *NRC acknowledges the importance of providing clear and consistent information. Because the comment was general in nature, no changes were made to the SEIS in direct response to this comment. Specific or detailed comments concerning clarity and consistency are addressed in Section B34.2, B34.3, and other sections of this appendix.*

Comments: MR009-051; MR009-113; MR018-085
Two commenters noted that tables in the SEIS text needed to be updated to reflect the most recent information available.

Response: *The NRC agrees with the responder, and tables in the SEIS were updated appropriately.*

B5.34.2 Technical Editorial

Comments: MR009-002; MR013-003
A commenter noted that the NRC SEIS states that the application is for a new "source material license." The applicant notes that the license would also need to authorize possession of byproduct material as defined in section 11e.(2) of the AEA. Another commenter requested the license not be referred to as a "source material license" but rather a "uranium recovery license" or a "combined source and 11e.(2) byproduct material license."

Response: *Per NRC regulations in 10 CFR Part 40, the applicant is issued a "source material license." No changes were made to the SEIS in response to this comment.*

Comments: MR009-007; MR018-017
Two commenters noted that the draft SEIS referred specifically to the applicant including "two deep disposal wells," and that the applicant may drill two or four deep disposal wells, depending on which formation is used.

Response: *NRC acknowledges the comment and has revised the language in the Executive Summary appropriately that two or four wells would be drilled.*

Comments: MR009-066, MR009-250
One commenter notes that the applicant no longer plans to utilize anhydrous ammonia at the site, and requests that reference to it in the SEIS be removed.

Response: *NRC acknowledges the comment, and references to anhydrous ammonia have been removed from the SEIS.*

Appendix B

Comment: MR009-085
One commenter requested that reference to "water being pumped from a different aquifer" be removed from the SEIS text, as the applicant would remove this option in the revised license application.

Response: NRC acknowledges the comment; the reference has been removed from the SEIS, as requested.

Comment: MR009-106
One commenter pointed out that heap leach mining does not necessarily require removal of the ore from the ground, but it may take place in situ.

Response: The NRC staff acknowledge this comment. The conventional use of the term "heap leaching" is usually in reference to ores that have been extracted from the ground. ISL is the term that is conventionally used to describe leaching in place. The purpose of the discussion of heap leaching in the SEIS is for comparison with ISR proposed at the Moore Ranch site. Consequently, no modifications have been made to the SEIS beyond this response.

Comment: MR009-109
One commenter was concerned with a reference to "hazardous or mixed waste," with respect to 11e.(2) byproduct material.

Response: NRC acknowledges the comment, and the reference to "hazardous or mixed waste" has been removed from the SEIS.

Comment: MR009-110
A commenter noted that the capital cost for mechanical evaporation would be approximately four times greater than those for deep well disposal, not greater than that for an ISR facility.

Response: The commenter is correct. Section 2.2.5 of the final SEIS has been revised in response to this comment.

Comment: MR009-127
One commenter noted that some minor differences in acreage were present between the Environmental Report and the SEIS.

Response: NRC acknowledges that there are slight differences, but because these are simply due to rounding approximations, no changes were made to the SEIS or the tables therein in response to this comment.

Comment: MR009-132
One commenter stated that the latitude and longitude of the Moore Ranch project were stated incorrectly.

Response: The applicant confirms a latitude and longitude for the Moore Ranch project of 43°34'12.83" and 105°50'49.72".

Comment: MR009-150
A commenter noted that throughout Chapter 4, NRC concludes the analysis of impacts for each resource area and life cycle phase of the project by noting that site-specific conditions are

Appendix B

comparable to those described in the GEIS. It would be helpful for the public if the NRC cited the specific section of the GEIS.

Response: At the introduction to each life cycle phase, NRC refers the reader to the particular section of the GEIS that addresses each phase. The comment does not provide any new and significant information; therefore, no change was made to the SEIS in response to this comment.

Comment: MR009-158
One commenter noted that the annual production of yellowcake was listed incorrectly at 40,000 pounds per year in the SEIS, rather than 4 million pounds per year.

Response: NRC acknowledges the comment and has made the appropriate correction in the SEIS.

Comment: MR009-161
One commenter indicated that the chemical waste shipments were identified as a transportation impact during aquifer restoration, and there are no chemical waste shipments associated with the Moore Ranch Project.

Response: The NRC staff have reviewed the discussion of transportation impacts during aquifer restoration in the SEIS. The SEIS has been modified to the more general term "waste shipments" instead of "chemical waste shipments."

Comment: MR009-179
A commenter stated that the text in 4.5.2.1.2.2 regarding excursions and groundwater quality should be rewritten to reflect the restoration standards as specified in 10 CFR Part 40, Appendix A. The commenter further noted that the standard is actually baseline or MCLs, whichever is higher, or ACLs.

Response: The comment is noted. Section 4.5.2.1.2.2 of the final SEIS has been revised in response to this comment.

Comment: MR009-219
A commenter stated that NRC should note that the applicant "would be required," rather than "would likely be required" to stop work and assess resources prior to continuing with construction, per a license condition.

Response: The commenter is correct. Section 4.9.1.1 has been revised in response to this comment.

Comment: MR009-255
One commenter suggested replacing "radioactive waste" with the more specific term "11e.(2) byproduct material."

Response: The NRC agrees with the commenter. The term "radioactive waste" has been replaced with the more specific term "byproduct material" in the SEIS.

Appendix B

Comment: MR009-256
One commenter suggested that referring to reverse osmosis and IX waste stream as "highly contaminated" was inconsistent with the typical concentrations for deep well injection provided by the applicant.

Response: SEIS Section 4.14.1.1.2 has been revised to clarify the discussion of the IX and reverse osmosis waste streams, in response to this comment.

Comment: MR009-262
One commenter noted that the discussion of cumulative impacts incorrectly stated the status of conventional uranium projects in the vicinity of the proposed Moore Ranch Project.

Response: The NRC staff have reviewed discussion of cumulative impacts, and the SEIS has been modified to clarify the status of conventional uranium projects in the vicinity of the Moore Ranch Project.

Comment: MR009-289
A commenter stated that NRC cannot say that surface water "could" potentially be impacted by an increase in sediment yield within "Unavoidable Adverse Environmental Impacts" for Water Resources.

Response: The comment is noted. The text within Table 8-1 has been revised, in response to this comment.

Comments: MR009-092; MR009-144; MR009-253
Several commenters suggested revising the text in various locations to clarify meaning.

Response: The NRC acknowledges the suggestions, and the SEIS has been modified accordingly to address these concerns.

Comments: MR009-031; MR009-033; MR009-083; MR009-116; MR009-121; MR009-133; MR009-136; MR009-137; MR009-138; MR009-139; MR009-166; MR009-172; MR009-173; MR009-187; MR009-190; MR009-223; MR009-224; MR009-229; MR009-242; MR009-251; MR009-244; MR009-247; MR009-257; MR009-264; MR009-276; MR009-283; MR009-285
One commenter pointed out several typographical or minor errors in technical data.

Response: The NRC acknowledges the comment and has corrected any errors that were found in the technical data.

Comment: MR009-226
A commenter noted that the phrase "an approved site reclamation plan" should be replaced by the phrase "an NRC-approved decommissioning plan."

Response: The commenter is correct. The applicant will have a site-wide general decommissioning plan and initial cost estimate by the end of plant operations. Prior to any decommissioning activities taking place, the applicant would need to submit a detailed plan to the NRC 12 months prior to decommissioning, per 10 CFR 40.42. The suggested change has been made to SEIS Section 4.10.1.4.

Appendix B

Comment: MR009-243
A commenter pointed out a typographical error on page 4-72, lines 20 and 21 of the SEIS.

Response: The commenter is correct. NUREG–1910, Table 4.2-2 shows Crow Butte offsite dose as 0.317 mSv [31.7 mrem] and Irigaray as 0.004 mSv [0.4 mrem] just as the SEIS does. The text in Section 4.13.1.2.1 of the final SEIS was revised in response to the comment.

Comment: MR013-004
One commenter stated that the final SEIS should indicate that the terms "ISL" and "ISR" can be used interchangeably.

Response: SEIS Section 1.1 states that for purposes of the SEIS, "in-situ recovery" or ISR is synonymous with "in-situ leach" or ISL. Because the SEIS discussion already addresses this issue, no changes were made to the SEIS in response to this comment.

Comments: MR018-021; MR018-022
A commenter identified two pages in the draft SEIS where subsurface depth intervals for deep injection wells did not correspond stratigraphically to the geologic formations being discussed.

Response: Sections 2.1.1.1.6.2 and 3.4.1 of the draft SEIS described that the Lance Formation and Fox Hills Sandstone were present at the Moore Ranch site at a subsurface depth of 1,128 to 2,286 m [3,400 to 7,500 ft]. However, this depth interval would also include the Tullock Member of the overlying Fort Union Formation, which is not part of the proposed deep injection zone. The SEIS was corrected to indicate that the Lance Formation and the Fox Hills Sandstone occur at the Moore Ranch site at a subsurface depth interval of approximately 1,615 to 2,286 m [5,300 to 7,500 ft].

Comment: MR018-025:
One commenter suggested deleting "an average flow rate of 643 L [170 gal] over 9 years" from SEIS page 4-81, line 29 and replacing it with "500 gal/min for 10 years."

Response: The text in question is not a flow rate and based on the GEIS should be edited to be "an average flow rate of 1700 L/min [450 gal/min] over a 9 year operating period". The SEIS text has been modified to reflect this correction.

Comments: MR018-075; MR018-079
The commenter noted that page 3-50, line 36 of the SEIS refers to the "Midwest Industrial Landfill" and that page 4-52, line 5 refers to a "county landfill." The commenter points out that the landfill referenced is the Midwest-Edgerton Landfill that is a municipal landfill.

Response: The reference to the landfill on pages 3-50 and 4-52 has been corrected to refer to the Midwestern Edgerton Municipal Landfill.

Comments: MR001-004; MR009-064; MR009-068; MR017-015; MR018-086
Several commenters discussed the quality of the figures in the SEIS. One commenter stated that the clarity of the visual figures and graphics in the draft SEIS was inadequate. This same commenter noted interactions with the NRC to obtain revised figures prior to the end of the comment period, and that only four revised figures were posted on the NRC website prior to the closure of the public comment period. Another commenter stated that the scale of the SEIS maps were inadequate for evaluation of surface resource impacts at the site-specific level.

Appendix B

Another commenter noted that not all figures were up to date with the most recent applicant information.

Response: *The NRC staff reviewed all figures in the SEIS and determined that many warranted revision. The NRC staff opted to revise and improve the following figures: 2-1, 2-2, 2-3, 2-5, 3-1, 3-2, 3-3, 3-5, 3-6, 3-7, 3-8, 3-9, 3-10, 3-11, 3-12, 4-1, and 6-1. Prior to the publication of the final SEIS, all revised figures were posted on the NRC website. Four of these revised figures were posted on the NRC website prior to the closure of the public comment period on March 3, 2010. In response to these comments, the figures identified in this response were revised to improve the quality. Figure 2-1 was updated to match the most recent applicant information submittal in response to these comments.*

Comment: MR002-008

One commenter suggested that the readability of SEIS Section 3.6.3 would be improved if the section was reorganized using the protection status for the subcategories rather than the species.

Response: *NRC acknowledges that information can be organized in different manners. NRC opted to organize the information by species and provided Table 3-8 which provides information columns describing protection status by species. Because the organization in the SEIS is considered appropriate, no changes were made to the SEIS in response to this comment.*

Comment: MR009-026

One commenter noted that in the Executive Summary, the discussion of ore production zone was out of place in a discussion about liquid waste disposal.

Response: *The NRC acknowledges the comment, and has removed the sentence from the Executive Summary for clarity.*

Comments: MR009-052; MR009-069; MR013-005

Two commenters recommended that references to the proposed action as "mining" should be replaced with the term "milling."

Response: *NRC agrees that the proposed action at Moore Ranch should be described as milling or rather than mining. NRC staff reviewed the discussions of the proposed action in the draft SEIS and corrected instances where the substitution of the term "milling" for "mining" was warranted.*

B5.34.3   Regulatory Editorial

Comment: MR008-001

One commenter felt that the language in the recommendations section seems full of negatives as opposed to more clearly written positive statements.

Response: *The NRC acknowledges the comment.*

Comment: MR008-002

The commenter suggests that the NRC staff describe how the safety review is integrated with the environmental review.

Appendix B

Response: As stated in Section 1.4.5, the NRC safety staff evaluate whether the applicant proposed action can be accomplished in accordance with the applicable provisions of 10 CFR Part 20 and 10 CFR Part 40, Appendix A. The SER evaluates the applicant proposed facility design, operational procedures, and radiation protection program to ensure that the applicable requirements in 10 CFR Part 20 and 10 CFR Part 40 would be met by the applicant. These facets of the safety review are integral to shaping the environmental review in the SEIS. The comment does not present any new or significant information; therefore, no change was made to the SEIS text in response to this comment.

Comments: MR009-004; MR013-006

Two commenters noted that NRC should use the terms "proposed" and "potential" when referring to the proposed action and the impacts analyzed.

Response: The NRC agrees with the comment, and has added the terms "proposed" and "potential" when referring to the site and proposed action throughout the SEIS, as appropriate.

Comments: MR009-042; MR009-114

A commenter noted that the NRC preliminary conclusion seems to be a negative endorsement of the analysis completed in the SEIS. NRC has concluded that virtually all of the impacts associated with the proposed Moore Ranch Project would be small and that there would be no long-term effects. Assuming there are no significant safety issues identified in the SER that would require denial of a license, then issuance of a license is a reasonable action by the NRC. The commenter also noted that the Preamble to 10 CFR Part 51 states that NEPA reviews do not rule out actions that have potentially significant impacts. So, where the NEPA review identified SMALL impacts and an SER with license conditions demonstrate adequate protection of public health and safety and the environment, NRC has to issue the license.

Response: Per 10 CFR 51.71(f), the draft environmental impact statement will include a preliminary recommendation by the NRC staff in response to the proposed action. This recommendation is based on the NRC staff analysis of all relevant information and review of all other Federal permits, licenses, approvals, and other entitlements that must be obtained to implement the proposed action. This preliminary decision will be reached after considering the environmental effects of the proposed action and reasonable alternatives, and after weighing the costs and benefits of the proposed action. The NRC preliminary conclusion is based upon the regulatory requirements specified in 10 CFR Part 51 and §§ 51.75, 51.76, 51.80, 51.85, and 51.95. The NRC staff preliminary conclusion states, "Should safety issues mandate otherwise, that the environmental impacts are not so great as to make issuance of a source material license an unreasonable licensing decision. If the proposed action presents certain safety issues that make licensing unreasonable, then the license will not be granted." The comment provides no new and significant information; therefore, no change was made to the SEIS text.

Comment: MR009-053

One commenter noted that NRC should reference reports submitted by the applicant as an aid to consultations.

Response: The NRC is not required to list every report completed by the applicant. The applicant did conduct surveys of the project area, and these reports have been reviewed by the NRC and SHPO. Section 1.7.2 has been revised to include additional consultation efforts to date and the SHPO determination that the project would have no effect on historic properties.

Appendix B

Comment: MR009-065
A commenter stated that NRC should specifically reference any secondary containment on all facilities, including berms or curbs, or both. This should be referenced throughout the SEIS.

Response: The SEIS (specifically Section 2.1.1.1.2.1) provides the reader with a general description of the Central Plant facility, its dimensions, and a description of activities that would occur within that building and its immediate environs. Some clarifications were made in the SEIS to reference secondary containments.

Comment: MR009-236
One commenter requested that, for consistency, NRC add a statement to the end of Section 4.11.1.2 that no new or significant information was identified by the NRC staff.

Response: The NRC acknowledges the comment and has added the requested statement to the final SEIS.

Comment: MR009-254
A commenter stated that the NRC should note that the requirement for a disposal agreement to be in place will be pursuant to a license condition.

Response: The comment is noted. Section 4.14 has been revised, in response to this comment.

Comment: MR009-271
A commenter noted that the statement is not accurate. "The sampling would be conducted in accordance with a standard operating procedure reviewed by the NRC staff." The commenter states that a full description of the sampling process was provided in an RAI response, dated July 11, 2008, [See section 5.12(f)]. The commenter states that since the NRC had no further questions or comments, this is not an open item.

Response: Section 6.2.5 has been clarified. While the NRC does not approve standard operating procedures, NRC inspectors can review procedures during site inspections.

Comment: MR009-274
A commenter stated that the NRC should make it clear that analysis of the full Guideline 8 parameters is only required for the first two sample events.

Response: WDEQ does not identify the number of Guideline 8 samples that must be obtained. NRC would require that the number of samples of Guideline 8 parameters be adequate to determine the baseline conditions. No changes have been made to the text to address this comment.

Comment: MR009-280
A commenter stated that NRC should add the text, "Under WDEQ requirements…" to the beginning of the sentence on page 6-6, line 14 to show interconnection between monitoring wells and the production patterns.

Response: The commenter is correct. Furthermore, Section 5.7.8.3(4) of NUREG–1569 states, "Once a wellfield is installed, it should be tested to establish that the production and injection wells are hydraulically connected to the perimeter horizontal excursion monitor wells

Appendix B

and hydraulically isolated from the vertical excursion monitoring wells." Section 6.3.1.2 has been revised in response to this comment.

Comment: MR013-001
One commenter stated that the existing SEIS language concerning the preliminary recommendation on issuing a license was inadequate and should be rephrased to provide a clear understanding that the environmental review has resulted in a finding that the license should be issued.

Response: *As described in SEIS Section 1.6.1, the NRC licensing process includes a detailed technical review of the Moore Ranch ISR Project license application, which is comprised of both a safety review and an environmental review. These two reviews are conducted in parallel (see GEIS Fig. 1.7-1). The environmental review is conducted in accordance with the regulations in 10 CFR Part 51. The focus of the safety review is to assess compliance with the applicable regulatory requirements in 10 CFR Part 20 and 10 CFR Part 40, Appendix A. The NRC staff reviewed the SEIS language concerning the preliminary recommendation and determined that the text was consistent with the NRC licensing process. Because the SEIS is considered appropriate, no changes were made to the SEIS in response to this comment.*

Comment: MR018-042
A commenter noted that Antelope Creek is protected through a rulemaking process. The commenter states that the term "suitable" gives the reader the impression that there have been onsite water quality assessments that have determined the suitability for all uses (recreation, other aquatic life, wildlife, agriculture, industry, and scenic value), which may or may not be the case. Please replace "suitable" with "protected."

Response: *The comment is noted. Section 3.5.1.4 of the final SEIS has been modified in response to this comment.*

Comment: MR018-045
A commenter suggested that the NRC add a sentence to Section 4.5.1.1: "Authorization from the Wyoming DEQ could be required when filling or crossing wetlands."

Response: *The comment is noted. Section 4.5.1.1 of the final SEIS has been revised, in response to this comment.*

APPENDIX C
ALTERNATE CONCENTRATION LIMITS

ALTERNATE CONCENTRATION LIMITS

In-Situ recovery (ISR) facilities operate by first extracting uranium from specific areas called wellfields. After uranium recovery has ended, the groundwater in the wellfield contains constituents that were mobilized by the lixiviant. Licensees shall commence aquifer restoration in each wellfield soon after the uranium recovery operations end (NRC, 2008b). Aquifer restoration criteria for the site-specific baseline constituents are determined either for each individual well or as a wellfield average.

NRC licensees are required to return water quality parameters to the standards in 10 CFR Part 40, Appendix A, Criterion 5B(5). As stated in the regulations: "5B(5)—At the point of compliance, the concentration of a hazardous constituent must not exceed—(a) The Commission approved background concentration of that constituent in the groundwater; (b) The respective value given in the table in paragraph 5C if the constituent is listed in the table and if the background level of the constituent is below the value listed; or (c) An alternate concentration limit is established by the Commission."

For an alternate concentration limit (ACL) to be considered by the NRC, a licensee must submit a license amendment application to request an ACL. In this ACL license amendment request, the licensee must provide the basis for any proposed limits including consideration of practicable corrective actions, that limits are as low as reasonably achievable (ALARA), and information on the factors the Commission must consider. The NRC will establish a site-specific ACL for a hazardous constituent as provided in paragraph 5B(5) if the NRC finds the proposed limit as ALARA, after considering practicable corrective actions, and determining that the constituent will not pose a substantial present or potential hazard to human health or the environment as long as the ACL is not exceeded.

To determine if the ACL does not pose a potential hazard to human health or the environment, NRC performs three risk assessments (NRC, 2003a). The first is a hazard assessment which evaluates the radiological dose and toxicity of the constituents in question and the risk to human health and environment. The second is an exposure assessment to examine the existing distribution of hazardous constituents, as well as potential sources for future releases and the potential consequences associated with the human and environmental exposure to the hazardous constituents. The last assessment is a corrective action assessment which evaluates (1) all applicant proposed corrective actions; (2) the technical feasibility of each proposed corrective actions; (3) the costs and benefits associated with each proposed corrective action; and (4) the preferred corrective action to achieve the hazardous constituent concentration which is protective of human health and the environment.

To perform these assessments, the NRC staff uses a rigorous review process. Licensees must provide a comprehensive ACL amendment that addresses groundwater and surface water quality and expected impacts on human health and the environment. Such information required in an amendment request pursuant to 10 CFR Part 40, Appendix A, Criterion 5B(6) includes the following factors:

- Potential adverse effects on groundwater quality, considering the following:
    - The physical and chemical characteristics of the waste in the licensed site including its potential for migration
    - The hydrogeologic characteristics of the facility and surrounding land
    - The quantity of groundwater and the direction of groundwater flow

- - The proximity and withdrawal rates of groundwater users
  - The current and future uses of groundwater in the area
  - The existing quality of groundwater, including other sources of contamination and their cumulative impact on the groundwater quality
  - The potential for health risks caused by human exposure to waste constituents
  - The potential damage to wildlife, crops, vegetation, and physical structures caused by exposure to waste constituents
  - The persistence and permanence of the potential adverse effects.

- Potential adverse effects on hydraulically connected surface water quality, considering the following:
  - The volume and physical and chemical characteristics of the waste in the licensed site
  - The hydrogeologic characteristics of the facility and surrounding land
  - The quantity and quality of groundwater, and the direction of groundwater flow
  - The patterns of rainfall in the region
  - The proximity of the licensed site to surface waters
  - The current and future uses of surface waters in the area and any water quality standards established for those surface waters
  - The existing quality of surface water including other sources of contamination and the cumulative impact on surface water quality
  - The potential for health risks caused by human exposure to waste constituents
  - The potential damage to wildlife, crops, vegetation, and physical structures caused by exposure to waste constituents
  - The persistence and permanence of the potential adverse effects.

Although state "class of use" standards are not recognized in NRC's regulations as restoration standards, these standards may be considered as one factor in evaluating ACL requests for ISR facilities located in Wyoming. Furthermore, in considering ACL requests, particular importance is placed on protecting underground sources of drinking water (USDWs). The use of modeling and additional groundwater monitoring may be necessary to show that ACLs in ISR wellfields would not adversely impact USDWs. It must be demonstrated that the licensee has attempted to restore hazardous constituents in groundwater to background or a maximum contaminant level—whichever level is higher.

Before an ISR licensee is allowed to extract uranium, the EPA under 40 CFR Part 146.4 and in accordance with the Safe Drinking Water Act must issue an aquifer exemption covering the portion of the aquifer in which the uranium-bearing rock is located. The EPA cannot exempt the portion of the aquifer unless it is found that "it does not currently serve as a source of drinking water" and "cannot now and will not in the future serve as a source of drinking water". Due to these criteria, only impacts outside of the exempted aquifer are evaluated. In most cases, the water in aquifers adjacent to the uranium ore zones does not meet drinking water standards. The staff will not approve an ACL if it will impact any adjacent USDWs. Therefore, the impact of granting an ACL request is SMALL.

Further guidance for the review of ACLs for ISR facilities is being developed in a revision of NUREG–1569 (NRC, 2003a). Existing guidance for the review of ACLs for conventional mills is in NUREG–1620, "Standard Review Plan for the Review of a Reclamation Plan for Mill Tailings Sites Under Title II of the Uranium Mill Tailings Radiation Control Act of 1978." (NRC, 2003b).

REFERENCES

10 CFR Part 40. Appendix A. *Code of Federal Regulations*, Title 10, *Energy*, Part 40, Appendix A, "Criteria Relating to the Operations of Uranium Mills and to the Disposition of Tailings or Wastes Produced by the Extraction or Concentration of Source Material from Ores Processed Primarily from their Source Material Content."

40 CFR Part 146. *Code of Federal Regulations*, Title 40, *Protection of Environment.* Part 146, "Underground Injection Control Program: Criteria and Standards."

NRC (U.S. Nuclear Regulatory Commission), 2003a. NUREG–1620 "Standard Review Plan for the Review of a Reclamation Plan for Mill Tailings Sites under Title II of the Uranium Mill Tailings Radiation Control Act of 1978," Final Report, Washington, D.C. NRC 2003

NRC, 2003b. NUREG–1569, "Standard Review Plan for *In-Situ* Leach Uranium Extraction License Applications." Final Report. Washington, DC: NRC. June.

APPENDIX D

NONROAD COMBUSTION ENGINE EMISSIONS ESTIMATES

NONROAD COMBUSTION ENGINE EMISSIONS ESTIMATES

D1 INTRODUCTION

The primary nonradiological emissions from *in-situ* recovery (ISR) facilities include diesel combustion engine emissions from construction equipment (including drilling rigs) and fugitive dust emissions from vehicular travel on unpaved roads (NRC, 2009, Section 2.7.1). This appendix provides estimates of the expected nonroad combustion engine emissions from the proposed action. Fugitive dust emissions are discussed in the supplemental environmental impact statement (SEIS) and, therefore, it is not discussed further in this appendix.

NRC has previously evaluated combustion engine emissions associated with ISR facilities in prior licensing actions (NRC, 1997; 2004) and has characterized the potential impacts to air quality as minor. The drilling rigs that are used during construction of these facilities, in particular, are not presently subject to State of Wyoming new-source emissions permitting, and applicants that propose facilities in attainment areas (i.e., areas in compliance with ambient air quality standards) are not presently required by the state to document their emissions from these sources. Similarly, the U.S. Nuclear Regulatory Commission (NRC) has not routinely requested detailed nonradiological emissions information from applicants. As a result, existing information pertaining to ISR construction emission activities is limited. Nonetheless, to address recent concerns expressed in public comments on the draft SEIS about potential air quality impacts (EPA, 2010), representative emissions estimates are calculated in this appendix.

Based on the similarities in design and construction of ISR facilities and the nature of associated nonradiological emissions, the nonroad combustion engine exhaust calculations in this section are based on a combination of proposal-specific and other representative information that the staff considers adequate to support a conservative emissions screening analysis. The current calculations incorporate the best available information provided by the applicant for the proposed action; representative information provided by other NRC applicants as applicable; and emissions factors provided by the U.S. Environmental Protection Agency (EPA). Mobile road (vehicle) combustion emissions were not calculated here because these engine emissions are controlled at the source by mandated emission control technology, and the magnitude of proposed road vehicle activity is small relative to existing road traffic (Section 4.3).

The calculations in this appendix were conducted to support the NRC evaluation of potential environmental impacts to air quality from the proposed action.[1] These calculations are provided to meet NRC obligations, pursuant to the National Environmental Policy Act of 1969 as amended, to more completely disclose the potential environmental impacts from the proposed action. While NRC is responsible for assessing the potential environmental impacts from the proposed action, NRC does not have the authority to develop or enforce regulations to control nonradiological air emissions from equipment used by licensees. This authority rests with the State of Wyoming Department of Environmental Quality (WDEQ). To ensure the air quality of

---

[1] The analysis in this appendix is based upon the applicant's original schedule provided in its license application (EMC 2007, Figure 1.8-1) which included the development of three wellfields. The applicant subsequently revised its application in 2008 to combine wellfields 2 and 3 into one wellfield which is now wellfield 2. The analysis presented in this section is still applicable.

Appendix D

Wyoming is adequately protected, in addition to addressing all NRC regulatory requirements that address radiological emissions, NRC applicants and licensees must also comply with all applicable State and Federal air quality regulatory compliance and permitting requirements.

D2 NONROAD DIESEL COMBUSTION ENGINE EXHAUST EMISSIONS CALCULATION METHODS

D2.1 Well Drilling Emissions Calculations

ISR facilities are constructed using commonly available construction equipment, including truck mounted or mobile drilling rigs (NRC, 2009). Based on past estimates (NRC, 2004), the NRC staff expect well drilling activities will represent the majority of combustion engine emissions during the construction period. Emissions from diesel combustion engines, including drilling rigs, that the staff evaluated for potential impacts to air quality include nitrogen oxides (NO_x), carbon monoxide (CO), sulfur oxides (SO_x), particulate matter (PM_{10}), formaldehyde, volatile organic compounds (VOC), and carbon dioxide (CO_2) emissions were calculated to support the NRC staff evaluation of greenhouse gas emissions in the SEIS.

Diesel emissions were estimated using emission factors provided by EPA. Emissions factors provide the ratio of the mass of a pollutant emitted to the atmosphere by a source engine to the level of activity of the emission source (Eastern Research Group, 1996). The level of activity of the emission source in an emission factor is represented by power output (in horsepower-hours) or fuel use represented by heat energy of combusted fuel in million British Thermal Units (MMBtu). Emission factors were developed by the EPA for different engine classes, based on their review of a variety of engine test data (EPA, 1996, 2004). Currently available EPA documentation of emissions factors for diesel combustion engines include AP-42 (EPA, 1996) and a more recent update of emissions factors for the EPA NONROAD model (EPA, 2004). The former reference (AP-42) is recognized by WDEQ as a source for emissions factors that may be used to estimate emissions from drilling rigs (WDEQ, 2010) while the NONROAD model factors represent a more current data source. For the following calculations, the emissions factors from AP-42 (EPA, 1996) were used. The updated emissions factors for the NONROAD model are considered for context in the discussion of the calculated emissions results.

The WDEQ provided methods for calculating emissions from drilling rigs based on fuel use (WDEQ, 2010). The WDEQ calculation methods are from worksheets they have provided to minor oil and gas emitters in a proposed ozone non-attainment area in southwestern Wyoming (Finley, 2010). These methods were adapted to the current analysis and are summarized by the following equations

$$E_{tot,r,i} = F_{tot,r}\, HC_{fuel}\, EF_i\, U_{conv} \tag{D-1}$$

where

$E_{tot,r,i}$ — annual total emissions for drilling rig type r and pollutant i [tons/yr]
$F_{tot,r}$ — annual fuel use for drilling rig type r [gal/yr]
HC_{fuel} — heat content of diesel fuel [Btu/gal]
$EF_{i,r}$ — emission factor for pollutant i from drilling rig type r [lb/MMBtu]
U_{conv} — unit conversion [MMBtu/1E+6 Btu][ton/2000lb]

Appendix D

and

$$F_{tot,r} = \sum DT_{n,r} \, FC_r \qquad (D-2)$$

where

DT$_n$ — duration of drilling for individual well n [hr]
FC$_r$ — hourly fuel consumption for drilling rig type r [gal/hr]

Input parameters for well-drilling equipment diesel emission calculations are provided in Tables D2–1, D2–2, and D2–3. Proposed drilling activities include (i) drilling water wells for wellfield operations and associated monitoring and (ii) drilling deep disposal wells for disposal of operational liquid wastes. Water well drilling is expected to involve truck-mounted drilling equipment that requires, on average, 12 hours of drilling per well and consumes approximately 2.5 gal of diesel fuel per hour (Cash, 2010). These operational water wells would be drilled to the depth of the ore body {approximately 76 to 91m [250 to 300 ft]} (EMC, 2007). Deep disposal well drilling would go to a greater depth (several thousand feet) relative to the water wells. Such drilling requires a more powerful drilling rig that consumes more fuel than a water well drilling rig. Completion of one deep well has been estimated by an NRC applicant to take approximately 528 hours of drilling and consume about 56 gal of diesel fuel per hour (Cash, 2010). The applicant proposes to drill four deep wells, although the schedule for drilling has not

Table D2–1. Well Drilling Input Parameters for Emissions Calculations			
Parameter	Symbol	Value	Remarks
Duration of drilling activities for 481 water wells [hr]	$\sum_n DT_{n,r}$	5532	Staff estimate for drilling one wellfield based on average per well drill time provided by an applicant* and the proposed number of wells for Wellfield #1
Hourly fuel consumption for truck-mounted drilling rig [gal/hr]	FC_{r_water}	2.5	Provided by an applicant*
Annual fuel use for truck mounted water well drilling rigs [gal/yr]	F_{tot,r_water}	13,830	Staff calculated from drilling duration and hourly fuel consumption
Duration of drilling activities for 2 deep waste disposal well [hr]	$\sum_n DT_{n,r}$	1056	Double the value provided by an applicant* for drilling 1 deep well
Hourly fuel consumption for deep well rig [gal/hr]	FC_{r_deep}	56.25	Provided by an applicant*
Annual fuel use for deep well drilling rig [gal/yr]	F_{tot,r_deep}	59,400	Staff calculated from drilling duration and hourly fuel consumption
Heat content of diesel fuel [Btu/gal]	HC_{fuel}	137,000	Value from EPA AP-42†

*Cash, 2010.
† EPA, 1996.

been provided. Because the proposed wellfield development is phased over a period of years (EMC 2007, Figure 1.8-1), for the calculation of annual emissions, the drilling of one wellfield and two deep wells in the first year is assumed. To account for the differences in the two proposed drilling operations, emissions calculations were conducted for each type of drilling activity (i.e., water wells, deep disposal wells). Input parameters for each activity are provided in Table D2–1.

Table D2–2. Emissions Factors (EF$_i$) for Uncontrolled Diesel Industrial Engines (lb/MMBtu)	
Pollutant	Value
Nitrogen Oxides (NO$_x$)	4.41
Carbon Monoxide (CO)	0.95
Sulfur Oxides (SO$_x$)	0.29
Particulate Matter (PM$_{10}$)	0.31
Carbon Dioxide (CO$_2$)	164
Formaldehyde	0.00118
Volatile Organic Compounds	0.35
Source: EPA, 1996.	

Table D2–3. Emissions Factors (EF$_i$) for Large Stationary Diesel Engines (lb/MMBtu)	
Pollutant	Value
Nitrogen Oxides (NO$_x$)	3.2
Carbon Monoxide (CO)	0.85
Sulfur Oxides (SO$_x$)	1.01
Particulate Matter (PM$_{10}$)	0.10
Carbon Dioxide (CO$_2$)	0.85
Formaldehyde	7.89E-5
Volatile Organic Compounds	0.09
Source: EPA, 1996.	

D2.2 Construction Equipment Emissions Calculations

In addition to the use of drilling rigs, proposed wellfield construction involves the use of common diesel powered construction equipment that will also contribute to air emissions. Emissions from this equipment were calculated using emission factors based on power output and operating time using the following equation

$$E_{tot,r,i} = HP_r \, OT_r \, EF_i \, U_{conv} \qquad (D-3)$$

where

$E_{tot,r,i}$ — annual total emissions for construction equipment type r and pollutant i [tons/yr]
HP_r — engine horsepower rating for construction equipment type r [hp]
OT_r — operating time for construction equipment type r [hr/yr]
EF_i — emission factor for pollutant i for diesel industrial engines [lb/hp-hr]
U_{conv} — unit conversion [ton/2,000lb]

Input parameters used in the construction equipment emissions calculations, including the types of equipment the applicant could use during the construction period, operating times for this equipment, and applicable emission factors, are provided in Tables D2–4 and D2–5. The information in Table D2–4 summarizes detailed equipment emissions information an applicant

(Cash, 2010) voluntarily submitted to the WDEQ to support a survey of small emitters. Table D2–5 lists the applicable power-output-based emissions factors for diesel industrial engines.

Table D2–4. Horsepower (hp) and Operating Times (hr/yr) for Diesel Construction Equipment		
Equipment	Horsepower (HP_r)	Operating Time (OT_r)
Lull 944E Telehandler	110	450
John Deere 710J Backhoe	126	360
John Deere 410 Backhoe	66	240
Truck	250	90
John Deere Loader	200	50
Scraper	600	50
Blade	300	38
Caterpillar D8 Dozer	321	12
Source: Cash, 2010.		

Table D2–5. Emissions factors (EF_i) for Uncontrolled Diesel Industrial Engines (lb/hp-hr)	
Pollutant	Value
Nitrogen oxides (NO_x)	0.031
Carbon monoxide (CO)	0.00668
Sulfur oxides (SO_x)	0.00205
Particulate matter (PM_{10})	0.00220
Carbon dioxide (CO_2)	1.15
Formaldehyde	0.00000826
Volatile organic compounds	0.00247
Source: EPA, 1996.	

D2.3 Reclamation Equipment Emissions Calculations

Emissions during the construction period are expected to bound annual emissions from operations and aquifer restoration phases because the use of diesel powered equipment during those phases is much less than during construction (NRC, 2004). Construction equipment use during decommissioning and reclamation (hereafter reclamation) is expected to be similar to the construction phase (NRC, 2004) because many aspects of reclamation, in effect, are the reverse of the activities conducted during construction. During construction, well-drilling and facility construction activities predominate, while during reclamation, diesel equipment is used for other activities, such as well plugging and abandonment, equipment removal, and land reclamation. The applicant has planned a two-year period for the reclamation of each wellfield (EMC, 2007, Figure 1.8-1).

Emissions for diesel equipment used for reclamation activities were calculated using the same methods as in Section D2.2 for construction equipment (Eq. D–3), although input parameters were revised for equipment horsepower and operating times to reflect available information on the proposed reclamation activities. The NRC staff identified the most detailed and complete information on proposed equipment and activities in the surety estimate for the proposed facility (EMC, 2007). Based on the types of equipment identified in the surety estimate, specific equipment models were selected by reviewing documentation of commonly used reclamation equipment provided by WDEQ (2009).

Equipment horsepower information for specific models was obtained from manufacturer documentation. A few items of equipment were only generally described in the surety estimate as truck or tow vehicles for well-abandonment activities, and these were assumed to be rated at 250 horsepower. Operating times for each item of equipment were derived from detailed information and assumptions on specific reclamation activities discussed in the surety estimate (EMC, 2007), including building demolition floor removal, pipeline removal, well abandonment, and reclamation of disturbed surface areas such as wellfields, facilities areas, and access roads. Equipment usage for backhoe and trackhoe equipment was explicit in the surety information, and operation times are expected to be the most reliable estimates of the equipment evaluated in this analysis. Other equipment was not explicitly called out for specific activities (e.g., dozer, dump truck, scaper, motor grader), so operating times were estimated based, in part, on assumptions about which reclamation activities would utilize the equipment (e.g., the motor grader was assumed for road grading and grading cleared foundation areas, the dozer was assumed to be involved in ripping packed land surface areas; the dump truck was assumed for transporting excavated contaminated soils; and the scraper was assigned the same hours as for the construction work discussed in Section D2.2). Information on equipment productivity, such as grading or ripping rates and payload amounts, were obtained from the aforementioned WDEQ documentation (WDEQ, 2009). The resulting equipment and operation times are provided in Table D2–6. The emissions factors used in the calculations are provided in Table D2–5.

D3 RESULTS AND DISCUSSION

The estimated annual emissions from well drilling and construction equipment are provided in Tables D3–1 and Table D3–2. The total estimated annual emissions from both calculations combined are provided in Table D3–3. The combined results for drilling and construction equipment show CO_2 and NO_x annual emissions are the highest of the pollutants evaluated. For well-drilling equipment, the rigs used for two deep disposal wells generated higher annual

Table D2–6. Horsepower (hp) and Operating Times (hr) for Diesel Reclamation Equipment		
Equipment	Horsepower (HP_r)	Operating Time (OT_r)
Heavy Truck	250	1280
Caterpillar 430E Backhoe	101	950
Caterpillar 320DL Track hoe	148	780
Lull 944E Telehandler	110	450
New Holland 545D Tractor	63	170
Dump Truck	250	60
Caterpillar 657G Scraper	600	50
Caterpillar D9 Dozer	474	50
Caterpillar 16H Motor Grader	265	5
Source: Derived by staff from information in the following references: (1) EMC, 2007. (2) Cash, 2010. (3) WDEQ, 2009.		

Appendix D

emissions estimates when compared to the emissions from drilling all of the water wells for a single wellfield (481 wells). This result is explained by the larger, more fuel-consuming, engine used by a deep well rig in comparison to the smaller water well rig and the long drilling time

Table D3–1. Calculated Annual Emissions From Well Drilling Activities* (tons/yr)							
Drilling Activity	NO_x†	CO†	SO_2†	PM_{10}†	CO_2†	Formaldehyde	VOC†
Operational wellfield (water well) drilling	4.2	0.9	0.27	0.27	160	0.0011	0.34
Deep well drilling	13	3.5	0.20	0.41	670	0.00032	0.37
Total	17	4.4	0.47	0.68	830	0.0014	0.71

*Includes drilling and construction of the first proposed wellfield, and two deep disposal wells
†NO_x = nitrogen oxides; CO = carbon monoxide; SO_x = sulfur oxides; PM_{10} – particulate matter; CO = carbon dioxide; VOC = volatile organic compounds

Table D3–2. Calculated Annual Emissions From Construction Equipment (tons/yr)*							
Equipment	NO_x†	CO†	SO_2†	PM_{10}†	CO_2†	Formaldehyde	VOC†
Lull 944E Telehandler	0.77	0.17	0.05	0.05	28	0.00020	0.06
JD 710J Backhoe	0.70	0.15	0.05	0.05	26	0.00019	0.06
JD 410 Backhoe	0.25	0.05	0.02	0.02	9.1	0.000065	0.02
Truck	0.35	0.08	0.02	0.02	13	0.000093	0.03
JD Loader	0.16	0.03	0.01	0.01	5.8	0.000041	0.01
Scraper	0.47	0.10	0.03	0.03	17	0.00012	0.04
Blade	0.18	0.04	0.01	0.01	6.6	0.000047	0.01
CAT D8 Dozer	0.06	0.01	0.0039	0.0042	2.2	0.000016	0.0047
Total	2.9	0.63	0.19	0.19	110	0.00077	0.23

*Includes equipment used to support drilling and wellfield development operations
†NO_x = nitrogen oxides; CO = carbon monoxide; SO_x = sulfur oxides; PM_{10} – particulate matter; CO = carbon dioxide; VOC = volatile organic compounds

Table D3–3. Total Calculated Annual Emissions From Drilling and Construction (tons/yr)*							
Activities	NO_x†	CO†	SO_2†	PM_{10}†	CO_2†	Formaldehyde	VOC†
Well drilling and construction	20	5.0	0.66	0.87	940	0.0022	0.94

*Includes drilling and construction of the first proposed wellfield, and two deep disposal wells. Results are the sum of results from Tables D3–1 and D3–2
†NO_x = nitrogen oxides; CO = carbon monoxide; SO_x = sulfur oxides; PM_{10} – particulate matter; CO = carbon dioxide; VOC = volatile organic compounds

per well required for deep drilling. For example, the two deep well-drilling activities combined are estimated to emit 76 percent of the annual NO_x drilling emissions, compared water well-drilling of 481 wells that represents 24 percent of the annual drilling NO_x emission total. The magnitude of the calculated construction equipment emissions is small relative to the results for drilling activities. The total construction equipment emissions of NO_x are 14 percent of the total annual NO_x from all activities included in the calculations while drilling activities constitute the remaining 86 percent of the total emissions.

Appendix D

The emissions estimates are expected to be conservative because they are based on emissions factors applicable to engines that have no pollution controls. Table D3–4 provides a comparison of the EPA AP-42 factors (EPA, 1996) that were used for the calculations in this appendix with updated emission factor values (EPA, 2004) that are based on more recent data

Table D3–4. Effect of Using Updated Emissions Factors That Account for Pollution Controls			
Pollutant	1996 Uncontrolled Emission Factor for Diesel Industrial Engines	2004 Updated Value (Tier 1 Controlled 300-600 HP Diesel Engines)	Reduction Ratio (Updated/ Uncontrolled)
Nitrogen oxides (NO_x)	0.031	0.0132	0.42
Carbon monoxide (CO)	0.00668	0.00288	0.43
Particulate matter (PM_{10})	0.00220	0.00044	0.20
Volatile organic compounds	0.00247	0.000446	0.18
Source: EPA, 2004; EPA, 1996.			

that apply to engines with pollution controls (Tier 1 representing the first phase of standards that were mandated by the Federal government in four phases of increasing limits). Table D3–4 shows that calculated emissions estimates for NO_x and CO would be reduced, approximately, by a factor of 2 and PM_{10}, and VOC emissions by a factor of 5 if the updated emission factors were used. Because the actual equipment that would be used is uncertain, the assumption of an applicant using older, uncontrolled engines bounds the emissions should older equipment be selected for this work and also provides margin in the estimates if newer equipment is selected.

The results of emissions calculations for the reclamation activities are provided in Table D3–5. Because wellfield reclamation is planned to take two years per wellfield using a phased approach, the annual emissions for years that do not overlap with any of the other scheduled wellfield reclamation activities would be approximately half of the values listed in Table D3–5. According to the applicant schedule, in two of the four years allocated for wellfield reclamation, two wellfield's reclamation activities overlap (EMC, 2007, Figure 1.8-1) and, therefore, the annual emissions during these years would equal the values listed in Table D3–5. Because many of the total values in Table D3–5 are less than the emissions calculated for construction (by no more than a factor of 2.5), the magnitude of annual reclamation emissions are less than the annual emissions calculated for wellfield construction, by approximately a factor of about two to five for several of the pollutants evaluated, depending on the specific year evaluated.

Cumulative emissions for the proposed action were also approximated using the calculated results and the number of wellfields proposed by the applicant. For the purpose of this analysis, cumulative emissions are the total lifecycle emissions from all phases of the proposed action. Because the principal diesel emissions from the proposed action are associated with equipment used for constructing and decommissioning the project, the analysis focuses on the emissions from those phases. The initial calculated annual emissions in Table D-3 apply to constructing a single wellfield with two deep disposal wells. Assuming these emissions are representative of the construction of the other wellfields to be developed, these results scale with the total number of wellfields and deep wells that are proposed. Because the applicant's initial proposal includes phased construction of three wellfields (EMC, 2007, Figure 1.8-1) and each of these wellfields

Appendix D

would require reclamation once operations and aquifer restoration are completed, the cumulative emissions were conservatively approximated by calculating a wellfield weighted-sum of emissions for the planned life of the facility. This was done by computing a weighted sum of the (pollutant-specific) total construction phase water well drilling emissions from Table D3–1, the total construction equipment emissions from Table D3–2, and the total reclamation emissions contributions from Table D3-5 where each contribution to this sum is weighted by a factor of 3 (the total number of planned wellfields for this proposed action). Because the total deep-well drilling emissions from Table D3-1 are independent of the water wellfields, the results for these emissions (for two deep wells) are doubled and added to the weighted sum result to complete the cumulative emissions totals that are provided in

Table D3–5. Calculated Diesel Equipment Emissions (tons) From Reclamation of One Wellfield* and the Central Processing Facilities*							
Equipment	NO$_x$†	CO†	SO$_2$†	PM$_{10}$†	CO$_2$†	Formaldehyde	VOC†
Heavy Truck	4.97	1.07	0.33	0.35	180	0.0013	0.40
Caterpillar 430E Backhoe	1.49	0.32	0.098	0.10	55	0.00040	0.12
Caterpillar 320DL Track hoe	1.79	0.39	0.12	0.13	66	0.00048	0.14
Lull 944E Telehandler	0.77	0.16	0.05	0.054	28	0.00020	0.061
New Holland 545D Tractor	0.17	0.036	0.011	0.012	6.2	0.000044	0.013
Dump Truck	0.24	0.052	0.016	0.017	9.0	0.000065	0.020
Caterpillar 657G Scraper	0.47	0.10	0.031	0.033	17	0.00012	0.037
Caterpillar D9 Dozer	0.37	0.079	0.024	0.026	14	0.000098	0.029
Caterpillar 16H Motor Grader	0.021	0.0044	0.0014	0.0015	0.76	0.0000055	0.0016
Total	10	2.2	0.68	0.72	380	0.0027	0.82

*The applicant plans reclamation of a single wellfield over a 2-year period, so annual emissions would be approximately half the values listed in this table when only a single wellfield is planned to be reclaimed in 1 year.
†NO$_x$ = nitrogen oxides; CO = carbon monoxide; SO$_x$ = sulfur oxides; PM$_{10}$ – particulate matter; CO = carbon dioxide; VOC = volatile organic compounds

Table D3–6. Estimated Cumulative Emissions of the Proposed Action* (tons)							
Activities	NO$_x$†	CO†	SO$_2$†	PM$_{10}$†	CO$_2$†	Formaldehyde	VOC†
Well drilling and construction of three wellfields and up to four deep wells and reclamation of all wellfields and facilities	78	18	3.8	4.4	3300	0.014	4.9

*The planned duration of the proposed action represents a phased construction, operation, aquifer restoration, and reclamation schedule for each wellfield and central plant over a period of 12 years.
†NO$_x$ = nitrogen oxides; CO = carbon monoxide; SO$_x$ = sulfur oxides; PM$_{10}$ – particulate matter; CO = carbon dioxide; VOC = volatile organic compounds

Appendix D

Table D3–6. The cumulative results are conservative, in part, because the (factor of 3) multiplier overcounts the contribution from the plant facilities decommissioning that is included in each wellfield reclamation emissions estimate.

D4 REFERENCES

Cash, J., 2010. "Air Quality Spreadsheet." E-mail communication (April 29) to A. Bjornsen, NRC. Casper, Wyoming: UR-Energy USA Inc.

Eastern Research Group, 1996. "Report on Revisions to 5th Edition AP-42, Section 3.3 Gasoline and Diesel Industrial Engines". Morrisville, North Carolina: Eastern Research Group. September.

EMC (Energy Metals Corporation), 2007. "Application for USNRC Source Material License Moore Ranch Uranium Project Campbell County, Wyoming." Technical Report, Vol. 1, Vol. 3, Appendix D. Casper, Wyoming: Energy Metals Corporation US. September.

EPA (U.S. Environmental Protection Agency), 2010. "Re: NUREG–1910, Supplements 1, 2, and 3 Draft SEIS for Three Wyoming Uranium ISR Projects: Lost Creek ISR Project CEQ# 20090425; Moore Ranch ISR Project CEQ# 20090421; Nichols Ranch ISR Project CEQ# 20090423." Letter (March 3) from C. Rushin to M. Lesar, NRC. Denver, Colorado: EPA, Region 8.

EPA, 2004. "Exhaust and Crankcase Emission Factors for Nonroad Engine Modeling—Compression-Ignition." EPA420–P–04–009, NR–009c. Washington DC: EPA. April.

EPA, 1996. "Compilation of Air Pollutant Emission Factors, Volume 1: Stationary and Point Area Sources." AP-42, Fifth Edition: Chapters 3.3 and 3.4. Washington DC: EPA. October.

Finley, D.A., 2010. "RE: 2009 Annual Minor Source & 2010 Winter (February and March) Upper Green River Basin Oil and Gas Emissions Inventories." Open letter (January 13) to all operators in proposed ozone non-attainment area in Southwestern Wyoming. Cheyenne, Wyoming: WDEQ, Air Quality Division.

Lost Creek ISR, LLC, 2008. "Lost Creek Project, South Central Wyoming, Technical Report, Application for US NRC Source Material License (Docket No. 40-9068)." Rev. 1, Vols. 1–4. Littleton, Colorado: Lost Creek ISR, LLC. March.

NRC, 2009. NUREG–1910, "Generic Environmental Impact Statement for In-Situ Leach Uranium Milling Facilities." Washington, DC: NRC. May.

NRC, 2004. "Environmental Assessment for the Operation of the Gas Hills Project Satellite *In-Situ* Leach Uranium Recovery Facility." Docket No. 40-8857. Washington, DC: NRC. January.

NRC, 1997. NUREG–1508, "Final Environmental Impact Statement To Construct and Operate the Crownpoint Uranium Solution Mining Project, Crownpoint, New Mexico." Washington, DC: NRC. February.

WDEQ (Wyoming Department of Environmental Quality), 2010. "Emissions Inventory Forms: 2009 Annual and 2010 Winter Upper Green River Basin Emissions Inventory, Winter Drill Rig

Emission Inventory." Cheyenne, Wyoming: WDEQ, Air Quality Division. <http://deq.state.wy.us/adq/ei.asp> (10 May 2010).

WDEQ, 2009. "Guideline No. 12: Standardized Reclamation Performance Bond Format and Cost Calculation Methods." Cheyenne, Wyoming: WDEQ, Land Quality Division. October.